PROCEEDINGS OF THE INTERNATIONAL CONFERENCE
ON HYDROLOGY AND WATER RESOURCES,
NEW DELHI, INDIA, DECEMBER 1993

VOLUME 1

SURFACE-WATER HYDROLOGY

Water Science and Technology Library

VOLUME 16/1

PROCEEDINGS OF THE INTERNATIONAL CONFERENCE ON HYDROLOGY AND WATER RESOURCES, NEW DELHI, INDIA, DECEMBER 1993

Volume 1

SURFACE-WATER HYDROLOGY

edited by

VIJAY P. SINGH

Department of Civil and Environmental Engineering, Louisiana State University, Baton Rouge, U.S.A.

and

BHISHM KUMAR

National Institute of Hydrology, Roorkee, India

SPRINGER SCIENCE+BUSINESS MEDIA, B.V.

A C.I.P. Catalogue record for this book is available from the Library of Congress

ISBN 978-94-010-4174-4 ISBN 978-94-011-0389-3 (eBook)
DOI 10.1007/978-94-011-0389-3

Printed on acid-free paper

To

Mother India

TABLE OF CONTENTS

SECTION 6: STOCHASTIC MODELING OF STREAM FLOW

SECTION 7: FLOOD FREQUENCY ANALYSIS

SECTION 8: DROUGHT ANALYSIS

PREFACE

Water is vital to life, maintenance of ecological balance, economic development, and sustenance of civilization. Planning and management of water resources and its optimal use are a matter of urgency for most countries of the world, and even more so for India with a huge population. Growing population and expanding economic activities exert increasing demands on water for varied needs--domestic, industrial, agricultural, power generation, navigation, recreation, etc. In India, agriculture is the highest user of water.

The past three decades have witnessed numerous advances as well as have presented intriguing challenges and exciting opportunities in hydrology and water resources. Compounding them has been the growing environmental consciousness. Nowhere are these challenges more apparent than in India. As we approach the twenty-first century, it is entirely fitting to take stock of what has been accomplished and what remains to be accomplished, and what accomplishments are relevant, with particular reference to Indian conditions.

India has had a long tradition of excellence spanning over several millennia in construction of water resources projects. She has produced civil engineers of highest quality, some of whom rank among the best in the world. In keeping with India's tradition of honoring distinguished engineers at their retirement, it was decided to hold an International Conference on Hydrology and Water Resources, December 20-22, 1993, in New Delhi, to celebrate the retirement of Professor Satish Chandra, Director of the National Institute of Hydrology at Roorkee, who has had a long and distinguished career spanning over three decades in hydrologic education, research, and development. Through this forum, it was hoped to bring together from all over the world researchers, educationists, and practitioners to New Delhi, India, to present the results of their latest work, to share their experiences, to interact with Indian water resources professionals, and to, hopefully, set the stage for future course of hydrology and water resources.

The objectives of this conference were therefore (1) to assess the current state of the art of hydrology and water resources, (2) to discuss applicability of hydrologic and water resources models, (3) to enhance interdisciplinary interaction, (4) to discuss research needs in hydrology and water resources, and (5) to determine the directions for further research.

We received an overwhelming response to our call for papers. The subject matter of the conference was divided into 36 major topics encompassing virtually all facets of hydrology and water resources. Each topic comprised an invited state-of-the-paper and a number of contributed papers. These contributions provided a natural blend to evolve a synthesized body of knowledge on that topic. Extended abstracts of all the invited and contributed papers were assembled in a pre-conference proceedings volume. This helped stimulate discussion and exchange of ideas during the conference.

The papers presented at the conference were reviewed by the Organizing Committee. As a result, some of the papers did not pass the review and were therefore eliminated from inclusion in the final proceedings. The papers contained in this book, SURFACE-WATER HYDROLOGY, represent one part of the conference contributions.

The other parts are embodied in three separate companion books, SUBSURFACE-WATER HYDROLOGY, WATER-QUALITY HYDROLOGY, and WATER RESOURCES PLANNING AND MANAGEMENT, which are being published simultaneously. Arrangement of the contributions in these books under four different titles was a natural consequence of the diversity of topics presented in the papers. These books can be treated almost independently, although some overlap may exist between them.

This book contains eight sections encompassing major aspects of surface-water hydrology. Each section usually starts with invited state-of-the-art papers, followed by contributed papers. Beginning with investigation of hydrologic processes in the first section, the papers go on to discuss hydrological measurements in Section 2, dynamics of overland flow in Section 3, modeling basin response in Section 4, flood routing in Section 5, stochastic modeling of stream flow in Section 6, flood frequency analysis in Section 7, and drought analysis in the concluding Section 8.

The book will be of interest to researchers as well as those engaged in practice of Civil Engineering, Agricultural Engineering, Hydrology, Water Resources, Earth Resources, Forestry, and Environmental Sciences. The graduate students as well as those wishing to conduct research in water resources planning and management will find this book to be of particular value.

We wish to take this opportunity to express our sincere appreciation to all the members of the Organizing and Advisory Committees, the Louisiana State University administration, and the Indian Association of Hydrologists. Numerous other people contributed to the conference in one way or another, and a lack of space does not allow us to list all of them by name here. We would, nonetheless, like to single out Mr. S. S. Kanwar, who served as Assistant Conference Organizing Secretary, for his inordinate assistance and effort without which the conference would not have come to fruition. Numerous other people contributed to the conference to varying degrees. The authors, including the invited keynote speakers, contributed to the conference technically and made the conference what it was; and this book is largely a result of their efforts. The session chairmen/co-chairmen administered the sessions in a positive and professional manner. We owe our sincere gratitude to all of these individuals.

If the success of a conference is measured in terms of the quality of participants and presentations, then most people would agree that this was a very successful conference. A large number of internationally well-known people, who have long been recognized for their contributions and have long been at the forefront of hydrologic and water resources research, came to participate in the conference. More than 20 countries, covering five continents and most of the countries of the world active in research in hydrology and water resources, were represented. It is hoped that long and productive personal associations and friendships will develop as a result of this conference.

V. P. Singh
Conference Chairman
B. Kumar
Organizing Secretary

ACKNOWLEDGMENTS

The International Conference on Hydrology and Water Resources was sponsored and co-sponsored by a number of organizations. The sponsors supported the conference financially without which it might not have come to fruition. Their financial support is gratefully acknowledged. The co-sponsors extended their support by announcing the conference through their journals, periodicals, transactions, newsletters, or magazines. The publicity so gained helped with increased participation in the conference, and is sincerely appreciated. The following is a list of symposium sponsors and co-sponsors.

Sponsors

Wallac Oy
Finland

Pharmacia United Limited
New Delhi, India

Electrotek Instruments Inc.
Madras, India

Elron Instruments Company Pvt. Ltd.
New Delhi, India

Department of Power
Govt. of Arunachal Pradesh
Itanagar, India

Environmental Measurements & Control
Cochin, India

AIC Agro Instruments (P) Ltd.
Calcutta, India

Spectrum Associates
New Delhi, India

Anubhav Printers & Packers
Roorkee, India

Ajay Printers
Roorkee, India

Co-Sponsors

American Geophysical Union
USA

International Association for
Hydraulic Research
The Netherlands

International Water Resources Association
USA

Soil and Water Conservation Society
USA

Association of American Geographers
USA

International Statistical Institute
The Netherlands

SECTION 1:

INVESTIGATION OF HYDROLOGIC PROCESSES

PROCESS STUDIES IN FOREST HYDROLOGY: A WORLDWIDE REVIEW

Peter F. Ffolliott
School of Renewable Natural Resources, College of Agriculture
University of Arizona
Tucson, Arizona 85721
USA

and

Kenneth N. Brooks
Department of Forest Resources, College of Natural Resources
University of Minnesota
St. Paul, Minnesota 55108
USA

ABSTRACT. Studies of hydrological processes help provide a fundamental understanding of precipitation-runoff relationships and the effects of altering vegetation and other watershed characteristics on the hydrological cycle. Such an understanding is necessary if we are to improve our ability to manage water and the multiple resources on watersheds to meet resource needs without adversely affecting the environment. Process studies in forest hydrology can focus on cause and effect relationships that complement larger scaled catchment experiments, and often are needed to help interpret the results of such experiments. In some cases the results of process studies help identify the need for larger scale catchment experiments. Furthermore, process studies often provide the basis for developing mathematical expressions needed to simulate or model the hydrological effects of modifying forested watersheds. Examples of process studies in temperate regions, humid tropical and sub-tropical forests, and drylands of the world are presented in this paper to illustrate their hydrological value. Studies of interception, evapotranspiration, and infiltration are reviewed. The role of forest hydrology process studies in improving operational watershed management is stressed.

1. Introduction

Recent critical reviews of hydrological science have stressed the need for improving the knowledge and understanding of basic processes if the science of hydrology is to advance (Klemes 1986, National Resources Council 1991). Because of the serious hydrological questions and issues associated with forested watersheds and resource management of these lands, it is imperative that we understand better the hydrological processes of these systems.

An important step in testing of a hypothesis in forest hydrology can involve process studies

V. P. Singh and B. Kumar (eds.), Surface-Water Hydrology, 1–18.

in which "cause-and-effect" relationships can be examined using small, homogeneous plots which are isolated under controlled conditions. Such plot studies have been used in water balance investigations, precipitation-runoff evaluations, and analyses of the effects of vegetative management practices on water regimes. Sometimes, the results of such studies indicate the need for larger scale catchment experiments. In other instances, the results from paired watershed or catchment experiments indicate that process studies are needed to interpret the results of catchment responses to various management treatments.

Information obtained from hydrological process studies is essential in the development of simulation models that are capable of predicting the hydrological effects of various forest and other resource management schemes. In some instances, a conceptual model can be formulated and used to help identify those hydrological processes that are not well understood. This model, therefore, can form the basis for determining process studies that are needed. A research program then can become an iterative one in which process studies are carried out, relationships developed and tested using hydrological models that are then applied to entire watershed systems, and then new process studies identified that are needed to further improve the hydrological models. The key here is that models should not be developed in isolation of field studies.

A review of process studies that have been conducted in temperate regions, humid and sub-tropical forests, and drylands of the world is presented in this paper to illustrate the "hydrological value" of these investigations. Water balance components considered in this review are interception, evapotranspiration, and infiltration. However, it is not a purpose of this paper to compare the different types of instruments, equipment, and methods that have been used in the reported process studies, a subject in itself and well worth pursuing at another time. In fact, instruments, equipment, and methods are mentioned only incidentally in the paper. The results presented should be viewed as a "representative sample" of studies to indicate the relative magnitudes and variabilities of the hydrological processes of interception, evapotranspiration, and infiltration in forests, woodlands, and adjacent vegetative types of the broadly-defined geographical-climatological regions considered. Importantly, the findings obtained from these studies have been, and continue to be extrapolated, largely with the use of mathematical models, to larger water supply basins and incorporated into operational watershed management practices.

2. Interception

In most water balance studies, interception is considered to be a storage term, to be subtracted from the total precipitation when estimating the amount of precipitation available to either replenish soil water deficits or become surface, sub-surface, or groundwater flow. The magnitude of interception loss is related, in general, to the physical character of the precipitation (that is, whether the precipitation falls as rain or snow), the size of the precipitation event, the type of forest overstory (broad-leaved or needle-leaved) species and total surface area of plant leaves (Kittredge 1948, Zinke 1967, Pereira 1973). The magnitude of annual interception can be substantial for forests and represents a loss of water from watersheds that can in some instances exceed transpiration losses. The following discusses interception for different climatological regimes.

2.1 TEMPERATE REGIONS

A few studies conducted in the USA are reviewed to illustrate the relative magnitudes of rainfall interception in temperate forest communities. In the eastern region of the USA, for example, interception of rainfall by broad-leaved trees, in both the growing and dormant seasons, averages about 10% of the rainfall event (Helvey and Patric 1965). Interception of rainfall in the coniferous forests of the southern part of the USA averages about 15% of total rainfall (Roth and Chang 1981). Interception storage of rainfall in the coniferous forests that characterize the northern region of the USA is between 5 and 15% for rainfall events less than 25 mm, and between 15 and 30% when the rainfall event exceeds this amount (Fox 1985). Depending upon the forest density conditions, between 10 and 25% of the rainfall was intercepted in the coniferous forests of the Rocky Mountain region (Aldon 1960). Interception in the coniferous forests of the Pacific Northwest varies with the size of the rain storm, ranging from 100% in storms less than 1.5 mm to 10 to 15% in storms greater than 75 mm (Rothacher 1963).

Canopy interception in coniferous forests in the temperate region of Canada ranged from 15 to nearly 40% annually, depending largely upon the nature of the precipitation events and density of the forest cover (Plamondon et al. 1984). Interception of rainfall in deciduous forests stands in Hungary averages about 25% of the annual precipitation, varying between 30 and 40% in the growing season and between 20 and 30% in the dormant season (Fuhrer 1981).

Studies of rainfall interception in other temperate regions of the world have been summarized in various publications in eastern Europe (Molchanov 1960), Great Britain (Rutter et al. 1971), the (then) Federal Republic of Germany (Weihle 1984), and New Zealand (Blake 1975). From these and other studies, it generally has been concluded that, hydrologically speaking, the interception of rainfall represents a transient form of water storage in the vegetative cover of a catchment. In more recent years, therefore, research interests have centered on the relative contributions of direct evaporation and transpiration to the overall water losses of wetted foliage.

Effects of snowfall interception are more difficult to analyze than interception of rainfall, because neither the initial amount of snowfall nor the water content of the accumulated snow on foliage can be measured accurately (Pereira 1973). Snow builds up rapidly on the foliage of forest overstories, especially on needle-leaved conifers. However, it subsequently can be blown off by wind, dislodged by falling snow from other branches, lost through mass release, washed off by rainfall, melted as snow, or vaporized directly into the atmosphere (Satterlund and Haupt 1970).

Most of the process studies of snowfall interception have been based on the comparisons of snowpacks in forested areas and adjacent clearings. Measurements made in these studies, conducted in the "snow zones" of North America, Europe, and Japan, have shown a range of greater snowpack water equivalents in the clearings than in the trees (Costin et al. 1961, Miller 1964, Satterlund and Eschner 1965).

The ultimate deposition of intercepted snowfall also has been the subject of process studies, employing time-lapse imagery and meteorological records in the western region of the USA. From these studies, it was suggested that most of the intercepted snow eventually reaches the ground (Hoover and Leaf 1967, Tennyson et al. 1974). The important processes of snow removal were wind erosion and snowmelt with subsequent dripping and freezing in the snowpack on the ground. However, more recent research on the Fraser Watershed in Colorado indicated that interception losses of snow appear to be a major factor in explaining differences in snow accumulation between small patch cuts, thinned and unthinned stands of lodgepole pine (Meiman 1987). Whether snowfall interception is a significant loss or not in the water balance, the deposition of the snow on the ground results generally in a redistribution of the snowpack which can affect the amount of water yield from snowmelt.

2.2 HUMID TROPICAL AND SUB-TROPICAL FORESTS

The proportion of rainfall that is intercepted by forests in humid and sub-tropics is variable, depending largely upon the composition and structure of the forests and character of the rainfall event. Estimates of interception are available from India, Australia, Southeast Asia, East Africa, and Costa Rica to illustrate this variability.

Near Dehradun, India, a number of interception studies have been carried out in forest plantations of various tree species compositions (Ghosh et al. 1982, Chandra 1985). Interception in coniferous forest plantations ranged from 20 to 25% of the rainfall amount. In broad-leaved forest plantations, interception of rainfall ranged from 20 to 40%, depending upon the tree species. Studies on the interception in eucalyptus forest plantations indicate that the maximum interception of rainfall approaches 50% of the rainfall. In excess of 25% of the rainfall was reported intercepted in a young-growth plantation of *Cedrus deodara* in Himachal Pradesh, India (Singh et al. 1983). Interception losses were less in the wet season than in the dry season. Minimum interception losses of up to 20% occurred in the month of maximum rainfall, with maximum interception of 70% in the month of lowest rainfall.

Observations in Australia indicate that interception in eucalyptus plantations represent about 10 to 15% of the total rainfall (Schofield 1984). Other Australian investigators also found that interception in eucalyptus was about 10% of the rainfall, and 1% of the rainfall became stemflow (Prebble and Stirk 1980). Interception storage was 2 mm, and rainfall amounts of 7 mm were required to initiate stemflow.

Interception of rainfall in the forests of Thailand ranges from about 65% of the total rainfall amount in natural teak forests, to nearly 10% of the rainfall in the hill-evergreen forests, to less than 5% in the dry-evergreen forests (Chunkao et al. 1971, Tangtham 1981). In the lowland rain forests of Peninsular Malaysia, the interception of rainfall is 20 to 25% of the total rainfall amount (Manokaran 1979). On the average, interception of rainfall in the humid tropical forests of the Philippines is between 60 and 70% of the total rainfall (Rarcos 1979). Measurements of interception from a secondary lowland tropical rainforest in West Java, Indonesia, indicated that 21% of the total rainfall was intercepted (Calder et al. 1986). Total calculated evapotranspiration from the site from August 1980 to July 1981 was 1481 mm, of which 595 mm was interception loss.

Studies conducted in East Africa have shown that the interception of rainfall is dependent largely upon the physical character of the rainfall. From the results of these studies, it appears that light rainfalls of small drop size can be intercepted totally, while heavy rainfalls of large drop size rapidly saturate a forest canopy (Pereira 1973). This apparently is a hydrological process largely independent of the character of the tree foliage. It is possible to illustrate the above point through previously reported comparisons of rainfall interception patterns in a tropical bamboo forest and a plantation of cypress trees near the equator in East Africa, and in a natural forest of broad-leaved trees in the temperate region of the eastern USA (Pereira 1952). Graphical representations of the relationships between the percent of rainfall intercepted by these forest canopies and size of a rainfall event basically were similar in shape and magnitude. In this example, therefore, the hydrological importance of rainfall interception by these forest overstories is explained in terms of the patterns of rainstorms and similar for many species of trees.

Throughfall, the difference between precipitation and the sum of interception and stem flow, has been measured in a mature lowland forest and a one-year-old successional area in Costa Rica. Throughfall, which was highly variable from day-to-day and point-to-point on both sites studied, averaged 50% of the rainfall amounts in the forest and in excess of 65% in the successional area (Raich 1983). Canopy interception was generally one-third higher in the forest than in the adjacent successional area.

2.3 DRYLANDS

Interception is related to precipitation and canopy density in the sparsely- stocked pinyon-juniper woodlands of the southwestern part of the USA. Depending upon the density of the canopy in the coniferous trees, between 5 and 10% of the rainfall is intercepted (Skau 1964a). It has been assumed that most of the intercepted precipitation is evaporated into the atmosphere, as the amount of subsequent stemflow is negligible. A study of rainfall distribution in the oak woodlands of the southwestern region of the USA reported that up to 70% of the late summer-early fall rainfall is intercepted directly on the canopies of the trees (Haworth and McPherson 1991). Throughfall varied from 100% (all trees, large storms) to about 30% (large trees, small storms).

Interception of rainfall is correlated with leaf area of *Larrea tridentata* shrubs in the western region of the USA (Tromble 1983). Loss of rainfall to interception is nearly 25% in stands of 30% crown cover. Canopy interception is nil for rainfall events that are less than 5 mm.

Measurements of rainfall interception by herbaceous vegetation adjacent to woodlands in the dryland regions of the USA are difficult to obtain, although several attempts have been made to obtain these measurements in grassland communities (Couturier and Ripley 1973). In one study, however, it was discovered that the water storage capacity of grasses in the drylands of the western USA was proportional to the average height and percent of ground that is covered by the grasses (Crouse et al. 1966). Furthermore, the total interception loss for a storm was related to the water storage capacity and number of showers in the storm. Information obtained from studies such as this have been utilized in developing hydrological models of the surface runoff processes on sparsely vegetated watersheds (Amisial et al. 1969). These models generally include an input of water storage capacities for the herbaceous vegetation.

2.4 COMMENTS ON INTERCEPTION STUDIES

Interception process studies from around the world indicate that there is considerable variability in results. As indicated by Zinke (1965), much of this variability is due to a lack of standardized instrumentation and documentation. For interception studies to be of value, the measurements in the field must be made using standard procedures and, therefore, related to the vegetative characteristics of the forest stands for which they are determined. Furthermore, the interactions of vegetative and storm characteristics for various vegetative systems needs to be quantified, so that generalized models can be developed and used to extrapolate process study results to the field.

3. Evapotranspiration

Evaporation from soils, plant surfaces, and water bodies, together with water losses through plant leaves, are considered collectively as evapotranspiration (ET). Therefore, the component of interception that evaporates from plant foliage is part of total evapotranspiration. The ET component of a water balance is a topic of interest to forest hydrologists because it affects the water yields from a watershed and is influenced by the land management practices on the catchment (Brooks et al. 1991). ET can represent a high percentage of the annual precipitation, approaching 100% on some sites. Changes in vegetative cover that reduce ET will increase streamflow runoff and groundwater recharge. Increases in ET generally have the opposite effect.

Although we recognize that the process of evapotranspiration controls the hydrological responses of watersheds to snowmelt and rainfall, we still understand little about the process itself and the various feedback mechanisms that control evapotranspiration in natural environments (Morton 1990). Much of what we do understand has originated from carefully designed and controlled plot studies usually with complex microclimatological instrumentation and indirect measurements of the process itself. This review of evapotranspiration studies gives some indication of what has been done in the past, but it should be emphasized that this area of hydrological research is in dire need of more basic, fundamental research that examines both small scale and large scale processes.

3.1 TEMPERATE REGIONS

The more elaborate methods of measuring ET, such as the operation of lysimeters or measurement of the flux of water vapor by instruments mounted above a forest canopy, are important in basic hydrological research. However, these methods can have limitations in terms of practical interpretation and extrapolation (Pereira 1986). Partly because of instrumental difficulties and the more general problem of experimental control, ET studies in temperate and other "geographical - climatological" regions have been carried out traditionally on small, carefully delineated, homogeneous plots (Ward 1971). To illustrate this approach, ET, calculated in terms of the differences between soil moisture in the spring and fall, adjusted for summer precipitation, averaged 488, 379, and 226 mm on small plots in the western United States occupied by aspen, spruce, and grassland, respectively (Brown and Thompson 1965). These ET values, which represent values for the growing season only, approximated 90, 70, and 50% of the annual precipitation at these sites, respectively.

A plot study in the western USA was used to estimate ET differences among bare soil, herbaceous vegetation, and a mixture of broad-leaved trees and herbaceous cover (Johnston 1970). Four-year average ET losses for these vegetative communities, expressed on an annual basis, were 285, 390, and 535 mm, respectively, representing nearly 20, 30, and 40% of the annual precipitation. The results from this and similar plot studies (Croft and Monninger 1953) helped to quantify the differences in ET among cover types and, as a result, indicate the potentials for increasing water yields through manipulations of vegetative cover. Nevertheless, it generally is difficult to extrapolate the results obtained by these methods to a watershed management basis. In many instances, therefore, the best estimates of the vegetative effects on actual ET are derived from catchment experiments and energy budget assessments.

Tritium tracers were used to measure evapotranspiration in a coniferous forest stand in the (then) German Democratic Republic (Simon et al. 1985). It was determined from these measurements that evapotranspiration was about 85%, and surface runoff and groundwater recharge nearly 15% of the almost 700 mm of precipitation measured in the year of study. Annual water consumption in the stand was estimated to be 825 m^3/t of dry biomass in the forest overstory.

Because of the absence of practical methods of measuring actual ET in the field, it often becomes necessary to utilize an index of potential evapotranspiration (PET), and then relate PET to the available water on a watershed. In taking such an approach, however, a researcher must realize the inherent problems associated with the PET methods of measurement, and also have detailed knowledge of soil-water characteristics and plant responses to soil water conditions.

Some of the weaknesses of the PET concept have been described in a review of evaporation and transpiration experiments with forest vegetation (Calder 1982). Early work in PET implied that all wetted vegetative surfaces experienced similar ET amounts, with the observed differences attributed to albedo. However, subsequent experimental work suggests that ET losses can vary significantly among vegetative cover types, the interception losses in forests can exceed the estimated PET values obtained (for example, from Penman's equation), and some forest overstories, well supplied with water, transpire significantly less water than estimated through energy budget assessments. Due to departures from the earlier PET concepts, therefore, alternative relationships to estimate PET have been synthesized (Monteith 1965), and these relationships then employed to estimate PET values on a watershed basis in temperate regions.

3.2 HUMID TROPICAL AND SUB-TROPICAL FORESTS

Only a few researchers have undertaken ET studies within humid and sub-tropical forest communities. Nevertheless, it can be assumed that there is considerable variation in ET values among forests of different compositions and structures, and among sites within specific forest overstories. Examples of this variation are indicated by the results of process studies conducted in India, Southeast Asia, Australia, and East Africa.

Studies of the consumptive use of water by coniferous tree species in India, with lysimeter measurements of water use taken from September to the following June, indicated consumption of approximately 535 mm (Chandra 1985). Consumptive use of water by broad-leaved tree species ranged from 1,145 to 2,705 mm. In excess of 5,525 mm of water was consumed by eucalyptus tree species.

Average values of ET from humid and sub-tropical forests, based on only a few meteorological measurements, are reported to range between 1,200 and 1,500 mm annually (Penman 1970). However, estimates of actual ET in a number of humid and sub-tropical forests in Southeast Asia generally are lower than this range of average ET values. In the dry-evergreen forests of northeastern Thailand, for example, estimated ET values approximate 950 mm annually (Sabhasri et al. 1970). ET in the hill-evergreen forests of northern Thailand is about 1,035 mm, or nearly 50% of the rainfall amounts in the region (Suwanarat 1981). Measurements of annual ET in the rain forests of West Malaysia were 1,065 mm, while ET measurements in areas where rain forests and rubber plantations are found in intermixture were 995 mm (Low and Goh 1972). In the Philippines, on areas characterized by forests of coniferous tree species, agricultural production, and grazing by livestock, the estimated ET values ranged from 835 to 850 mm (Esteban 1979).

In one application of a water budget analysis, both actual ET and PET values were estimated on a monthly basis for a forested watershed in northern Thailand (Brooks et al. 1991). As expected from this analysis, the monthly ET values differed largely in relation to the annual precipitation regime in the area. Monthly estimates of actual ET ranged from 10 to 115 mm, corresponding to 85% and more of the monthly rainfall. Such an approach should be viewed as providing only rough estimates of evapotranspiration and are not equivalent to detailed process studies.

Annual evapotranspiration from the ground flora, litter, and soil in a jarrah forest in southwestern Australia was nearly 30% of the precipitation in daylight hours and 35% of the precipitation in the night (Greenwood et al. 1985). Other investigators in Australia, who measured evapotranspiration rates on lysimeter plots, reported that the water loss was about three times greater when a pine canopy was wet than when it was dry (Holmes and Olyszyezka 1982).

Penman's equation to estimate PET has been tested in the tropics of East Africa and, in general, shown to produce "good results" on forested watersheds (Pereira 1986). For a continuous forest canopy, freely supplied with water, the primary change from an open-water calculation of PET is an increase of the reflection coefficient in the energy budget and an allowance for a greater roughness of the canopy surface in the ventilation estimate. In addition, there is an altitude effect, which is considered in the calculation of PET for the area.

3.3 DRYLANDS

On most drylands, ET represents the largest loss of water from land surfaces, with surface runoff and infiltration relatively minor losses. Selected process studies from different regions of the world report actual ET losses in relation to various plant communities found in dryland environments.

In studying the conversion of shrub oak communities to perennial grass and forb species to improve the production of forage in the drylands of California, USA, no residual moisture was found in the soils on plots supporting a shrub oak cover or on plots converted from oak to grasses and forbs, suggesting little change in ET values as a result of the conversion treatment (Pillsbury et al. 1963). Measurements of soil moisture changes under natural and cleared stands of pinyon-juniper woodland species in the southwestern part of the USA indicated that the clearing of these woodlands for grazing land improvement had little effect on surface water yields, insofar as these water yields were influenced by soil water storage (Skau 1964b). In another study, relative quantities of water that were evapotranspired during a growing season by 12 plant communities on grazing lands in the western region of the USA increased as the percent of bare soil decreased (Branson et al. 1976). The amounts of water loss by ET annually increased from about 60 mm with 75% of bare soil, to nearly 130 mm with 30% of bare soil.

Many tree species survive in dryland environments by controlling their transpiration rates through adjustment of the stomatal openings in their leaves. When the stomata are closed, transpiration virtually ceases. To illustrate this point, in a study of acacia in Australia, actual ET losses declined as soil moisture stresses increased (Slayer and Mabbutt 1964). Similar results with other plant species generally have been obtained elsewhere.

A number of studies have been carried out in the dryland regions of the world to estimate PET, and then relate PET to the available water in a locale to estimate ET (Jones et al. 1981). In one such study, daily ET was estimated to be 3.7 mm in a stand of phreatophytes in Saudi Arabia. In general, these studies have shown that the ET component of water budgets in dryland environments amounts to a relatively high proportion of the rainfall but a low proportion of PET.

3.4 COMMENTS ON EVAPOTRANSPIRATION STUDIES

Little progress has been made in evapotranspiration research over the past few decades (Klemes 1986, Morton 1990). This observation is particularly true on forested environments. The need for new and innovative research on evapotranspiration is apparent. Until we better understand evapotranspiration in forests and other vegetative systems, and develop better means of measuring or estimating evapotranspiration in the field, our overall understanding of the science of forest hydrology will continue to be constrained. More and better process studies are needed that can improve our understanding and ultimately lead to the development of more physically-based models for field application.

4. Infiltration

Once rainfall reaches the ground, it moves into the soil or flows over the surface of the soil. The process by which water enters the soil surface is, by definition, infiltration. Plant material and the litter accumulations on the soil surface, which influence the rate of infiltration, are viewed as two hydrologically distinct layers. There is an upper layer of leaves, stems, and other undecomposed plant material, and a lower layer of decomposed plant material that "behaves" hydrologically like mineral soil. The maximum rate at which water can enter a soil surface is the infiltration capacity. Estimates of infiltration, both cumulative infiltration in time and the infiltration capacity, are obtained from infiltrometer measurements and infiltration equations.

4.1 TEMPERATE REGIONS

There have been a number of process studies utilizing infiltrometer measurements to estimate infiltration in temperate regions. From these studies, estimates of infiltration capacity and, in many instances, surface erodibility have been obtained. In one study, conducted in the coniferous forests of the Pacific Northwest in the USA, it was determined that infiltration capacities and surface erodibility values on logged areas did not differ significantly from the values obtained on unlogged sites (Johnson and Beschta 1980). However, areas that were disturbed by skid trails, cable log paths, and where logging residues had been piled and burned were characterized by reduced infiltration capacities and increased surface erodibility values. Similar results have been obtained from other studies in the same region (Meeuwig 1971, Johnson 1978).

Humid temperate forests with deep soils generally have high infiltration capacities, and are usually much greater than most rainfall intensities that are experienced (Hewlett and Troendel 1975). Therefore, when undisturbed, soils in these forests rarely experience overland flow. Infiltration rates under such conditions do not follow the "classical" infiltration curves.

As might be expected, the "classical" infiltration curve that is described in most textbooks, has been characterized repeatedly by infiltrometer studies in many types of soils and conditions of vegetative cover. In the case of agricultural soils and soils under livestock grazing conditions, infiltration curves in the temperate regions have been approximated by several equations (Horton 1940, Philip 1957, Holtan 1971). However, forested soils generally exhibit considerable variability in infiltration and, therefore, do not conform necessarily to the smooth curves characterized by these equations. A simplified approach for quantifying infiltration, and an approach suitable for agricultural soils, soils under grazing conditions, and forested soils, has been employed to obtain estimates of the relatively constant rates (or final rate) of infiltration in time for a variety of soil-vegetation-land use categories (Gray 1970).

Process studies using infiltrometers are too numerous and site specific to elaborate upon in this paper. However, it is important to state that measurements from infiltrometers, in addition to being of hydrological value in themselves, frequently are a basis to "localize" the inputs to the more general infiltration equations.

Land management practices affect infiltration capacities by altering the soil moisture contents and other soil characteristics. To estimate the magnitude of this effect, studies on the hydrological impacts of land management practices on infiltration have been the focus of many research efforts in temperate regions. Two special considerations are described below.

Soil frost, common in the cold, continental climates of the world, influences the infiltration rate and other hydrological properties of a soil. The occurrence and depth of soil frost is affected by the vegetative cover, soil texture, depth of litter and, when present, the depth of snow. In a specific example, the different cover conditions found in the coniferous forests in the northern region of the USA affected both the occurrence and depth of frost during the winter months (Weitzman and Bay 1963). The removal of forest overstories by timber harvesting or through silvicultural treatments results in more frequent and deeper occurring soil frosts. Furthermore, the occurrence of a "concrete" frost is more prevalent in soils that are non-forested than in soils with forest cover. As this term implies, infiltration rates for concrete frost conditions are minimal.

Hydrophobic soils present another special case. Most hydrophobic soils repel water as a result of organic, long-chained hydrocarbon substances coating the soil particles. As a result, water "beads up" on the soil and will not penetrate readily. Hydrophobic soils under oak shrub vegetation in the coastal mountain ranges along the Pacific Ocean in the western region of the USA have received considerable study (DeBano 1981). Fires, which occur frequently in this area, intensify the hydrophobic condition, and, apparently, volatilize organic substances and drive the water repellent layer deeper into the soil. The resulting layer restricts water movement into and through the soil and, therefore, inhibits infiltration. Fires also have been reported to promote water repellency in soils of forests and grasslands.

4.2 HUMID TROPICAL AND SUB-TROPICAL FORESTS

Infiltration rates of humid and sub-tropical soils generally are high in comparison to the infiltration rates of soils in temperate regions (nonforested) and many drylands. These relatively high infiltration rates usually are attributed to the more favorable structure and greater stability of aggregates in humid and sub-tropical soils. Infiltrometer measurements in forests have verified this important hydrological difference.

As points of reference, an initial infiltration rate in excess of 1,115 mm/hr, with a constant infiltration rate of 280 mm/hr, has been measured in the hill-evergreen forests of northern Thailand (Chunkao and Naksiri 1976). Soils under dry dipterocarp forests, again, in northern Thailand, are characterized by lower infiltration rates, averaging 195 mm/hr (Inthasothi and Chunkao 1976). In the Philippines, extensive hillside cultivation and inappropriate farming practices have resulted in a lowering of infiltration rates in comparison to the conditions observed on areas in original forest covers (Bala et al. 1982). In specific, infiltration rates on sites that are burned annually ranged from 45 to 70 mm/hr.

Infiltration studies have been carried out in India to determine the effects of land management practices on this important hydrological parameter. Infiltration rates in eucalyptus plantations generally were greater than those recorded in natural grasslands and on lands under terraced cultivation (Ghosh et al. 1982). In Puerto Rico, measured infiltration rates for a variety of soils on forested watersheds ranged from a low of 240 to a high of 133,150 mm/hr (Bonnet 1968).

4.3 DRYLANDS

Infiltration rates in natural plant communities in the drylands of the USA have been measured by numerous investigators to provide baselines against which the effects of land management practices can be analyzed (Branson et al. 1981). Different plant compositions, comparisons on soils with and without plant covers, and different "range conditions" have been the focus of these infiltration studies.

Because livestock grazing is a major land use on drylands in the USA, the influence of grazing on infiltration has been studied extensively. In the southwestern parts of the USA, infiltration on sites that are brush-dominated was greater than on either grazed sites or ungrazed sites that are grass-dominated (Tromble et al. 1974). Similar results were reported on drylands in the Rocky Mountain region (Lusby et al. 1971) and the Great Basin region (Branson et al. 1981).

Fire is another common phenomenon in many dryland ecosystems of the USA and, therefore, the influence of fire on infiltration also has been the subject of investigation. For example, relative infiltration rates generally are higher on burned areas immediately after a fire, and then this condition persists one year later in many cases (Scott 1956). However, repeated annual burning in the shrub oak communities of California reduced the average infiltration rates (Rowe 1948). The response to fire of the infiltration capacity for soils generally varies as a function of the number of annual burns, litter depth, and soil texture (Settergren 1967).

Infiltration rates on an afforested site in Morocco were greater than on sites that were moderately and heavily grazed by livestock (Berglund et al. 1981). In comparison to the moderately and heavily grazed sites, the afforested site represents a successful rehabilitation effort to improve infiltration conditions. Consequently, afforestation seems to be an effective watershed rehabilitation technique in Morocco.

4.4 COMMENTS ON INFILTRATION STUDIES

Infiltrometer measurements and infiltration equations have been used to estimate infiltration capacities for many situations throughout the world. It has been shown that land management practices, for example, the harvesting of forest overstories and livestock grazing, affect these estimates of infiltration capacities by changing soil moisture contents. The results obtained from infiltrometer measurements largely represent "point-estimates" in that they are generated from measurements taken on relatively small plots. It is necessary, therefore, to consider the problems of inherent variabilities on watershed lands to extrapolate these results to a larger scale.

5. Summary and Recommendations

Process studies on interception, evapotranspiration, and infiltration are important to understanding the hydrological regimes of watershed lands. It has been documented in the literature, as shown by the review presented in this paper, that the hydrological effects of forests, woodlands, and other vegetation vary greatly with geographical-climatological regions, vegetative types, and watershed features. The results of these studies, therefore, should not be applied directly to specific sites without further investigation. Nevertheless, these results can provide "qualitative" insights to the effects of watershed management practices on a water balance and, as a consequence, precipitation-runoff relationships in a general context.

Many of the environmental issues of concern at the end of this century encompass questions of various scales that will only be answered when we improve the status of our knowledge of hydrological processes. The cumulative effects of human activities on watersheds are a focus of much current research in forest hydrology, and will require process studies. The restoration, protection and management of riparian systems requires that we learn more about the hydrological functions of wetland systems. Furthermore, research on hillslope processes and water movement through the vadose zone will promote better management to reduce detrimental effects of human activities on runoff, sediment flow, and non-point source pollution. Even though considerable research has focused on evapotranspiration, it still remains the major hydrological process that we understand the least, and a critically important process that cannot be measured or estimated with any degree of confidence. These few examples of research needs indicate that many challenges remain in forest hydrology. We need to respond to these challenges with well designed research projects that can yield better models and methods that advance hydrological science.

Hydrological process studies need to be designed on the basis of an overall conceptual models in which the objectives of the study are articulated clearly. Long-term studies that provide an iterative process of field - process studies, development of a working model, followed by testing of the model on field data, model adjustments, and the identification of needed (further) process studies are needed. There must be a feed-back of field study results to model development, testing, and modification. Conversely, models help to identify the key questions that need to be addressed and then point to the gaps in our knowledge that can only be filled by process studies in many instances. With the development of a more comprehensive research approach, we should be able to solve the mysteries of hydrological science that have evaded us for so long.

6. References

Aldon, E. F. 1960. Research in the ponderosa pine type. In: Progress Report on Watershed Management Research in Arizona. USDA Forest Service, Rocky Mountain Forest and Range Experiment Station, Fort Collins, Colorado, USA.

Amisial, R. A., J. P. Riley, K. G. Renard, and E. K. Israelson. 1969. Analog computer solution of the unsteady flow equation and its use in modeling the surface runoff process. Utah Water Resources Laboratory, Report PRWG 38-2, Logan, Utah, USA.

Bala, U. T., L. P. Biete, R. F. Comandante, A. R. De La Druz, R. Gabutan, M. I. Hernandez,

C. T. De Jesus, V. T. Manarpaac, A. G. Reforma, E. M. Unite, Jr., and D. Urmatan. 1982. Management of watersheds in the Philippines. Paper Presented at the Regional Training Course on Watershed Resources Management and Environmental Monitoring, College, Laguna, Philippines.

Berglund, E. R., A. Ahyoud, and M. Tayaa. 1981. Comparison of soil and infiltration properties of range and afforested sites in northern Morocco. Forest Ecology and Management 3:295-306.

Blake, G. J. 1975. The interception process. In: Chapman, T. G., and F. X. Dunin. (eds.) Prediction in catchment hydrology. National Symposium on Hydrology, Australian Academy of Science, pp. 59-81.

Bonnet, J. A. 1968. Relative infiltration rates of Puerto Rico soils. Journal of Agriculture 52:233-240.

Branson, F. A., G. F. Gifford, K. G. Renard, and R. F. Hadley. 1981. Rangeland hydrology. Kendall-Hunt Publishing Company, Dubuque, Iowa, USA.

Branson, F. A., R. F. Miller, and I. S. McQueen. 1976. Moisture relationships in twelve northern shrub communities near Grand Junction, Colorado. Ecology 57:1104-1124.

Brooks, K. N., P. F. Ffolliott, H. M. Gregersen, and J. L. Thames. 1991. Hydrology and the management of watersheds. Iowa State University Press, Ames, Iowa, USA.

Brown, H. E., and J. A. Thompson. 1965. Summer water use by aspen, spruce, and grassland in western Colorado. Journal of Forestry 63:756-760.

Calder, I. R. 1982. Forest evaporation. In: Proceedings of Canadian Hydrology Symposium, Fredericton, New Brunswick, Canada, pp. 173-193.

Calder, I.R., I.R. Wright, and D. Murdiyarso. 1986. A study of evaporation from tropical rain forest - West Java. Journal of Hydrology 86:13-31.

Chandra, S. 1985. Small watersheds and representative basin studies and research needs. In: Proceedings of the National Seminar on Watershed Management, Dehradun, India, pp. 63-71.

Chunkao, K., and C. Naksiri. 1976. Infiltration capacity of natural hill-evergreen forest at Doi, Chiengmai, Kog-Ma Watershed. Faculty of Forestry, Watershed Resources Bulletin 25, Kasetsart University, Bangkok, Thailand.

Chunkao, K., N. Tang Tham, and S. Ungkulpakdikul. 1971. Measurements of rainfall in early wet season under hilt and dry-evergreen, natural teak, and dry-dipterocarp forests of Thailand. Faculty of Forestry, Forest Resources Bulletin 10, Kasetsart University, Bangkok, Thailand.

Costin, A. B., L. W. Gay, D. J. Wimbush, and D. Kerr. 1961. Studies in catchment hydrology in the Australian Alps - preliminary snow investigations. CSIRO Division of Plant Industries, Technical Paper 16.

Couturier, D. E., and E. A. Ripley. 1973. Rainfall interception in mixed grass prairie. Canadian Journal of Plant Science 53:659-663.

Croft, A. R., and L. V. Monninger. 1953. Evapotranspiration and other water losses on some aspen forest types in relation to water available for stream flow. American Geophysical Union Transactions 34:563-574.

Crouse, R. P., E. S. Corbett, and D. W. Seegrist. 1966. Methods of measuring and analyzing rainfall interception by grass. International Association of Scientists in Hydrology, Bulletin 11:110-120.

DeBano, L. F. 1981. Water repellent soils: A state-of-the-art. USDA Forest Service, General Technical Report PSW-46.

Esteban, I. D. 1979. Forestry as a major land use type in Agno River Basin. Paper Presented at the Regional Training Course on Integrated River Basin Management, College, Laguna, Philippines.

Fox, S. J. 1985. Interception-net rainfall relationships of red pine stands in northern Minnesota. Unpublished Plan B Paper, University of Minnesota, St. Paul, Minnesota, USA.

Fuhrer, E. 1981. Interception measurements in beech stands. Erdeszeti Kutatasek 74:125-137.

Ghosh, R. C., O. N. Kaul, and B. K. Subba Rao. 1982. Environmental effects of forests in India. Indian Forest Bulletin No. 275.

Gray, D. M. 1970. Handbook on the principles of hydrology. Water Information Center, Port Washington, New York, USA.

Greenwood, E. A. N., L. Klein, S. D. Beresford, G. G. Watson, and K. D. Wright. 1985. Evaporation from understory in jarrah (*Eucalyptus marginata*) forest, southwestern Australia. Journal of Hydrology 80:377-349.

Haworth, K., and G. R. McPherson. 1991. Effect of *Quercus emoryi* on precipitation distribution. Journal of the Arizona-Nevada Academy of Science, Proceedings Supplement 1991:21.

Helvey, J. D., and J. H. Patric. 1965. Canopy and litter interception by hardwoods of eastern United States. Water Resources Research 1:193-206.

Hewlett, J.D., and C.A. Troendle. 1975. Non-point and diffused water sources: a variable source area problem. In: Watershed Management, Symposium Proceedings, Logan, Utah. American Society of Civil Engineers. pp. 21-46.

Holmes, J. W., and B. Olyszyezka. 1982. Evaporation of surface water from the foliage of an isolated pine tree. In: First National Symposium on Forest Hydrology, Barton, ACT, Australia, pp. 7-11.

Holtan, H. N. 1971. A formulation of quantifying the influence of soil porosity and vegetation on infiltration. In: Proceedings of the Third International Seminar for Hydrology Professors, West Lafayette, Indiana, USA, pp. 228-239.

Hoover, M. D., and C. F. Leaf. 1967. Process and significance of interception in Colorado subalpine forest. In: Proceeding of the International Symposium on Forest Hydrology, Pennsylvania State University, Pergamon Press, New York, USA, pp. 212-222.

Horton, R. E. 1940. An approach to the physical interpretation of infiltration capacity. Soil Sciences of America 5:399-417.

Inthasothi, S., and K. Chunkao. 1976. Infiltration capacity of soils in various types of land use at Mae Talai Watershed, Chiengdao Kog-Ma Watershed. Faculty of Forestry, Forest Resources Bulletin 26, Kasetsart University, Bangkok, Thailand.

Johnson, M. G. 1978. Infiltration capacities and surface erodibility associated with forest harvesting activities in the Oregon Cascades. M.S. Thesis, Oregon State University, Corvallis, Oregon, USA.

Johnson, M. G., and R. L. Beschta. 1980. Logging, infiltration capacity, and surface erodibility in western Oregon. Journal of Forestry 78:334-337.

Johnston, R. S. 1970. Evapotranspiration from bare, herbaceous, and aspen plots: A check on a former study. Water Resources Research 6:324-327.

Jones, K. R., O. Berney, D. P. Carr, and E. C. Barrett. 1981. Arid zone hydrology for agricultural development. FAO Irrigation and Drainage Paper No. 37.

Kittredge, J. 1948. Forest Influences. McGraw-Hill Book Company, New York, New York, USA.

Klemes, V. 1986. Dilettantism in hydrology: transition or destiny? Water Resources Research 22: 177S-188S.

Low, K. S., and K. C. Goh. 1972. Water balance studies and implications on water resource utilization in West Malaysia. Journal of Tropical Geography 35:153-175.

Lusby, G. C., V. H. Reid, and O. D. Knipe. 1971. Effects of grazing on the hydrology and biology of the Badger Wash basin in western Colorado, 1953-1966. U.S. Geological Survey, Water-Supply Paper 1532D.

Manokaran, N. 1979. Stemflow, throughfall, and rainfall interception in a lowland tropical rain forest in Peninsular Malaysia. The Malayan Forester 42:174-201.

Meeuwig, R. O. 1971. Infiltration and water repellency in granitic soils. USDA Forest Service, Research Paper INT-111.

Meiman, J. R. 1987. Influence of forests on snowpack accumulation. In: Management of Subalpine Forests: Building on 50 years of Research. USDA Forest Service Technical Report RM-149, Ft. Collins, Colorado.

Miller, D. H. 1964. Interception processes during snowstorms. USDA Forest Service, Research Paper PSW-18, 24 p.

Molchanov, A. A. 1960. The hydrological role of forests. Academy of Sciences of the USSR, Institute of Forestry, Moscow, Russia.

Monteith, J. L. 1965. Evaporation and the environment. Society of Experimental Biology 19:205-234.

Morton, F.I. 1990. Studies in evaporation and their lessons for the environmental sciences. Canadian Water Resources Journal 15:261-286.

National Research Council. 1991. Opportunities in the hydrologic sciences. National Academy Press, Washington, D.C.

Penman, H. L. 1970. The water cycle. Scientific American 223:99-108.

Pereira, H. C. 1952. Interception of rainfall by cypress plantations. East African Agricultural Journal 18:1-4.

Pereira, H. C. 1973. Land use and water resources. Cambridge University Press, London, England.

Pereira, H. C. 1986. The management of tropical watersheds. In: Strategies, approaches, and systems in integrated watershed management. FAO Conservation Guide, No. 14, pp. 10-32.

Philip, J. R. 1957. The theory of infiltration: The infiltration equation and its solution. Soil Science 83:345-357.

Pillsbury, A. F., J. O. Osborn, and P. E. Pelishek. 1963. Residual soil moisture below the root zone in southern California watersheds. Journal of Geophysical Research 68:1089-1091.

Plamondon, A. P., M. Prevost, and R. C. Naud. 1984. Rainfall interception in a fir/white birch stand in the Montmorency Forest. Canadian Journal of Forest Research 14:722-730.

Prebble, R. E., and G. B. Stirk. 1980. Throughfall and stemflow on silverleaf ironbark (*Eucalyptus melanophloia*) trees. Australian Journal of Ecology 5:419-427.

Raich, J. W. 1983. Throughfall and stemflow in mature and one-year-old wet tropical forest. Tropical Ecology 24:234-243.

Rarcos, R. S. 1979. Critical ecological consideration in river basin management in Southeast Asia. Paper Presented at the Regional Training Course on Integrated River Basin Management,

College, Laguna, Philippines.

Roth, F. A., and M. Chang. 1981. Throughfall in planted stands of four southern pine species in east Texas. Water Resources Bulletin 17:880-885.

Rothacher, J. 1963. Net precipitation under a Douglas-fir forest. Forest Science 9:423-429.

Rowe, P. R. 1948. Influence of woodland chaparral on water and soil in central California. State of California, Department of Natural Resources, Division of Forestry, Unnumbered Publication.

Rutter, A. J., K. A. Kershaw, P. C. Robins, and A. J. Morton. 1971. A predictive model of rainfall interception in forests - derivation of model from observations in a plantation of Corsican pine. Agricultural Meteorology 9:367-384.

Sabhasri, S., K. Chunkao, and C. Hgampongsai. 1970. The estimation of evapotranspiration of the old clearing and the dry-evergreen forest, Sakaerat, Nakorn Rachasima. Faculty of Forestry, Kasetsart University, Bangkok, Thailand.

Satterlund, D. R., and A. R. Eschner. 1965. The surface geometry of a closed conifer forest in relation to losses of intercepted snow. USDA Forest Service, Research Paper NE-34.

Satterlund, D. R., and H. F. Haupt. 1970. The deposition of snow caught by conifer crowns. Water Resources Research 6:649-652.

Schofield, N. J. 1984. A simulation model predicting winter interception losses from reforestation stands in southwest W. A. Australia. Australian Forest Research 14:105-127.

Scott, V. H. 1956. Relative infiltration rates of burned and unburned upland soils. Transactions of the American Geophysical Union 37:67-69.

Settergren, C. D. 1967. Reanalysis of past research on effects of fire on wildland hydrology. Missouri Agricultural Experiment Station, Research Bulletin No. 954.

Skau, C. M. 1964a. Interception, throughfall, and stemflow in Utah and alligator juniper cover types of northern Arizona. Forest Science 10:283-287.

Skau, C. M. 1964b. Soil water storage under natural and cleared stands of alligator and Utah juniper in northern Arizona. USDA Forest Service, Research Note RM-24.

Simon, K. H., G. Ritter, and K. Mraz. 1985. Water balance for a spruce stand in the central Bohemian uplands. Beitrage fur die Forstwirtschaft 19:65-69.

Singh, R. P., K. C. Sharma, H. N. Mathur, M. K. Gupta, and A. K. Gupta. 1983. Interception studies in *Cedrus deodara* London plantation in Himachal Pradesh. Indian Forester 109:261-266.

Slayer, R. O., and J. A. Mabbutt. 1964. Hydrology of arid and semi-arid regions. Handbook of

Hydrology, Section 24, McGraw-Hill Book Company, New York, New York, USA.

Suwanarat, R. 1981. Water balance in the hill-evergreen forest, Doi Pui, Chiengmai. Master's Thesis, Kasetsart University, Bangkok, Thailand.

Tangtham, N. 1981. Ecological aspects of watershed resources management in humid tropical ecosystems. Paper Presented at the Regional Training Course on Watershed Resources Management and Environmental Monitoring, Chiangmai Mai, Thailand.

Tennyson, L. C., P. F. Ffolliott, and D. B. Thorud. 1974. Use of time-lapse photography to assess potential interception in Arizona ponderosa pine. Water Resources Bulletin 10:1246-1254

Tromble, J. M. 1983. Interception of rainfall by creosotebush (*Larrea tridentata*). In: Proceedings of XIV International Grassland Congress, Lexington, Kentucky, USA, pp. 373-375.

Tromble, J. M., K. G. Renard, and A. P. Thatcher. 1974. Infiltration for three rangeland soil-vegetation complexes. Journal of Range Management 27:318-321.

Ward, R. C. 1971. Small watershed experiments - an appraisal of concepts and research developments. University of Hull, Occasional Papers in Geography No. 18.

Weihle, J. 1984. Wetting and interception in beech and spruce stands. Allgemeine Forst und Jagdzeitung 155:241-252.

Weitzman, S., and R. R. Bay. 1963. Forest soil freezing and the influence of management practices, northern Minnesota. USDA Forest Service, Research Paper LS-2.

Zinke, P. J. 1967. Forest interception studies in the United States. In: Proceedings of the International Symposium on Forest Hydrology, Pennsylvania State University, Pergamon Press, New York, USA, pp. 137-162.

MODELLING EVAPOTRANSPIRATION OF DUNE VEGETATION

P.J.M. de Laat and P. Varoonchotikul[1]
International Institute for Infrafructural,
Hydraulic and Environmental Engineering (IHE)
POB 3015
2601 DA Delft
THE NETHERLANDS

ABSTRACT. The Penman-Monteith-Rijtema approach for cropped surfaces is used to estimate evapotranspiration of natural dune vegetation. Use is made of drain outflow data from lysimeters in Castricum, The Netherlands. One of the lysimeters is bare, another one is covered with dune vegetation. The lysimeter with bare soil is used to calibrate parameters for soil evaporation and to verify the soil physical data. The vegetation parameters were calibrated and verified for the lysimeter covered with dune vegetation. Subsurface flow and root water uptake were simulated with the model for unsaturated flow above a shallow water-table, MUST. This model is based on a pseudo steady-state solution of the flow equation. Results show the applicability of the method to estimate evapotranspiration of natural dune vegetation.

1. Introduction

Dune areas in The Netherlands have been used since the middle of the last century for the extraction of groundwater to supply the densely populated Western part of the country with drinking water. The importance of the dunes for the water supply has stimulated hydro(geo)logical research in this area from the very onset of the extractions. In 1940 four lysimeters were constructed in the dunes to study the evapotranspiration from the area for different types of land use. The lysimeters have been continuously monitored and the data were extensively used for research (Hiege, 1987). So far no attempt has been made to simulate evapotranspiration and unsaturated flow in the lysimeters. However, with the application of transient groundwater flow models in the dune area there is a need for a more detailed deterministic approach to the simulation of hydrological processes in the unsaturated region in relation to evapotranspiration.

2. Lysimeter data

Four lysimeters were constructed in 1940 in the dunes along the North Sea coast near Castricum, 30 km NW of Amsterdam. Each lysimeter sizes 25x25 m^2 with a depth of 2.50 m. At the bottom a drainage system is installed, consisting of small drainage channels at a depth of 2.25 m, leading to a central collection reservoir which is connected to a storage tank where the drain water is weekly measured. Each lysimeter has a different type of land use: I bare soil, II shrubs (Sea Buckthorn), III coniferous trees and IV deciduous trees. Lysimeters I and II are surrounded by natural dune vegetation and the other lysimeters within a distance of at least 30 m by the same type of trees. As the type of land use in the dunes is predominantly shrubs and bare soil, data pertaining to lysimeters I and II were used for this study.

[1] At present: Royal Irrigation Department, Bangkok, Thailand

V. P. Singh and B. Kumar (eds.), Surface-Water Hydrology, 19–27.

The lysimeters are filled with sand from local, older dune deposits. In 1948 soil moisture characteristics were measured at two different depths. These data appear very similar to the pF-curve of Coarse Sand, which is one of the soils in the international soil series presented by Rijtema(1969). The moisture characteristic and hydraulic conductivity relation of this soil have been used in this study, as it appeared impossible to obtain permission for additional soil physical measurements in the lysimeters. According to the American Soil Taxonomy System the soil could be classified as a Typic Quartzipsamment.

Lysimeter II is covered with natural vegetation, which is predominantly Sea Buckthorn (Hippophae Rhamnoides). The shrub is tolerant to a salty environment, strong winds and droughts. It is a thorny, deciduous plant up to nine metres in height and it effectively suppresses most other plant growth in the dunes. Most of the roots are found in the top ten to twenty centimetres, but some of the rhizome-like roots may penetrate to depths of at least 1.2 m. No measurements of actual rooting depth are available. The current height of the shrub in lysimeter II was found to be 3.0 m.

Daily meteorological data covering a period of five years (1984-1988) were used for this study. Precipitation data were obtained from a local ground level rain gauge and data on air temperature, sunshine duration, wind speed and relative humidity were provided by the Royal Netherlands Meteorological Institute (KNMI) for the synoptic station De Kooy which is also near the coast at a distance of 50 km to the North. The data show annual precipitation amounts varying from 790 to 912 mm, where 1986 is relatively dry and 1987 relatively wet.

3. Model for unsaturated flow

Transient vertical flow in the soil is simulated with the Model for Unsaturated flow above a Shallow water Table, MUST (DeLaat, 1985). The model is based on a unique, pseudo steady-state, solution of the flow equation, which proved to be a very convincing and efficient method for simulating flow above a shallow water table. A concise description of the method is found in DeLaat (1992), a summary is given below.
The soil profile is schematised into a root zone with a constant depth and a subsoil. The upper boundary condition consists of a flux across the soil surface which is governed by the precipitation and the evapotranspirative demand from plants and atmosphere, and the water pressure in the root zone. The actual flux is computed from an iterative solution of the modules for evapotranspiration and unsaturated flow.
The lower boundary of the unsaturated zone is chosen at a fixed level just below the lowest water table depth. The flux across this boundary is also found in an iterative way by solving the modules for unsaturated flow and drainage in turn.
The module for unsaturated flow simulates the transient vertical transport in the subsoil using a succession of steady flow situations. Before the start of a simulation run, soil moisture content- and water pressure profiles are computed for a large number of steady flow situations using the soil moisture characteristics and hydraulic conductivity relations that apply for the various layers. For given upper and lower boundary fluxes the steady flow situation in the subsoil may than be found from interpolation in these curves in combination with the continuity equation.
For capillary rise the assumption of steady flow is seriously violated if the flux across the

lower boundary is large in the downward direction, so that the actual soil moisture profile has a more elongated shape than the assumed steady-state profile. Therefore, the water table drawdown is recalculated assuming steady flow in the lower part of the subsoil corresponding to the flux across the lower boundary. If this yields a lower water table depth, a percolation profile develops, simulating the elongation of the steady-state profile for capillary rise. The non-steady flow situation is then approached by a superposition of the two soil moisture profiles for steady flow.

4. The evapotranspiration module

The equation for the evaporation from a wet crop (Monteith, 1965) may be written similar to the formula of Penman for open water

$$E_{wet} = C \frac{sR_n + c_p\rho_a(e_s - e_a)/r_a}{(s+\gamma)L} \quad (cm.d^{-1}) \tag{1}$$

where C is a conversion factor to change the units for the evaporation rate from $kg.m^{-2}.s^{-1}$ to $cm.d^{-1}$ (the symbols are explained in the appendix). When using (1) for cropped surfaces it should be noted that the albedo of the vegetation (say 0.24) differs from open water (0.06) and that the diffusion resistance r_a is larger due to the roughness length of the vegetated surface. Taking into account the evaporation of intercepted water E_i, Rijtema(1965) has shown that the equation for the transpiration of a crop takes the form

$$E_T = \frac{s+\gamma}{s+\gamma(1+r_c/r_a)} (E_{wet} - E_i) \quad (cm.d^{-1}) \tag{2}$$

In the evapotranspiration module the aerodynamic resistance r_a is for all types of land use, except for grass and forest, estimated with the following formula

$$r_a = \frac{1}{k^2 u(2+d)} [\ln(2/z_o)]^2 \quad (s.m^{-1}) \tag{3}$$

where the roughness length z_o and the zero plane displacement d are taken equal to 10, respectively 70 % of the crop height. The value for u(2+d) at the standard model height (2+d) is as follows estimated from the wind speed u_s measured elsewhere over short grass

$$u(2+d) = c_u\, u_s \quad (m.s^{-1}) \tag{4}$$

where the correction factor c_u is calculated in relation to the height HU at which the windspeed is measured.

$$c_u = \frac{\ln(60/0.03)}{\ln(HU/0.03)} \frac{\ln(2/z_o)}{\ln(60/z_o)} \quad (-) \tag{5}$$

The canopy resistance r_c is an empirical function of the moisture potential in the leaves h_L as shown in figure 1. The pressure head h_L is computed as

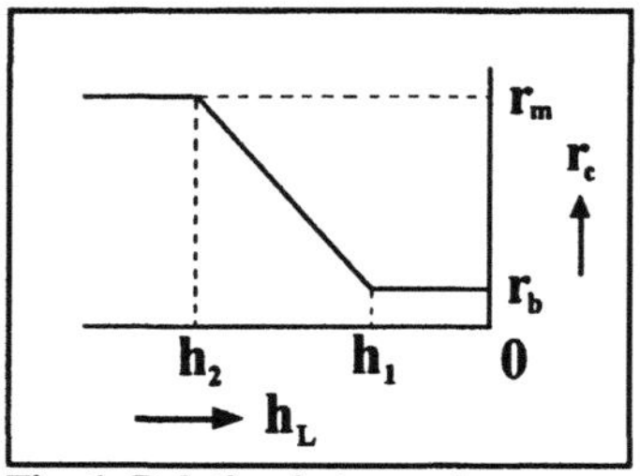

Fig. 1 Relation between the canopy resistance r_c and the leaf water potential h_L

$$h_L = h_s - 3E_T\,[R_{pl}+b/K(h_s)] \quad (cm) \tag{6}$$

where the geometry factor of the root system b follows from

$$b = B\,(R_{pl}/D_r) \quad (cm) \tag{7}$$

The above equations show in combination with figure 1 the dependence of the actual transpiration E_T on the hydraulic conductivity of the soil ($K(h_s)$) and the pressure head in the root zone (h_s). The computation of E_T requires an iterative solution of the equations (2) and (6).
The model uses an interception reservoir to simulate the interception of precipitation by the canopy. Each day rainfall is recorded the interception reservoir fills up to a maximum, plant-dependent, value S_{max}. The rate of evaporation from the reservoir E_i equals E_{wet}, hence

$$E_i = \min(E_{wet}, S_a/\Delta t) \quad (cm.d^{-1}) \tag{8}$$

where S_a is the actual amount stored in the interception reservoir.
The evaporation of bare soil E_s is estimated from the Penman open water evaporation E_{pen} as follows

$$E_s = \alpha_s\,E_{pen} \quad (cm.d^{-1}) \tag{9}$$

where the relation between the parameter α_s and the pressure head in the root zone h_s is expressed by the formula

$$\alpha_s = 1 - \frac{\log(1-h_s)}{\log(16000)} \quad (-) \tag{10}$$

For soil water potentials exceeding a certain maximum value ($h_s > h_{max}$), the value for h_s in (10) is set equal to h_{max} in order to limit the maximum evaporation rate from bare soil. Penman open water evaporation E_{pen} is computed according to (1) with the net radiation estimated for an albedo of 0.06 and with the aerodynamic resistance computed for an open water surface from (see e.g. De Bruin, 1981)

$$r_a = \frac{245}{0.54u(2)+0.5} \quad (s.m^{-1}) \tag{11}$$

The fraction of the soil covered by the vegetation S_c finally determines the relative contribution of the components that comprise the total actual evapotranspiration

$$E = S_c(E_T+E_i)+(1-S_c)E_s \quad (cm.d^{-1}) \tag{12}$$

5. Calibration of physical parameters

The model MUST was first applied to the lysimeter with bare soil in order to calibrate the parameters for soil evaporation and to verify the soil physical data. The average drain outflow from lysimeter I appears to be equal to 1.86 mm/d. This value is taken as the initial flow situation at the start of the simulations. MUST includes a drainage module which has been used to schematize the drainage system of the lysimeter. The depth of the 'root zone' is taken as one centimetre.

Simulations were carried out for the five year period 1984-1988 using a time step of one day (1820 time steps). The results were evaluated in two different ways. The total observed and simulated volumes drained from the lysimeter are compared in a parameter RV, the Relative Volume percentage. RV is computed as the ratio of the simulated and observed volumes multiplied by hundred. For the second assessment use is made of the model efficiency RE as defined by Nash(1970). For a simulated drainage outflow rate which is constant and equal to the average observed value, RE = 0 % and for a perfect fit RE = 100 %.

The selected soil type, Coarse Sand, appeared to gave the best results. Other soil types show hydraulic conductivity properties resulting in large evaporation rates and consequently in simulated drainage values which are tens to hundreds of millimetres less than observed.

The only parameter to be calibrated is the maximum pressure head of the soil moisture in the root zone h_{max} which determines the maximum value of α_s in (10). Calibration showed that h_{max} = -10 cm, limiting the maximum evaporation rate from bare soil to 75 % of the open water evaporation E_{pen}.

The simulated and observed drainage rates (weekly averages) for lysimeter I are presented in figure 2.

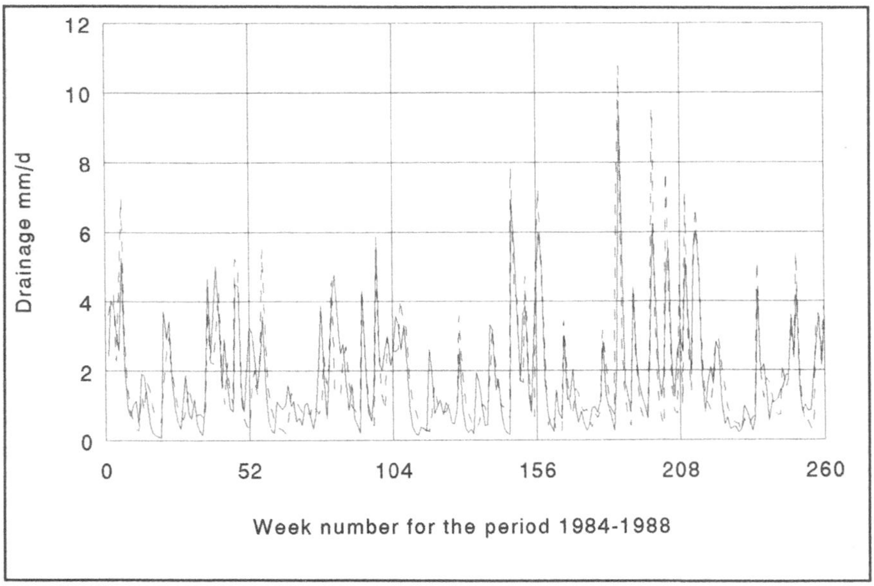

Fig. 2. Simulated (—) and observed (---) drain outflow rates from lysimeter I

The response time of the system is generally less than one day. The time interval over which precipitation is measured does not exactly coincide with the monitoring interval of the drain outflow which partly explains that the simulated and observed outflow peaks are often slightly out of phase. This adversely affects the model efficiency: RE = 69 %. Neglecting the change in the amount of water stored over the five year period in lysimeter I, the evaporation is simply found as the difference between the observed precipitation and the drain water outflow. Table 1 shows that the evaporation as obtained from the water balance is almost exactly equal to the simulated value, which indeed equally applies to the drainage volumes: RV = 99.7 %.

	Observed	Simulated
Precipitation	4337 mm	4337 mm
Drainage	3396 mm	3387 mm
Evaporation	941 mm	939 mm

Table 1. The components of the water balance for lysimeter I for 1984-1988.

6. Calibration and verification of vegetation parameters

The rooting depth D_r of Sea Buckthorn and the change of the fraction of soil covered by the vegetation S_c are taken from Roger(1961). Hence, for the rooting depth a value of 1.2 m is used while S_c is set equal to 0.2 in winter and 1.0 in summer. The increase of S_c takes place during March-April and the reduction during October-November.

The height of the vegetation in lysimeter II appeared to be 3.0 m. The calibration is then restricted to seven parameters which govern the actual transpiration of the crop: albedo, interception reservoir capacity, plant resistance for liquid flow and the four parameters which determine the relation between canopy resistance and leaf water potential (fig. 1). For the most important agricultural crops in The Netherlands the crop parameters have been determined (see e.g. DeLaat, 1985). These values were used as initial parameter values in the calibration procedure.

The calibration was carried out over the three year period 1984-1986. The daily simulated drainage outflows were aggregated into 156 weekly values for comparison with the observed data. The result of the calibration is given below.

Albedo	= 0.24
Maximum interception reservoir capacity S_{max}	= 0.20
Resistance for liquid flow in the plant R_{pl}	= 3000 days
Basic canopy resistance r_b	= 45 s.m^{-1}
Maximum canopy resistance r_{max}	= 350 s.m^{-1}
Parameter h_1 in the relation $r_c(h_L)$	= -14000 cm
Parameter h_2 in the relation $r_c(h_L)$	= -25000 cm

For verification the two year period 1987-1988 was used. Figure 3 shows the simulated and observed drainage rates for the verification period. The excellent fit is also expressed by the high value for the model efficiency RE = 90.5 %. The simulated and observed volumes are almost identical: RV = 99.0 %.

For comparison with lysimeter I, the water balance for lysimeter II is presented for the same time period in table 2.

	Observed	Simulated
Precipitation	4337 mm	4337 mm
Drainage	1712 mm	1694 mm
Evaporation	2625 mm	2629 mm

Table 2. The components of the water balance for lysimeter II for 1984-1988.

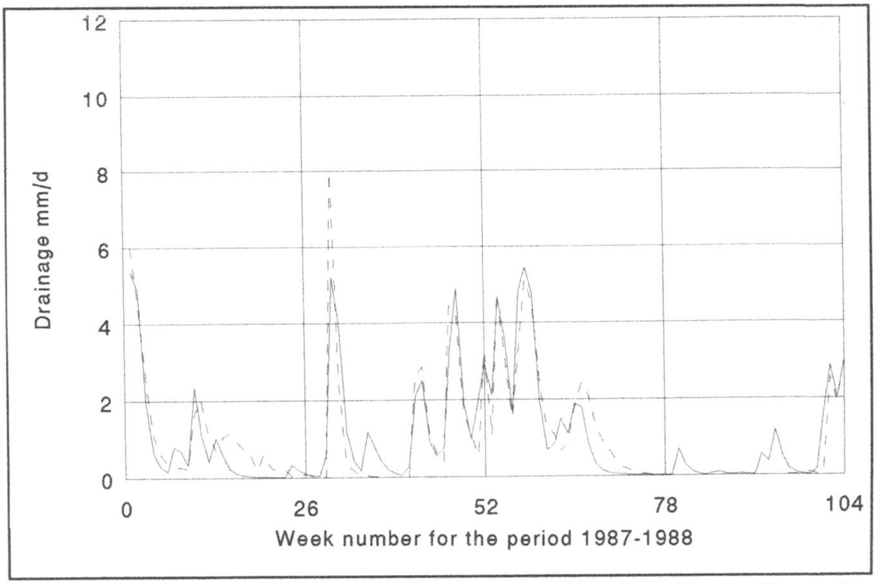

Fig. 3. Simulated (—) and observed (---) drain outflow rates from lysimeter II (verification period)

7. Discussion

The Penman-Monteith-Rijtema approach to the evapotranspiration of cropped surfaces appears to be applicable to natural dune vegetation. The parameter values that describe the relation between the leaf water potential and the canopy resistance are found to be in the same range as those for agricultural crops such as cereals, potatoes and sugar beets. The accuracy of the vegetation parameters as determined in this study is difficult to assess, as there is some uncertainty about the actual rooting depth D_r and the soil physical parameters. A sensitivity analysis showed that a difference of 20 cm in D_r affects the actual evapotranspiration over the calibration (3 year) period by approximately 9

mm/year.
The ability of the model for unsaturated flow and evapotranspiration MUST to simulate accurately the flux at the lower boundary of the unsaturated region in the presence of a shallow water table is clearly demonstrated. During some time steps a large difference may occur between simulated and observed drain outflow due to a slight phase shift. The difficulty in accurately simulating the response time of the lower boundary flux to rainfall excess is well-known. It explains why models for unsaturated flow using a given water table depth as lower boundary condition often produce unrealistic drain outflow rates.

References

Bruin, H.A.R. de, 1981. The determination of (reference crop) evapotranspiration from routine weather data. Comm. Hydrol. Research TNO, The Hague. Proceedings and Informations 28: 25-37.

Hiege, W., 1987. Inventarisatie van de gegevens van het lysimeterstation Castricum 1941-1985. Rapport No. 23, Vakgroep Fysische Geografie en Bodemkunde, Rijksuniversiteit Groningen. 111 p.

Laat, P.J.M. de, 1985. MUST-A Simulation Model for Unsaturated Flow. Report Series No. 16. IHE, Delft, 91 p.

Laat, P.J.M. de, 1992. MUST, A Pseudo Steady-State Approach to Simulating Flow in Unsaturated Media. ICID Bull. 41(2): 49-60.

Monteith, J.L., 1965. Evaporation and Environment. Proc. Symp. Soc. Exp. Biol. 19: 205-234.

Rijtema, P.E., 1965. An analysis of actual evapotranspiration. Agric. Res. Rep. 659, Pudoc, Wageningen, 107 p.

Rijtema, P.E., 1969. Soil moisture forecasting. ICW nota 513, Wageningen, 18 p.

Roger, J.A., 1961. The Autecology of Hippophea Rhamnoides L. Ph.D. thesis, University of Nottingham.

Appendix

Symbol	Definition	Units
b	geometry factor of root system	cm
B	empirical constant (B = 0.04) in (7)	$cm^2.d^{-1}$
C	conversion constant (C = 8640)	m^3.kg-1
c_p	specific heat of air at constant pressure (c_p = 1004)	$J.kg^{-1}.K^{-1}$
c_u	correction factor for wind velocity	-
d	zero-plane displacement	m
D_r	thickness of root zone	cm
e_a	actual vapour pressure at 2 m height	mbar
e_s	saturation vapour pressure at 2 m height	mbar
E	total evapotranspiration flux	$cm.d^{-1}$
E_s	soil evaporation flux	$cm.d^{-1}$
E_T	transpiration flux	$cm.d^{-1}$
E_i	evaporation flux of intercepted water	$cm.d^{-1}$
E_{wet}	evaporation of a wet surface	$cm.d^{-1}$
h_1,h_2	parameters in relation $r_c(h_L)$, see fig. 1	cm
h_L	leaf water potential	cm
h_s	soil water potential	cm
h_{max}	maximum value for h_s in (10)	cm
HU	height at which wind speed is measured	m
k	von Karman constant (k = 0.41)	-
K	hydraulic conductivity	$cm.d^{-1}$
L	latent heat of vaporization ($L = 2.451 \times 10^6$)	$J.kg^{-1}$
r_a	aerodynamic resistance	$s.m^{-1}$
r_b	basic canopy resistance	$s.m^{-1}$
r_c	canopy resistance	$s.m^{-1}$
r_m	maximum canopy resistance	$s.m^{-1}$
R_n	net radiation	$W.m^{-2}$
R_{pl}	crop resistance for liquid flow	d
s	slope of the temperature-saturation vapour pressure curve	-
S_a	actual storage interception reservoir	cm
S_c	fraction of soil covered by the crop	-
$u(x)$	wind speed at height of x metres	$m.s^{-1}$
u_s	wind speed observed at a height of HU metres	$m.s^{-1}$
z_o	roughness length	m
α_s	reduction factor soil evaporation	-
γ	psychrometric constant (γ = 0.67)	$mbar.K^{-1}$
ρ_a	density of air (ρ = 1.2047)	$kg.m^{-3}$

PALEOHYDROLOGY OF QUATERNARY RIVER SYSTEMS OF NORTH BIHAR PLAINS, INDIA

RAJIV SINHA*
Department of Earth Sciences, University of Cambridge
Cambridge CB2 3EQ, England

ABSTRACT The surface impressions on the satellite images, combined with field investigations and subsurface data, have been used to infer the paleohydrologic conditions in parts of the world's largest tract of Quaternary sedimentation. The present-day set-up in the north Bihar plains, eastern India consists of fan and interfan rivers and it appears likely that their relative positions have kept changing. The paleochannel traces seen on the satellite images as well as in the field bear testimony to the migration trends and paleohydrologic conditions of the region. From a sedimentologist's point of view, the floodplain deposits provide information about the morphology and hydraulic regime of the rivers that transported them. Variations in the grain size of the sediments may therefore be indicative of fluctuations of speed, and/or strength in the transporting flows The present study attempts to draw important inferences about the channel pattern changes, channel belt migration trends, paleoflow conditions and depositional history of the Quaternary sediments in the north Bihar plains.

Introduction

The plains of north Bihar are the areas of extensive Quaternary sedimentation. The surface of the plains is marked by a dense network of river systems which form alternating fan and interfan areas. These rivers have been undergoing significant changes in terms of their spatial distribution, channel morphology, hydrology, and sediment characteristics (See Gohain & Parkash, 1990; Mohindra et al, 1992; Sinha, 1992; Sinha & Friend, in press). These changes are recorded on the floodplains of the rivers as paleochannels, cut-offs, ox-bow lakes and varied sequences of depositional features which provide important clues about the paleohydrology of the river systems. The analysis of hydrological and sediment data of the rivers over the years help derive the trend of variation of discharge and sediment load parameters. The characteristics of alluvial deposits viz. their grain size distribution and bedding structures have been shown to be important indicators of paleoflow conditions.

General Set-up of the Area

The study area forms a part of the Ganga basin in the north Bihar plains. The topmost materials in the basin are the unconsolidated deposits of the major present-day river systems, the Ganges and its tributaries. These deposits are together classed as *alluvium*, belonging broadly to the Quaternary. The plains of north Bihar are bounded in the north by the Himalayan Frontal Thrust and by the Peninsular hills in the south. Detailed discussions of topography and drainage pattern of the area is available elsewhere (see Geddes, 1960; Gole & Chitale, 1960; Singh, 1971). The Ganges forms the axial drainage in the Ganga basin generally flowing parallel to the trend of the basin but located well to the south, against the Peninsular hills. The present paper is concerned with the tributary rivers of the Ganges (Fig. 1) that have built the alluvial plains of north Bihar with sediments that originated in the

* Presently at National Institute of Hydrology, Roorkee-247667, INDIA

V. P. Singh and B. Kumar (eds.), Surface-Water Hydrology, 29–41.

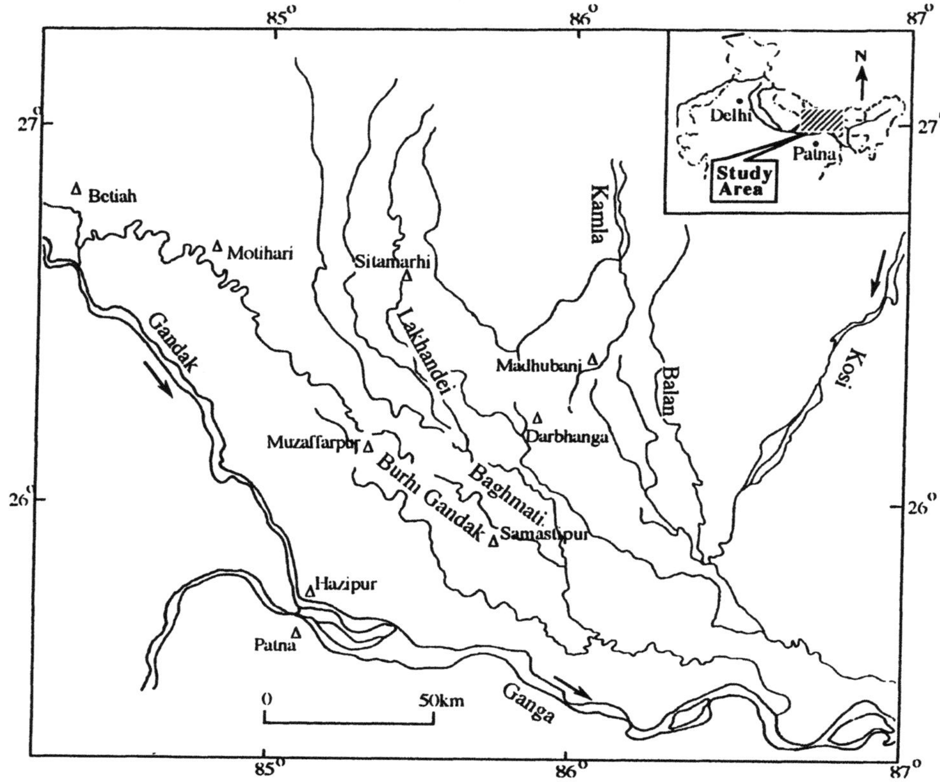

Figure 1. Major drainage of the north Bihar plains; the Ganges flows axially, the Gandak and the Kosi are the major fan rivers and the interfan area is drained by the Burhi Gandak, Baghmati and the Kamla-Balan rivers and their tributaries.

Himalayas. The Gandak and the Kosi are the two major rivers draining the plains which have formed 'megafans' (Gohain and Parkash, 1990; Mohindra et al, 1992). The interfan area is drained by smaller rivers such as the Burhi Gandak, the Baghmati and the Kamla-Balan rivers which are interconnected through minor streamlets.

A cursory study of the drainage alignment (Fig. 1) leads one to visualize two major directions: (i) N-S to NNE-SSW and (ii) WNW-ESE to E-W. While the latter corresponds to the major basinal axis of the Ganga basin, the former directly reflects the sediment input from the Himalayas. The upper catchments of most of these river basins lie in the Himalayan mountain region whereas the lower catchments in alluvial plains where they form the left bank tributaries of the Ganges. In the monsoon period, the interfluve areas are partly covered with flood water due to spills from the rivers on either side and it is almost impossible to delineate the separate courses of the rivers.

Morphological Analysis of Paleochannels

The examination of satellite images as well as field observations suggest that all the rivers of the study area have undergone a wide range of morphological changes (Fig. 2).

A close look at the Gandak - Burhi Gandak interfluve on the satellite images reveals a series of paleochannel traces suggesting a major westward shift of the Gandak river (Mohindra et al, 1992). The most likely mechanism for channel shift seems to be avulsion i.e. the sudden abandonment of a part or whole of a meander belt of a stream for some new course at a lower level on the floodplain (Allen, 1965). Apparently, the Burhi Gandak developed as the remnant drainage of the Gandak as evidenced by a very wide river valley, illogical with respect to the present-day configuration of the Burhi Gandak river. After the major avulsion, the Burhi Gandak river has been changing its course locally (eastward) through minor avulsions and in that process has undergone significant changes in its morphology. In the upstream reaches, extremely sinuous paleochannels and very tight meander cut-offs (Fig. 2) defying the present-day morphology of the river support this observation. A positive build-up of sediments in the adjoining Gandak megafan area in the west might have caused the eastward avulsion of the Burhi Gandak river. Similar avulsive systems have also been described by Richards et al (in press) for the adjoining UP plains.

Cut-offs are formed by catastrophic abandonment of a meander loop and are also one of the major mechanisms by which a river shortens its course. In the reach AB (Fig. 2) of the Burhi Gandak river, the measurements of channel length from toposheets have shown that the cut-offs have caused a decrease in channel length by about 25% between the years 1928 and 1986 (Sinha, 1992).

The paleochannels of the Baghmati river on the satellite image suggest a gradual westward migration of the river. The 'underfit' condition of a few paleochannels of the Baghmati river (Fig. 3) indicates that some cause has operated to reduce the stream size and meander size below their former values. The most plausible cause of underfitness in this case seems to be channel migration through avulsion. It seems likely that following avulsion the paleochannel is receiving much less water and sediment (mainly through monsoonal flow) than the original channel.

The paleochannels of the Kamla - Balan rivers also suggest westward migration of the river but perhaps there has not been any significant change in their morphology.

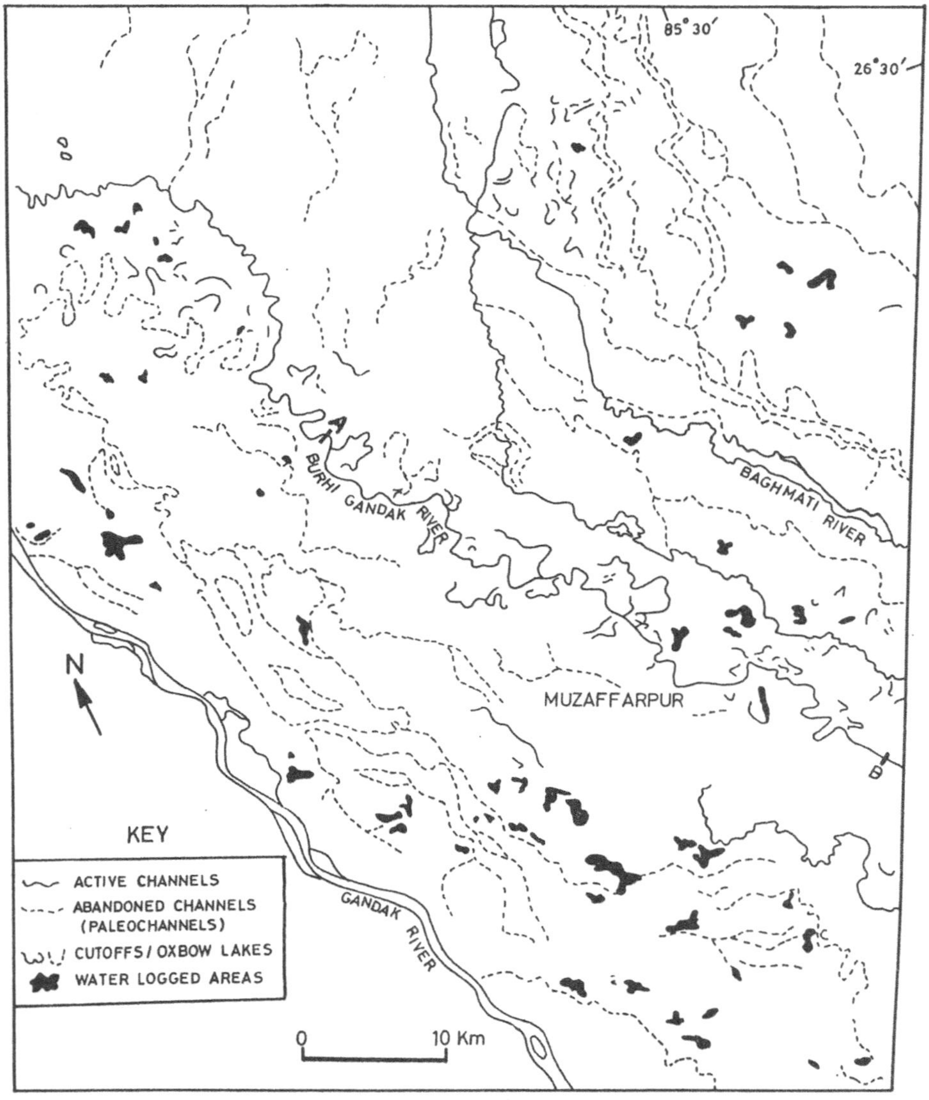

Figure 2. Paleochannel morphology, migration trends and other geomorphological features, as seen on the satellite image. AB is the reach in which the Burhi Gandak shows a decrease in channel length by about 25%.

Figure 3. Field photograph showing an underfit channel (paleochannel) of the Baghmati river; the underfitness implies that the water and sediment flow in the channel was much larger than at present.

Quantity and Mode of Water and Sediment Transport

Water and sediment transport behaviour of north Bihar rivers have been studied by analyzing discharge and sediment load data for the last 10 years (1980-89). The variations in discharge-sediment data are reflected in the floodplain morphology and deposits. The data therefore help in a better understanding of floodplain features and vertical sequences.

The average monthly discharge (averaged over 10 years) plot (Fig. 4a) clearly highlights the point that the fan rivers have much higher discharge values throughout the year than the interfan rivers. The contribution of snowmelt in case of fan rivers causes an early rise of discharge (in May) whereas the interfan rivers show the typical monsoonal rise of the discharge curve. The values of peak discharge (highest discharge observed in a year) are extremely fluctuating for all the rivers (Fig. 4b), presumably in response to variation in monsoon and due to tributary influence. In general, the values for the fan rivers are higher and show a larger variation as compared to the interfan rivers.

The analysis of sediment load data has indicated that the major fan rivers are characterised by low percentage of silt-clay in their suspended load whereas the smaller interfan rivers have high concentration of 'fines' in their suspended load (Sinha and Friend, in press). This is reflected in the *sandy* and *muddy banks* of the fan and interfan rivers respectively.

The amount of sediment transported varies with flow discharge (Fig. 4c & d) and the patterns of variation in case of fan and interfan rivers are distinctly different. The fan rivers show a very rapid increase in sediment load after a certain value of discharge (about 5000

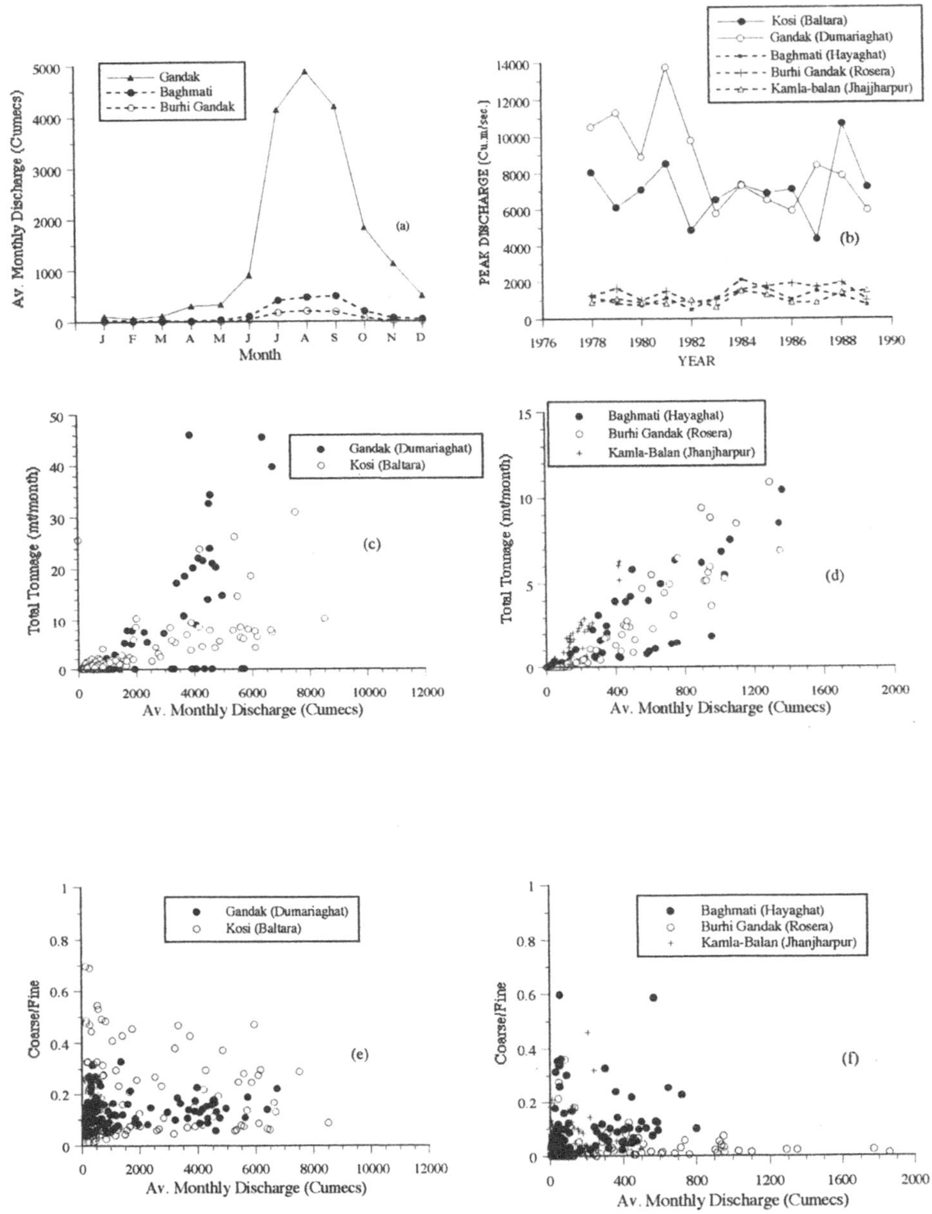

Figure 4. Hydrology and sediment characteristics of the present-day river systems of north Bihar; (a) Av. monthly discharge variation; (b) Peak discharge variation ; (c) Av. monthly discharge vs total tonnage for fan rivers at downstream stations; (d) Av. monthly discharge vs total tonnage for interfan rivers at downstream stations; (e) Av. monthly discharge vs. coarse-fine ratio of suspended sediments for fan rivers at downstream stations; (f) Av. monthly discharge vs. coarse-fine ratio of suspended sediments for fan rivers at downstream stations.

cumecs). This discharge may be some kind of a threshold value after which a large amount of sediment from bars and banks go into suspension thereby increasing the total suspended load to a great extent. On the contrary, the interfan rivers show a steady increase in suspended load with discharge; the degree and extent of variation are different in different rivers. The plot of coarse to fine ratio with discharge (Fig. 4e & f) clearly illustrates that the increase is mainly in terms of fine sediments (wash load). The rather stable coarse-fine ratio of fan rivers reflects the ready and uniform availability of coarse and fine material. On the other hand, the coarse-fine ratio tends to decrease in case of interfan rivers reflecting the increase in fine sediments with discharge perhaps due to erosion of muddy banks. The most usual effect of this seasonal variation is seen when the river overtops its banks and deposits a *fining upward sequence.*

Alluvial Deposits as indicators of Paleoflow conditions

The alluvial deposits can easily be interpreted in terms of morphology and hydrology of the rivers that deposited them. Grain size has long been regarded as one of the most important indicator of transporting flow. Several theoretical models (e.g. Allen, 1985) predicted that larger grains require stronger flows to move them. Schumm (1960, 1977) derived his 'M' index as a measure of grain size and correlated it with width/depth ratio of the channel cross sections which in turn would reflect flow conditions. A few recent publications (Friend, 1993; Friend and Sinha, in press) have also demonstrated that variations of grain size of the deposits can be interpreted in terms of flow characteristics and resulting channel morphology.

SHALLOW VERTICAL SEQUENCES

The vertical sequences of the alluvial deposits can reveal information about their depositional regime, particularly in terms of their proximity to the channels and hence they indirectly help to deduce flow conditions. Some of the typical sequences from the north Bihar plains studied through shallow excavations are shown in Figure 5 and interpretations about their depositional history follows :

(i) Predominantly sandy horizons capped by muddy layer (Fig. 5a) may be interpreted as *point bar* or *channel plug* deposits. The sandy horizons represent channel deposition under strong flow conditions whereas the muddy layer indicate slow siltation of the suspended sediments under low flow conditions.

(ii) Interbedded sequences of muddy and sandy horizons (Fig. 5b) may be classified as *natural levees* which indicate alternating high and low flow conditions and close proximity to the channels (channel margins).

(iii) Minor to very minor sand bodies interbedded in an otherwise muddy matrix are *crevasse* deposits (Fig. 5c) and they form beyond the main channel at high stage when excess water leaves the channel through isolated low sections or breaches in the natural levees.

(iv) Predominantly muddy (at least 70% and upto 100%) deposits constitute the *overbank* deposits (Fig. 5d) and are deposited away from the active channel in the floodbasin areas through settling of suspended fine grained sediments from overbank flows.

BEDDING STRUCTURES

Bedding structures on the point bar sections provide valuable information about the depositing flow. Deposition at the point bar is very rapid. Mostly, small ripple bedding is observed, sometimes associated with horizontal bedding. Two point bar sections from the

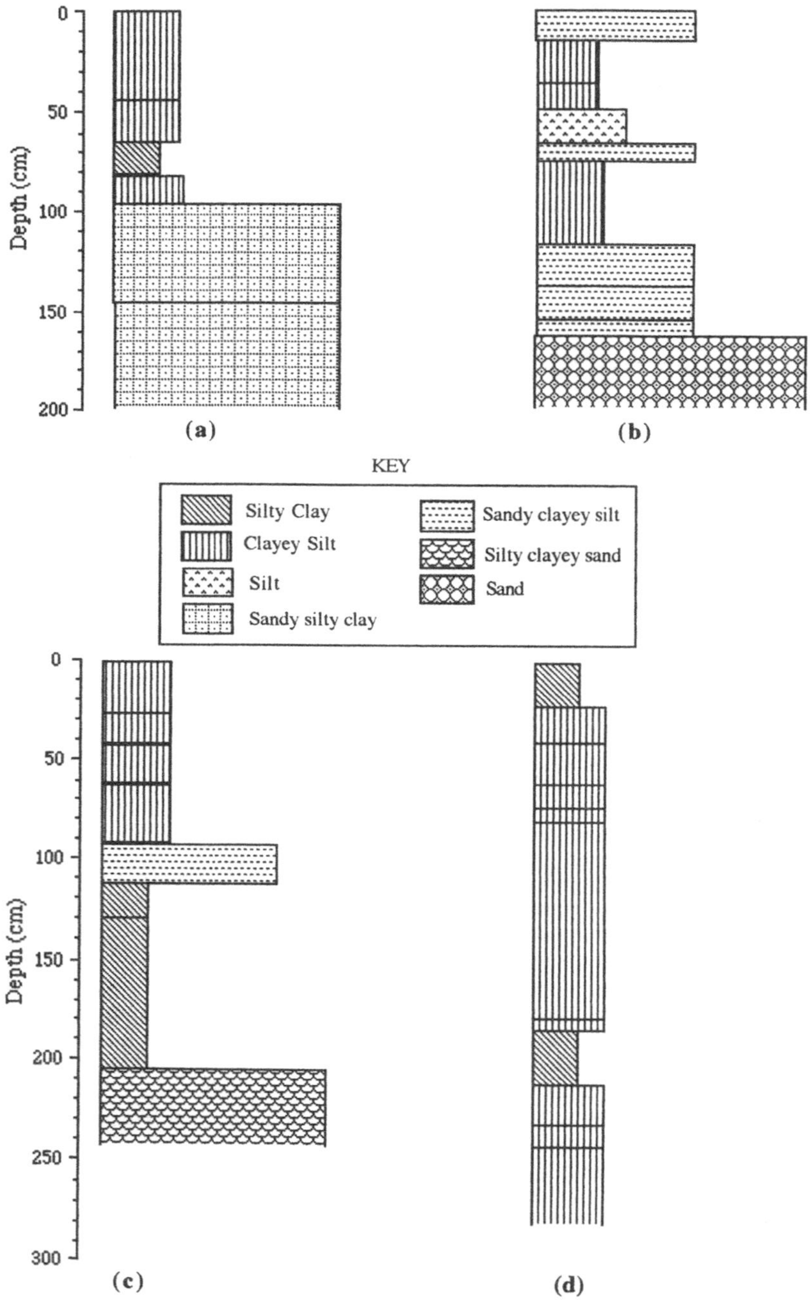

Figure 5 Vertical sequences depicting the flow conditions of the river that deposited them, (a) Typical fining upward point bar sequence from the Burhi Gandak floodplain, (b) Alternate sandy- muddy natural levee sequence from the Baghmati floodplain, (c) Sandy patches in a muddy sequence (crevasse deposits) from the Burhi Gandak floodplain, (d) Predominantly muddy overbank sequence from the Kamla-Balan floodplain (see text for interpretation)

north Bihar plains shown in Figure 6a and 6b illustrate the utility of bedding structures to infer depositional flows.

Figure 6a shows a point bar section in the Burhi Gandak floodplain, near Muzaffarpur in which *small ripple bedded units* are interbedded with horizontal bedded layers. The horizontal units show an erosional contact at the top, indicating thereby that they were eroded before the ripple-bedded units were deposited. The lower contact of the horizontal units seems to be non-erosional. It is well established, through hydrodynamic studies, that small ripples form in lower flow regimes, with Froude number < 1 (Simons et. al., 1965). On the other hand, horizontal laminations can be produced in several ways (Reineck & Singh, 1973). However, this particular association of ripple and horizontal laminations with an erosional contact indicates deposition in a pulsating environment where lower-flow regime conditions alternated with upper-flow regime conditions in quick succession. Much of the horizontal bedding may have been produced by the deposition of suspended fine sand due to a decrease in turbulence, or to fluctuations in the current velocity. A major part is deposited during the last phase of a receding flood when velocity and competence of the flow are reduced. This may have resulted in the deposition of suspended sand in the form of horizontal laminations.

Figure 6b shows an excellent example of *climbing ripple laminations* at a point bar section in the Baghmati floodplain near Minapur, in which small scale ripple-bedded units show a strong climbing upward trend. The development of climbing ripple laminations requires that abundant sediment is continually available to a current so that the ripples are built upward in an overlapping series rather than merely migrating in a forward direction. Most rivers in the study area have a very high suspended load with a high concentration of fine sand, especially in the midstream to downstream reaches (Sinha and Friend, in press). When much sediment is available, especially in suspension, the extra sand quickly buries and preserves the original rippled layers and a series of superimposed rippled layers are formed that eventually result in climbing-ripple lamination. In the upper part of the sequence, the climbing ripple laminations are followed by very small ripple cross-bedding which may reflect the fluctuations in the flood phase causing change in the depositional energy and the resulting bedforms.

BORE HOLE DATA AND MULTICYCLIC DEPOSITION

Apart from the analysis of shallow excavations, deep borehole data can give information about the long-term depositional cycle in the region. Frequent avulsion by fan as well as by interfan rivers builds a case for *multicyclic sedimentation* in the area. Each cycle may be thought of as starting with a sandy sequence (representing channel) and ending with a muddy sequence (representing overbank environment). This interpretation would probably hold true for the greater part of the thick alluvial fill in the plains, as is demonstrated by deep borehole data available from the north Bihar area. Figure 7 shows three borehole sequences of upto 120 metres at sites near Muzaffarpur. Some general interpretations from these sequences are as follows :

(i) Sequence I consists of alternations of thick sandy and muddy units and may be interpreted as an *alternation of channel and overbank environments*, the latter being predominant.

(ii) Sequence II consists of predominantly thick sandy units in the lower portion whereas the upper portion has relatively thin sandy layers with thicker mud layers in between. This may be interpreted to have started with *channel-fill* sediments, passing on to *channel margin (levee)* and then to *overbank* deposits.

Figure 6(a). Field photograph showing a point bar section on the Burhi Gandak floodplain; small ripple units interbedded with horizontal layers indicate deposition in pulsating environment.

Figure 6(b). Field photograph showing a point bar section on the Baghmati river floodplain; climbing ripple laminations have formed due to the availability of much sediment available in suspension.

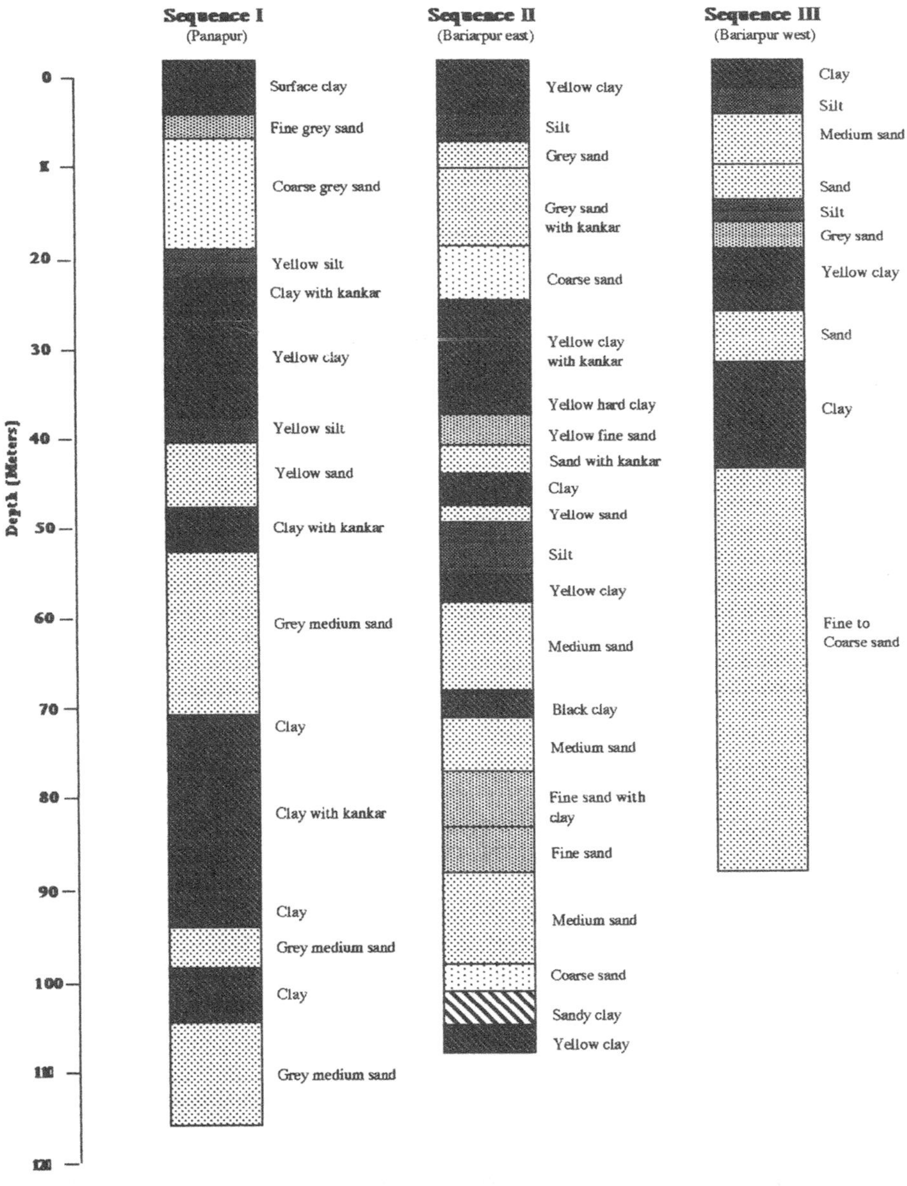

Figure 7 Deep borehole data from the north Bihar plains near Muzaffarpur illustrating multicyclic sedimentation sequence I consisting of alternate sandy and muddy layers may be interpreted as an alternation of channel and overbank environments the latter being predominant sequence II is interpreted to have started with channel fill sediments predominantly sandy passing on to channel margin and overbank sequence with more muddy layers sequence III is predominantly sandy with a cover of sand silt alternations and may represent a typical channel deposit with a cover of levee deposit

(iii) Sequence III is predominantly sandy in most parts, except for the upper 25 meters consisting of alternations of sandy and silty layers. This may represent a typical *channel-fill* deposit with a cover of *levee* deposit.

The borehole data provide valuable information about the dimension of the sandbodies below the plains. Most of these sandbodies may be classified as sand sheets; the thickness of the individual sandbodies indicated by the sequences I, II and III ranges from at least 1.6m and up to over 50m. The thick sandbodies (>20m) may be the result of vertical stacking of sand sheets due to reoccupation after avulsion.

Summary and Conclusions

The present paper illustrates the importance of paleochannel morphology to deduce the paleohydrology of the river systems of the region. The distribution of paleochannel traces on the satellite images and their field settings help in the reconstruction of the drainage system, migration trends, and causes of variations in water and sediment discharge parameters. All rivers of north Bihar show evidences of migration through avulsion and the most likely mechanism of avulsion seems to be sedimentological readjustments or tectonic disturbances. Channel migration has also resulted in underfitness of channels in certain reaches.

It is also established that the characteristics of the alluvial deposits provide valuable information about the depositional regime and paleoflow conditions. The deposits may be classified in terms of their grain size distribution in a vertical sequence which in turn would indicate the proximity to the channel. Bedding structures observed on the point bar sections may indicate particular flow conditions prevailed at the time of deposition.

Acknowledgements

The study presented in this paper forms a part of author's PhD Thesis at the University of Cambridge, England. The author is thankful to the Nehru Trust for Cambridge University, New Delhi and Cambridge Commonwealth Trust, Cambridge for providing a research studentship for three years at Cambridge. For hydrological data of the study area, the help of Central Water Commission, Govt. of India is sincerely acknowledged. Prof. B. Parkash, University of Roorkee and Dr. Sudhir Kumar, Scientist, National Institute of Hydrology, Roorkee provided useful comments on the paper.

References

Allen, J. R. L. 1965. A review of the origin and characteristics of recent alluvium sediments. Sedimentology, **5**, p 89-191

Allen, J. A. (1985). Principles of physical sedimentology. George Allen & Unwin, London.

Friend, P. F. (1993). Control of river morphology by the grain size of the sediment supplied. Sedimentary Geology, **85**, p 171-177.

Friend, P. F. a. Sinha, R. (in press). Braiding and Meandering Parameters. Journal of Geological Society, London (Braided River Conference), edited by J. Best and C. S. Bristow.

Geddes, A. (1960). The alluvial morphology of the Indo-gangetic plain: its mapping and geographical significance. Institute of British Geographers, Pub. No. **28**, p 253-278.

Gohain, K., and Parkash, B. (1990). Morphology of Kosi Mega fan. In *Alluvial fans : A Field Approach.* (eds A. H. Rachocki and Church. M.), p 151-178, John Wiley and Sons Ltd.

Gole, C. V., and Chitale S. V. (1966). Inland delta building activity of Kosi river. Journal of the Hydraulics Division, Proceedings American Society of Civil Engineers, **92**, p 111-126.

Mohindra, R., Parkash, B. and Prasad, J. (1992). Historical Geomorphology and pedology of the Gandak megafan, middle Gangetic plains, India. Earth Surface Processes & Landforms, **17**, p 643-662.

Reineck, H. E. and Singh, I. B. (1973). *Depositional sedimentary environments.* Springer-Verlag, Heidelberg.

Richards, K. S., Chandra, S. and Friend, P. F. (in Press) . Avulsive channel systems : Characteristics and examples. Journal of Geological Society, London (Braided River Conference), edited by J. Best and C. S. Bristow.

Schumm, S. A. (1960). The shape of alluvial channels in relation to sediment type. In *Erosion and sedimentation in a semiarid environment* (eds S. A. Schumm), p 17-30, Washington, Geological Society of America Professional Paper.

Schumm, S. A. (1977). *The Fluvial System.* John Wiley and Sons.

Simons, D. B., Richardson, E. V. and Nordin, C. F. (1965). Sedimentary structures generated by flow in alluvial channels. In Primary sedimentary structures and their hydrodynamic interpretation (eds G. V. Middleton), pp 34-52, Society of Economic Paleontologists & Mineralogists Special Publication.

Singh, R. L. 1971. India-A Regional Geography. National Geographic Society of India, Silver Jublee Publication, Varanasi.

Sinha, R. (1992). Quaternary alluvial sedimentology of the Gandak-Kosi Interfan, North Bihar, India. Unpublished PhD Thesis, University of Cambridge, England.

Sinha, R. and Friend, P. F. (in press). River systems and their sediment flux, Indo-Gangetic plains, north Bihar, India. Sedimentology.

Modelling the Effects of Climate Change on the Hydrologic Response of a Mountain Watershed

Ashutosh S Limaye[1], Alok K. Sikka[2], B. Rajagopalan[3], and J. Paul Riley[4]

Abstract

Climate changes are likely to have serious impacts on water supply and demand in arid and semi-arid basins. We consider CO_2 - induced changes in vegetation and the changes in temperature and precipitation scenarios in this analysis. In this paper, we present application of a distributed parameter model to simulate the effects of climate and vegetation changes on the hydrologic response of Chalk Creek watershed in the Weber River basin in Utah. Leaf area index (LAI) is used as a measure of forest structure to quantify energy and mass exchange. Hydrologic response unit (HRU) is used to partition the watershed into spatially distributed units. Sensitivity analysis assumed changes in climate and plant variables over credible ranges of change based on the GCMs output and current literature. For the range of scenarios studied, the results suggest greater effect of warming on runoff than for cooling. A greater sensitivity of annual runoff to precipitation than to temperature was in agreement with the results of other studies done else where. The results, in general, indicate a marked shift in the timing and seasonality of runoff.

Introduction

A growing body of scientific opinion predicts a continued global climate change with regional imbalances over the next century as a result of increasing concentrations of "greenhouse" gases in the atmosphere. Water supply, quality, and demand may be affected by climate changes in arid and semi-arid regions. These changes may have ramifications for making decisions on water allocations and water rights in the future. For this reason, a quantitative assessment of climate change impacts on water resources assumes an urgent priority in the semi-arid western United States. Water resources of the Weber River basin in the semi-arid zone of the western United States are extensively developed, with the flow being

---

[1]Graduate Research Assistant, Civil and Environmental Engineering, Utah State University, Logan, UT 84322-4110 (801) 750-4011

[2]Forest Sciences Lab, Oregon State University, Corvallis, OR 97330-7501 (503) 750-7409

[3]Graduate Research Assistant, Utah Water Research Laboratory, Utah State University, Logan, UT 84322-8220 (801) 750-3151

[4]Professor, Civil and Environmental Engineering, Utah State University, Logan, UT 84322-4110 (801) 750-2783

V. P. Singh and B. Kumar (eds), Surface-Water Hydrology, 43–52.

regulated by the five major reservoirs. In past no study has been done in this basin to assess the water yield under projected climate change. We recently started hydrologic modeling studies at the utah State University (USU) to examine these impacts.

Coarse resolution and simplified hydrologic parameterization in the general circulation models (GCMs) make the direct use of their outputs for hydrologic interpretation a risky proposition at the watershed scale. Until the spatial resolutions and hydrologic parameterizations of GCM's are improved, the hydrologic effects of climate change can best be examined by using watershed hydrologic models. However, to realistically represent the effects of climate changes on watershed response, hydrologic models need to have a relatively high degree of spatial resolution and also be capable of simulating the effects of changes in watershed vegetation which might result from climatic changes, including elevated levels of CO_2 in the atmosphere. This raises the question as to how well existing hydrologic models reflect climate-vegetation-evapotranspiration (ET) and hydrologic interactions. For this reason, models which do not consider the effects of vegetative changes and which contain averaging effects from large spatial variations of the physical and meteorological variables can produce inaccurate estimates of runoff volumes or streamflows.

A spatially distributed estimation procedure which includes a realistic representation of vegetative changes is needed to better understand and model the changes in hydrologic responses of watersheds as a result of climate and CO_2 induced vegetation changes. A recent attempt was made at the USU (Sikka, 1993) to develop such a model after incorporating modifications, to explicitly link vegetation with hydrology, in the PRMS model of USGS (Leavesley et al., 1983). The model can simulate the effects of climate change and vegetation change (LAI, stomatal conductance) scenarios on hydrology of a vegetated mountain watershed. The model gave encouraging results in the Causey watershed of the Weber River basin. We are applying the model to different upper, unregulated catchments to model the response of the basin as a whole.

This paper presents the application and results of this distributed parameter model to the Chalk Creek subwatershed in the Weber basin to simulate the effects of climate and vegetation change scenarios on the runoff characteristics of the watershed.

Methods of Analysis

Hydrologic Model

It is a deterministic, physical process based, distributed parameter daily hydrologic simulation model. The model uses daily data of precipitation, temperature, and solar radiation. The development of a distributed parameter evapotranspiration model for vegetated mountain watersheds and modification of an existing hydrologic model is described in Sikka (1993). The watershed is divided into spatial subunits, called hydrologic response units (HRUs), to provide a distributed parameter capability. Leaf area index (LAI) is used as a measure of forest structure to quantify energy and mass exchange. LAI is changed over the growing season. Water and energy balances are computed daily for each HRU and the weighted sum of their response on a unit area basis produces the overall response of the watershed. The model needs site variables, initial conditions of state variables, and a set of topogrphic, meteorologic, and soils, and vegetation parameters. Parameters generally relate to particular watershed characteristics or conditions. They can be measured or derived empirically from historic data and/or relevant information from the literature. The model, which is based on the mass water balance equation, treats the watershed system as a series of linear and nonlinear reservoirs whose outputs are combined to produce total system response in terms of mean daily streamflow.

For the purpose of this study, vegetation has been broadly categorized as a) evergreen/conifers; b) deciduous, c) shrubs and grasses, and d) mixed vegetation in an HRU. The model treats vegetation canopy as being proportional to the LAI (the area of leaves per unit area of ground, taking one side of leaf into account, that is, the projected area). The uniform depth of canopy defined by the LAI can be divided into a number of horizontal layers. Evaporation from the soil and plant canopy, and transpiration from plants is modeled separately to compute actual evapotranspiration.

Available radiation energy (incoming shortwave radiation) is partitioned in such a way that potential radiation energy is available first for evaporating intercepted water, with the remaining energy being divided between soil evaporation and transpiration based on the LAI and a species dependent light extinction coefficient. The soil moisture model consists of two layers, namely, the top 15 cm layer and from that level to the rooting depth. Soil evaporation is assumed to take place only from the upper layer, with the potential soil evaporation being limited by the available soil water in this layer. The soil profile depth is user defined and depends upon the effective root zone depth of the pre-dominant vegetation in the HRU. Transpiration is assumed to take place from the entire soil profile depth depending on the relative moisture availability in the upper and lower soil layers.

Actual transpiration is computed using the Penman-Monteith (PM) equation. To account for soil moisture depletion, the PM equation is linked to a soil moisture balance model so that the canopy resistance term is adjusted for the available soil moisture. The canopy resistance term is then sequentially corrected for humidity, temperature, and radiation. Soil moisture is updated at the end of each day by accounting for ET, surface runoff, infiltration, and percolation. Daily infiltration is computed as the difference between daily precipitation and surface runoff. Surface runoff is computed using the contributing area concept (Dickinson & Whiteley, 1970 and Hewlett and Nutter, 1970). The runoff and snowmelt components of the Precipitation Runoff Modelling System (PRMS) are used in the model (Leavesley et al., 1983).

Leaf area index (LAI) is kept constant during the non active transpiration period and it is changed during the active transpiration period. The beginning and end of the active transpiration period are determined on the basis of the accumulated weekly degree-day index From the first day of the active season until full leaf, the LAI increases linearly from an initial value to a user defined peak. It remains constant at the peak value until the beginning of leaf color change in the fall, and thereafter it decreases to the initial LAI value at the end of growing season.

Watershed Partitioning into HRUs

Watershed characteristics such as elevation, aspect, slope, vegetation, and soils were used in partitioning the Chalk creek watershed into HRUs. Vegetation map of the watershed was obtained from the Department of Fisheries and wildlife, USU. Using quad maps composites for the watershed; vegetation map; and soils information from the reports of USDA Soil Conservation Service, hydrologic response units (HRUs) were delineated. For a given HRU, single value of a parameter is used.

Model Calibration

The model was calibrated for the Chalk Creek watershed in northern Utah using precipitation and runoff data for the 1989 water year. Precipitation and temperature data from Coalville station were used. Meteorologic parameters such as maximum and minimum monthly temperature lapse rates, precipitation correction factor were estimated using the data from nearby stations. Soils parameters were largely obtained based on SCS Reports. The Thematic Mapper (TM) data was used for computing Normalized Difference Vegetative Index (NDVI). Relationships between NDVI and leaf area index (LAI), available for broad vegetation types were used for computing LAI (Sikka, 1992). Some parameter values obtained from from the literature. Spatially varying parameters were obtained for each HRU.

The runoff volumes and the hydrograph shape were used in the calibration process. The matching of streamflow was first done for water balance volumes and then followed with fitting the timings of flows. It was also ensured that the simulated outputs of intermediate processes such as soil moisture, snowmelt timing and snowpack water equivalent make a reasonable sens. The simulated daily hydrograph matched well with the observed hydrograph, with the daily and monthly coefficients of determination (r^2) being 0.88 and 0.97 respectively. The parameters values which were established by the calibration procedure were used for conducting the sensitivity analyses involving various climate change scenarios.

Sensitivity Analyses

Scenarios of changes in temperature, precipitation, LAI, and stomatal conductance were used as inputs to the watershed hydrologic model to assess the potential impacts of climate change on runoff from the Chalk Creek watershed. Because GCMs are currently not able to provide realistic meteorological data at the basin scale, assumed scenarios were used for this study. Extreme climate and vegetation change scenarios are based on conditions which reflect a doubling of the atmospheric CO_2 content. The values for scenarios are estimated from available GCM output information and data from current literature (Flaschka, Stockton, and Boggess 1987, Gleick 1987, Schaake 1990, Rosenberg et al. 1990, and Running and Nemani 1991). The temperature (T), precipitation (P), LAI (L), and stomatal conductance (CC) changes for each scenario are given in Table 1.

Unlike most of the conventional studies, the temperature changes were not applied uniformly throughout, but instead are based on a disaggregation approach. Maximum canopy conductance value was changed as a parameter for different simulation runs to represent change in stomatal conductance due to elevated levels of CO_2. Model runs were made for single and combined scenarios.

Results and Conclusions

The effects of assumed climate change scenarios on annual runoff are given in Table 2. The magnitudes of changes in annual runoff for the assumed scenarios ranged from a decrease in annual runoff of 33 percent (T+4°F, P-10%) to an increase in runoff of 27 percent (P+10). Annual runoff decreased by 6, 11, and 25 percent for temperature increases of 2°, 4°, and 6° F, respectively, with no change in precipitation. The effect on runoff is greater for warming than for cooling. This result is explained by the aridity of the basin. Warming combined with an increase in stomatal resistance and LAI produced a decrease in annual runoff. The changes in ET produced by warming largely occur during late spring and summer, and thus have little effect on annual runoff dominated by snowmelt. A greater sensitivity of annual runoff to precipitation than to temperature agrees with the results of recent studies conducted in the Colorado River basin (Nash Gleick 1991, and Schaake 1990).

The effect of the assumed scenarios on the timing of daily runoff is illustrated in Figures 1 and 2. Warming in general causes peak runoff to occur earlier in the year. Temperature increases of 4° and 6° F alone and in combination with other changes cause the peak runoff to occur in April rather than in May, while a temperature increase of 8° F shifts the peak runoff backwards to March from May. The magnitude of the peak is reduced in all the cases except those with increased precipitation. The results, in general, indicate a marked shift in the timing and seasonality of runoff with a significant increase in winter and early spring runoff followed by a decrease in late spring and summer runoff. Scenarios with increased precipitation result in corresponding increases in the annual runoff. However, spring and summer runoff are decreased because of early snowmelt.

The results of this study indicate that significant hydrologic impacts would result from climate change, but the impacts appear to be less severe than the results reported by others for regions in western United States (Nash & Gleick 1991, Schaake 1990, and Flaschka et al. 1987). The study provides useful preliminary results of the hydrologic implications of climate change on the Chalk Creek subwatershed. However, these results are based on model calibrated with limited data and further research is in progress for evaluating the potential impacts of climate change on water yield in the entire Weber River basin.

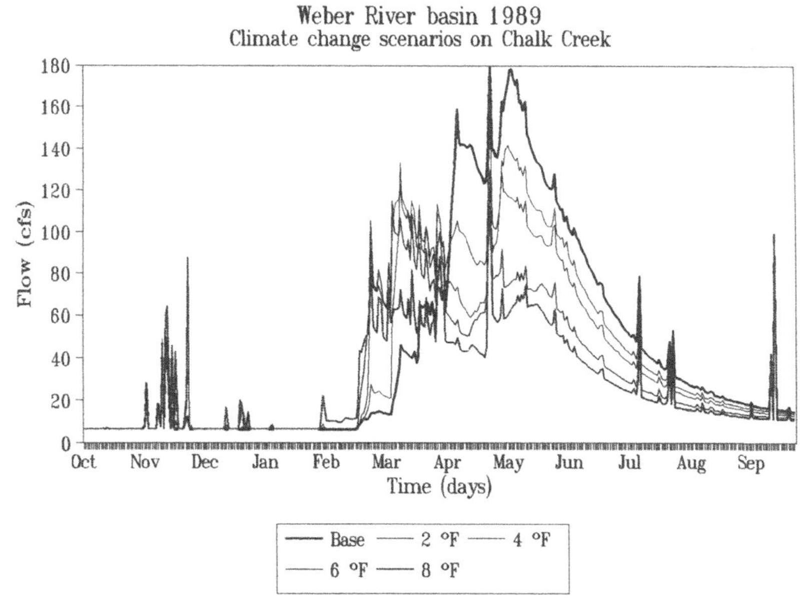

Figure 1: The effects of climate change temperature scenarios on Chalk Creek runoff.

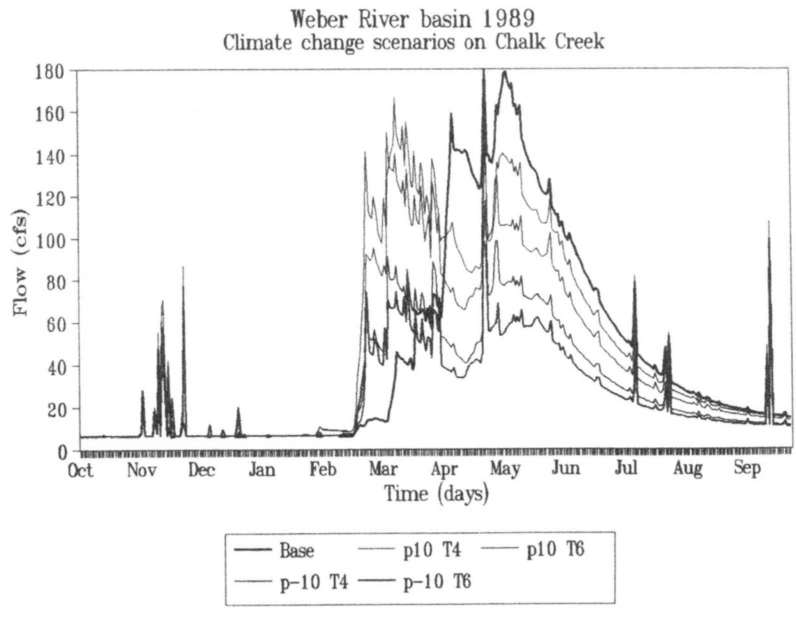

Figure 2: The effects of climate change precipitation and temperature scenarios on Chalk Creek runoff.

Table 1: Climate change and CO_2-induced vegetation change scenarios used.

Scenarios	Temperature change (° F)	Precipitation change (%)	Leaf-area index (%)	Stomatal Cond. (%)
Base	0	0	0	0
T 2	2	0	0	0
T 4	4	0	0	0
T 6	6	0	0	0
T 8	8	0	0	0
p 5	0	5	0	0
p 10	0	10	0	0
p 15	0	15	0	0
p -5	0	-5	0	0
p -10	0	-10	0	0
p -15	0	-15	0	0
p10 T4	4	10	0	0
p10 T6	6	10	0	0
p-10 T4	4	-10	0	0
p-10 T6	6	-10	0	0
C-2 L15 T4	4	0	15	-20
C-2 L15 P10 T4	4	10	15	-20
C-2 L15 P-10 T4	4	-10	15	-20
C-3 L25 T6	6	0	25	-30

Table 2: The effects of climate change scenarios on flows from Chalk Creek watershed.

Scenarios	Flows from Chalk Creek (in CFS)	Flow reduction from Chalk Creek as % of flow at base scenario.
Base	43.93	1
T 2	41.09	6.46
T 4	38.82	11.63
T 6	32.84	25.24
T 8	27.68	36.99
p 5	49.9	-13.59
p 10	56.02	-27.52
p 15	63.56	-44.68
p -5	37.46	14.73
p -10	32.48	26.06
p -15	28.31	35.56
p10 T4	46.95	-6.87
p10 T6	42.41	3.46
p-10 T4	29.35	33.19
p-10 T6	25.32	42.36
C-2 L15 T4	40.09	8.74
C-2 L15 P10 T4	49.94	-13.68
C-2 L15 P-10 T4	30.62	30.29
C-3 L25 T6	35.63	18.89

References

Dickinson, W. T., and Whiteley, H. (1970). "Watershed areas contributing to runoff." IASH Pub. No.96, 1.12-1.28.

Flaschka, J. M., Stockton, C. W., and Boggess, W. R. (1987). "Climatic variation and surface water resources in the Great basin region." Water Resources Bulletin, 23(1), 47-57.

Gleick, P. H. (1987). "Regional hydrologic consequences of increases in atmospheric CO_2 and other trace gases." Climatic Change, 10, 137-161.

Gunderson, R. W., Riley, J. P., and Sikka, A. K. (1992). "An interactive graphics fuzzy clustering procedure for partitioning a watershed into hydrologic response units." Abstract in AGU Fall meeting, Eos, Transactions, 73(43), 109.

Hewlett, J. D., and Nutter, W. L. (1970). "The varying source area of streamflow from upland basins." Paper presented at Symposium on Interdisciplinary aspects of Watershed Management, Montana State University, Bozeman, Aug. 3-6, 1970.

Kruse, E. G., and Haise, H. R. (1974). "Water use by native grasses in high altitude Colorado meadows." ARS-W-6, USDA.

Leavesley, G. H., Lichty, R. W., Troutman, B. M., and Saindon, L. G. (1983). "Precipitation Runoff Modelling System - users manual." USGS Water Resources Invest. Rep. 83-4238.

Nash, L. L., and Gleick, P. H. (1991). "Sensitivity of streamflow in the Colorado basin to Climatic changes." J. Hydrology, 125, 221-241.

Rosenberg, N. J., Kimball, B. A., Martin, P., and Cooper, C. (1990)."From climate and CO_2 enrichment to evapotranspiration. In Climate Change and U.S. Water Resources, P.E. Waggoner, (Ed.), John Wiley and Sons, 151-176.

Running, S. W., and Nemani, R. R. (1991). "Regional hydrologic and carbon balance responses of forests resulting from potential climate change." Climatic Change.

Schaake, J. C. (1990). " From climate to flow." In Climate Changeand U.S. Water Resources, P.E. Waggoner, (Ed.), John Wiley and Sons, 177-206.

Sikka, A. K. (1993). "Modelling the influence of climate change on evapotranspiration and water yield." Unpublished Ph.D. Dissertation submitted to Utah State University, Logan Utah.

OPERATIONAL HYDROLOGY ASPECTS OF DESERTIFICATION

ARIE BEN-ZVI
Israel Hydrological Service
P.O. Box 6381
Jerusalem, 91063
ISRAEL

ABSTRACT. Operationl hydrology deals with methods for providing the professional background for collection, storage and dissemination of information about the state of water resources and for preparation of hydrologic predictions and forecasts. Provision of this background is difficult when dealing with desertification, which is a gradual transition of an area from a semi-arid into an arid state. As the precipitation and water flow magnitudes, for either state, are highly variable, distinction between the two states requires a delicate analysis. Owing to the difficulties in operating stations in arid and arid-prone areas, the quality of data does not reach the standards which are usually met in more humid areas. Yet, in consideration of the deep consequences of desertification, its assessment is often requested before sufficient data are available. The wide-range fluctuations in magnitudes of hydrologic variables might cause then variations in the assessment results.

1. Introduction

1.1 DEFINITIONS

Desertification is the process leading to intensification or extension of desert conditions, where vegetation is scarce or absent because of deficient rainfall or edaphic aridity (after U.N., 1977a). The process is involved with a reduction of soil moisture, air humidity and plant biomass, which cause, among others, changes in the hydrological properties of the area.

Operational hydrology is the branch of hydrology which deals with the collection, transmission, processing, storage and dissemination of data on hydrologic variables; the preparation of predictions and forecasts on the state of hydrologic variables; and the development of methods, networks and instrumentation for these purposes (after WMO, 1993).

V. P. Singh and B. Kumar (eds.), Surface-Water Hydrology, 53–60.

1.2 SCOPE

This article deals with methods and means to monitor and assess hydrological changes which are associated with desertification. Its main body consists of descriptions of the hydrological aspects of desertification, and of the relevant activities in the field of operational hydrology. It describes the links between these subjects, but does not include quntitative criteria and models. In the present level of knowledge, and with respect to the needs of expected readers, such an elaboration might be misleading.

2. Hydrological Aspects of Desertification

2.1 THE PARTICULAR HYDROLOGY OF DESERTS

Deserts are characterized by deficiency of moisture with respect to the needs of a perennial or of a seasonal vegetation. The potential rate of evaporation is high, while the precipitation events are rare and of short duration. As a result, the moisture which is supplied by the precipitation suffices to support an occasional short living vegetation only. A permanent vegetation may be found at sparse points of favorable micro-conditions, or at oases replenished through exotic rivers or groundwater flow. In general, the desert landscape is bare, or almost bare, of organic materials. The absence of vegetal cover on the surface, and of organic material in the soil, encourages evaporation and results in a low resistance to flow of water and to wind.

Except for areas covered by sand dunes, the infiltration capacity of desert soils is low. Consequently, surface runoff is initiated from rainfalls of low depths. Owing to the low resistance, the concentration of runoff into the channels is fast. In most cases, the channel beds are composed of granular materials, in which the rate of infiltration is high. As a result, replenishment of groundwater resources in the deserts is concentrated under the surface channel network or under sand dune formations. Yet, owing to the low depth of rainfall, the deficiency in soil moisture, and the granular texture of the channel beds which enables evaporation even from deep layers, only small volumes of water replenish the aquifers.

Resulted from the low rate of infiltration, and the low resistance to flow, the peak discharges which are generated in arid areas are much higher than those generated in semi-arid or in humid areas. Owing to the low volumes of most of the rainfall events in the desert, and to the high rate of infiltration in the channels, a large portion of the runoff volume, and a large number of the runoff events, do not reach the drainage bases of the rivers. Thus, the mean volume of runoff per unit area decreases as the drainage area increases. The combination of high discharges and low volumes of runoff results in a rapid rise and a fast decline of runoff events in the desert. The moisture which is left behind the flow, in the soil and in the channel beds, evaporates during the long time interval till the commencement of a subsequent flow event. By that, vacant voids are prepared for absorption of future abstractions.

The high discharges, which occur in the desert, obtain high erosion capacities, during both the sheetflow and the channel flow phases. Heavy loads of sediments are transported in the rivers. The decline of the discharge along the flow path, through infiltration and routing effects, causes a reduction in the carrying capacity of the water, which results in a deposition of sediments along the channels. Heavy depositions are left along reaches of high infiltration rates and at sites of sharp decline in the flow velocity. These are located upstream from constrictions, at entrances into flat valleys, and at outlets into lakes and seas. The deposited sediments interefere with the flow of subsequent runoff events and cause the rivers to slow down, meander and distribute.

The low volumes of runoff, and the high rate of evaporation, encourage formation of closed lakes in the topographical depressions. These lakes may be perennial, seasonal, or of occasional appearance. Springs which emerge in the desert lose much waters along their surface flow paths. Many of them vanish after a short distance of travel on the surface. Shallow groundwater sources lose water by evaporation through pores in the soil and through vegetal transpiration. In some cases, the entire flow of potential springs is abstracted before it can emerge out of the ground. The dissolved solids, transported by the inflowing waters, remain at the evaporative spots and in the closed lakes. The accumulating salinity reaches, in many cases, levels which are detrimental to the residing fauna and flora. As a result, some of the waters which are available in the desert are too saline for regular uses.

2.2 DESERTIFICATION TRENDS

The areas prone to desertification lie, in most cases, in semi-arid regions which are adjacent to arid regions. The environmental humidity in the semi-arid areas is sufficient to sustain a seasonal, and some perennial, vegetation of a relatively dense distribution. The existence of this vegetation encourages infiltration, harnesses runoff discharges and wind speeds, and protects the soil against erosion. These contribute to the environmental moisture, which, in turn, helps sustain the vegetation. This cyclic effect may be considered as a natural self-guarding mechanism of the vegetation. The more extensive and denser is the distribution of vegetation, the more efficient is the mechanism.

The rate of solar radiation, which reaches semi-arid areas, is about as high as that reaching arid areas. The difference between the two types of areas is mainly in the precipitation and in the factors governing evapotranspitation. When these attain a trend towards a drier state, the vegetation in the semi-arid area is exposed to a more intensive stress and its density declines. The remaining population can protect less against evaporation from bare soils, and harness less the actions of water and wind. In general, the safeguard mechanism becomes less effective. If the trend continues, the biological environment might collapse and the area would become arid.

The vulnerability to desertification is nonuniformly distributed within a semi-arid area. It is higher at sub-areas located more closely to arid areas, where precipitation is lower and evaporation is higher. Particulrly vulnerable are sub-areas located in the downwind direction from arid areas, at sunny-side aspects of the terrain, and where the soil can retain less moisture.

2.3 ROLE OF CLIMATIC FLUCTUATIONS

Magnitudes of climatic variables change periodically and geographically. In addition, they fluctuate temporally and spatially. The amplitude of fluctuation in the precipitation is high at areas where the mean depth is low. Fluctuations in the precipitation and in the evaporation inflict fluctuations in the environmental moisture. When a drier phase occurs, the vegetation is destroyed and the desert expands. Then flooding and erosion develop and soil is lost. On the contrary, when a wetter phase occurs, the vegetation recovers and a soil is generated. In this manner, the desert borders fluctuate continuously in a transgression onto, and in a regression from, the adjacent semi-arid areas. Yet, as the destructive processes develop faster than the constructive ones, frequent climatic fluctuations around a stable average state would generally result in a desertification process.

In case that the desertification or the rcovery process developes gradually, the flora, fauna and the people can migrate to avert risk or to take advantage of improving conditions. However, in case of a rapid change, there is no sufficient time for modifications, and the results might be fatal, or the opportunity might be lost. As the civilization developes, the adjustment of people to changes becomes more complicated and slower. Although the risk associated with small fluctuations is reduced, the society at large becomes more vulnerable to large fluctuations.

2.4 DESERTIFICATION DUE TO MAN'S ACTIVITY

Many activities of man might enhance desertification. Particularly dangerous are those that destruct vegetation, enhance erosion or increase salinity. Frequently cited examples are of overgrazing, non-conservative agricultural practices, deforestation, minning, massive construction and warfares. Industry might release harmful contaminants to the atmosphere or through its sewage and gaebage, irrigated agriculture might drain saline water, and many of the other activities might have detrimental effects on the biological environment. Regulation of rivers might restrict too strongly the flow downstream and endanger there the potential use of water for human and for biological purposes. A number of case studies on man induced desertification are described by the UN (1977b).

2.5 ANTI-DESERTIFICATION ACTIVITIES

The activities against desertification are centered around the increase of environmental moisture and the encouragement of vegetal population. The best known action is of agricultural irrigation by waters diverted from surface and subsurface sources. In some cases the sources are very distant from the area of application. Other actions are afforestation, soil conservation, grazing control, wind breaking and water harvesting from impervious surfaces and from minor watersheds. An ancient technology, which takes advantage of the runoff generated in arid areas, is through damming of the rivers by a large number of low earth dams that cause the water to spread over wide areas, reduce its velocity, deposit silt, and infiltrate into the soil at desired locations. A modern source of water for irrigation is from municipal sewage of cities located in or near to the desert. A systhematic review of the measures, suitable to combat desertification, has been published by WMO (1985).

3. Operational Hydrology Activities Related to Desertification

3.1 REQUIREMENTS

The main objective of operational hydrology with respect to desertification, is to provide means for detection and assessment of probable negative processes and of results of actions taken to combat desertification. The particular activities stem from the regular activities of operational hydrology and strengthen them where needed. Attention is called here for this strengthening.

3.2 DATA COLLECTION NETWORKS

Detection of desertification, estimation of its rate of progress, as well as assessment of actions taken to combat it, should be based upon data recorded at the area prone to desertification and at adjacent desert areas. Noting the high temporal and the spatial fluctuations of the variables, a dense network which produces accurate data is required. Achievment of this goal is particularly difficult in arid areas.

The network should cover the principal variables, such as precipitation, evaporation, surface runoff, springflow, groundwater, lakewater, transported sediments, dissolved solids, etc. Data are required on the quantity and the quality of these variables. The measurments should cover all the major sources of water as well as those minor sources which might be sensitive to probable changes due to desertification. In many cases, these minor variables shed a bright light on the investigated problem.

The various aspects of network design are discussed by WMO (1993). In addition to the regular requirements, the network should be strengthened with respect to the particular aspects of desertification. These are concluded in additional stations at areas prone to desertification, at areas where measures are taken to combat desertification, and at adjacent desert areas. Another conclusion is through an inclusion of secondary variables which are sensitive to desertification. The strengthened network should be operated very carefully because the application of such data is more delicate than most other applications of hydrologic data.

3.3 PROCESSING

Owing to the difficulties associated with construction and operation of hydrometric networks in arid areas, the acquired data might be of low quality. Yet, due to the scarcity of data on arid areas, those records are highly valuable. Processing of records obtained at the early stages should not be deferred until a better knowledge is gained. The benefits of early warnings are, in many cases, higher than the cost of risking subsequent revisions. A proper processing should include tolerance margins which are wider than those of regular hydrologic works. The actual accuracy of the data would be taken into consideration in the quantitative hydrologic works.

3.4 EVALUATION AND PREDICTION

Conclusions about an ongoing desertification process, or about measures taken to combat it, should be based upon data recorded in the field. Yet, these conclusions are seldom self-evident from the raw data. Processing, evaluation, prediction and forecast are always required in order to transform the data into workable conclusions.

Desertification is a process of change of a semi-arid environment into an arid one. The hydrologies of these environments are similar to each other and the change occurs gradually. Therefore, a distinction between the two states is difficult. Most of the common techniques for hydrologic analysis have been developed with respect to regimes prevailing in regions which are not prone to desertification. In addition, the desire to take measures against desertification at early stages of its development, encourages an early preparation of assessments. The data, which are available at that time, are short and less accurate than those provided for regular hydrologic works. A serious analysis of such data, with the aim to arrive at properly conclussive results, calls for the talents of highly skilled hydrologists.

Some compensations for the shortness in data series can be made through a simultaneous examination of different variables. In addition to the regular intercomparison of statistical parameters of flow variables, the analysis may include an examination of changes in the relationship between rainfall and runoff, steepness of the hydrograph, transmission losses along the channels, chemical composition, sediment transportation, etc.

Following the analysis, predictions should be prepared on the probable routes of development of the desertification process, and on the probable states of important variables. Forecasts on some of the more probable routes may be issued. It is advisable to express the results in clear quantitative terms and to specify the uncertainties which are involved with the predictions and forecasts. These uncertainties are wider than those confronted with in regular hydrologic works.

3.5 DISSEMINATION

The concluded predictions and forecasts should be reported to the responsible authorities. Yet, before the submission of such a report, it is advisable to undertake a peer review of the data and methods by the professional community. Such a review contributes, in many cases, to the quality of reports, particularly when dealing with difficult subjects, data of low accuracy, new techniques and wide uncertainties.

The report to the responsible authorities should be styled clearly and conclussively, because the addressies are not professional in the subject matter. However, the style should not be overdeterministic. It should rather describe the various possibilities for future development of the desertification process and the expected effects of proposable actions. It should also describe the occurrence likelihood of possible altrnatives. The descriptions should be phrased in simple quantitative terms. A chapter on further data collection, analyses and subsequent revisions of the conclusions should also be included.

4. Conclusions

4.1 GENERAL ASPECTS

Areas prone to desertification are, usually, deficient of moisture to a degree which is not as severe as that prevailing in desert areas. The depth and distributions of precipitation in these areas are highly variable, and as a result, the geographic locations of the borders between the arid and the semi-arid areas fluctuate back and forth. Therefore, the desertification process does not develop monotoneously, but it rather appears as a trend in the climatic and hydrologic fluctuations. Provision of means for detection and quantitative estimation of this trend is the major goal of operational hydrology programs taken with respect to desertification.

In view of the serious effects of desertification, the scope of activities of the operational hydrology programs should be widely extended into the areas prone to desertification, as well as into the desert. Data and interim reports should be publicly disseminated right from the early stages of the work. The needs for early actions require an early formulation of conclusions. The methods for preparation of the background for these conclusions ought to be selected with respect to the availabilty of data and skills. When the responsible authorities are aware of the seriousness of desertification it is easier to improve the means allocated for these activities.

4.2 COMMUNICATION

Detection of desertification, and prediction of its probable developments, require skills of professional hydrologists. Planning and implementation of actions taken to combat desertification require skills of professionals in other fields. Good communication procedures between the hydrologists and the other professionals would contribute to the efficiency of the operational hydrology activity.

The actions to combat desertification are composed of field and relief activities. The field activity calls upon operational hydrology for the assessment of its results. The relief activity requires predictions and forecasts of probable developments of the process. Owing to the natural fluctuations, and to the needs for early warnings and assessments, the hydrologic conclusions cannot be accurate. Predictions, forecasts and assessments issued on later times might contradict those that had been issued on earlier times.

Viewing the dangers involved with desertification, and the needs for early conclusions, the fluctuations found in the hydrologic conclusions should be tolerated by the other professionals. The relief activities should be planned flexibly enough to stand various deviations of the real situation from the previously forecasted one. Likewise, the field activities should stand fluctuations in the assessment of the desertification process and of the measures taken to combat it. A good communication between the different groups of people, which are involved with the different activities, would ensure easy adaptations to the fluctuating hydrologic conclusions.

5. Acknowledgement

This article is based upon a report prepared for the World Meteorological Organization when the author served in the capacity of a Rapporteur on Drought and Desertification. Permission of WMO for this publication is acknowledged.

6. References

UN (1977a). 'World map of desertification: explanatory notes'. UN Conf. on Desertification, Nairobi, Kenya.

UN (1977b). 'Synthesis of case studies of desertification'. UN Conf. on Desertification, Nairobi, Kenya.

WMO (1985). Use of Meteorological Climatological and Hydrological Data in the Combat against Desertification. Special Environmental Rep., pt. I, WMO, Geneva Switzerland.

WMO (1993). Guide to Hydrological Practices, 5'th ed., Geneva, Switzerland.

SPATIAL ESTIMATION TECHNIQUES FOR PRECIPITATION ANALYSIS - APPLICATION TO A REGION IN INDIA

BALAJI RAJAGOPALAN, ALOK K. SIKKA, DAVID S. BOWLES AND
ASHUTOSH S. LIMAYE
Utah Water Research Laboratory
Utah State University
Logan, Utah 84322-8200
U.S.A.

ABSTRACT. Precipitation data from northern India was analyzed using three different spatial estimation techniques, viz. Kriging, Lowess and Smoothing Spline ANOVA. Annual precipitation was considered as a function of latitude and longitude. Various quantitative measures were considered for comparison for these techniques.

1. Introduction

Spatial estimates of precipitation are required for modeling hydrologic and ecologic processes. With the advancement of remote sensing and geographic information systems (GIS), application of spatialy distributed hydrologic and ecologic models are gaining momentum. Spatially distributed estimates of precipitation are crucial for driving such models. spatial estimation of precipitation over a region from point observations is a chalenging task. The irregularity of sampling in space and substantial variability in precipitation with topography (i.e. non-stationarity) makes this task even more difficult in heterogenous regions with spase number of observation points.

Simple spatial techniques such as Thiessen polygons, polynomial trend surfaces, and inverse distance methods for precipitation data are commonly used. The weights assigned to each supporting data point in these methods are not necessarily optimal. In view of this, robust, powerful and optimal techniques are needed for spatial etimation of precipitation.

Kriging, a parametric regression procedure, is a widely used geostatistical technique for spatial estimation. It assumes a priori specification of the functional form of the parameter of interest. Its estimates suffer from second order stationarity assumptions and subjective way of fitting the variogram. The assumptions of kriging are most often unrealized in practice. Nonparametric and semiparametric techniques such as Splines, LOWESS (Locally Weighted Regression) and Kernel regression are capable of addressing most of the above problems. These techniques are locally adaptive and objective nature. Akin (1992), has extensively compared various spatial estimation techniques on a large number of theoretical and real data sets, and found that kriging was inferior to other methods on most of the data sets. Similar results were observed by Yakowitz and Szidarovszky (1985), who compared the theoretical properties of kriging and kernel function estimation. Satagopan and Rajagopalan (1993) while analyzing precipitation data from Columbia river basin (USA) also found that Splines and Lowess performed better than Kriging.

With this motivation at the background, we planned this study to examine the relative performance of the three methods viz., **Kriging, Smoothing Splines and Lowess** for spatial interpolation of precipitation in an area largely dominated by monsoon rainfall. We use the

V. P. Singh and B. Kumar (eds.), Surface-Water Hydrology, 61–75.

smoothing spline analysis of variance (SS-ANOVA) proposed by Gu and Wahba (1992). Brief discussion of the three methods is first presented, followed by a note on the study area, data set and the statistical model. Comparative results and discussion conclude the paper.

2.1 Kriging

Kriging is a parametric regression procedure due to Krige (1951) and Journel (1977). It has become synonymous with geostatistics over the last decade and represents the state of the art for spatial analysis problems. Issacs and Srivastava (1989) present a comprehensive and applied treatment of kriging, while Cressie (1991) provides a comprehensive treatment that covers much of the recent statistical research on the subject. Most of the work has been largely focused on Ordinary Kriging. The model considered is:

$$y = f(x) + \varepsilon \tag{1}$$

where the function f(x) is assumed to have a constant but unknown mean and stationary covariance, y is the observed vector and ε is a vector of i.i.d noise. Most often the assumptions for the function f are not satisfied, especially in the case of mountaneous precipitation. Consequently, Universal kriging, Co-kriging and the Intrinsic Random Function Hypothesis attempt to deal with non stationary situations, but fitting variograms in these cases is even more tenuous, which affects the estimates and are difficult to implement. Therefore, we restrict ourselves to Ordinary kriging in our data analysis. However, the general principle in all the various types of kriging is the same, which is described below. Cressie (1991), Journel (1989) and de Marsily (1986) have detailed discussions on the various types of kriging and their estimation procedures as applied to different situations.

Kriging is an exact interpolator at the points of observation, and at other points it attempts to find the best linear unbiased estimator (BLUE) for the underlying function and its mean square error (MSE). The underlying function f(x) is assumed to be a random function and f(x) and $f(x+\delta x)$ are dependent random variables. This leads to the ergodicity and stationarity assumptions. The kriging estimate f_k is formed as a weighted linear combination of the observations as

$$f_k(x_0) = \sum_{i=1}^{n} \lambda_{0i} y_i \tag{2}$$

The subscript k stands for kriging estimate. the weights are determined through a procedure that seeks to be optimal in a mean square error sense. The wrights relate to the distance between the point at which the estimate is desired and the observation points, and to the degree of covariance between the observations as a function of distance, called the variogram $\Gamma(h)$. The variogram is given as

$$\Gamma(h) = \mathrm{Var}(y) - \mathrm{Cov}(y(x), y(x+h)) \tag{3}$$

where h is the distance.

The wrights λ_{0i} are determined by solving the normal equations for kriging which are,

$$\sum_{j=1}^{n} \lambda_{0j}\Gamma(x_i-x_j) + \mu = \Gamma(x_i-x_o) \qquad i = 1,...,n \tag{4}$$

$$\sum_{i=1}^{n} \lambda_{0i} = 1 \tag{5}$$

Where μ can be interpreted as a Lagrange multiplier for satisfying the constraint that the weights sum to unity, in an optimazation problem formed for minimizing the mean squared error estimation. λ_{0i} s are obtained by solving the above two equations more like in a regression situation. the ideas of gaussian linear estimation are thus implicit in the kriging process.

The MSE of the estimator f_k is given by Cressie (1991) as

$$MSE(f_k) = \sum_{i=1}^{n} \lambda_{0i}\Gamma_f(\delta x_i + \mu) \tag{6}$$

where $\Gamma_f(x_i,x_j)$ is the semi variogram given by $Var(f(x_i) - f(x_j))/2$.

The above estimation procedure is under the presumption that the variogram is a known function, which in practice is never known *a priori*. In reality the observations are unequally spaced, and hence a direct estimate of $\Gamma(h)$ from the data is not feasible. Therefore, the data are grouped into distance categories and then a parametric function (e.g exponential or spherical) is fit to the estimated or raw variogram. This is called **variogram fitting** and is a central issue in kriging. fitting the variogram is the most difficult and important part of kriging, more so in case of nonstationarity. Lack of objective methods to fit the variogram, results in a poorly fit variogram and consequently, the estimates and the estimates of MSE are likely to be significantly in error.

Yakowitz and Szidarovszky (1985) argue that there is no consistent variogram estimator, even for the case where the data is noise-free. Wabha (1990) also shows tha tno consistent estimators of the variogram parameters from the data are readily available as part of the kriging estimation process. Journel (1989) has discussions on the demerits of kriging, and stresses that stationarity assumptions are made for ease of analysis and are not necessarily properties of the process studied. Akin (1992) has studied kriging with known data sets and also groundwater data. He found that kriging performed verey poorly in almost all the cases as compared to other techniques. Bowles et.al (1991) compared kriging with thin plate smoothing splines on precipitation data from mountaneous region and made similar inferences. This substantiates the arguement by Yakowitz and Szidarovszky (1985).

We have analyzed the precipitation data sets using Ordinary kriging on the public domain software GeoEAS, developed and used by government regulating agencies and consulting firms.

2.2 Smoothing Splines - ANOVA

Smoothing spline analysis of variance (SS-ANOVA) is a semiparametric procedure for fitting models. The model considered in this case is similar to the kriging model. The SS-ANOVA method decmposes the function f into various components like in any analysis of variance model i.e., the function f is split into main effects and interaction terms. This is very useful because one can find out how the observed data is affected by each variable.

Consider the model:

$$y_i = f(x_{1i}, x_{2i}, .., x_{ki}) + \varepsilon_i \quad i = 1,..,n \tag{7}$$

where $y_1, y_2, .., y_n$ are observations, f is the function to be estimated, $x_1, ..., x_k$ are variables such that the jth variable $x_j \in X^j$, some measurable space and $\varepsilon_1, .. \varepsilon_n$ are i.i.d and $\varepsilon_i \sim N(0, \sigma^2)$, σ^2 unknown. Usually the space considered is $X^j = [0,1]$. Whenever the variables are not in the range [0,1], we can rescale them to lie in this range. Wahba (1990), Gu(1989) and Gu and Wahba (1992) give an overview of the SS-ANOVA models. They discuss applications to polynomial splines, tensor product splines and thin plate splines. The SS-ANOVA model is described briefly in what follows.

The assumption in this model is $f \in H$, where H is a Hilbert space. The function f is required to be smooth in its domain with $f, f', .. f^{(m-1)}$ absolutely continuous where $f^{(i)}$ denote the ith derivate of f, $f^{(m)} \in L^2$ and $\int_0^1 f(t)dt = 0$. The space H is uniquely decomposed into a tensor sum as:

$$H = \{1\} \oplus \sum_{i} H_i \oplus \sum_{i<j} H_i \oplus H_j \oplus \ ... \tag{8}$$

Based on this representation of H, the function f is decomposed uniquely as

$$f = \mu + \sum_{i} f_i + \sum_{i<j} f_{ij} + .. \tag{9}$$

where μ is a constant, $f_i \in H_i$ are the main effect terms, $f_{ij} \in H_i \oplus H_j$ are the 2 factor interaction terms and so on. The space H is decomposed in such a way that the resulting subspaces are unique and orthogonal in the tensor product norm induced by the original inner products. This decomposition is very similar to the decomposition in any analysis of variabce problem. The Hilbert space H can further be decomposed as polynomial and smooth spaces. Let H_i have an orthogonal decomposition $H_{\pi i} \oplus H_{si}$, where $H_{\pi i}$ is the polynomial or parametric space and H_{si} is the smooth space.

The f_is satisfy

$$\int_0^1 f_i(x_i)d\mu_i = 0, \tag{10}$$

where μ_i is a lebesgue measure on [0,1]. f_{ij}s satisfy

$$\int_0^1 f_{ij}(x_i,x_j)d\mu_i d\mu_j = 0 \tag{11}$$

and so on. These are similar to the side conditions in any analysis of variance model. The SS-ANOVA procedure obtains f_λ as an estimate of f which minimizes the quantity

$$\frac{1}{n}\sum_{i=1}^{n} (y_i - f(x_{1i},..x_{ki}))^2 + \lambda[\sum_i \frac{1}{\theta_i} J_i(f_i) + \sum_{i,j} \frac{1}{\theta_{i,j}} J_{i,j}(f_{ij}) + ... \tag{12}$$

λ, θ_i, $\theta_{i,j}$,... are smoothing parameters. Js are smoothness penalty functionals. The polynomial space does not have any penalty. Kimeldorf and Wahba (1971) and Wahba(1990) discuss different penalty functionals for cubic spline models, thin plate spline models, tensor product spline models etc.

Gu and Wahba(1991) discuss the application of thin plate splines. They give the penalty function J_m^k for the thin plate spline as being invariant under rotation of the variables $x_1,...,x_k$, thereby making thin plate splines rotation invariant. Bayesian confidence intervals for the main effect and interaction terms can be obtained for a thin plate splines model, Gu and Wahba(1992). For further details on thin plate splines and its applications, refer to Gu and Wahba (1991).

Thin plate splines differ from polynomial and tensor product splines in the penalty function, reproducing kernels and span of polynomial spaces. The reproducing kernels of the smooth spaces are more abstract for thin plate spline models.

The publicly available code RKPACK by Gu (1989) enables one to fit tensor product splines, polynomial splines and thin plate splines for SS-ANOVA models. We have used the thin plate spline program in RKPACK for our data analysis.

2.3 Locally Weighted Regression (Lowess)

Locally weighted regression or Lowess is a local regression procedure that is a quick and versatile tool for exploratory data analysis and regression surface estimation. Cleveland et. al (1988) and Cleveland and Devlin (1988) developed Lowess. Under certain conditions, Muller (1987) shows an equivalence between lowess and nonparametric kernel regression estimation. The estimator proposed by them considers local linear or quadratic fits using k nearest neighbours of the point at which the estimate is desired. They consider the standard multivariate nonparametric regression situation defined through the general model

$$f_i = f(x_i, \beta) + \varepsilon_i, \quad i = 1, .., n \tag{13}$$

Where $f(x,\beta)$ is the smooth regression function or conditional expectation with parameter β and $\varepsilon_1,..., \varepsilon_n$ are i.i.d with $\varepsilon_i \sim N(0,\sigma^2)$, σ^2 usually unknown.

An estimate $\hat{f}(x)$ of f(x) is developed nonparametrically (in the global sense) by a weighted local linear or quadratic least squares regression in the neighbourhood of x devised by its k nearest neighbours. The quadratic model fitted locally is,

$$\hat{f}(x) = D\hat{\beta}$$

Where D is a design matrix and the coefficients $\hat{\beta}$ are determined as the solution to a weighted least squares problem defined as

$$\min_{\beta} \sum_{i \in k(x)} (y_i - f(x_i, \beta))^2$$

where k(x) is the index set of x_i that are the k nearest neighbours of x and $w_i(x)$ is a weight function defined as

$$w_i(x) = \frac{W(\rho(x,x_i))}{d_k(x)}$$

where $\rho(.)$ is the euclidean distance function, $d_k(x)$ is the euclidean distance from x to the kth nearest neighbour in x_i. W(.) is usually taken as the tricubic weight function.

The number of nearest neighbours, k, acts as a smoothing parameter. As k increases, bias increases, but variance decreases. The fit determined by Lowess depends on the choice of the number k of nearest neighbours and the order r of the fit. Cleveland et.al (1988) propose a graphical method, based on analyzing an M-plot for the selection of both these parameters. for further details we refer the reader to Cleveland et.al (1988).

For the present analysis, the span (fraction of sample used to compute local regression) and degree (order of local approximation 1=linear, 2=quadratic) were chosen by using the F statistic to compare alternate values of span and degree. A lowess surface is fit for a span and degree and the residual sum of squares is computed. the F-statistic using the proper number of degrees of freedom in each case is used to compare the residual sums of squares (RSS) at a significance level of 0.01. In case the residual sum of squares are significantly different (at the 0.01 level) the span/degree with the lower RSS is selected. Otherwise the span/degree with the lower number of equivalent parameters (higher degrees of freedom) is selected.

Cleveland (1979) suggests that local linear fitting produces good results, especially in the boundary region. akin (1992) has shown that the scheme works well even for a small number of independent variables. From his comparisons of lowess with kriging on a large number of known data sets he found that lowess performed very well in reproducing the known functions.

3. Study Area and Data Set

The study area covers mostly the northern region of India covering the states of Punjab, Rajasthan, Uttar Pradesh and Madhya Pradesh. It mostly covers major parts of Indus and Gaga basin. The major source of annual precipitation in this area is south west monsoon spanning between June to September. The area extends between 20° N to 34° N latitude to 72° E and 84° E longitude. The elevation of the stations selected in the study area, ranges from 50 m. to 3500 m. The area is characterized with a vast variation in topography, climate, and landcover. The topography map of the area is shown in Figure 1. The variation in topography with mountains could cause non-stationarity in the precipitation process.

The data consisted of average annual precipitation obtained from 137 precipitation gauges sprad over the entire region. The daily precipitation data was obtained from the National Centre for Atmospheric Research (NCAR), Boulder, Colorado, USA. Stations with a common period of record extending 19-- to 19-- were used to compute average annual precipitaion.

Model

The model considered in our study was,

$$y = \log(\text{precipitation}) = f(\text{easting, northing}) + \varepsilon$$

The precipitaion data had large variance, hence, the analysis was carried out in log transformed space. Besides, the basic requirement in any analysis is the assumption of constant variance for error. This assumption is violated usually when the response variable y follows a probability distribution in which the variance is functionally related to the mean. Especially in the case of precipitation data, the assumption of constant variance does not usually hold. Log transformation helps in such cases.

The model considered for SS-ANOVA was,

$$y = \mu + f_{\text{easting}} + f_{\text{northing}} + f_{(\text{easting,northing})} + \varepsilon$$

where, y = log(precipitation), μ is the constatnt term, f_{easting} denotes the effect of easting, f_{northing} denotes the effect of northing, $f_{(\text{easting,northing})}$ denotes the interaction between easting and northing and ε denotes i.i.d gaussian error. This model is very similar to any analysis of variance model.

4. Results and Discussion

Visual examination of the estimated surfaces are very helpful in assessing the performances of different estimators. However, measures of accuracy are also important in decision making as they are based on quantitative criteria. We estimated a few measures that quantify the local and global measures of fit, for comparint the three techniques. We estimated the following measures, maximum absolute deviation (MAD), mean absolute deviation (MNAD), residual sum of squares (RSS), scaled residual sum of squares ($SRSS_1$ and $SRSS_2$), and scaled variance of deviation (SVD).

The measures are defined as follows,

$$\text{MAD} = \max_{i=1,..,n} |y_i - \hat{y}_i|$$

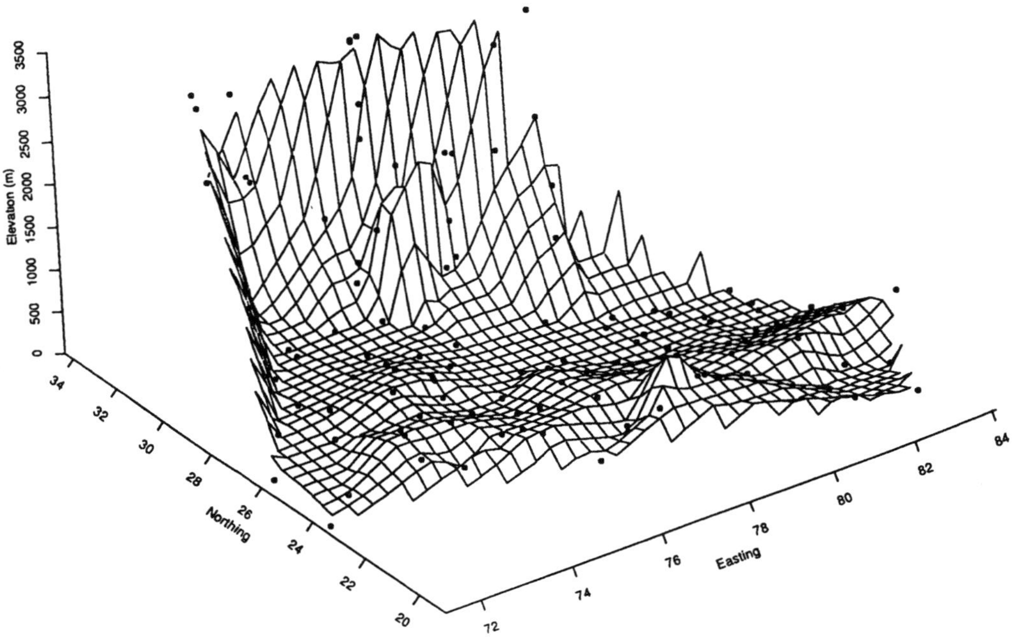

Figure 1: Surface plot of the elevation of the study area.

$$\text{MNAD} = \frac{1}{n}\sum_{i=1}^{n} |y_i - \hat{y}_i|$$

$$\text{RSS} = \sum_{i=1}^{n} (y_i - \hat{y}_i)^2$$

$$\text{SRSS}_1 = \frac{1}{n\,\text{var}(y)} \sum_{i=1}^{n} (y_i - \hat{y}_i)^2$$

$$\text{SRSS}_2 = \frac{\sum_{i=1}^{n} (y_i - \hat{y}_i)^2}{\sum_{i=1}^{n} \hat{y}_i^2}$$

$$\text{SVD} = \frac{\text{var}(\hat{y})}{\text{var}(y - \hat{y})}$$

where y denotes the response variable (log(precipitation)) and $\hat{y}$ denotes the estimated values.

MAD signifies how far away the estimate is from the true underlying function at the point where the fit is poorest, but does not ensure the best fit over the entire range of the function. SRSS_1 and SRSS_2 measure the nearness of fit to the true underlying function. They are suited to determine the better estimator where a good fit is desirable over the entire range of the function and to ensure no regions of relatively poor fit. SVD would return the squared signal to noise ratio, this also tells how well the techniques discriminate the noise in the function. Akin (1992) has detailed discussion on these measures.

Measure	Kriging	LOWESS	SS-ANOVA
MAD	2.866	2.960	2.748
MNAD	0.781	0.642	0.760
RSS	138.265	104.188	128.312
SRSS1	0.996	0.739	0.911
SRSS2	0.031	0.023	0.028
SVD	0.007	0.009	0.008

Table 1: Measures of comparison for the three methods considered

Figure 1 shows the topography of the area along with the location of the observation stations. Contour plot of the observed data is plotted in Figure 2a, this is using a simple bilinear interpolation. Figures 2b, 2c and 2d show the contour plots of the estimates from Kriging, SS-ANOVA and LOWESS respectively, the corresponding surface plots are plotted in Figures 3a, 3b, 3c and 3d. We can see from Figures 2b and 3b that the estimates from kriging are oversmoothed,

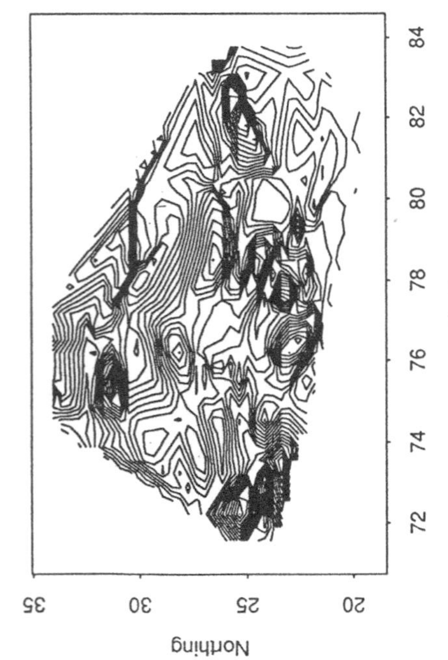

Figure 2a: Contour plot of Log of precipitation of the observed data.

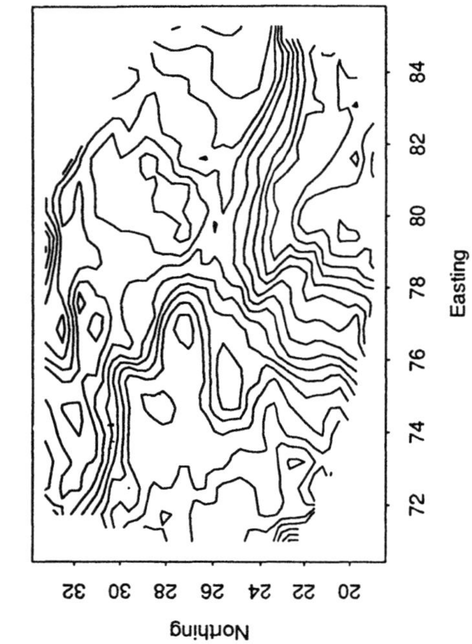

Figure 2b: Contour plot of estimates of Log of precipitation from Kriging.

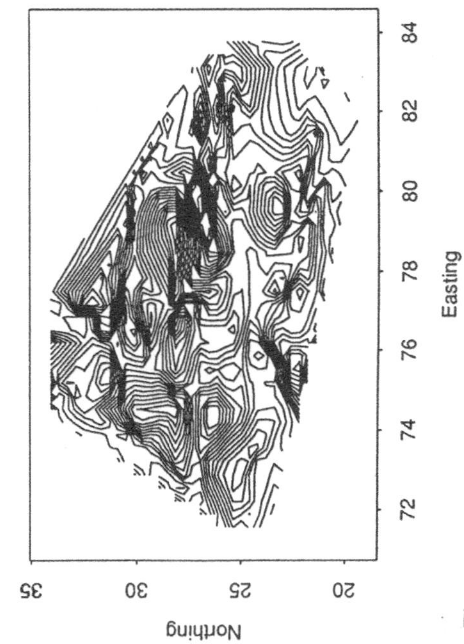

Figure 2c: Contour plot of estimates of Log of precipitation from SS-ANOVA.

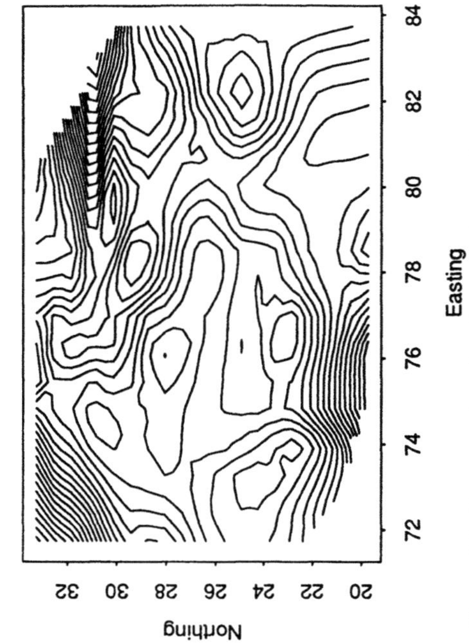

Figure 2d: Contour plot of estimates of Log of precipitation from LOWESS.

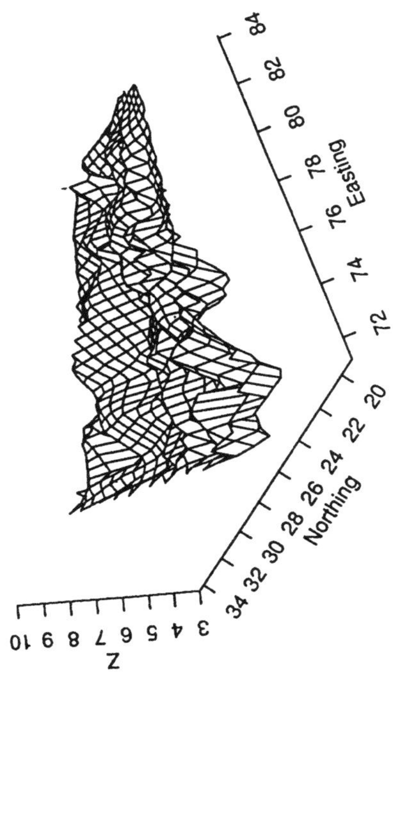

Figure 3a: Surface plot of Log of precipitation of the observed data.

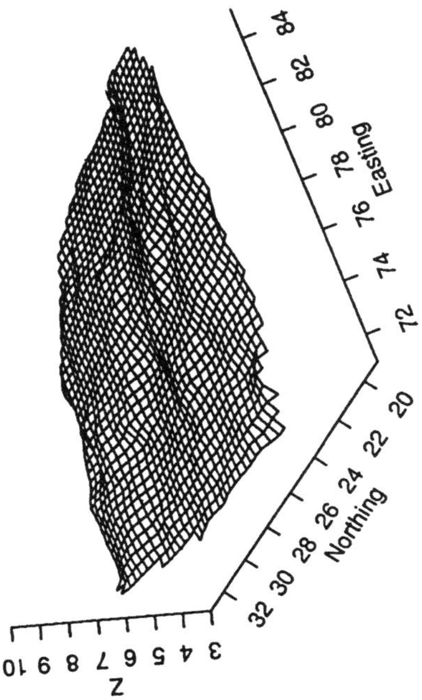

Figure 3b: Surface plot of estimates of Log of precipitation from Kriging.

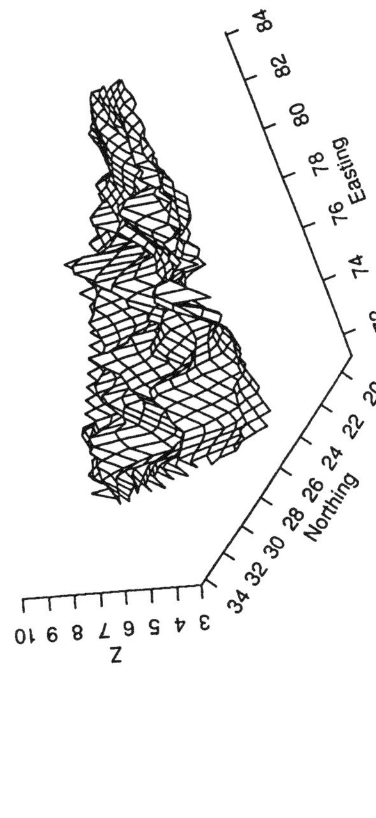

Figure 3c: Surface plot of estimates of Log of precipitation from SS-ANOVA.

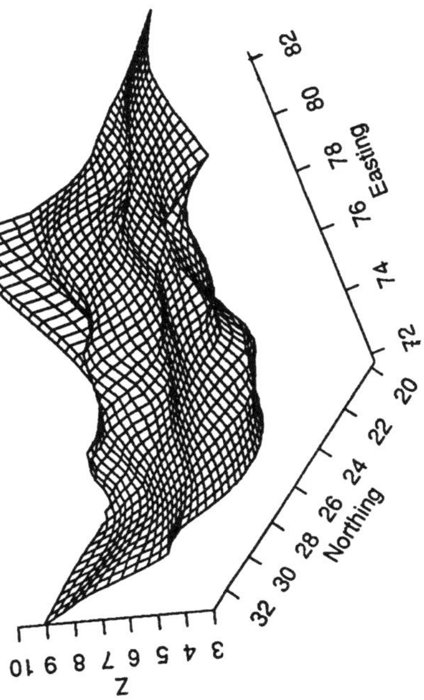

Figure 3d: Surface plot of estimates of Log of precipitation from LOWESS.

reproducing only the mean behaviour. Recall, that Kriging works under the assumption of stationarity of the underlying process. The situation that we have here, there is a good amount of nonstationarity due to large variation in the topography. In such cases the fitted variogram is poor (see Figure 4a) resulting in an oversmoothed estimates. LOWESS (see Figures 2d and 3d) and SS-ANOVA (see Figures 2c and 3c) do not suffer from the obvious shortcomings of Kriging (viz. stationarity assumption and arbitray variogram fitting). As a result they tend to perform better both in terms of smoothing as well as trying to admit nonstationarity. Another advantage with using SS-ANOVA is that, effects of various factors can be obtained (see Figures 4b, 4c and 4d). The effects of Easting, Northing and their interaction seem to be consistent with the elevation change across the area. Comparing the various measures for the three methods in Table 1 similar observations can be made.

The results suggest that for the data set analysed here LOWESS and SS-ANOVA perform better than Kriging. Weaker assumptions necessary for LOWESS and SS-ANOVA and its better performance provides attractive alternatives to Kriging, especially when the process is nonstationary.

Finally, it is worth noting that all interplators are smoothers. They cannot be expected to recover the finer features of data, or to be able to extrapolate to values outside the range of observations. While, Kriging tries to estimate parametric variograms and drift functions, which detracts from its use. We have presented the results from our exploratory data analysis. Future work is required in terms of more data analysis taking into account the elevation, to consolidate our preliminary findings.

Acknowledgements

We would like to thank Dr. Upmanu Lall for valuable suggestions and for providing us with relevant manuscripts. We also thank Jaya Mathangi Satagopan, Dept. of Statistics, Univ. of Wisconsin - Madison, for helping with running the software, RKPACK for SS-ANOVA.

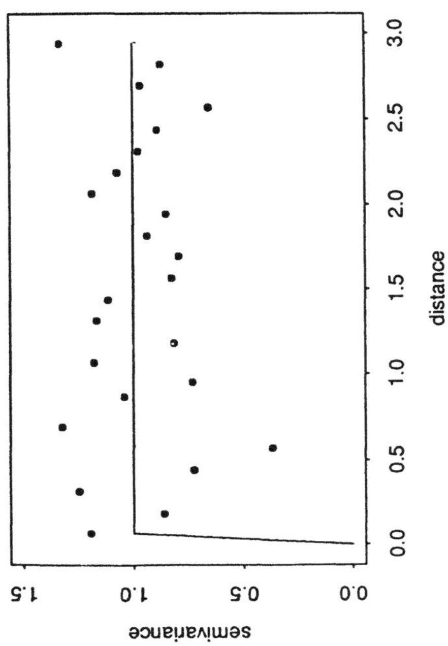

Figure 4a: Fitted variogram for Kriging. The dots are the data estimated values, the line is the fitted spherical variogram.

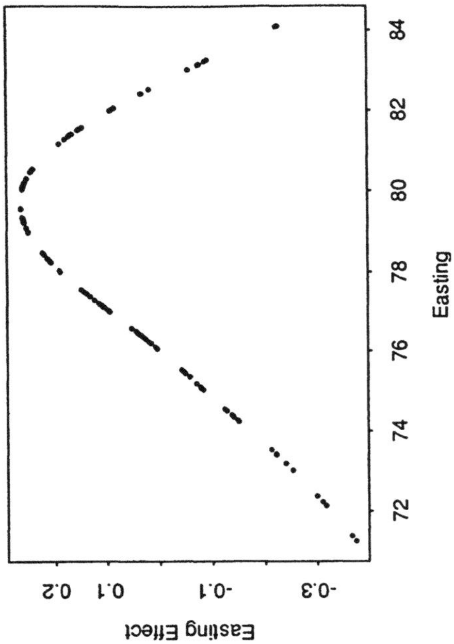

Figure 4b: Effect of Easting obtained from SS-ANOVA.

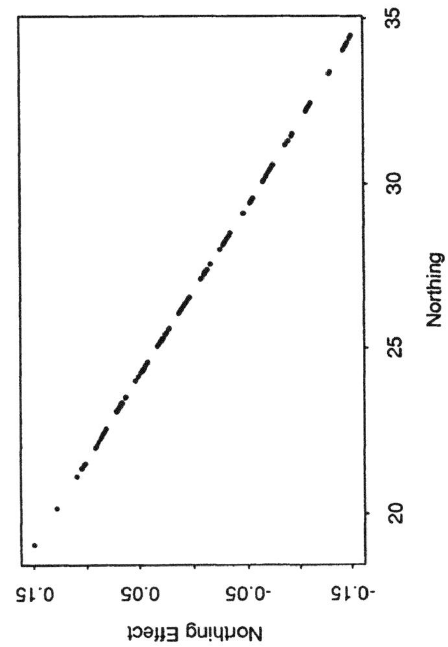

Figure 4c: Effect of Northing obtained from SS-ANOVA.

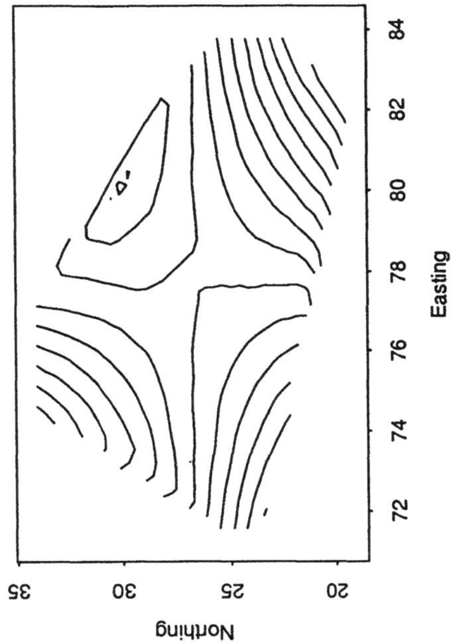

Figure 4d: Interaction effect obtained from SS-ANOVA.

References

Akin, O.O. (1992). A comparative study of nonparametric regression and kriging for analyzing groundwater contamination data. M.S. thesis, Dept. of Civil and Environmental Engg., Utah State University, Logan, Utah.

Bowles, D.S., Bingham, G.E., Lall, U., Tarboton, D.G., Al-Adhami, M., Jensen, D.T., McCurdy, G.D., Jayyousi, E.F. (1991). Development of mountain climate generator and snow pack model for erosion predictions in the western United states WEPP, Project report - III.

Cleveland, W.S. (1979). Robust locally weighted regression and smoothing scatter plots. Journal of American Statistical Association, 74(368).

Cleveland, W.S., and Devlin, S.J. (1988). Locally weighted regression: an approach to regression analysis by local fitting. Journal of American Statistical Association, 83(403).

Cleveland, W.s., Devlin, S.J. and Grosse, E. (1988). Regression by local fitting. Journal of econoomics, 37.

Cressie, N. (1991). Statistics for spatial data, John Wiley, New York.

de Marsily, G. (1986). Quantitative hydrogeology, groundwater hydrology for engineers. Academic press,

Gu, C. (1989). Rkpack and its applications: Fitting smoothing spline models. Technical report 857, University of Wisconsin - Madison.

Gu, C. and Wahba, G. (1991). Semiparametric ANOVA with tenson product thin plate splines.

Gu, C. and Wahba, G. (1992). Smoothing spline ANOVA with component-wise bayesian confidence intervals. Technical report 881(rev.), University of Wisconsin - Madison.

Issaks, E.H. and Srivastava, R.M. (1989). An introduction to applied geostatistcs. Oxford University Press.

Journel, A.G. (1977). Kriging in terms of projections. Journal of Mathematical Geology, 9(6) 563-586.

Journel, A.G. (1989). Fundamentals of geostatistics in five lessons. American Geophysical Union, Washington, D.C.

Kimeldorf, G., and Wahba, G. (1971). Some results on Tchebycheffian spline functions. Journal of Mathematical Analysis and Probability, 33.

Krige, D.G. (1951). A statistical approach to some mine valuations and allied problems in Witwaterstrand, unpublished M.s. thesis, University of Witwaterstrand, South Africa.

Muller, H.G. (1987). Weighted local regression and kernel methods for nonparametric curve fitting. Journal of American Statistical Association, 82(397).

Satagopan, J. and Rajagopalan, B. (1993), Comparing spatial estimation techniques for precipitation analysis. Paper presented at the International conference on stochastic and statistical methods in hydrology and environmental engineering, Univ. of Waterloo, Waterloo, Canada, June 1993.

Wahba, G. (1990). Spline models for observational data. SIAM series in Applied Mathematics.

Yakowitz, S. and Szidarovszky, R. (1985). A comparison of kriging with nonparametric regression methods. Journal of Multivariate Analysis, 16(1), 21-53.

SECTION 2:

HYDROLOGICAL MEASUREMENTS

DESIGN CRITERIA FOR THE IMPROVEMENT OF HYDROMETEOROLOGICAL DATA ACQUISITION SYSTEMS. AUTOMATION OF SICILY NETWORKS.

G. ROSSI
University of Catania
Institute of Hydraulics, Hydrology and Water Management
V. A. Doria, 6 - 95125 Catania
Italy

A. ANCARANI
Studio Associato G&A
V.G.Leopardi, 141 - 95127 Catania
Italy

A. JAKOMIN
Studio Associato G&A
V.G.Leopardi, 141 - 95127 Catania
Italy

ABSTRACT. The automation of present hydrometeorological networks requires not only an accurate design of all components of the data acquisition system, but also a detailed account of the necessary changes in the organizational structure and of the related training of personnel. Also an analysis of the economic feasibility of the modernization of the system is necessary. This paper describes a project to improve and automate the existing hydrometeorological networks in Sicily (Italy). The selection of the meteorological and streamflow stations to be included in the system is discussed. The main requirements for instruments for automatic measurements of the variables already under observation (rain, temperature, river stage) and of the new variables to be measured (air humidity, wind speed and direction, sun radiation, sea level and sea wave characteristics) are discussed.The system's architecture, which includes 103 gauging stations, two control centers and an automatic transmission line by radio with 20 repeaters is also presented, as well as the characteristics of the hardware and software equipment in the control centers. Some recommendations for construction and operation of the system in order to facilitate the success of the network modernization are issued. The criteria for identifying and evaluating costs and benefits of the project, and for carrying out cost-benefit analysis are illustrated.

1. Introduction

Recent advances in technology for data collection, transmission, storage and processing lead to a modernization of the existing hydrometeorological networks.

A significant difference exists between designing new automatic networks and designing the automation of existing systems for acquisition of hydrometeorological data. In the latter case, an accurate design of the network lay-out, of observation instruments and methods, of transmission lines and of the processing center is not sufficient. Great attention has to be also paid to the necessary changes in the organizational structure of the Hydrological Service and the related training of personnel. Moreover, a benefit-cost analysis is necessary in order to ascertain the economic feasibility of the proposed project to modernize the system.

V. P. Singh and B. Kumar (eds.), Surface-Water Hydrology, 79–93.

Three additional points should be considered carefully in planning the automation of an existing system:

i) the selection of the number and location of gauging stations to be linked through an automatic transmission system;
ii) the identification of the new variables and/or the new measurement points which can be required by actual and future demands for hydrometeorological information;
iii) a detailed analysis of the needs of network operation and maintenance during the initial years after construction has finished, including the sharing of the responsabilities between the Hydrological Service and firm in charge of the project.

This paper describes a project for the improvement and automation of some hydrometeorological networks in Sicily (Italy). It is part of a more complex Telematic Plan for Sicily, which has been coordinated by TELEINFORM on behalf of the Sicilian Regional Government with financial support from the National Agency for Southern Italy's Development.

At the time of the start of the project, computer technology was not widely utilized in the public administration of Sicily. The Telematic Plan was conceived in order to reduce the delay in computer application with respect to many other Italian regions. Standard criteria in data collecting, management and processing were adopted in order to facilitate data transfer from local offices to regional departments, between the same regional departments and between these departments and the national government.

This project concerns the automation of a part of the existing rainfall, temperature and streamflow gauge networks operated by the Hydrological Service.

In the following chapters, this project and some technical requirements regarding the automatic stations will be described, and some recommendations about assistance and maintenance during the start-up period after the network construction will also be outlined. Then the main benefit-cost analysis criteria will be summarized.

2. Design Criteria

2.1 PRESENT HYDROMETEOROLOGICAL NETWORKS AND AUTOMATION NEEDS

Sicily, is the largest island in the Mediterranean Sea (25.500 km^2). The Sicilian climate is semiarid, with rainfall concentrated in the autumn period, and a long summer period without rain when the temperatures are at their maximum. The average annual rainfall range is from a 400-500 mm minimum in the coastal plains (Catania, Trapani) to 1200-1300 mm maximum in the mountain areas of the island. Almost all rivers are short, with basin area less then 1000 Km^2, except for the three main rivers, Simeto (4186 Km^2), Salso (2002 Km^2) and Platani (1795 Km^2). Most of the rivers are ephemeral with short and heavy floods occurring during the October-March period.

Existing hydrometeorological networks include about 230 rain gauges, 70 temperature gauges, 20 snow depth gauges and 55 streamflow gauges. They are operated by the Regional Office of National Hydrological Service (H.S.), located in Palermo. Since 1917, when H.S. began its measurements, there has been an increase in the number of stations (above all for streamflow measurement), but the instruments are still of the paper-recording type.

There is a great need to improve the present gauging networks in order to minimize data losses caused by instrument damage and to make data collection and control and first processing faster and more reliable. Moreover, the automated networks set up the basis for new tasks to be developed in the future such as a systematic monitoring of water quality, a flood forecasting system in selected rivers, where it is possible, and an agrometeorological service.

Furthermore, the improved and automated hydrometeorological networks may have positive effects for:

- faster decision making processes both in the operation of existing water systems and in water resources planning);
- quick and objective identification of flood-prone or drought-prone areas in order to determine fiscal reductions and reimbursements;
- more timely research on the meteorology and hydrology of the island.

2.2 REQUIREMENTS OF THE DATA ACQUISITION SYSTEM

The following main requirements for an automatic hydrometeorological data acquisition system should be considered:

1) to achieve high reliability of the whole system through:
   - architecture reliability, i.e. providing redundancy of some components of the system in order to minimize data losses;
   - equipment reliability, especially in the instruments for data collecting;
   - transmission reliability, obtained through accurate tests (oriented to ascertain the radio conflicts with airport or other service transmission), and also through a certain redundancy in main repeaters (in order to maintain good transmission links also during extreme events);
2) to adopt an open architecture, in order to permit a further expansion both in the number of stations and in variables to be measured;
3) to choose the instruments which present minimum maintenance needs, also in very adverse climatic conditions; in particular, simple instruments constructed with hard and resistent materials, able to deter thefts and vandalism, should be preferred; in large networks,well tested instruments should be used instead of advanced devices lacking field check;
4) to choose friendly and simple user interfaces both in measurement station software and station management software in control centers;
5) to choose software for use by non skilled personnel:
   a) easy to use: all the parts of the control center software must provide a menu to guide the users also through help on line;
   b) including simple procedures for network configuration changes;
   c) including automatic procedures for data processing;
6) to adopt international standards for data collection, storage and processing in order to facilitate data transfer with other information systems.

2.3 SYSTEM ARCHITECTURE

The system for automatic acquisition of hydrometeorological data in Sicily was organized according to the scheme of fig. 1.

The size of the project was determined under the basic assumption that a gradual automation of the networks should be preferred in order to reduce both the complexity of

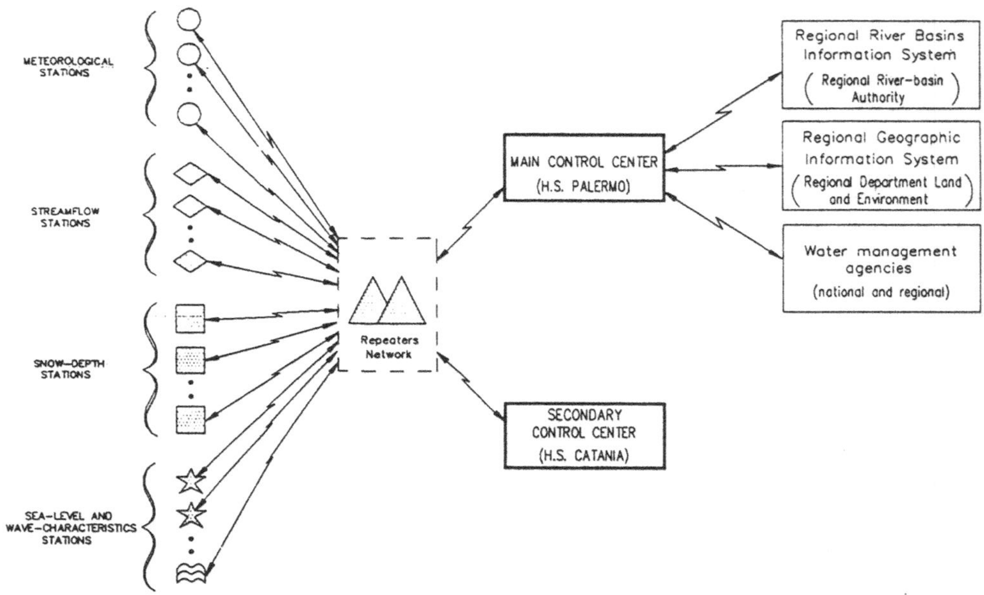

Figure 1. Scheme of the data acquisition system

the telemetering system and the impact on the organizational structure of the Hydrological Service. Infact, a limited number of stations to be linked through the automatic transmission system could facilitate the operation in the start-up period and H.S. staff acceptance. A detailed investigation on the optimal grid for measurement of meteorological variables will be carried out on the basis of the experience from the initial years in order to plan further development.

The system includes:

1) 103 stations for the measurement of hydrometeorological data, namely:
   - 22 existing streamflow stations to be automated;
   - 62 existing meteorological stations to be automated and expanded for the measurement of new variables;
   - 19 new stations (6 snow depth stations located in mountain areas, 6 sea level stations and 7 wave characteristics stations);
2) one main control center and one secondary control center, each one including hardware and software for data collection, storage and processing, for station management and communication with other Information Systems (Regional River-basin Authority, Regional Department of Land and Environment, etc.);
3) a multifrequency radio transmission network to connect stations and control centers.

The selected existing river gauging stations are those located in the rivers where streamflow knowledge is required to improve water use and water quality control and/or where a flood forecast system has to be implemented in the future.

The selection of the present meteorological stations to be automated was done by identifying the stations which have historical records over a long period and are representative of certain climatic areas on the basis of altitude, orientation, main wind direction, etc.

In order to improve knowledge about the snow cover in mountain areas both for water resources evaluation purpose and touristic information purposes, six new snow depth stations were introduced in the meteorological network.

The sea level automatic stations have been located in the main Sicilian ports. The sites for measurements of wave characteristics were defined in order to improve the available information for design of port protection works and sea erosion defences.

Figure 2 indicates the location of the automatic stations of the meteorological network and figure 3 shows the location of the river gauging stations.

Table I shows the distribution of instruments considered in the project into the main hydrological areas which include the basins of the rivers flowing into the Tyrrhenian Sea, the Channel of Sicily and the Jonian Sea, besides the small islands.

Table I Distribution of the instruments among hydrologic areas

MEASURED VARIABLE	HYDROLOGIC AREA			SMALL ISLANDS	TOTAL SICILY
	EAST	NORTH	SOUTH		
Rainfall	17	28	21	2	68
Temperature	17	28	21	2	68
Air humidity	7	8	9	-	24
Wind direction	7	11	7	-	25
Wind speed	7	11	7	-	25
Barometric pressure	2	2	2	-	6
Sun radiation	2	2	2	-	6
Snow-depth	2	4	-	-	6
Streamflow	5	7	10	-	22
Sea level	1	2	3	-	6
Wave characteristics	2	3	2	-	7

3. Design of the System Components

3.1 STATIONS

Each station is composed of measurement instruments, hardware and software for data collection storage and transmission, radio transmission device and supplementary equipment (solar panels and battery charger).

According to the requirements outlined in paragraph 2.2, instruments which functioned simply, had proven reliability, and limited needs of maintenance were preferred. Where it is possible, the automatic stations will use the same devices already operating (e.g. for rainfall and water stage measurement) in order to preserve the recording on paper during the initial period.

Table II shows the technical requirements of the selected instruments.

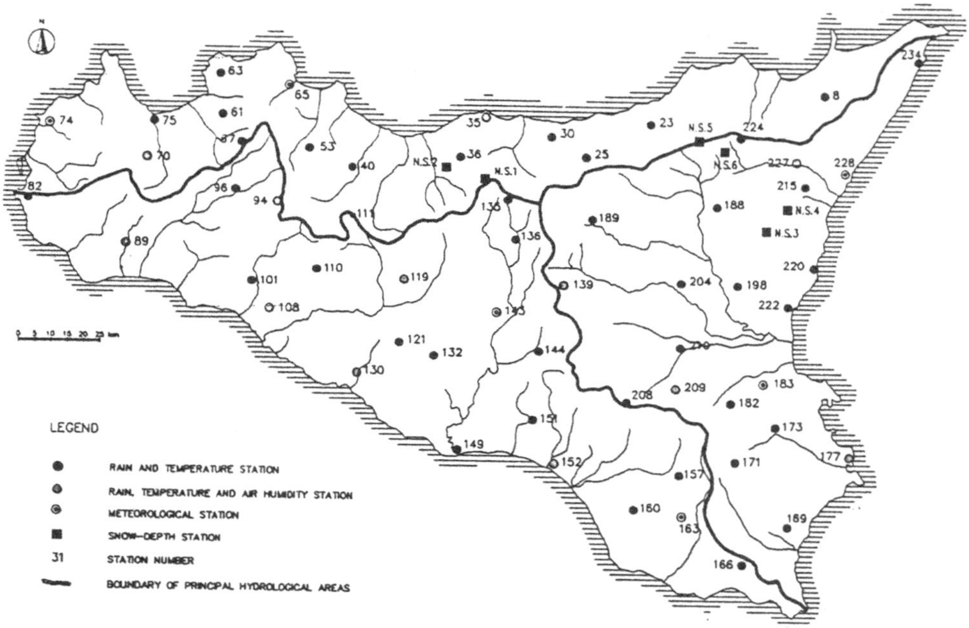

Figure 2. Stations of the meteorological network

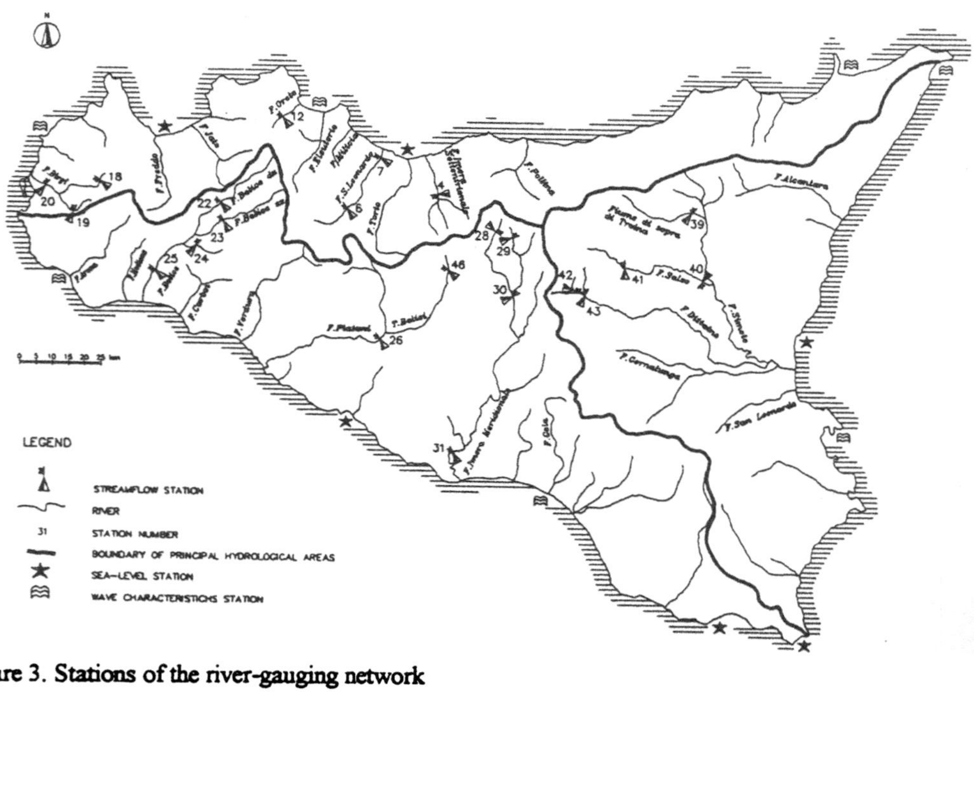

Figure 3. Stations of the river-gauging network

Table II Main instrument requirements

INSTRUMENT	RANGE	SENSIBILITY	REPEATEDLY	ACCURACY	CALIBRATION INTERVAL
Tipping bucket rain gauge	0/300 mm/h	1 impulse=0.2 mm	+/-0.25 mm/h at 60mm/h	+/-0.1 mm/h at 10 mm/h and 0.3 mm/h at 60 mm/h	2 years
Temperature sensor (platinum resistance)	-20/+50 °C	0.1 °C	+/-0.1 °C	+/-0.30 °C	2 years
Cup anemometer	0/160 km/h	0.1 km/h (threshold 1 km/h)	-	+/-0.25 km/h	no periodic calibration need
Wind direction sensor (potentiometer type)	0/360 °	+/- 1° (threshold 1 km/h)	-	+/- 2°max	no periodic calibration need
Relative humidity sensor (capacitive element)	0/100% R.H.	<0.01 % R.H.	+/- 1% R.H. at 25 °C	+/-2% R.H. at 25 °C	1 year
Barometric pressure sensor (piezoresistance type)	600/1100 hPa	<0.1 hPa	-	+/-0.2 hPa at 0<T<30 °C and 850<hPa<1050	no periodic calibration need
Solar radiation sensor (thermopile type)	0/500 W/sqm	negligible	-	+/- 15 W/sqm	2 years
Ultrasonic snow-depth gauge	1/7 m	0.1 cm	+/-0.6 cm	+/- 1/1000 of bottom scale	1 year
Ultrasonic water level gauge	0/10 m	0.1 cm	+/-0.6 cm	+/- 1/1000 of bottom scale	1 year
Pneumatic water level gauge	0/10 m	0.25 cm	+/-0.5 cm	+/- 1/1000 of bottom scale	1 year
Wave direction gauge	0/20 m	1 cm (resolution)	-	+/- 1/1000 of bottom scale	1 year
Electromagnetic current meter	0/150 cm/s	0.2 cm/s (resolution)	-	+/-0.2 cm/s	1 year

All functions in the station operation will be executed by a microprocessor unit. It will be able to handle one rain gauge or one water-stage gauge in addition to at least 16 sensors. The microprocessor unit will utilize both a RAM (Random Access Memory) buffer and an EPROM (Erasable Programmable Read Only Memory) solid memory. Collecting and recording frequency will be indipendent. Configuration parameters will be memorized on EEPROM (Electrically EPROM) 8 kByte memory, where the intervals for measuring andrecording for each sensor will be also stored. It is possible to specify if average, minimum or maximum value and standard deviation value would be recorded, both on RAM buffer and on EPROM memory.

Changing the parameters will be easily accomplished both from the station keyboard and from the control centers. A help menu will be available to users.

The station management software will be recorded on the EPROM memory (64 kByte, at least) . Local manual and automatic calibration will be available for each sensor. RAM memory capacity dedicated to recording software parameters and data storage, will be at least 24 kByte,

The EPROM recorder has been chosen both because it has no mechanical parts, and for its wide temperature range of operation (from -20 to +60°C). Since data are recorded in digital form, it permits transferring data easily to a computer unit and protects data from accidental damages.

Photovoltaic panels were chosen as a source of power for the stations. Only the meteorological stations in mountain areas will need a 220 V power connection for the operation of a hot rainfall sensor. In case of electric power failure, station operation will be uninterrupted for one year (with a 20 minute recorder frequency without any radio transmission).

The installed software in each station will be able to control the following functions:

- collect data at intervals either established in advance or requested by the control center;
- record all collected data or only selected ones;
- execute commands either established in advance or requested by the control center;
- guarantee the system from data error in case of power failure;
- automatically restart the system after a power failure;
- start automatic sensor calibration and verify whether it has been done correctly.

3.2 HARDWARE AND SOFTWARE IN CONTROL CENTERS

Automatic hydrometeorological networks will be operated in normal conditions from the principal control center, located in the Hydrological Service offices in Palermo. This center (in order to manage the information system) is composed of (fig.4):

1. one front-end unit and its peripherals (printer, radio and modem), to communicate with remote stations for the collection of raw data; because of the importance of the front-end for a regular operation of the system, another computer unit in hot stand-by will be installed in order to automatically substitute the principal unit in case of emergency, avoiding the failure risk. In normal activity, the principal front-end will operate as master unit. The slave unit does not call the stations and gets fresh information from the master unit only through the last summary row at the end of the call operation. In case of emergency, software procedures automatically change the control unit.
2. one Server unit and its peripherals (printer and scanner), both for connecting and handling the hardware units of the H.S. center, through a Local Area Network (LAN), and for communicating with other information systems through a Geographycal Area Network (GAN). The server unit will be also able both to manage the data base where raw data can be stored, modified and retrieved, and to validate data through an applicative software package for filling series, processing and print information; it will operate as LAN server, gateway for remote users (Regional River-basin Authority, Regional Department of Land and Environment, etc.), gateway for the present VAX system operating in Hydrologic Service offices, and process unit for applicative software.
3. some personal computers and their peripherals (printer and digitizer A3),both for processing in real-time raw data and validate data;

The front-end unit will be a UNIX graphic workstation, with one rewritable optical disk drive, and one DAT cartridge to improve recorder capacity. The server unit will be one RISC work station. The personal computers will be PC INTEL 80486 (33/66 MHz) using X-Window interface. They will also be used as UNIX graphic terminals.

Communications system architecture will be based on TCP/IP (Transmission Control Protocol/Internet Protocol).

The secondary control center, located in the Hydrological Service office in Catania, includes (fig.4):

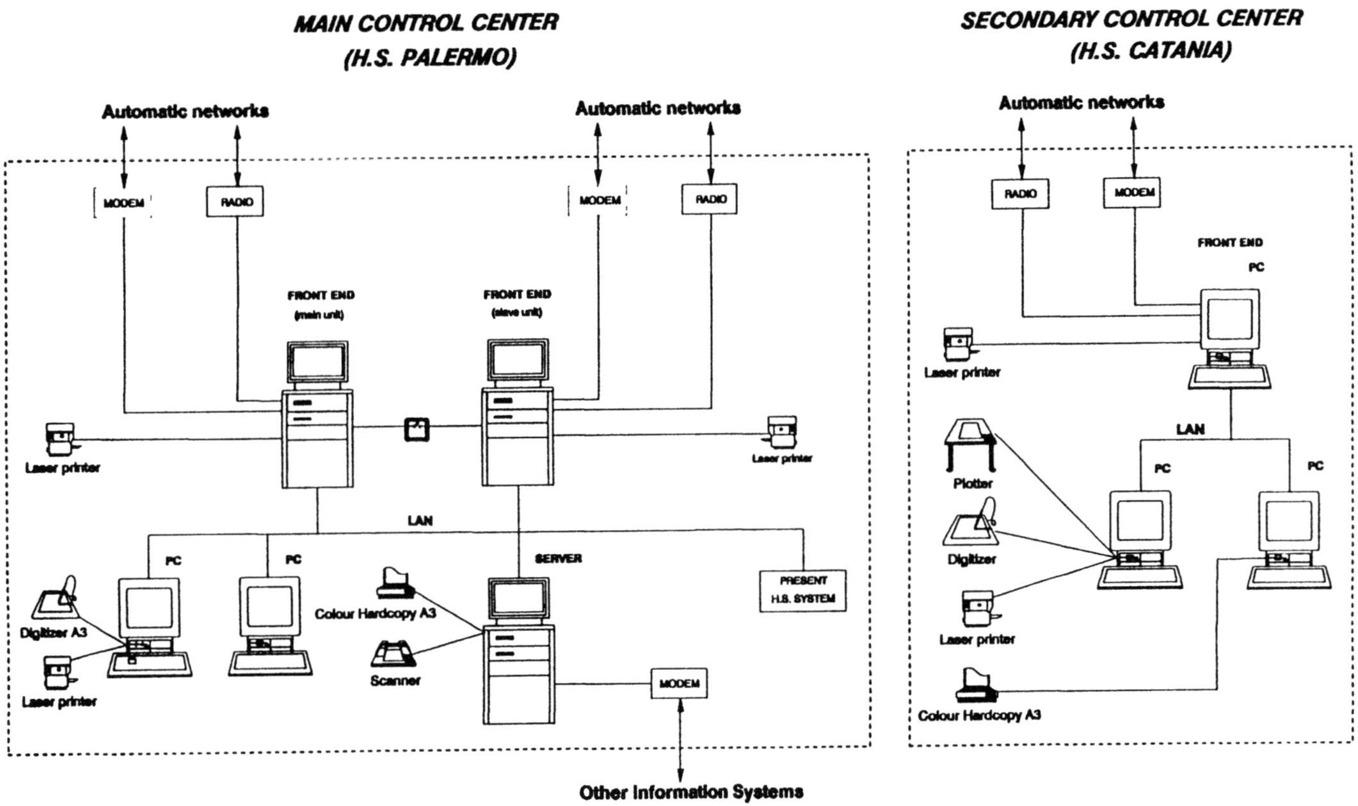

Figure 4. Architecture of the control centers

- one front-end unit (personal computer MS-DOS INTEL 80486-50 MHz), and its peripherals (printer and radio), able to handle a medium meteorological network (up to 500 sensors); the unit normally operates only in reception and it gets fresh information from the master unit in Palermo through the last summed data row at the end of the call operation. It will be able to substitute for the principal control center in case of emergency;
- two personal computers and their peripherals (printer, digitizer and plotter), able to manage and process information transmitted from networks; the characteristics are the same as the ones of the principal control center;
- processing software subsystem.

Some activities of the Hydrological Service will be automated by using specific software for:

i) automatic drawing of precipitation isolines for given interval and/or given duration by using the precipitation data observed in selected stations and computation of average precipitation over basins of interest;
ii) construction of the relationship between discharge and water stage (rating curve) to be selected among several interpolation equations obtained with the method of least squares and testing that the new experimental points fit well with the present curve to evaluate the necessity of an up-dating;
iii) automatic validation of raw data in each station, by checking whether the observed data are inside the confidence intervals computed on the basis of past measurement data obtained for the same time period in the station, and/or obtained by using the regression equation fitted to contemporary data in the station of interest and in closed stations.

3.3 DATA TRANSMISSION SYSTEM

The transmission system connecting automatic stations to control centers will be an UHF radio link, because even if there is a major investment cost, the benefits of this type of communication are very important, such as continous uninterrupted transmission, even in case of extreme conditions, independence from electrical or telephone networks, and lower up-keep costs.

The multifrequency radio system was chosen in order to take into account the particular physical characteristics of the Sicilian mountains. The system will permit many alternative links in case of emergency, will make collecting time faster, and could be used for vocal transmission, too. Each station could be used like a repeater.

The project includes 16 main repeaters, 4 secondary repeaters, and 12 stations used like repeaters (fig.5). Only one temperature and rain gauge station will be linked by commuted telephone line, because of long distance from any other station or repeater. The main repeaters will include two egual radio units in order to maintain a good level of safety. Secondary repeaters will include one radio unit. Every repeater will use a Yagi type antenna.

4. Construction, Operation, and Maintenance of the System

Figure 6 shows the time scheduling of the project realization. It is possible to distinguish four main parts:

1) Test of radio transmission: the firm in charge of the contract will perform first the radio transmission tests, under the guidance of the H.S. staff, in order to verify the design

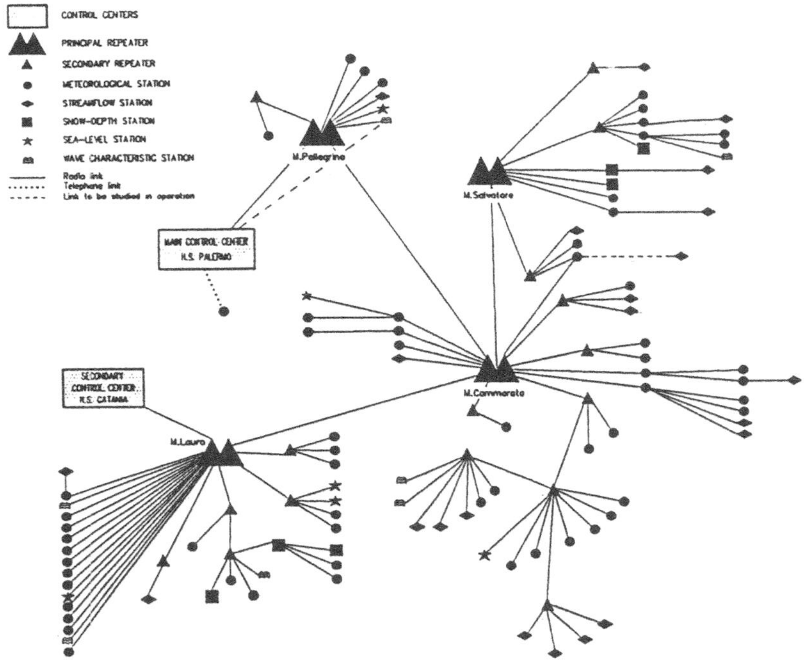

Figure 5. Data transmission scheme of proposed automatic networks in Sicily

ACTIVITY	1st year	2nd year	3rd year	4th year
Testing of radio transmission for each station				
Supplying of the equipment				
Installation of stations, repeaters and control centers				
Development of applicative software for the control centers				
Testing of the complete system				
Training courses for Hydrologic Service personnel				
Maintenance under guarantee and assistance				
Assistance and maintenance after final testing				

Figure 6. Time-scheduling of the proposed project

location of the stations;

2) Supply and installation of the equipment and development of the applicative software: the firm personnel will have to interact with the H.S. employees in order to make use of their experiences in hydrometeorological data collection;
3) Training courses for the H.S. personnel at different levels. The courses on operation, maintenance, and repair of electronic devices, on instruments, and on identification of most common causes of failure, as well as on programming and data input, will be oriented to updating some of the employees. More advanced courses on the UNIX and MS-DOS operative systems and applicative software will be directed to new personnel to be hired by H.S. These courses will be oriented to give the basis for a direct operation of the automatic networks and control centers by H.S. personnel;
4) Assistance and maintenance during the start-up period. It is possible to distinguish:
   a) the first year after construction has finished: in this period the firm will be responsible for the maintenance of the system and assistance to the H.S staff. In particular, it must eliminate any failure which will appear during the system operation with free replacement of the defective parts under guarantee, and with the replacement of the components which don't work well due to extreme events, under the financial coverage of a maintenance contract.
   b) the second and third years: it is evident that the difficulties in introducing computer and telemetering technologies increase if the system doesn't work well during the initial years after the start-up period. It is a crucial moment when some employees may look at the new instruments with suspect and want to return to the old techniques. So, it is very important that the project includes provisions for financing the maintenance of the system and assistance to the H.S. staff for two years more after the final check. During this period the H.S. staff will work together with the firm staff in order to acquire the necessary skill. Also, in the maintenance contract scheme the obbligation of providing the spare parts for six years was introduced in order to prevent troubles from the abandonement of the construction of the adopted measurement instruments.

In order to permit the H.S. staff to follow the development of the system, it is necessary to introduce in the construction contract some check-points.

In the project for Sicily hydrometeorological networks three check-points were considered as follows:

1) Preliminary check of the instruments: one station of each type must be available, inside the italian territory (e.g. at the production factory), to be tested under the supervision of the H.S. staff, in order to verify the respect of the technical requirements;
2) Check of the stations, repeaters, and control centers: each element of the system, after its installation, will be tested in order to ascertain the correct operation both of the single instruments and of the whole system;
3) Final check of the system: the complete system for data collection, transmission, and processing will be tested after the first year of operation also through a comparison between the automatically recorded data and manually collected data.

After each test with positive result the firm can require a part of the payment according to the contract's statements.

5. Benefit-Cost Analysis

Benefit-cost analysis was carried out in order to check the economic feasibility of the project.

The investment costs were computed on the basis of average prices for similar services developed in recent years. Since the hardware prices tend to decrease along with the increase of computational capability, their estimate can be considered to be conservative, inspite of the uncertainty about the time when the project will be funded and implemented.

The computed costs for operation during the start-up three year period include the costs connected with:

- the assistance and maintenance contract of the automatic system in the first year after the construction has finished, and before final check;
- the operation and maintenance contract during the 2-year period after the final approval;
- the replacement of equipment parts which fail due to exceptional conditions and therefore are not included in the guarantee;
- further administrative costs to be covered by the Hydrologic Service (hiring of new personnel, purchase of consumable materials, publicity programs).

The computed costs in normal operations (when H.S. staff will totally substitute firm staff) include the costs connected with:

- the maintenence contracts for automatic station networks (8% of the instruments cost), control center hardware (10% of the hardware cost), and control center software (5% of the software cost);
- other administrative tasks to be covered by H. S. (publicity programs, personnel updating, purchase of consumable materials).

The evaluation of benefits included financial revenues, economic monetary benefits and non-quantifiable benefits.

Financial revenues arise from a quicker data service and from an increased and more reliable availability of raw and processed data for private and public users and from consulting activities by the H.S. becoming possible as a result of the project. The unit prices were evaluated on the basis of the actual H.S. charge, and of a reasonable increase which takes into account the willingness to pay by users due to service improvement.

Economic monetary benefits were computed by evaluating the value of information which is made available by the automated network for water quality monitoring and for future agrometeorological service and flood forecasting systems.

Finally, the respect of international standards in data collecting, storing, transmitting and processing as well as the increasing of reliability, and the aid to decision making in water resources management, and to research activities, were considered as economic non quantifiable benefits.

The financial benefit-cost analysis, carried out considering a 10-year time period and a discount rate of 8%, gave a negative present net value, as expected given the prevaling character of public service of hydrometeorological data acquisition system; however, the financial revenues during the system life will cover operational costs. The economic benefit-cost analysis, including the monetary benefits computed for agrometeorology and flood forecasting (which cannot be considered financial revenues but are real benefical effects from a public point of view) gave positive results with a present net benefit value of about 5 billion Italian liras; the internal rate was 24,15%.

6. Conclusions

On the basis of the experience gained during the development of the project for automation of the hydrometeorological data acquisition system in Sicily, some reccomandations may be proposed in order to facilitate the success of modernizing the existing networks.

It is convenient to automate just a limited number of the existing stations not only because of the financial constraints, but above all in order to facilitate the gradual changes in the organization of the Hydrological Service. The selection of the present existing stations to be linked through the transmission lines has to be done by identifying the existing meteorological stations which have historical records over a long period and are representative of certain physiographic and climatic areas, and the streamflow stations which are located in the most important rivers either for water resources use or for flood control.

Moreover, increasing the exstension of the acquisition system to include new variables should be restricted only to those meteorological characteristics which have a direct relation to meeting the future needs of H.S. such as in agrometeorology, water quality control, and flood forecasting systems.

Among the technical requirements for system development there is the need for using instruments which have already been tested in field work (even if they may not represent the latest technology) and which operate simply and have easy maintenance. There is also a need for a certain redundance in the transmission lines and in the control center's computers.

The most frequent reasons for failure in the introduction of updated technologies in organizations working with traditional methods should be carefully avoided, namely:

1) the lack of acceptance of changing the way to operate by management staff; and
2) the lack of close cooperation between the firm which is in charge of the construction of the automatic system and the staff of the responsible organization.

The first step in introducing an updated technology is to persuade the service staff of its value in solving the main problems of the service. For this reason, it is very important for the H.S. staff to collaborate during the entire development of the project.

Particular care was devoted to persue good collaborative efforts between the firm and the H.S. staff in all phases of the project development, from the beginning of the construction to the end of the 3 year start-up period. For this reason, it is very important for the H.S. to appoint an executive of the staff to closely follow the development and the construction of the system. He will become the link between the firm and H.S. during the construction and start-up period, and could become the network's manager during its normal operation.

In the first year after the construction end, the firm which built the system has to be responsable for the maintenance of the networks and assistance to H.S. staff in order to minimize system failures, and to facilitate acquisition of necessary skills by the Hydrological Service.

Training courses in the operation of the automatic stations and the procedures for acquisition, storage and processing of hydrologic data in control centers are necessary for H.S. personnel.

Economic feasibility has to be evaluated in terms of benefit-cost analysis because many positive improvements in public administration efficiency and/or hydrologic research represent social benefits which cannot be accounted by strict financial analysis. However, it is also important that financial revenues should cover for normal operation costs.

Ackowledgements

We would like to thank TELEINFORM for permission to publish this paper. The writers are also grateful to the Palermo staff of the Hydrological Service for several cooperative discussions about the station choice and operation needs. Information about instrument characteristics and transmission devices was kindly provided by CAE, Bologna, SIAP, Bologna, and ECOMAR, Milan.

We are particulary grateful to the late Professor Mario Gramignani who developed the studies on the automation of hydrometeorological networks for Sicily, until his early departure.

References

- Starosolszky, O, (1987), Design and operation of hydrologic networks, in O. Starosolszky (ed.) Surface Hydrology, Water Resources Publication.
- World Meteorological Organization (1981), Guide to Hydrological Practices, vol. 1 Data acquisition and processing, Report no. 168, Geneva, Switzerland.
- Strangeways, I.C. (1985), Automatic weather stations, in J.C. Rodda (ed.), Facets of Hydrology, vol.II, John Wiley and Sons, Chichester, U.K..
- Moss, M.E. (1979), Some basic considerations in the design of hydrologic data networks, Water Resources Research, vol.15, n°6, pp.1673-1676.
- O'Connel, P.E. et al. (1979), A case study of rationalization of a rain gage network in Southwest England, Water Resources Research, vol.15, n°6, pp.1813-1822.
- Herschy, R.W. (1986), New Technology in Hydrometry, Adam Hilger, Bristol, U.K.

A NEW AUTOMATIC RAINFALL SENSOR

Bhishm Kumar * V.C.Goyal* Sandeep Gupta**

ABSTRACT

Rainfall Sensor/Recorder with the latest technique has been developed. The complete sensor is based on dual mechanism i.e., weighing and tipping bucket. The main data of rainfall is fed to data logger through load cell i.e., weighing mechanism. Apart from weighing mechanism a very special feature of this raingauge is automatic drainage without loss of data in this duration. This special feature is accomplished through tipping bucket and solenoid valve. The complete unit is to be operated through a data logger especially designed for this purpose. The latest technique used in this raingauge minimize the errors evolved at the time of drainage in any type of raingauge. The calibration of this type of raingauge is very simple and can be easily programmed through datalogger. The experiment carried out at NIH laboratory is reported.

INTRODUCTION

The types of raingauges widely used for the measurement of rainfall can be classified under three broad groups.[1-8]

1. Non-recording raingauge
2. Recording raingauge
3. Storage gauge.

The basic principle of each type of raingauge is same that the water is collected through a catch of some standard diameter and therefrom it is fed to some measuring mechanism. This is the measuring principle which categorize different types of raingauges. The different types of raingauges and their working principle have been reported in the following section of this paper.

TYPES OF RAINGAUGES

1. Non recording raingauge

The principle of working of the non recording raingauge is to collect the rainfall through an opening of regular dimensions and working out the depth of rainfall. But, for the purpose of refinement and to ensure uniformity of specifications and to maintain required degree of accuracy, raingauges are standardized for adoption. In India IMD is using the Symon's pattern raingauge of 5" diameter catch. The standard gauge of U.S. Weather Bureau consists of an 8" dia.(50.4sq.in area) receiver with knife edge top. The water is funneled into a measuring tube with 1/10th of the cross section of the receiver. This magnification helps in accurate measurement of depth.

---

* Scientist **Project Officer
National Institute of Hydrology, Roorkee - 247 667 (UP) India.

V. P. Singh and B. Kumar (eds.), Surface-Water Hydrology, 95–103.

2. Recording Raingauges

The recording raingauges are used to collect information on the intensity of rainfall in addition to amount of rainfall. These gauges are equipped with recording device which gives continuous record of depth of rainfall with respect to time in the form of a curve which when processed gives the following information.

(i) Total amount of rainfall
(ii) Times of start and cessation of rainfall
(iii) Duration of rainfall
(iv) Rainfall intensity.

Three types of recording gauges are widely used which can be distinguished by the mechanism adopted for measuring the depth of rain, since the time elements in all the three types have common features.

(i) Syphon type

The most widely used syphon type raingauge consists of a float chamber with a float connected in the bottom to a syphon chamber which houses the long discharge tube. The recording mechanism consists of a pen attached to the float, and the chart which is connected to the spring driven clock. The movement of the float is directly transferred to the chart through pen.

(ii) Tipping bucket type

The tipping bucket type of raingauge which is rapidly becoming popular consists of a tipping bucket , magnet and reed switch. The magnet is directly attached to the tipping bucket and the reed switch is positioned near to the magnet. The water from the catch is funneled into the tipping bucket and as soon as the bucket becomes full it tips to the opposite side. The movement of magnet alongwith tipping bucket causes the contact closure of reed switch, which is further recorded to some digital storage device.

(iii) Weighing type

In this type of raingauge the time element consists of a high quality clock, chart drum and set of gears. The gauge element consists of a spring or lever balance with a platform on which a bucket rests. The spring is connected to the recording pen by means of appropriate linkage. The compression of the spring and weight of water in the bucket are precisely calibrated. The incoming rain caught by the receiver is funneled into the bucket which compresses the spring. The compression of the spring is transmitted to the pen which covers a distance proportional to the depth of water received by the bucket. The gauges are generally equipped with reversal mechanism which reverses the direction of the pen after it touches the margin of the chart.

3. Storage gauges

In remote and inaccessible areas where it is not practicable to take observations every day or week, gauges with larger capacities called storage gauges are generally used. The observations may be taken after every month, season or year for which the instrument is designed. These are generally standard raingauges with bigger size of containers to hold more rain. The container is provided with some type of oil which can form a thin film on the surface of water to avoid evaporation losses.

The movement of a float, bucket, or weighing mechanism can also be converted into an electrical signal and transmitted by radio or wire to a distant receiver where recorders may be made from a number of rain recorders on data logging equipment.

PROBLEMS WITH EXISTING RAINGAUGES

Different kinds of errors are associated with the syphon type, tipping bucket type and even weighing type raingauges. For example, in syphon type raingauge, some amount of water is required to uplift the float which is evaporated if there will be the gap of longer period between two consecutive rains, which creates an error in the measurements. Secondly, accountability of rainfall is not possible if it occurs during the draining out of water, although, the time period of draining the water may be short, but, the error percentage will depend upon the intensity of rainfall. Similarly, the error percentage increases with the intensity of rainfall as due to insufficient capacity of bucket, the increased frequency of tipping , which increases the error.

In addition to the above, as India is a very big country and there is a great variation in the rainfall pattern. Thus, accordingly, raingauges of suitable specifications are required. Keeping in view the facts given above, there is a strong need to develop raingauges using advanced instrumentation techniques which should be free from all possible errors and should also satisfy the need of the region/area concerned.

Also, the advanced type raingauges which are being used for some specific studies, are required to be imported from foreign countries thereby causing a loss to our foreign currency and also difficulties are encountered in their maintenance. Moreover, they are less accurate in the regions where high intensity rainfall occurs.Therefore, the indigenous development of these equipment is also essential to save the foreign currency and to fulfill our needs.

India is such a big country that some parts of it experiences very less rainfall while other parts enjoy the full monsoon. Therefore a suitable mechanism is required to suit all the environments.

As mentioned earlier, whether it is tilting bucket type raingauge or siphon type, all have some inherent problems due to which any rain falling during the movement of the central mechanism i.e., during siphoning or during tilting of the bucket, is invariably lost. Keeping in view the above difficulties, it is desired to have a raingauge which should be free from the above stated problems and should suit to the Indian or similar environment.

Now a days, tipping bucket type raingauges are being used in most of the developed and developing countries. In India these are imported from foreign countries. However, very few firms have started manufacturing of these type of raingauges in India but these have not been standardized so far.

NEW TECHNIQUE FOR RAINFALL MEASUREMENT

The complete sensor is based on dual mechanism i.e., weighing and tipping bucket. The main data of rainfall is fed to data logger through load cell i.e., weighing mechanism. Apart from weighing mechanism, a very special feature of this raingauge is the automatic drainage without loss of data in this duration. This special feature is accomplished through tipping bucket and solenoid valve. The complete unit is to be operated through a data logger especially designed for this purpose. The complete assembly consists of following sections:

1. Outer container with 203.4mm dia and 450mm height.
2. Inner container of 175mm dia and 150mm height.
3. Load cell of capacity 10kg.
4. Solenoid valve 12V dc operated.
5. Base with platform/pipe mounting arrangement.
6. Tipping bucket mechanism.

The water from atmosphere is collected through catch of outer container and leads to inner container via tipping bucket. Tipping bucket is calibrated to collect 0.1mm equivalent of water before tipping. As soon as the bucket tips, the water goes to inner container where it is collected and weighed by load cell. When the inner container becomes full or after a prespecified time whichever comes earlier, the solenoid valve is operated to drain out the complete water from inner container. The data from load cell and tipping bucket is fed to datalogger and it is stored in a replaceable memory module.

SPECIFICATIONS

SENSOR	
Catch dia.	203.4 mm.
Overall height	475 mm.
Resolution	0.1 mm.
Material	Stainless steel

OPERATION

The raingauge is based on latest technology available in the world and it needs no human intervention for its operation after proper installation. The complete set-up is shown in figure 1.The replaceable memory module is to be changed after every one, two or three months depending upon the capacity of the memory module used and number of channels whose data is being stored in the memory module. The complete operation takes place in accordance with following steps.

1. The complete unit is resetted after every 24 hrs., if there is any amount of water in the container. Resetting involves the following works at the same time.

i) Data reading from the load cell is stopped.
ii) Solenoid valve is opened for drainage for 1.5 min.
iii) Tipping bucket counts are recorded in temporary memory RAM for 1.5 min.
iv) These counts are converted in mm of rain through software.
v). The initial reading is set to the mm of rain as recorded by tipping bucket.
vi) Further data is recorded from load cell and added to the initial reading.

2. Data from the load cell is first converted into mm. of rain through software and then stored in memory along with date and time after a predetermined time.

3. A separate memory module having capacity to store the data after a programmable time along with date and time up to three months can be changed if the presently connected module is full.

4. Regular check is made on the current reading of load cell. In case it reaches to weight, 8 cm. equivalent to the rain, the solenoid valve is operated immediately for drainage and complete resetting operation is performed i.e., from step 1 to 4, but this time, the reading from the load cell at the time of opening of the valve is added to the reading from tipping bucket counter and is considered as the initial reading of the main unit.

TESTING AND ACCURACY

The sensor was tested in laboratory using a rainfall simulator. For experiment, the water to the simulator was provided through a constant head tank and constant pressure in simulator was also monitored using a pressure probe with Universal Indicator. However, the required data logger as discussed above is under development, therefore a normal data logger for tipping bucket measurements and universal indicator for load cell measurements were used for the testing of rainfall sensor. The universal indicator gave the results directly in mm of rain as per details given in table 1 and shown in figure 2 at different simulated rain intensities. The tipping bucket operation also confirmed the same set of readings as shown in table 1.

As the rainfall of 8cm. equivalent of height can be stored in the container attached with the weighing sensor, therefore, the testing of the sensor was carried out for 6,11 and 20 minutes at rainfall 15,10 and 5 millimeter per minute respectively. This was repeated for five times in order to check the consistency in observations. However, a set of readings only for six minutes have been reported in this paper.

TABLE - 1

OBSERVATIONS OF WEIGHING/TIPPING SENSOR AT DIFFERENT TIMES

Elapsed time (min.)	Rainfall (mm) observed at intensities(mm/min.)		
	5	10	15
0	0	0	0
1	4.9	9.5	14.7
2	10.0	19.6	30.0
3	15.1	29.5	45.1
4	20.0	39.5	60.0
5	25.1	49.5	74.9
6	30.0	59.5	89.9
Manual readings	29.9	59.4	89.8

Manual readings were taken by draining out the water from inner container just after last reading shown in the table and then by measuring the volume.

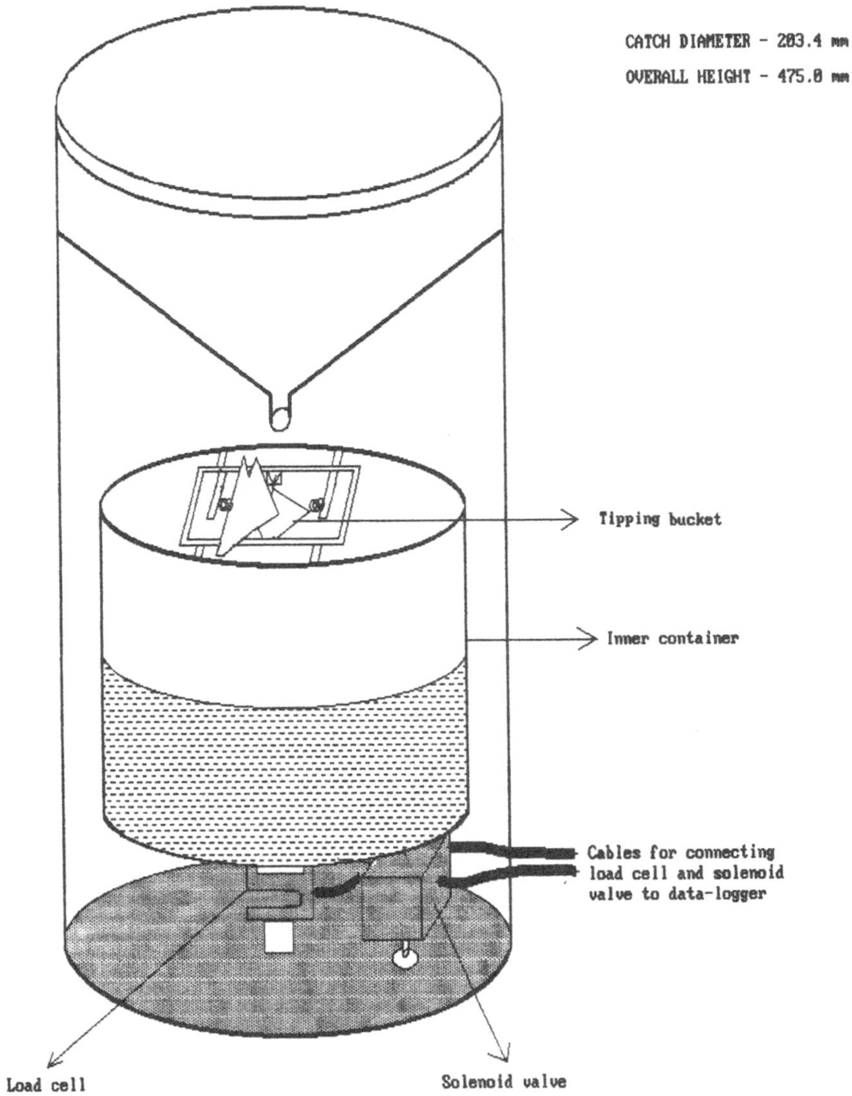

FIG-1 : AUTOMATIC RAINFALL SENSOR

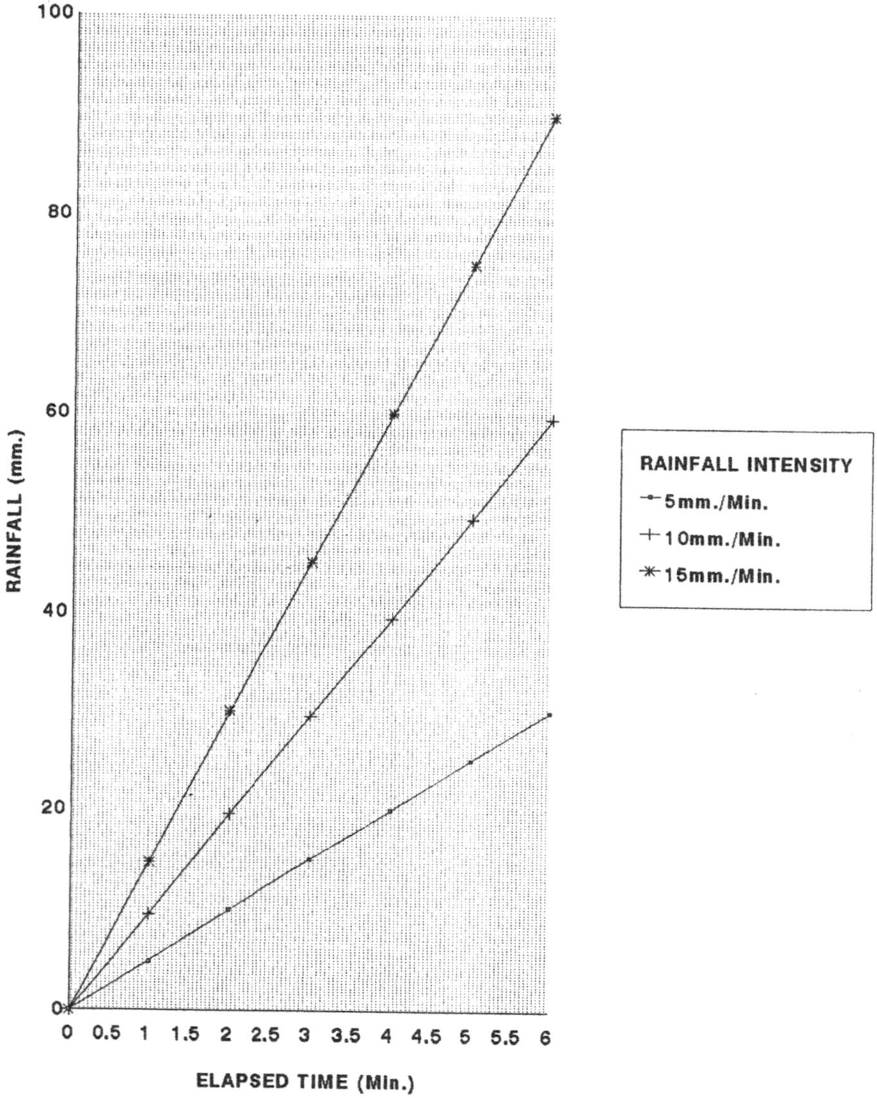

Fig.2: VARIATION OF RAINFALL PATTERN WITH

DIFFERENT SIMULATED RAINFALL INTENSITIES

ACKNOWLEDGEMENTS

Authors are very much thankful to Shri R.M. Sethi, Joint secretary and Shri Y.P. Yadav, deputy commissioner, Ministry of Agriculture, Government. of India for providing financial support to develop the automatic rainfall sensor. Thanks are also due to to Director, National Institute of Hydrology, Roorkee for allowing to work on this interesting and challenging task..

REFERENCES

1. Wilson, W.T. (1953) Discussion on paper 'Accuracy of determination of annual precipitation over a given area'. Trans. Amer. Geophys. Union 34 (1).

2. Huff, F.A. and J.C. Neill (1955),'Rainfall relations in a Urban Catchment ' Bull. Illinois State Water Survey no.44.

3. World Meteorological Organization (1968)

4. Indian Standard IS 5235-1969.

5. Wassi Ullah and others (1969) Adequacy of raingauges in small experimental watersheds.

6. Wassi Ullah and others (1972) Hydrological measurements for watershed research.

7. Seth S.M. and Ramashastri K.S. (1984-85),'Processing of precipitation data', report M-2 National Institute of Hydrology, Roorkee.

8. Training manual for hydrological and sediment monitoring of small water sheds, New Delhi- June 91.

AN AUTOMATIC ELECTRONIC INFILTROMETER

BHISHM KUMAR
National Institute of Hydrology, Roorkee - 247 667 (U.P), INDIA

SYNOPSIS

A new equipment for infiltration rate measurement has been developed which is very simple, accurate and fully automatic. The automatic electronic infiltrometer works on the constant head principle. Microprocessor (8085) has been incorporated with it so that the infiltration data can be stored in the memory module with a maximum of thousand readings of water level with time. The accuracy of the instrument is as per the Indian standards and its main advantage is its portability. A special constant head device has also been used to maintain a constant head. The infiltrometer has been so designed that one infiltrometer ring can be kept inside the other rings (due to varying diameter of inner and outer infiltrometer rings) in order to reduce the space requirement for the easy transportation to field. The microprocessor can be programmed in order to store the infiltration data at a desired interval from 1 to 99 minutes. It is easy to operate even by an unskilled worker, once installed at the experimental site. The microprocessor data can be transferred to personal computer and analyzed in the laboratory with ease.

INTRODUCTION

The flow of water through soil surface is called infiltration. Quantitatively, infiltration rate is defined as the

V. P. Singh and B. Kumar (eds.), Surface-Water Hydrology, 105–117.

volume of water passing into the soil per unit area per unit time and has the dimension of velocity.

There are two general approaches to determine the infiltration capacity of a soil cover and soil moisture complex. One of these is the analysis of hydrographs of runoff from natural rainfall on plots and watersheds. The other is the use of infiltrometers with artificial application of water to enclose sample areas. The brief description of types of infiltrometers and the automatic infiltrometer developed has been described in the following sections.

TYPES OF INFILTROMETERS

Infiltrometers may be broadly classified into two general categories.

(i) Single ring infiltrometer

(ii) Double ring infiltrometer

Single ring infiltrometers are out dated because of some of their demerits (discussed below) and double ring infiltrometers are generally used with new modifications for regulating the flow and recording data.

SINGLE RING INFILTROMETER

Single ring infiltrometer consists of a single tube about 22.5 cm diameter and 45 to 60 cm long which is driven into the ground at least to a depth upto which the water percolates during the experiment and thus no lateral spreading of water can occur.

The water added into the tube at regularly to maintain a constant water head is noted down.This data is used to determine the infiltration rate and to draw the infiltration rate curve. The affect of compression of the entrapped air is not accounted using infiltrometer.It disturbs the soil structure near the boundary. Results and conclusions are drawn from very small experimental area and cannot be applied to bigger areas or watersheds. However, the main drawback of this type infiltrometer is the loss of water due to horizontal flow.

DOUBLE RING INFILTROMETER

The most common type consists of two shallow concentric rings of sheet metal,usually ranging from 22.5 to 90 cm diameter.These rings are placed with their lower edges a few cm below the ground surface and with the upper portion projecting above. Water is now applied in both rings and is always kept at the same level in both the apartments. The function of the outer ring is to prevent the water spreading out from the inner space to a larger area,after penetrating below the bottom of the ring. From the rate at which water is added into the inner ring in order to maintain a constant water head, the infiltration rate and the infiltration capacity is determined..

Some other types .of infiltrometers are Self dispensing infiltrometer, Closed top infiltrometer, Automated tension infiltrometer, Suction crust infiltrometer, Rainfall simulators and Constant head permeameter which also work on the similar principle with only difference in maintaining the constant water level head and infiltration data acquiring technique.

PROBLEMS WITH EXISTING INFILTROMETERS

It is a well known fact that manual operation is associated with errors which depend upon the person concerned operating the system In existing infiltrometers, the constant water head is maintained manually which is very difficult and erroneous. Also, in double ring infiltrometers, the water level in both rings is to be kept at same level, which creates error if maintained manually. For operating these type of infiltrometers, some skilled workers are required. However, some of the infiltrometers which are technically very sound, the system of operation is either complicated and requires expertise in situ operation or their use in situ and transportation to field is rather difficult. Therefore, keeping in view of all the above problems, it is required to have an instrument which can be operated easily and automatically.

AUTOMATIC ELECTRONIC INFILTROMETER

The automatic electronic infiltrometer is a complete device to conduct infiltration studies in situ and once installed in the field, an unskilled worker can take care of it during the experimental work.

This system consists of the following components:

1) Two G.I. rings of appropriate dimensions
2) Two G.I. containers of appropriate diameter filled with water
3) Two solenoid valves, one each fitted with the water container
4) Two sets of electrodes, one each fitted with G.I. infiltrometer ring
5) One sensor (capacitance type) to be installed in the water container from which the water is supplied to the inner ring.
6) Microprocessor based programmable water level measuring device with digital display and memory module along with electronic circuitry to control the constant water need in the rings.
7) Two 12 Volts, maintenance free, batteries for the operation of solenoid valves and microprocessor based water level

measuring device separately.

The sensor which is an ordinary hollow rod with copper wire fitted at the centre, connecting the both ends, works as a water level sensor. It is placed in one of the container whose water level is to be sensed. The drop in the water level of the container, decreases the capacitance value of the sensor. Using this capacitance value the microprocessor based unit displays the water level in direct engineering units and stores this value along with time in the memory module at the preselected time intervals.

In order to save the power, the provision is made to switch off the display by pressing the key designated to 'RUN' the device. However, a sound beep is given by the equipment whenever the water level reading will be stored in the memory according to the preselected time-interval. At the time of beep, the display will automatically be switched on for three seconds so that one can note down the W.L. reading manually. The 'HOLD' key enables the user to see the current values at any instant. The data can be retrieved on Personal Computer by connecting it with the equipment through RS-232 interface.

SPECIFICATIONS

a) Electronic Device

The main features of the microprocessor based electronic device are described below:

- Programmable recording time ranging from 1 to 99 minutes.
- Seven segment eight digit LED display
- Memory module of capacity to store 1000 readings.
- Re-calibration of the system if water quality is changed.

- Resetting of any water.level to zero - as a reference.
- Keypad on front panel with seven keys for different operations.
- Presets for zero setting and calibration of the equipment on front panel.
- RS-232 port at one side of the equipment.

The complete unit is enclosed in a box of size 20 x 20 x 10 cm.

b) Infiltrometer rings and water containers

Different M.S. infiltrometer rings have been designed, with varying diameters for examples, the inner and outer sets of 10 gauge M.S. rings having the diameters 22.5 cm & 37.5 cm., 25 & 40 cm, 27.5 & 42.5 cm, 30 cm & 45 cm, 32.5 cm & 47.5 cm and 35 & 50 cm. Accordingly, the water containers of G.I. sheets of different diameters (equal to the diameter of infiltrometer rings) each 50 cm the height fitted with a transparent plastic scale have been designed. The water level reading in the container can be seen in the container and can be noted down manually if required for testing the performance of the system.

c) Water level sensor

Hallow aluminium tube of 1/2" diameter and 50 cm long, fitted with a copper wire inside.

d) Solenoid valve - 12 V D.C. operated with 1/2" opening.

e) Power supply - 12 V, 6/7 Amp. hour maintenance free battery

INSTALLATION

The infiltration rings are hammered in the ground in such a way that at least 20-25 cm of the lower part is went inside the soil strata. The inner ring is hammered first so that the soil strata, to be studied, may not be compacted due to hammering of outer ring.

The connection of the solenoid valve, electrodes and battery are made as per shown in the fig. 1. The water containers are filled with water and water level sensor is kept in one of the containers from which the water is to be supplied to the inner ring. The connection of the sensor are made with the electronic device as per shown in the diagram.

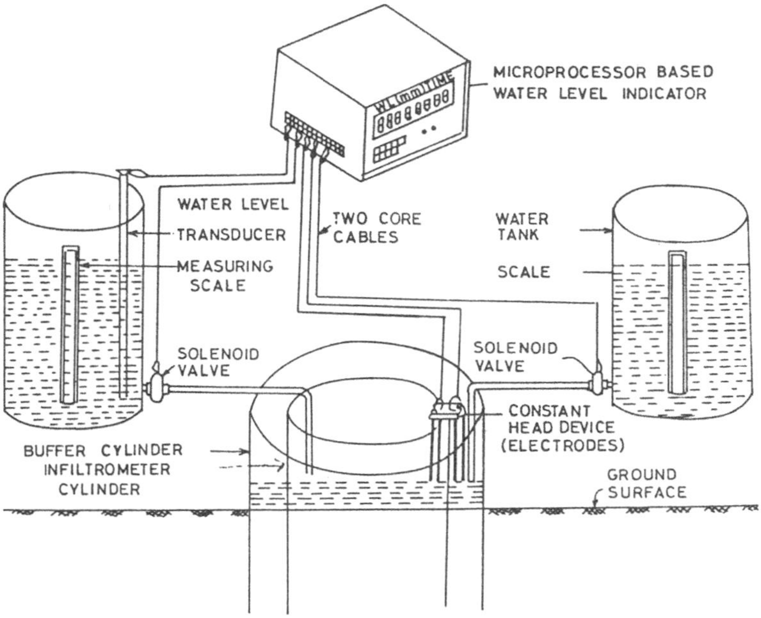

FIG. 1 MICROPROCESSOR BASED DOUBLE RING INFILTROMETER WITH CONSTANT HEAD DEVICE

OPERATION

After the proper installation of the equipment at the suitable site as described above. The equipment on display indicates 'nih' in the beginning. The equipment is now to be reset using the 'reset key' and the time interval is programmed using the lower side keys i.e. if the water level reading is to be taken after every 5 minutes, the all values in the time section of display are selected for zero except the first one which is made at 5. The time display is a 4 digits LED in which two digits are designated for minutes and the two for hours.The equipment can be programmed for a time interval from 1 to 99 minutes as required. After setting the time interval, the incremental key is again pressed,therefore,the display will again indicate'nih'. Now, the system is ready to run. But, before, starting the experiment, the water is filled-up in the containers and zero value is set in the left part of the display with the help of preset designated for zero setting. After this, the equipment is calibrated by taking out the sensor from the water container for a known height of the water column. The same value should be indicated in the display. If it is not indicating the same value, the water level value is set with the help of preset designated for calibration. The system is calibrated for the maximum possible change in water level in order to minimise the percentage of error in the measurements. After calibrating the system, the water level sensor is again kept in its original position and the water is filled-up in the infiltrometer rings from the external source in such a way that the upper surface of water should touch to the end of the electrodes. At the same time, the key pad 'run' should be pressed to start the final operation of the equipment. These all operations are done in such a way that the time clock, W.L. reading and infiltration of water in the rings start at the same time.

The equipment will now start functioning and the water level values along with time will be stored in the memory module at the pre selected time intervals. The solenoid valves will automatically operate to supply the water into the rings in order to maintain the constant head. Thus the water level will change in the water container.

After the specified period, the experiment can be closed and another experiment can be performed at some other site. The infiltration data can be later on retrieved on P.C. with the help of the interface available with the equipment.

PRECAUTIONS

The following precautions should be taken while operating and using the electronic infiltrometer:

i) The connection should be as prescribed and shown in the diagram (fig.1)

ii) The batteries should be fully charged while starting the experiment in the beginning and these should be recharged every night if the series of experiments are to be performed continuously for few days.

iii) The memory module can store maximum 1000 readings of water level along with time, therefore the data should be retrieved from the memory module on personal computer as soon as it is possible.

iv) The calibration of the equipment is essential, if the water quality changes which is quite possible in case of in situ studies at different places.

v) It is advisable to check the performance of the equipment by taking some manual reading of drop in water level using the scale fitted in the container.

FIELD TESTING

This portable infiltrometer has been tested at different sites in the field. The operation and use of this type of infiltrometer has been found quite satisfactory. However, the efforts are being made to make the zero setting and calibration of the equipment through software controlled keys and not by using the presets presently designated at the front panel. The experimental data of one of the sites is given in table 1 as an illustration and its plot has been shown in fig.2.

Table - I

Observations of Infiltration rate using automatic electronic infiltrometer developed at NIH, Roorkee, Uttar Pradesh, India

Time Elapsed (minutes)	Infiltration Rate (cm/hr)
5	2.5
10	2.2
15	2.0
20	1.8
40	1.2
50	1.0
70	0.8
145	0.8
200	0.8

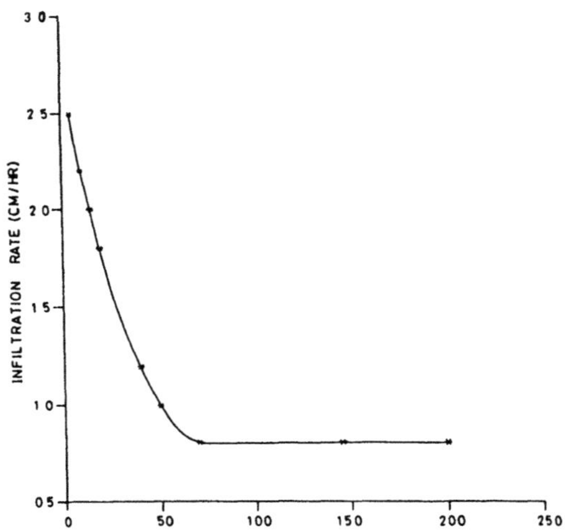

FIG 2 INFILTRATION RATE CURVE

REFERENCES

(1) Baver.L.D. 1933. Soil Erosion in Missouri. Missouri Agriculture Exp. Sta. Bull.(349) Page 266.

(2) Booltink.H.W.G, J.Bouma and D.Gimenez. March-April 1991. Suction Crust Infiltrometer for Measuring Hydraulic Conductivity of Unsaturated Soil near Saturation. Soil Science Society of America Journal. Vol. 55. Page 566-568.

(3) Diebold.C.H. 1951 Soil layers causing run-off from hard land wheat fields in Colarado and New Mexico. Journal of Soil and Water Conservation, 6, Page 202-209.

(4) Dixion.R.M.Deaign and Use of Close Top Infiltrometers. Soil Science Society of America Proc. Volume 39. No.4.August 1975.

(5) Duley.F.L.and L.L.Kelley. 1939. Effect of Soil Type, and Surface Conditions on Intake of Water. University of Nebraska Agri. Exp. Sta. Res. Bull. No. 112. Page 1-16.

(6) Fletcher.J.E. 1949. Some properties of water solutions that influence infiltration. Trans American Geophys Union (30) Page 548-554.

(7) Fletcher.J.E. 1960. Some Aspects of Plant Growth on Infiltration in the South West. AAAS. South Western and Rocky Mountain division, Alpine Texas.May 3. 1960.

(8) Green.R.E. 1962. Infiltration of Water into Soils as Influenced by Antecedent Moisture. Ph.D. Dissertation. Iowa State University, Ames, Iowa.

(9) Holtan.H.N. 1945. Time Condensation in Hydrograph Analysis.Trans.American Geophys Union.26. Page 407-413.

(10) Horner.W.W. and C.L.Lloyd. 1940. Infiltration Capacities as Determined from a Study of an Eighteen Month Record at Edwards-Ville, Illinois. Trans American Geophys Union. 21. Page. 522-541.

(11) Horton.R.E. 1933. The rate of infiltration in the Hydrologic cycle. Trans. American Geophys.Union.(14). Page 446-460.

(12) Horton.R.E. 1935. Surface Runoff Phenomena Hortan. Hydrological Laboratory. Paper No. 101.

(13) Horton.R.E. 1940 An approach toward a physical interpretation of infiltration capacity. Soil Science Society of America (Proceedings) 5, Page 693-711.

(14) Horton,Kaspor,and Ankery.M.D. Design for an Automated Tension Infiltrometer.Soil science Society of America Journal. Volume 52, No.3. May-June 1988.

(15) Jens.S.W. and M.B.Mcpherson 1964. Hydrology of Urban Areas. Handbook of Applied Hydrology (ed. V.T.Chow). Sec.20. McGraw Hill book co. New York. NY.

(16) Kostiakov.A.N. 1932. On the Dynamics of the Coefficient of Water Percolation in Soils and on the Necessity for Studying it from a Dynamic Point of View for Purposes of Amelioration.

(17) Lewis.M.R. and W.L.Powers 1938. A study of factors affecting infiltration. Soil Science Society of America (Proceedings) 3 Page 334-339.

(18) Linsley.R.K.; M.A.Kohlar. and J.L.H. Paulhus. 1949. Applied Hydrology. McGraw Hill book co. New York. N.Y. Page 689.

(19) Lull.H.W. 1964. Ecological and Silvicultural Aspects. A HandBook of Applied Hydrology. (ed.by V.T.Chow). Sec.6. McGraw Hill book co. New York, NY.

(20) Musgrave.G.W.and H.N.Horton 1964. Infiltration. Handbook of Applied Hydrology (ed. by V.T.Chow). Sec.12. McGraw Hill book company; New York. NY.

(21) Packer.P.E. 1951. An Approach to Watershed Protection Criteria. Journal of Forestry 49(9). Page 639-644.

(22) Parlange.J.Y. 1971.Theory of Water Movement in Soils. 2. One Dimensional Infiltration. Soil Science.111. Page.170-174.

(23) Philip.J.R. 1957a. Numerical Solutions of Equations of the Diffusion Type with Diffusivity Concentration Dependent. II Australian Journal of Physics.10. Page 29-42.

(24) Philip.J.R. 1957. Numerical Solution of Equations of the Diffusion Type with Diffusion Concentration Dependent, II. Australian Journal of Physics.

(25) Philip.J.R. 1973. Theory of Infiltration Advances in Hydroscience (ed. by V.T.Chow) Vol. 5. Academic press, New-York, N.Y. Page. 215-296.

(26) Philip.J.R. 1973. On Solving the Saturated Flow Equation.1. The Flux Concentration Relation. Soil Science. 116. Page. 328-335.

(27) Powell.G.M.and R.P.Beasley 1967 Effect of erosion on water infiltration rates. University of Missouri. Agri. Exp. Sta. Res. Bull. 922. Page 47.

(28) Rauzi.F. and C.L.Fly. 1968.Water Intake on Mid-Continental Range lands as Influenced by Soil and Plant Cover.USDA. Tech. Bull. No. 1390. Page 58.

(29) Richards.L.A. 1952. Report of the sub-committee on permeability and infiltration, committee on technology. Soil Science Society of America Journal Proc. 16(1). Page 85-88.

(30) S.Matulaand C.Dirsken. Jan-Feb 1989. Automated Regulating and Recording System for Ring Infiltrometers. Soil Science Society of America Journal.Volume 53.Page 299-305.

(31) Saxena.R.K. Jan-april 1992. Evolution of Soil Auto Infiltrometer. Journal of Water resources Society. Vol.12.No.1&2. Page 63-65.

(32) Schreiber.H.A.and D.R.Kincaid. 1967. Regression Models for predicting On-side Runoff from short duration Convective storms. Ware resources research.3 (2).Page 389-395.

(33) Self Dispensing Infiltrometer.Page 59. Hydrological Investigations.soil and Water Conservation Research.

(34) Sharda.V.N.and Sita Ram Singh July 1992. Infiltration and Ground Water Recharge. Water Science Education Series..No.3.

(35) Sharp.A.L.and H.N.Hortan. 1940.A Graphical Method of Analysis of Sprinkled Plot Hydrographs.trans. American Geophys. Union.21. Page 558-570.

(36) Sharp.A.L.and H.N.Hortan. 1942. Extension of Graphic Method of Analysis Sprinkled Plot Hydrographs to the Analysis of Hydrographs of Control Plots and Small Homogeneous Watersheds.Trans.American Geophys Union.23. Page 578-593.

(37) Th.M.Boers, F.J.M.P.Van Deurzen, L.A.A.J.Eppink, R.E.Ruytenberg 1992. Comparison of Infiltration Rates Measured with an Infiltrometer, Rainulator and a Permeameter for Erosion Research in S.E.Nigeria. Soil Technology. Vol. 5., Page 13-26.

(38) Trimble.G.R.1958. How Type of Forest Effects Infiltration, Journal of Soil and Water Conservation.13. Page. 81-82.

(39) Willis.W.S. 1965. Use of rainfall Simulator in a Laboratory Study of Infiltration Rates of Bare Soils.

SECTION 3:

DYNAMICS OF OVERLAND FLOW

MODELING SURFACE RUNOFF WITH KINEMATIC, DIFFUSION, AND DYNAMIC WAVES

V. M. PONCE
Department of Civil Engineering
San Diego State University
San Diego, California 92182 USA

ABSTRACT. Surface runoff can be modeled deterministically with kinematic, diffusion, or dynamic waves. Kinematic waves convect runoff, but do not diffuse it. Diffusion waves convect runoff under a small but perceptible amount of diffusion. Dynamic waves convect runoff and strongly diffuse it. Kinematic and dynamic waves are well established in hydrologic engineering research and practice. Diffusion waves, however, have only recently been accepted in practice, as shown by their inclusion in Version 4.0 of HEC-1: Flood Hydrograph Package (Hydrologic Engineering Center, 1990). Given the current emphasis on computational modeling of surface runoff, a review of these three models is warranted at this time. This paper focuses on the following: (1) definition and properties of kinematic, diffusion, and dynamic waves, (2) role of numerical diffusion in kinematic wave modeling, (3) nature of kinematic shock, and (4) role of Vedernikov number in surface runoff modeling.

1. Introduction

The modeling of surface runoff using physically based methods with the aid of a computer is well established in research and practice. Any such modeling uses either kinematic, diffusion, or dynamic waves, with papers continuing to appear in the literature describing applications of the three models. However, there is still some confusion as to their applicability, as demonstrated by a recent paper by Tayfur et al. (1993), where a comparison is made of results obtained by the three models using a very steep slope (S_o= 0.086), while disregarding the effect of time-of-rise t_r. The comparison showed essentially no difference between results obtained with any of the three models.

Kinematic and dynamic waves have been established since the 1960's, first in research, and then in practice. Diffusion waves, however, have remained within the realm of research until only recently, when they were incorporated into Version 4.0 of HEC-1: Flood Hydrograph Package (Hydrologic Engineering Center, 1990). Given the current emphasis on computational models of surface runoff, for both water quantity and water quality applications, a review of the three models is warranted at this time. Accordingly, this paper focuses on the following: (1) definition and properties of kinematic, diffusion, and dynamic waves, (2) role of numerical diffusion in kinematic wave modeling, (3) nature of kinematic shock, and (4) role of Vedernikov number in surface runoff modeling.

V. P. Singh and B. Kumar (eds.), Surface-Water Hydrology, 121–132.

2. Kinematic Waves

The mathematical theory of kinematic waves is attributed to Lighthill and Whitham (1955), although its origin can be traced back to the end of the last century, specifically to Kleitz and Seddon (Chow, 1959). In the past three decades, the application of kinematic wave theory to overland flow and streamflow has gained considerable momentum (Wooding, 1965; Woolhiser and Liggett, 1967; Hydrologic Engineering Center, 1990). There is a substantial body of knowledge about the kinematic wave, and papers continue to appear in the literature describing the model's capabilities and limitations (Hromadka and DeVries, 1988; Ponce, 1991a). There is still, however, some misunderstanding about the precise role of kinematic waves in surface runoff modeling (Woolhiser, 1992).

A kinematic wave can be defined in three ways. First, a kinematic wave transports mass, in contrast to the inertial wave, the so-called "gravity" wave of classical mechanics, which transports energy. Second, a kinematic wave is characterized by a one-to-one relationship between discharge and stage. Third, a kinematic wave accounts for only gravitational and frictional forces, and neglects the forces arising from the flow depth gradient (dy/dx, in which y = flow depth, and x = distance along the channel) and inertia.

The above three definitions are related. To put them in the proper perspective, we note that Lighthill and Whitham (1955), in laying the foundations of kinematic wave theory, subtitled their paper *Flood Movement in Long Rivers*. Flood waves transport mass; kinematic waves also transport mass. However, while flood waves are kinematic in nature, not all kinematic waves are flood waves. To clearly distinguish between flood waves and kinematic waves, we examine Lighthill and Whitham's subtitle further. The implication is that the kinematic wave is applicable only to long rivers! If this were the case, the kinematic wave could not be applied to steep mountain streams, which are relatively short compared to most alluvial rivers. In practice, however, kinematic waves are seen to be applicable to both 'short' mountain streams and 'long' alluvial rivers.

The resolution of this conflict was made possible by Ponce and Simons (1977), who identified the parameter describing the applicability of kinematic waves. They showed that the latter is controlled, not by the 'length' of the river, or by the length L of the shallow wave, but rather by the dimensionless ratio L_o/L, where L_o is a reference channel length, defined as

$$L_o = d_o / S_o \tag{1}$$

with d_o= average flow depth, and S_o= channel bed slope. In general, a wave is kinematic if the dimensionless wave number

$$\sigma = (2\pi/L)L_o \tag{2}$$

is sufficiently small for a given Froude number

$$F_o = u_o/(gd_o)^{1/2} \tag{3}$$

The reference channel length L_o is readily obtained, provided an average flow depth can be established, which is usually the case. The same does not follow for the wavelength L, which needs to be converted to the temporal domain for practical use. Since

$$L = cT \tag{4}$$

where c is the wave celerity and T is the wave period, the ratio L_o/L can be expressed as:

$$L_o/L = d_o/(cTS_o) \tag{5}$$

Seddon's law (Seddon, 1900; Chow, 1959) can be used to express the kinematic wave celerity c in terms of the mean flow velocity:

$$c = (5/3)\, u_o \tag{6}$$

applicable to Manning friction in hydraulically wide channels. The ratio L_o/L is now expressed as:

$$L_o/L = (3/5)\,(d_o/u_o)\,/\,(TS_o) \tag{7}$$

Furthermore, the flood wave period T can be expressed in terms of the more familiar time-of-rise t_r, or time-to-peak of the flood hydrograph. Assuming for simplicity

$$T = 2t_r \tag{8}$$

then

$$L_o/L = (3/10)\,(d_o/u_o)\,/\,(t_rS_o) \tag{9}$$

In nature, while d_o and u_o are usually restricted within a narrow range, t_r and S_o tend to vary within a broad range. In fact, the flood wave time-of-rise t_r can be as short as 5 to 15 minutes in small steep catchments, and as long as 3 to 6 months in large catchments of mild relief. To give an extreme example, the time-of-rise of the Upper Paraguay River at Porto Murtinho, Brazil (at the outlet of the huge *Pantanal* of Mato Grosso), is approximately 6 months. The channel bed slope S_o typically varies between $S_o = 0.1$ (or steeper) in some mountain stream situations, and $S_o = 0.00001$ (or milder) in certain deltaic and tidal settings. Thus, in general, the ratio L_o/L is inversely related to the product t_rS_o. For a given Froude number, the larger the value of t_rS_o (and therefore, the smaller L_o/L), the more kinematic the wave is.

In light of the preceding considerations, the meaning of Lighthill and Whitham's subtitle is now fully elucidated: The adjective 'long' should be construed to refer to a small L_o/L or large t_rS_o. The latter implies that either t_r or S_o, or both, should be large. Experience reveals that in nature these two parameters are not likely to be large simultaneously. Either the time-of-rise t_r is long (as in a large catchment of mild-relief), or the channel bed slope S_o is steep (as in a mountain stream, or in overland flow in a typical small-catchment setting), but usually not at the same time. This behavior confirms the wide range of field situations in which the kinematic wave is applicable: for *both* steep and mild catchments, and *both* fast-rising and slow-rising hydrographs, provided the product t_rS_o is sufficiently large.

Ponce et al (1978) developed a criterion for the applicability of kinematic waves in surface runoff. The criterion states that for a shallow water wave to be kinematic, it should satisfy the following dimensionless inequality:

$$N = t_r S_o (u_o / d_o) > 85 \tag{10}$$

The larger the value of N, the more kinematic the wave is. For instance, if t_r = 6 hr, S_o = 0.01, mean velocity u_o= 2 ms^{-1}, and flow depth d_o= 1 m, it follows that N = 432 > 85, confirming that this wave is kinematic. According to the definitions of kinematic wave, this wave: (1) will transport mass, (2) will describe a one-to-one relationship between discharge and stage at any cross-section, and (3) the forces arising from the flow depth gradient and inertia will be so small so as to be negligible compared to the gravitational and frictional forces.

3. Diffusion Waves

The specification of a one-to-one relationship between discharge and stage, a key trait of the kinematic wave, imposes a significant physical and mathematical constraint: The wave cannot diffuse. It can convect downstream and transport mass in the process, but it cannot dissipate its discharge or stage. This limitation of the kinematic wave is grounded in its formulation: The neglect of the flow depth gradient and inertia terms results in a first-order partial differential equation governing the motion. This equation cannot describe diffusion, since the latter is a second-order process. From a physical perspective, the one-to-one stage-discharge relationship implies that wave diffusion is clearly out of the problem, since the latter is caused by the presence of a loop (however small!) in the stage-discharge rating.

Since in nature there exist shallow water waves which do diffuse, although in small amounts, the theory of kinematic waves is incomplete without a means of incorporating this important diffusion mechanism. Lighthill and Whitham (1955) clearly saw this when they suggested the extension of kinematic waves to the realm of *diffusion waves*, i.e., of kinematic waves that incorporate a small amount of diffusion. To accomplish this, the formulation of kinematic waves is modified to include the flow-depth gradient term, while still excluding the inertia terms. This significant extension allows the description of looped stage-discharge ratings, and consequently, of the diffusion of kinematic waves, now properly, diffusion waves. To summarize, diffusion waves are still kinematic in nature; they still transport mass; however, unlike kinematic waves, diffusion waves have the capability to undergo small but perceptible amounts of physical diffusion.

This physical diffusion is confirmed by theory and experience. As long as the flow depth gradient is not negligible, it will produce a looped stage-discharge rating for every shallow wave, which will in turn cause the wave to dissipate as it travels downstream. In practice, as the channel bed slope S_o decreases (as the flow moves from mountain streams to alluvial rivers), the friction slope S_f decreases accordingly (as channel roughness typically decreases in a downstream direction), and the flow depth gradient becomes increasingly too important to be disregarded. Intuitively, while kinematic waves are seen to apply to mountain streams, diffusion waves are seen to apply to valley streams and alluvial rivers. A rule of thumb validated by experience, says that if the channel bed slope is greater than 1 percent (S_o > 0.01), the wave will most likely be kinematic, will feature a one-to-one relationship between discharge and stage, and will not diffuse appreciably. If the channel bed slope is less than 1 percent, the wave may not be kinematic; it may be a diffusion wave. If so, it will feature a looped stage-discharge rating and show a small but perceptible amount of diffusion.

Ponce et al (1978) have presented a criterion for the applicability of diffusion waves. The criterion states that for a shallow wave (whether flood wave or overland flow wave) to be a diffusion wave, it should satisfy the following dimensionless inequality:

$$M = t_r S_o (g/d_o)^{1/2} > 15 \tag{11}$$

For instance, if $t_r = 6$ h, $S_o = 0.001$, and flow depth $d_o = 1$ m, it follows that $M = 67.6 > 15$, confirming that this wave is a diffusion wave. This wave will have the following properties: (1) it will transport mass, like the kinematic wave, (2) it will diffuse appreciably, unlike the kinematic wave, (3) it will describe a looped stage-discharge relationship at any cross-section, and (4) the force arising from the flow depth gradient can no longer be neglected.

It is noted that in the example of the previous section, had the channel bed slope been $S_o = 0.001$, then $N = 43.2$, and the wave would not have qualified as a kinematic wave. However, in the example of this section, if the slope is $S_o = 0.01$ instead, then $M = 676$, and the wave would still qualify as a diffusion wave. It is concluded that while the kinematic wave model does not apply to diffusion waves, the diffusion wave model does apply to kinematic waves. In other words, the theory of diffusion waves can properly describe both kinematic and diffusion waves.

4. Dynamic Waves

Dynamic waves approach the problem of shallow water wave propagation in its most general form, i.e., by considering all forces, including gravitational, frictional, flow-depth gradient, and inertial force. This leads to a set of two partial differential equations of continuity and motion, the Saint Venant equations, for which there is no known complete analytical solution. There are, however, several incomplete analytical solutions of the Saint Venant equations (Lighthill and Whitham, 1955; Dooge, 1973; Ponce and Simons, 1977). In particular, the linear solution of Ponce and Simons is significant because it gives great insight into the behavior of shallow water waves, including kinematic, diffusion, dynamic, and inertial waves. (It is noted that in classical mechanics, the inertial waves are commonly referred to as "gravity" waves. This is the source of some confusion, since the gravity force is not included in the formulation of inertial waves).

The work of Ponce and Simons (1977) is summarized as follows:

1. The dynamic wave lies towards the middle of the spectrum of dimensionless wave numbers ($10^0 < \sigma < 10^2$), while kinematic/diffusion waves lie to the left ($10^{-2} < \sigma < 10^0$), and inertial waves to the right ($10^1 < \sigma < 10^4$).
2. In the stable flow regime, i.e., for Vedernikov number $V < 1$, the dynamic wave shows

very strong diffusive tendencies. The Vedernikov number is the ratio of relative kinematic wave speed to relative Lagrange wave speed (Craya, 1952; Ponce, 1991b).

3. At the threshold of flow instability ($V = 1$), the Seddon and Lagrange speeds (Chow, 1959) are the same, and kinematic, dynamic, and inertial waves have the same celerity and lack diffusion.
4. In the unstable flow regime ($V > 1$), kinematic, dynamic, and inertial waves have a tendency to amplify during propagation.

The findings of Ponce and Simons (1977) pose an interesting question which helps place the nature of shallow water waves in the proper perspective. Kinematic waves, and by extension diffusion waves, lying to the left of the wave number spectrum, transport mass. On the other hand, the inertial waves, lying to the right of the wave number spectrum, transport energy. What, then, do dynamic waves transport, since they lie in the middle of the wave number spectrum? Mass or energy? The logical answer is that they transport *both*. Therein lies the reason for the markedly strong dissipative tendencies of dynamic waves. Shallow water waves can transport mass and energy simultaneously *only* at the expense of wave diffusion. In the stable flow regime ($V < 1$), the more dynamic a wave is, the more strongly dissipative it is (Ponce et al, 1978). At the threshold of flow instability ($V = 1$), dynamic waves lose their ability to dissipate, and their properties coalesce with those of kinematic and inertial waves.

The preceding discussion raises the following question: If the dynamic wave is so strongly dissipative in most cases of practical interest, is it worth attempting to compute it? Would it not dissipate shortly after it is generated, with its mass going to join the underlying larger, kinematic/diffusion, wave? Or, can it be tracked downstream as it propagates? If so, at what characteristic speed? Seddon's or Lagrange's? A more practical question is: If the dynamic wave is so strongly dissipative, could it be properly construed as a flood wave? Lighthill and Whitham (1955) put it very aptly when they stated (op. cit., p. 293): "Under the conditions appropriate for flood waves... the dynamic waves rapidly become negligible, and it is the kinematic waves, following at lower speed, which assume the dominant role."

In summary, dynamic waves do not apply to floods in steep mountain streams. There is still the unresolved question of whether dynamic waves apply to the routing of flood waves in a typical valley setting. Dam-break flood waves notwithstanding, perhaps the only clear statement that can be made today is that the dynamic wave applies to tidal flow and similar situations where there is a significant downstream control of the flow.

5. Role Of Numerical Diffusion In Kinematic Wave Modeling

If kinematic waves cannot diffuse, why is it that numerical models of kinematic waves are able to show some wave diffusion? The resolution of this paradox lies in the conversion of a partial differential equation into a finite difference equation in a computer model.

This conversion can only be done at the expense of introducing an error. This error is a function of the grid size (Δx and Δt) and tends to disappear as the grid size is progressively refined. In flood routing, the error that creeps into a typical computation using finite differences manifests itself as numerical diffusion and numerical dispersion effects. These effects are the direct result of specifying a discrete space-time domain, and are not necessarily related to the physical diffusion and dispersion which are inherent in the nature of flood waves.

Numerical diffusion arises because the calculated wave amplitude is smaller than the physical wave amplitude. Numerical dispersion arises when the calculated wave celerity is different from the physical wave celerity (Ferrick et al, 1984). In conventional finite difference shallow wave models, the aim is to minimize numerical diffusion and dispersion by choosing a grid size sufficiently small to drive these errors to inconsequential amounts. Then, the convection and diffusion of the shallow wave can be properly described by the numerical model.

Unfortunately, not all finite difference kinematic wave models have sought to minimize numerical diffusion and dispersion. Often, a finite difference kinematic wave model has inadvertently used the numerical diffusion as a way of showing a certain amount of "physically realistic" diffusion in the calculated results (Li et al, 1975; Curtis et al, 1978). Cunge (1969) demonstrated that finite difference schemes of the kinematic wave equation introduce varying amounts of numerical diffusion and dispersion. The latter interfere with the physical effects, modifying them (Abbott, 1976; Ponce, 1991a). Thus, a finite difference kinematic wave model may show some diffusion, the amount being a function of the grid size and weighting factors used in discretizing the terms of the kinematic wave equation. The fact that this diffusion is artificial and intrinsically related to the grid size can be readily demonstrated by solving the same problem several times, each time halving the spatial increment Δx and temporal increment Δt. Carried to the practical limit, this test leads to the eventual disappearance of the numerical diffusion in question, with the result approaching the analytical solution of the kinematic wave, which is nondiffusive.

In summary, if the kinematic wave is solved properly, achieving the complete elimination of numerical diffusion and dispersion, the model user can only hope to describe kinematic waves, but not diffusion waves. If the problem does have some physical diffusion, the latter would be entirely missing from this approach. Conversely, if the kinematic wave is solved improperly, introducing numerical diffusion and dispersion by the choice of grid size, there is no guarantee that these will be related to the diffusion and dispersion, if any, of the physical problem. Any arbitrary choice of grid size will cause some numerical diffusion and/or dispersion, and since the latter are unrelated to the physical problem, the solution degrades accordingly.

Fortunately, there is a way out of this difficulty. Cunge (1969) and others (Natural Environment Research Council, 1975; Ponce and Yevjevich, 1979) have shown that the numerical diffusion and dispersion of finite difference kinematic wave models can be managed. There is a way to optimize the numerical diffusion while minimizing the numerical dispersion, to make the method and its inherent errors work for us instead of against us.

By a careful match of physical and numerical diffusion, the finite difference kinematic wave model can reproduce *both* kinematic and diffusion waves, in the methodology referred to as the Muskingum-Cunge method. This is a variant of the Muskingum method (Chow, 1959) in which the parameters K and X are calculated directly, based on hydraulic data (channel friction, bed slope, and cross-sectional characteristics), instead of indirectly, based on hydrologic data (storage-weighted flow relations). The Muskingum-Cunge method was first applied to open channel flow, and later to overland flow (Ponce, 1986). Extensive tests have shown that the method holds promise for overland flow, since unlike conventional finite difference kinematic wave models, the Muskingum-Cunge model is essentially independent of the grid specification.

6. Nature Of Kinematic Shock

Kinematic waves lack physical diffusion. However, kinematic waves are nonlinear, a property which gives them the inherent tendency to change their shape upon propagation: either steepen or flatten, depending on the stage relative to the channel cross-section (flow inbank or overbank). Under the right set of circumstances, a kinematic flood wave can steepen to the point where it becomes for all practical purposes a "wall of water." (In overland flow situations, the "wall of water" would be a small discontinuity in the water surface profile). This is the kinematic shock, i.e., a kinematic wave that has steepened upon propagation to the point of being nearly discontinuous.

Contrary to conventional wisdom, there is no physical unreality about the kinematic shock. If the steepening tendency is allowed to continue unchecked, the kinematic shock will form in due time. Diffusion, however, acts to counteract the steepening tendency. Therefore, in cases where diffusion, either physical or numerical, is present, the development of the kinematic shock is likely to be arrested. This explains the pervasive presence of kinematic shocks in analytical solutions of the kinematic wave. On the other hand, kinematic shocks are often absent from finite difference kinematic wave models, particularly from those that have appreciable amounts of built-in numerical diffusion.

Ponce and Windingland (1985) have clarified the conditions under which the kinematic shock is likely to develop. Based on theoretical considerations supported by extensive numerical experiments, they established the following conditions for kinematic shock development:

1. The wave must be kinematic, i.e., it must have negligible physical diffusion. Diffusion tends to counteract the development of the shock.
2. The ratio of base-to-peak flow Q_b/Q_p must be small, with zero as the lower limit, such as in the case of an ephemeral stream (recall the flash floods occurring on dry river beds).
3. The channel is (a) hydraulically wide, i.e., of nearly constant wetted perimeter, to allow the wave steepening to progress unchecked by the cross-sectional shape; and (b) sufficiently long to allow enough time for the shock to develop.

4. The flow is at high Froude number, within the stable flow regime ($V < 1$). The higher the Froude number within the stable flow regime, the smaller the physical diffusion, and the more likely the shock can continue to develop unchecked. In the limit, as the Vedernikov number approaches 1 (and the Froude number approaches 1.5, for hydraulically wide channels with Manning friction), diffusion vanishes as the flow reaches the threshold of instability.

In practice, all four conditions may prevail at the same time in a given situation. Whether a kinematic shock will form will depend on the strength of any one condition, or, if more than one is present, on their combined strength. For instance, an analytical solution of the kinematic wave in an overland flow plane satisfies conditions 1 and 3(a), and maybe even 3(b) if the plane is long enough. The case of a flash flood in an ephemeral stream in an arid or semiarid region satisfies condition 2, and probably even 3(a), 3(b), and 4. The fact that kinematic shocks are not a common sight in nature points to the practical difficulty of satisfying all or most of these conditions at the same time.

Condition 1 is satisfied in channels where the product $t_r S_o$ is large. Condition 2 is satisfied in ephemeral streams. Condition 3(a) is satisfied in inbank flow in wide rectangular channels, but not if the flow goes overbank, since in this case the wetted perimeter would cease to be nearly constant. Condition 3(b) is dependent of the catchment's physiography, geology, geomorphology, and drainage density. The longer a stream, uninterrupted by lateral inflow at tributary confluences is, the better the chances for the shock to develop. Condition 4 is dependent on the channel aspect ratio, boundary friction, and presence or absence of riparian vegetation. Jarrett (1984) has noted that high-Froude-number flows are rare in natural streams. So Condition 4 is more likely to be the exception rather than the rule.

It is noted that kinematic shocks, particularly those associated with flash floods, are very difficult to document precisely, given the obvious likelihood of bodily harm and possibly even death for those daring enough to attempt it. Hjalmarson (1985) has documented the flash flood of July 26, 1982, in Tanque Verde Creek, east of Tucson, Arizona, in which the lives of eight unsuspecting bathers were claimed. This flood was in all likelihood a kinematic shock. Kinematic shocks and flash floods are associated with one or more of the following: (1) intense cloud bursts, (2) a semiarid or arid region, (3) a steep, ephemeral stream, (4) a low-friction channel (in both bed and banks), and (5) a catchment with relatively long stream channels.

7. Role Of Vedernikov Number In Surface Runoff Modeling

As pointed out by Hayami (1951) in his classical paper on diffusion waves, the hydraulic diffusivity is the controlling physical parameter in diffusion waves. The hydraulic diffusivity is:

$$\nu = q_o/(2S_o) \qquad (12)$$

in which q_o= reference (average) unit-width discharge, and S_o= channel slope. Therefore, the amount of diffusion that a flood wave undergoes during propagation is directly proportional to the unit-width discharge and inversely proportional to the channel bed slope. In other words, the steeper the channel slope, the lesser the amount of flood wave diffusion. In the limit, for a sufficiently steep channel, the diffusion disappears and the flood wave becomes a kinematic wave. Hayami's hydraulic diffusivity is properly a kinematic hydraulic diffusivity

$$\nu_k = q_o/(2S_o) \tag{13}$$

because it lacks inertia altogether. It is strictly applicable to flow well within the stable regime, i.e., for small Vedernikov numbers, in the range $0 < V < 0.25$.

By including inertia in the formulation, Dooge (1973) and Dooge et al (1982) have extended the concept of hydraulic diffusivity to the realm of dynamic waves. This leads to the concept of dynamic hydraulic diffusivity (Ponce, 1991a; 1991b):

$$\nu_d = (1 - V^2)\, q_o/(2S_o) \tag{14}$$

Unlike its kinematic counterpart, the dynamic hydraulic diffusivity is also a function of the Vedernikov number. As the Vedernikov number $V \rightarrow 0$, as with low-Froude-number flows: $\nu_d \rightarrow \nu_k$. Conversely, as $V \rightarrow 1$: $\nu_d \rightarrow 0$, and wave diffusion vanishes. The latter process can not be simulated with Eq. 13. It follows that ν_d applies through a wider range of flow conditions than ν_k (in the range $0 < V < 1$) . Since ν_d does not significantly complicate the expression for hydraulic diffusivity, it should be the preferred way of modeling surface runoff with diffusion waves. In practice, since wave diffusion is usually small, the dynamic contribution to wave diffusion turns out to be also small (Perumal, 1992).

The inclusion of the Vedernikov number in the expression for hydraulic diffusivity has the additional advantage that it can account for channels of arbitrary cross-sectional shape, i.e., those other than hydraulically wide. Taken to the limit, i.e., for the *inherently stable channel* (Liggett, 1975; Ponce, 1991b), the Vedernikov number is zero ($V = 0$), and the dynamic hydraulic diffusivity (Eq. 14) reduces to its kinematic counterpart (Eq. 13). It is seen that in this case, the flood wave attenuation is governed by the kinematic hydraulic diffusivity, for all values of discharge or stage.

8. Summary

Kinematic waves are useful tools in applied hydrology. With the applicability issue now clearly settled, they should continue to be used in the future. When the extension is made to diffusion waves, the applicability issue is no longer a serious roadblock. The strongly diffusive dynamic wave has yet to show that it is needed in a typical application of surface runoff modeling.

Caution should be exercised when applying the kinematic wave to overland flow and channel flow in the context of a numerical computer model, since the presence of uncontrolled numerical diffusion and dispersion may degrade the accuracy of the computation.

The Muskingum-Cunge method holds particular promise, given its demonstrated grid independence. The use of a dynamic hydraulic diffusivity in lieu of its kinematic counterpart will ensure that the dynamic component of surface runoff, however small, is being accounted for.

Research is needed into the nature of kinematic shock and its applicability to the study of flash floods. The conditions for shock development having been clearly identified, the focus should now shift to the development of a regional hazard rating for flash floods. This should be based on: (1) climatic and precipitation patterns, (2) geology, geomorphology, and catchment drainage density, and (3) stream channel slope, boundary friction, and cross-sectional shape.

9. References

Abbott, M. B., 1976. Computational Hydraulics: A Short Pathology. *Journal of Hydraulic Research,* 14:271-285.

Chow, V. T., 1959. Open-Channel Hydraulics. McGraw-Hill, New York.

Cunge, J. A., 1969. On the Subject of a Flood Propagation Computation Method (Muskingum Method). *Journal of Hydraulic Research,* 7:205-230.

Curtis, D. C., R. L. Mitchell, and J. C. Schaake, Jr., 1978. A Deterministic Runoff Model for Use in Flash Flood Planning. Conf. of Flash Floods: Hydrometeorological Aspects, American Meteorological Society, May 2-3, Los Angeles, Calif.

Dooge, J. C. I., 1973. Linear Theory of Hydrologic Systems. *Technical Bulletin* 1468, 327 pp., U.S. Department of Agriculture, Washington, D. C..

Dooge, J. C. I., W. B. Strupczewski, and J. J. Napiorkowski, 1982. Hydrodynamic Derivation of Storage Parameters in the Muskingum Model. *Journal of Hydrology,* 54:371-387.

Ferrick, M. G., J. Bilmes, and S. E. Long, 1982. Modeling Rapidly Varied Flow in Tailwaters. *Water Resources Research,* 20:271-289.

Hayami, S., 1951. On the Propagation of Flood Waves. Bulletin, Disaster Prevention Research Institute, Kyoto University, 1:1-16.

Hjalmarson, H. W., 1984. Flash Flood in Tanque Verde Creek, Tucson, Arizona. ASCE *Journal of Hydraulic Engineering,* 110:1841-1852.

Hromadka, T. V., and J. J. DeVries, 1988. Kinematic Wave and Computational Error. ASCE *Journal of Hydraulic Engineering,* 116:207-217.

Hydrologic Engineering Center, U.S. Army Corps of Engineers, 1990. HEC-1, Flood Hydrograph Package, User's Manual. Version 4.0, Sept.

Jarrett, R. D., 1984. Hydraulics of High-Gradient Streams. ASCE *Journal of Hydraulic Engineering,* 110:1519-1539.

Li, R. M., D. B. Simons, and M. A. Stevens, 1975. Nonlinear Kinematic Wave Approximation for Water Routing. *Water Resources Research*, 11:245-252.

Liggett, J. A., 1975. Stability, Chapter 6 in Unsteady Flow in Open Channels, K. Mahmood and V. Yevjevich, editors, Water Resources Publications, Fort Collins, Colorado, 259-281.

Lighthill, M. J., and G. B. Whitham, 1955. On Kinematic Waves. I. Flood Movement in Long Rivers. Procceedings, Royal Society of London, Series A, 229:281-316.

Natural Environment Research Council, 1975. Flood Studies Report, Vol. III: Flood Routing Studies. London, England.

Perumal, M., 1992. Comment to "New Perspective on the Vedernikov Number." *Water Resources Research,* 28:1734.

Ponce, V. M., 1986. Diffusion Wave Modelling of Catchment Dynamics. ASCE *Journal of Hydraulic Engineering,* 112:716-727.

Ponce, V. M., 1991a. The Kinematic Wave Controversy. ASCE *Journal of Hydraulic Engineering*, 117:511-525.

Ponce, V. M., 1991b. New Perspective on the Vedernikov Number. *Water Resources Research,* 27:1777-1779.

Ponce, V. M., R. M. Li, and D. B. Simons, 1978. Applicability of Kinematic and Diffusion Models. *Journal of the Hydraulics Division,* ASCE, 104:353-360.

Ponce, V. M., and D. B. Simons, 1977. Shallow Wave Propagation in Open Channel Flow. *Journal of the Hydraulics Division,* ASCE, 103:1461-1476.

Ponce, V. M., and D. Windingland, 1985. Kinematic Shock: Sensitivity Analysis. ASCE *Journal of Hydraulic Engineering,* 111:600-611.

Ponce, V. M., and V. Yevjevich, 1978. Muskingum-Cunge Method with Variable Parameters. *Journal of the Hydraulics Division,* ASCE, 104:1663-1667.

Seddon, J. A., 1900. River Hydraulics. Transactions, ASCE, 43:179-229.

Tayfur, G., M. L. Kavvas, and R. S. Govindaraju. (1993). Applicability of St. Venant Equations for Two-dimensional Overland Flow Over Rough Infiltrating Surfaces. ASCE *Journal of Hydraulic Engineering,* 119:51-63.

Wooding, R. A., 1965. A Hydraulic Model for the Catchment-Stream Problem. *Journal of Hydrology,* 3:254-267.

Woolhiser, D. A., and J. A. Liggett, 1967. Unsteady One-dimensional Flow Over a Plane--The Rising Hydrograph. *Water Resources Research,* 3:753-771.

Woolhiser, D. A., 1992. Closure to "The Kinematic Wave Controversy," by V. M. Ponce, ASCE *Journal of Hydraulic Engineering,* 118:1337-1339.

COMPUTATIONAL ASPECTS IN KINEMATIC MODELLING

BERNHARD H SCHMID & WOLFGANG SUMMER
Technical University Vienna
Institute of Hydraulics, Hydrology and Water Resources Management
Karlsplatz 13/223
A-1040 Vienna, Austria

ABSTRACT Mathematical and computational properties of kinematic wave models in hydrology are reviewed and discussed Special attention is paid to questions relating to the formation and propagation of shocks as well as the presence of numerical diffusion in certain finite difference schemes As shocks and shock structures (the latter obtained from higher-order "parent" models) travel with the same celerity, locally increased error in the vicinity of a solution discontinuity does not invalidate the kinematic model Care must be taken, however, in the evaluation of numerical results, as parasitic oscillations may develop Numerical diffusion inherent in several numerical schemes should either be suppressed or matched to a previously determined amount of physical diffusion The practical usefulness of kinematic models in hydrology is discussed, and so are certain recent modelling attempts aiming beyond the micro-catchment scale Finally fields of future research are indicated

1. Introduction

Mathematical models resting on kinematic wave theory are in widespread use in hydrology As many conceptual models can claim little, if any, physical basis and fully rigorous approaches are often too demanding in man-power and data requirements, a kinematic model may frequently offer an attractive alternative midway between these two types Since the mathematical concept of a kinematic wave arises whenever a process is described by a continuity equation together with a one-to-one relationship between the dependent variable concerned (e g water-depth) and an associated flux quantity (e g discharge), it is not surprising that this kind of model can be found in many different, and apparently not very closely related subfields of hydrology This point may be illustrated by the following list of subfields together with a few sample references (which do not even remotely pretend to name all contributions of significance)

- overland flow (Smith and Woolhiser, 1971, Singh, 1976a, Govindaraju et al , 1992, Saquib and Kavvas, 1992, Yomota and Islam, 1992) and transport (Brazil et al , 1979)
- watershed hydrology (Singh, 1975, Rovey et al , 1977)
- urban storm drainage including certain cases of closed conduit flow (Stephenson and Meadows, 1986, Chen and Wong, 1987) as well as kinematic options provided by several software packages currently in use (e g "MOUSE" distributed by the Danish Hydraulic Institute)

V P Singh and B Kumar (eds) Surface-Water Hydrology 133–147

- porous media flow (Beven, 1982, Smith, 1983, Schmid and Gutknecht, 1988) and transport (Charbeneau, 1984)
- soil erosion (Hjelmfelt et al , 1975, Shirley and Lane, 1978, Lane and Shirley, 1982, Rose et al , 1983, Singh and Prasad, 1982)
- sediment transport (Kluwick, 1977)
- pollution routing in rivers (Koussis et al , 1983)

Obviously, it is worthwhile to occupy oneself with this type of model To obtain meaningful results, however, attention must be paid to the following points

a In nearly all cases the kinematic wave arises out of some simplification of a more complex process so that a range of validity of the kinematic model has to be specified and observed This is an aspect to be judged by direct reference to the physics of the process involved, e g by considering more accurate higher-order differential equations, as was done for overland flow modelling by Woolhiser and Liggett (1967), Morris and Woolhiser (1980) and Viera (1983)

b A purely kinematic model does not account for any effect in the nature of diffusion, which is why shocks are particularly likely to appear This not very endearing quality is, in itself, independent of any particular solution technique that might be used, but, on the other hand, may influence the choice of computational method

c The outcome from numerical computations should not be confused with the properties of the purely kinematic wave, especially if the approach selected introduces an appreciable amount of numerical diffusion, whether controlled or not

Much of the above, clearly, involves issues regarding the underlying mathematics, both with respect to model properties and computational techniques It is these aspects that will be considered subsequently here Unless stated otherwise, the hydrological framework used in the derivations will be that of overland flow modelling

2. Mathematical Properties of the Purely Kinematic Wave

As mentioned before, a kinematic wave arises out of a continuity equation,

$$\frac{\partial A}{\partial t}+\frac{\partial Q}{\partial x}=R \tag{1}$$

(e g with A as the cross-sectional area of flow, Q discharge and R a source/sink term) and a unique relationship of the form

$$Q=Q(A,x,t) \tag{2}$$

For the case of overland flow over a plane of constant width this reduces to

$$\frac{\partial h}{\partial t}+\frac{\partial q}{\partial x}=i_e \tag{3}$$

and

$$q = q(h,x,t) \tag{4}$$

where q denotes the discharge per unit width, h the water-depth and i_e the rate of rainfall excess

In most situations of hydrologic relevance explicit dependency of q on t need not be considered, which leaves $q = q(h,x)$ Assumption of hydrologically homogeneous units (for instance, the individual planes in the context of a kinematic cascade) lets eqn (4) finally become $q = q(h)$, which is usually written in power-law form

$$q = a\,h^m \tag{5}$$

with a, m as parameters Substitution of eqn (5) into eqn (3) yields

$$\frac{\partial h}{\partial t} + c(h)\frac{\partial h}{\partial x} = i_e \tag{6}$$

with $c(h) = a\,m\,h^{m-1}$ denoting kinematic wave celerity It is eqn (6) that will be discussed below rather than the more general form eqn (1) with eqn (2), as the computational aspects of interest here are essentially the same, but can be presented more clearly

Eqn (6) is a quasi-linear first-order partial differential equation of the hyperbolic type with one family of characteristics To define a well-posed problem one initial condition and one (upper) boundary condition is required No lower boundary condition need, and indeed, may be given, which makes it clear that kinematic models cannot be used for problems involving impoundment, as no downstream control is accounted for

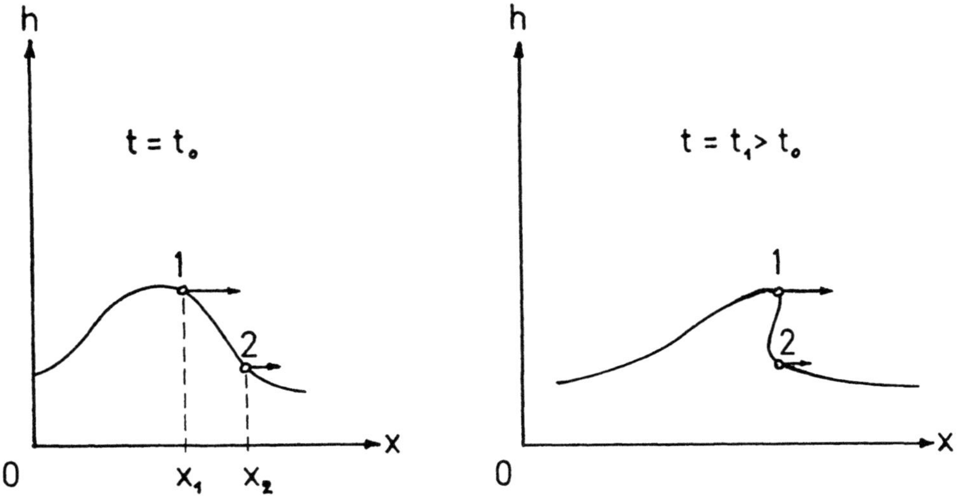

Figure 1 Breaking of a kinematic wave - schematic sketch

Since celerity c is in most cases a monotonously increasing function of h, the kinematic wave has a steepening tendency, which may eventually lead to the development of shocks (Fig 1)

Mathematically, a shock is generated as characteristics belonging to the same family (there is only one family in this context) intersect (Fig 2)

This results in a jump appearing in the solution, which is therefore no longer continuous At the location of the shock the partial derivatives of the governing differential equation become undefined, a difficulty the mathematicians circumvented by introducing so-called weak solutions (see e g Smoller, 1983) obtained from a corresponding integral equation A practical result to be used in shock routing is the jump condition

$$c_S = \frac{\Delta q}{\Delta h} \tag{7}$$

originally known from gas dynamics as the Rankine-Hugeniot equation This follows from the above-mentioned theory of weak solutions, but can also be derived by means of a continuity argument (Whitham, 1974, Kluwick, 1977)

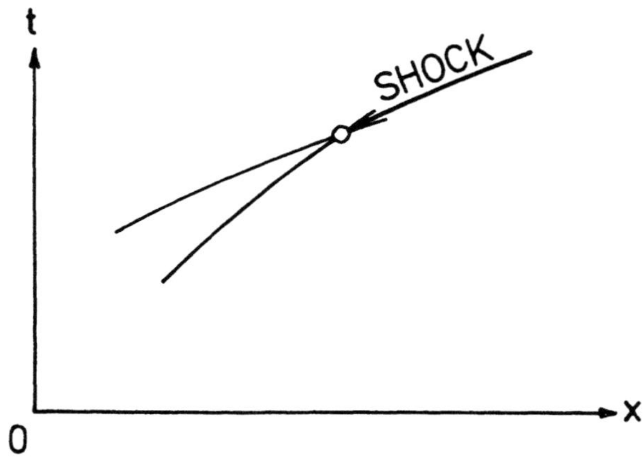

Figure 2 Shock formation viewed as intersection of characteristics

The question as to what a kinematic shock means from the physical point of view has repeatedly been discussed by various researchers Contributions by Ponce (1991), Schmid (1990d) as well as Ponce and Windingland (1985) may be mentioned as fairly recent references The following points are of particular significance in this context

The kinematic wave is an approximate model displaying a strong tendency towards shock development A priori this is a model property which may or may not reflect physical reality What really happens at the location of the shock cannot be judged from the kinematic equation alone, but an appropriate higher-order "parent" equation (e g diffusion wave, de St Venant equations) must be referred to This is exactly what was done by Lighthill and Whitham (1955) and Kluwick (1977) These studies showed that there may be either a comparatively sharp, but continuous transition or, depending on the Froude-number, a continuous part plus a hydraulic jump The latter, naturally, also involves a discontinuity, but this need not match in size with the kinematic shock (Fig 3)

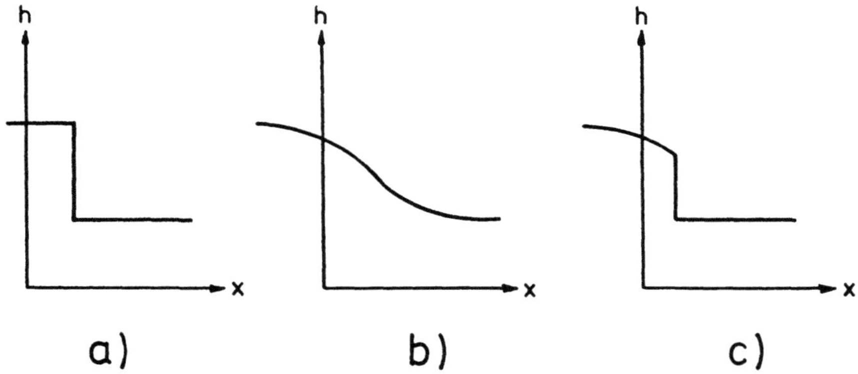

Figure 3 Kinematic shock and possible physical manifestations (adapted from Kluwick, 1977)

The practically most important result is, however, that in both cases the "shock structure" as determined from the higher-order equations travels with the same celerity as the kinematic shock so that the increased approximation error keeps close to the vicinity of the shock and does not compromise the rest of the kinematic solution It is, therefore, not correct to say that the kinematic wave approximation "breaks down" when a shock develops

3. The Kinematic Cascade

As kinematic modelling is probably most popular in the hydrology of small catchments, it is appropriate to devote one section of this paper to a concept that originated from this field, the kinematic cascade The idea to remove the severe geometric restrictions then associated with kinematic overland flow models by treating a series of n successive planes (Fig 4) was first introduced by Brakensiek (1967)

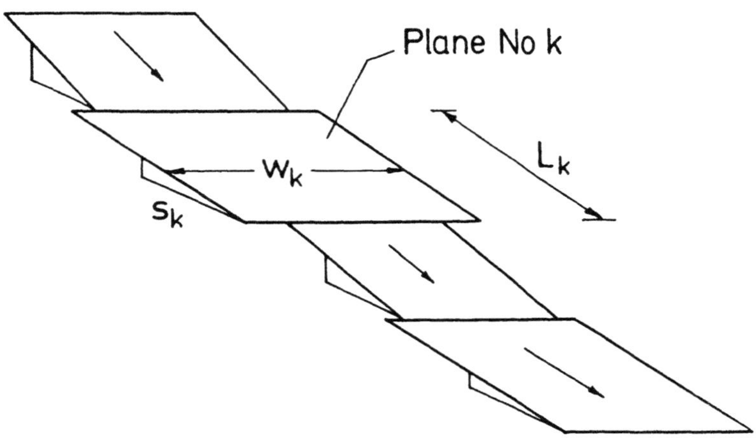

Figure 4 Example of a cascade comprising 4 planes

Subsequent pioneering work by Kibler and Woolhiser (1970, 1972) helped clarify the underlying mathematical issues, particularly with regard to shock formation and shock propagation Since then, the concept has received attention from quite a number of scientists, among them Borah et al (1980) and Hairsine and Parlange (1986)

Returning to eqn (6) and rewriting it as

$$\frac{\partial h_k}{\partial t} + a_k\ m\ h_k^{m-1} \frac{\partial h_k}{\partial x_k} = i_{e,k} \tag{8}$$

with k as a plane index (no 1 is the uppermost) a kinematic cascade may be defined by

$$Q(x_1 = 0, t) = 0 \tag{9}$$
$$Q(x_k = 0, t) = Q(x_{k-1} = L_{k-1}, t) \tag{10}$$

together with suitable initial conditions Denoting (constant) plane width by w_k, eqn (5) can be written to yield

$$Q_k = w_k\ a_k\ h_k^m \tag{11}$$

The exponent m depends on the flow regime (laminar/turbulent) and is assumed to be the same throughout, although one must be aware that this will be very nearly, but not quite strictly true

If the initial condition is a dry surface (or constant water-depth), it can be shown that the first and uppermost plane will be free from shocks for $i_{e\,1}$ depending on t only, but not on x_1 (for proof see e g Schmid, 1990d)

For the case of a rainfall excess rate constant in space and time, Kibler and Woolhiser (1972) derived the following shock formation criterion

$$P_s = \frac{w_{k-1}\ a_{k-1}}{w_k\ a_k} > 1 \tag{12}$$

The shock develops instantaneously This criterion works well, if it is kinematic flow on impermeable surfaces that is to be modelled Within natural or, at least, not completely sealed watersheds, the rate of rainfall excess can no longer be assumed constant in time, even if the rainfall intensity is, since

$$i_e(t) = r(t) - f_r(t) \tag{13}$$

with f_r denoting time-dependent infiltration rate

In contrast to kinematic flow over impermeable planes, shock formation on infiltrating surfaces is strongly influenced by the respective times of ponding, which mark the onset of overland flow From literature (Smith and Parlange, 1978, may serve as just one example) a variety of methods are known of how ponding time t_p can be determined Subsequently, it will be assumed that reasonably accurate values of t_p have been obtained somehow

To derive a generalized shock formation criterion, we consider two adjacent infiltrating planes of numbers k-1 and k, resp The following two cases may be encountered

a) $t_{p,k-1} < t_{p,k}$ This case is characterized by immediate shock formation at the upper end of plane k

b) $t_{p,k-1} > t_{p,k}$ For clarity of presentation, it is useful to adopt the following definitions here

- A - characteristic starts from the line $t = t_{p,k}$ (dash-dotted x_k - parallel in Fig 5)
- B - characteristic emanates from the t- axis ($x_k = 0$) and starts at time $t = t_{B,0}$ and with $h = h_{B,0}$

A shock may form, if

$$h_{B,0} + \int_{t_{B,0}}^{t_S} i_{e,k}(t)dt > \int_{t_{p,k}}^{t_S} i_{e,k}(t)dt \tag{14}$$

Rearrangement of terms finally yields the generalized relationship desired (Schmid, 1990a)

$$\frac{h_{B,0}}{\int_{t_{p,k}}^{t_{B,0}} i_{e,k}(t)dt} > 1 \tag{15}$$

Inequality (15) is a necessary criterion, which becomes a sufficient one, if plane k can be assumed to be "long enough" As it should be, the Kibler-Woolhiser criterion, inequality (12), is obtained as a special case of the above, generalized version

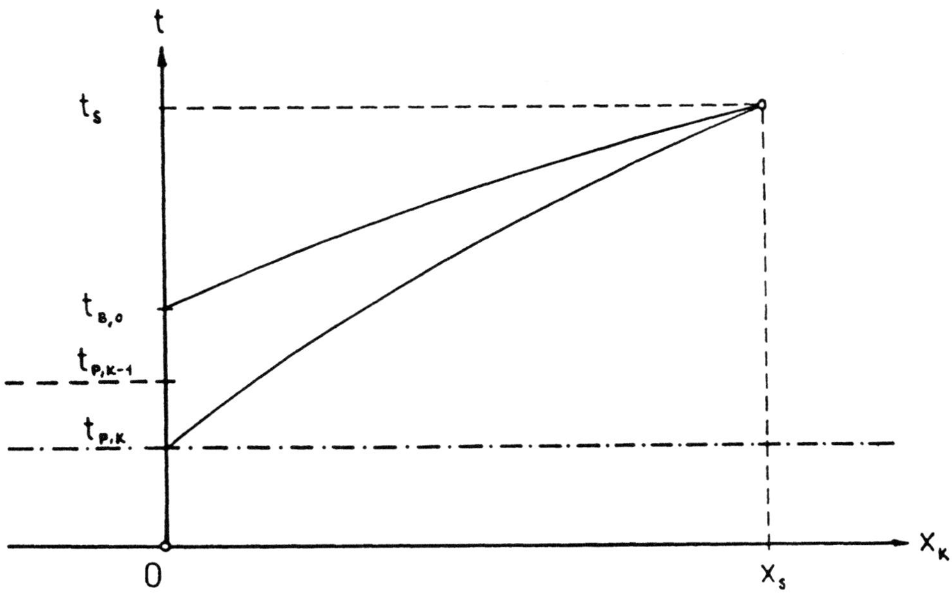

Figure 5 Picture of characteristics / derivation of generalized shock formation criterion

4. Some Remarks on Solution Methodology

As is usual with partial differential equations, methods of solution can be grouped into the two classes of analytical and numerical techniques Whereas early work concentrating on highly idealized flow conditions still permitted authors to derive and use comparatively simple analytical solutions (e g Wooding, 1965a, b), later attempts at modelling increasingly complex situations more and more called for numerical analysis In this context it should be stated that - in contrast to some genuinely non-linear equations known from mathematical hydraulics - the quasi-linear kinematic wave equation in principle permits analytical treatment of many rather complicated problems However, as can be seen from the example of shock routing along a cascade comprising more than three planes, with several first-order shocks developing and later merging to form higher-order shocks, the ensuing task may, even if it remains still feasible, become extremely laborious It is, therefore, justified to turn to numerical approaches

A large amount of papers on the numerical solution of the kinematic wave equation has been published to date Most of the techniques proposed are finite difference methods or else are in some way characteristics-based The former group will be discussed separately in the subsequent section, while the latter may frequently be regarded as semi-analytic in character (e g Eggert, 1987) There are also reports on Finite Element models (e g Blandford and Meadows, 1990, Goodrich et al , 1991), but this has not yet become a very widespread approach When combined with the possibilities provided by modern geoinformatics (digital elevation models, geographic information systems, etc), the FE-method may, however, help advance our ability to build 2D rainfall-runoff models Future experience gained on a technical scale will show what can actually be achieved in that way

5. On Difference Methods and the Philosophy of Kinematic Modelling

Using difference methods one first covers the domain within which a solution is to be obtained with a (mostly rectangular) grid and then proceeds to express partial derivatives approximately by means of finite space and time increments, resp Usually, there is a huge variety of possibilities of how this can be done, but only few schemes out of the conceivable multitude prove practically operational There are several properties that indicate the potential usefulness of a difference algorithm, the two most important issues being consistency and convergence

Consistency may be described as a measure of agreement in the limit between corresponding differential and difference equations This analysis yields the order of truncation (step) error and tells the user how fast that error will vanish if the increments go to zero Assuming that the grid is fine enough, a higher-order scheme will be preferable to a lower-order one (for practical comparisons of schemes see e g Singh, 1976b)

Convergence denotes the property that in the course of grid refinement the exact solution of the difference equation(s) gets arbitrarily close to the exact solution of the differential equation Direct proof of convergence is a task usually left to mathematicians, if required at all Such proof is in nearly all cases of practical interest replaced - consciously or not - by the invocation of Lax's equivalence theorem This theorem states that, given a properly posed linear initial value problem and a consistent difference approximation, stability is the necessary and sufficient criterion of convergence It should be noted that the theorem in this form rigorously applies only to linear problems Practical experience has shown that "a limited degree of nonlinearity" does usually not upset the validity of the theorem However, the development and propagation of shocks is a pronouncedly non-linear phenomenon, whence it must - regretfully - be inferred that Lax's theorem does not tell us what happens in the immediate vicinity of solution discontinuities And indeed, it is this point that will have to be addressed again later on

Another property of finite difference techniques that is of particular interest in kinematic modelling is a tendency of many such schemes to introduce a certain amount of numerical diffusion As mentioned before, the purely kinematic wave does not possess any diffusive terms, whereas higher-order "parent" equations (diffusion wave, de St Venant equations) do With regard to numerical diffusion there are two courses open to the modeller, each of which stems from a logically consistent line of reasoning

Once a kinematic model has been chosen, it may be argued that the purpose of the numerical approach is to approximate the exact solution of the given differential equation(s) as closely as possible Numerical diffusion then is simply a nuisance to be suppressed, which can be done with the aid of adequate second- or higher-order schemes The probably most popular technique in accordance with this aim is the explicit second-order Lax-Wendroff scheme, which was applied with success to a variety of problems in watershed hydrology (e g Smith and Woolhiser, 1971, Singh, 1976a) Difficulties encountered in cases affected by shock development were reported by Croley and Hunt (1981) They found that in such instances physically meaningless multiple peaks may be obtained numerically, which appear stable nevertheless In this context it should be remembered that - as indicated before - Lax's equivalence theorem does not hold in the vicinity of solution discontinuities (Abbott, 1979) Thus, whenever shocks are encountered, difference solutions, though derived from a consistent and stable scheme, need not necessarily converge towards the exact solution as space and time increments approach zero Therefore, numerical solutions in the vicinity of a shock must be viewed with caution It shall be added, however, that this is not a unique experience restricted to kinematic modelling, but very similar pitfalls are encountered in other fields where discontinuous solutions of hyperbolic differential equations play some role Gas dynamics and hydraulics of rapidly varying open channel flows may serve as just two examples In the majority of hydrologically significant cases, shocks are mostly weak and parasitic oscillations remain small so that numerical results are usually not invalidated It is, however, sound policy to take a look at the computed hydrograph instead of just accepting the maximum value sought automatically by a program

A second way to look at numerical diffusion is with the intention of controlling it in such a manner that it corresponds to the physical diffusion inherent in the process studied Attempts at quantification of the amount of physical diffusion will, of course, not rely on the purely kinematic wave (which yields none), but on some higher-order "parent" model This is, roughly, the underlying idea of the "matched diffusivity approach" which, in a sense, extends kinematic wave theory into the realm of diffusion waves, and was first introduced in the meanwhile classic paper by Cunge (1969) and further developed since by a number of researchers As examples, contributions by Ponce and Yevjevich (1978), Ponce (1986), Koussis (1976) and Holden and Stephenson (1988) shall be mentioned

The derivation of method, which is not repeated here, starts from a weighted four-point scheme (Fig 6)

$$\langle\frac{1}{c}\rangle\frac{\Theta(Q_j^{n+1}-Q_j^n)+(1-\Theta)(Q_{j+1}^{n+1}-Q_{j+1}^n)}{\Delta t}+\frac{Q_{j+1}^{n+1}-Q_j^{n+1}+Q_{j+1}^n-Q_j^n}{2\Delta x}=i_e\,w \qquad (16)$$

with < > denoting cell mean, c celerity, w plane width and i_e rate of rainfall excess The weighting parameter Θ is adjusted so that certain expressions obtained for physical and numerical diffusion, resp , match Rearrangement of terms yields the following Muskingum-type structure (Holden and Stephenson, 1988)

$$Q_{j+1}^{n+1} = C_0 \; i_e \; w \; \Delta x + C_1 \; Q_j^n + C_2 \; Q_j^{n+1} + C_3 \; Q_{j+1}^n \tag{17}$$

to be applied together with appropriate relationships for C_0 through C_3 Eqn (17) with the coefficients C_k (k=0 3) derived from eqn (16) has come to be known as "Muskingum-Cunge Method" (Weinmann and Laurenson, 1979) or "Fractional Scheme" (Koussis et al , 1983) Working on the differential equation

$$\frac{1}{c}\frac{\partial Q}{\partial t} + \frac{\partial Q}{\partial x} = q \tag{18}$$

with q denoting a "lateral inflow" term here, Koussis (1976) showed a theoretically more precise way of solution by carrying out discretization in space only and then solving the ensuing ordinary differential equation for assumed linear variation of Q over the time interval Δt The result of these operations has the same form as eqn (17) before, but with different expressions for the coefficients as well as the weighting parameter Θ This approach has been termed "Generalized Kinematic Model" (Weinmann and Laurenson, 1983), "Exponential Scheme" (Koussis et al , 1983) or "Muskingum-Koussis Method" (Schmid, 1990b) In many cases, both of the above methods yield very similar results Applied to shock-affected kinematic cascades, however, the Muskingum-Cunge ("MC") Method was shown to perform better than the Muskingum-Koussis Method, the latter being a less robust algorithm (Schmid, 1990b)

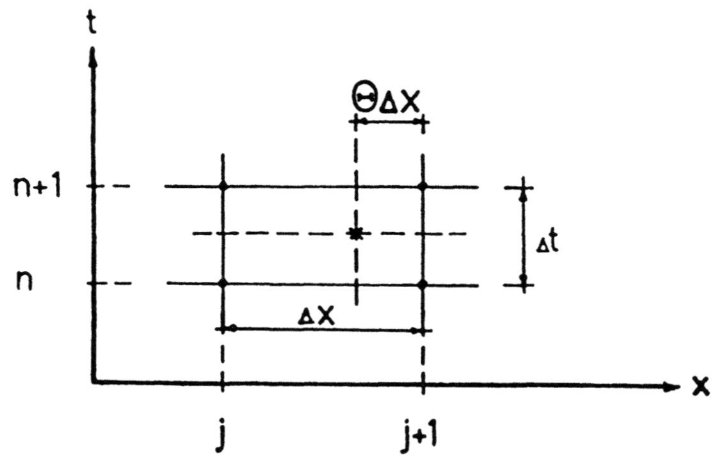

Figure 6 Illustration of 4-point scheme

Regarding the mathematical properties of the MC-scheme, the original paper by Cunge (1969) provides the following statements

- The scheme is stable for $\Theta \leq \frac{1}{2}$ and unstable for $\Theta > \frac{1}{2}$
- Wave attenuation is zero for $\Theta = \frac{1}{2}$ and increases to a maximum value as Θ approaches zero
- Phase error usually remains within acceptable limits, provided that $\Theta \leq \frac{1}{2}$ and Courant-number ≥ 1

The behaviour of the scheme when applied to kinematic cascades with and without reasonably strong shocks was investigated in some detail by the senior author of this contribution (Schmid, 1990 b, c) The study demonstrated the adequacy of MC-routing for turbulent flows with and without shocks and for laminar flow without shocks The method is, however, not recommended for the numerical simulation of laminar, shock-affected kinematic flow

Finally, it shall be added that, beyond the diffusion wave analogy, kinematic wave theory may also be extended to include dynamic effects Notable work in this respect was done, among others, by Dooge (1973) and Ponce (1990)

6. Practical Usefulness of Kinematic Models and Certain Trends of Development

Having reviewed the underlying theory, it is now appropriate to turn to the more practical question of applicability and usefulness The more so, as there has been some controversy on this matter only fairly recently (Dawdy, 1990, Goldman, 1990, Hromadka and De Vries, 1988, Merkel, 1990, Unkrich and Woolhiser, 1990, Woolhiser and Goodrich, 1990, Ponce, 1991)

In this context it seems necessary to distinguish between scientific applications and studies carried out by model developers ("taylored models") on one hand and kinematic options embedded in generally available hydrologic software packages on the other Whereas the scientific model developer is usually familiar with the background of theory and experience reported in literature, the "general" program user quite frequently is not In the latter case, the success of the kinematic modelling effort, therefore, depends on the quality of manuals and on-screen guidance at least as much as on the properties of the algorithm employed It is certainly no easy task to write a kinematic subroutine that rules out all possible misuse As a consequence, it is by no means justified to reject the kinematic approach as a whole, just because one particular program either performs poorly or has given wrong answers due to incorrect handling More effort will probably be required in the development of programs for general use to better define the conditions under which a kinematic option may be applied, and to make this transparent to the user

Apart from mistakes on the part of model developers and/or users, the kinematic approach is, of course, not the final answer in hydrology, as also remarked by Stephenson and Meadows (1986), but a convenient and powerful tool in the hands of an experienced hydrologist, if, and only if its applicability has been duly ascertained For overland flow, criteria by Woolhiser and Liggett (1967), Morris and Woolhiser (1980) and Viera (1983) have been mentioned previously here The ranges of applicability become somewhat wider as diffusion wave or dynamic wave extensions are used However, no downstream control is accounted for and, therefore, such a boundary must not be of significance in a problem treated with the aid of kinematic waves Finally, one must remain aware of the model's tendency towards shock development and of how the numerical scheme chosen will react

The practical adequacy of the approach, naturally, also depends on the scale of the problem at hand The kinematic model is primarily a tool for application to small catchments Ponce's (1991) figure of the upper limit being some 2 5 km^2 catchment area is probably still a fair estimate at the present date The increasing use of modern geographic information systems (GIS) and remote sensing technology, however, makes it seem reasonable to expect that this practical upper limit may be distinctly higher by the end of this decade Concepts aiming beyond the hillslope and micro-catchment scales are being developed and/or refined A Finite Element based approach combining kinematic waves with triangular irregular network (TIN) data has already been mentioned (Goodrich et al , 1991) One further method that shall be cited here is that of "hydrodynamic averaging", which originated from soil erosion research Emphazising the case of bare, rilled soil surfaces exposed to rain (e g freshly cut hillslopes adjacent to motorways),

Govindaraju and Kavvas (1992) introduced the concept of the ergodic length scale, characterizing the hillslope rill geometry in terms of a limited number of parameters, besides the approach to combine rill flow dynamics with sheet flow dynamics (Kavvas and Govindaraju, 1992) The stochastic flow dynamics is averaged leading to extended overland flow equations - based on the kinematic wave approximation for sheet flow over interill areas as well as rill flow The approach has the advantage of avoiding explicit modelling of flows over each interill area as well as in each rill separately

Current research pursued, among others, by the co-author of this contribution, aims at an extension of the concept from the hillslope scale to the catchment scale Means provided by modern GIS-technology as well as satellite imagery are hoped to permit progress to be achieved in this context

7. Conclusions

Many significant mathematical and computational aspects associated with kinematic models centre around two key issues numerical diffusion and shock routing There are two logically consistent ways of dealing with the former problem to suppress or to control numerical diffusion Methods in agreement with these two lines of reasoning have been discussed With regard to shocks it may be stated that these non-linear phenomena do not invalidate the kinematic wave model outside a fairly narrow space interval of transition Finite difference schemes applied to the problem should, however, be carefully analyzed for their behaviour in the vicinity of such discontinuities

A third key issue has arisen due to the desire for more strongly physically based modelling beyond the micro-catchment scale on one hand, and the availability of modern geographic information systems and remote sensing technologies on the other Some of the approaches that show promise in this context have been mentioned and described

With a view to future numerical work, attention shall be drawn to the fact that one of the best-known (first-order) equations in hydraulics and fluid mechanics, the advection equation, represents a special kinematic wave Although the wealth of concepts developed to deal with advection-dominated problems (where shocks are absent) may not be unreservedly transferable to kinematic wave problems (with a tendency to form shocks), it seems probable that some cross-fertilization may take place One notable effort was made by Koussis et al (1983), who applied the matched diffusivity technique to problems of pollution routing in rivers Moving in the opposite direction, certain high-accuracy fixed-grid characteristics methods might be applied to kinematic wave equations with success Besides that, schemes with notable shock-capturing ability are known from the theory of rapidly varying open channel flows, but have not been used in a kinematic framework yet It is impossible to foresee which of the possible candidate schemes will finally turn out best - the ongoing search for more robust, more accurate and faster algorithms, however, is, and will remain, for some time to come, a very interesting one

8. References

Abbott, M B , 1979 *Computational Hydraulics* Pitman Publishing Ltd , London

Beven, K , 1982 On subsurface stormflow Predictions with simple kinematic theory' *Water Resour Res* , **18**(6) 1627 - 1633

Blandford, G E and Meadows, M E , 1990 'Finite element simulation of nonlinear kinematic surface runoff' *J Hydrol* , **119** 335-356

Borah, D K , Prasad, S N and Alonso, C V , 1980 'Kinematic wave routing incorporating shock fitting *Water Resour Res* , **16**(3) 529-541

Brakensiek, D L , 1967 'Simulated watershed flow system for hydrograph prediction A kinematic application' *Proc Int Hydrol Symp* , Fort Collins, Collerado, Pap No 3

Brazil, L E , Sanders, T G and Woolhiser, D A , 1979 'Kinematic parameter estimation for transport of pollutants in overland flow' In *Surface and Subsurface Hydrology* (H J Morel-Seytoux, editor), Water Res Publ 555-568

Charbeneau, R J , 1984 Kinematic models for soil moisture and solute transport *Water Resour Res* , **20**(6) 699-706

Chen, C -N and Wong, T S W , 1987 Comparison of kinematic wave and rational methods for site drainage design *Proc XXII Congress IAHR*, Lausanne, Techn Session D 264-269

Croley, II , T E and Hunt, B , 1981 'Multiple-valued and non-convergent solutions in kinematic cascade models *J Hydrol* , **49** 121-138

Cunge, J A , 1969 'On the subject of a flood propagation method (Muskingum method)' *J Hydr Res* , **7**(2) 205-230

Dawdy, D R , 1990 Discussion of 'Kinematic wave routing and computational error', *J Hydr Engrg* , ASCE, **116**(2) 278-280

Dooge, J C I , 1973 'Linear theory of hydrologic systems' *Techn Bulletin No 1468*, USDA Agricultural Research Service, Washington, D C

Eggert, K G , 1987 'Upstream calculation of characteristics for kinematic wave routing' *J Hydr Engrg* , ASCE, **113**(6) 743-752

Goldman, D , 1990 Discussion of 'Kinematic wave routing and computational error , *J Hydr Engrg* , ASCE, **116**(2), 280-282

Goodrich, D C , Woolhiser, D A and Keefer, T O , 1991 'Kinematic routing using finite elements on a triangular irregular network *Water Resour Res* , **27**(6) 995-1003

Govindaraju, R S and Kavvas M L , 1991 Modeling the erosion process over steep slopes approximate analytical solutions *J Hydrol* , **127** 279-305

Govindaraju, R S and Kavvas M L , 1992 'Characterization of the rill geometry over straight hillslopes through spatial scales *J Hydrol* , **130** 339-365

Govindaraju, R S , Kavvas M L and Tayfur, G , 1992 'A simplified model for two dimensional overland flows *Advances in Water Resources*, **15**, 133-141

Hairsine, S Y and Parlange J -Y , 1986 Kinematic shock waves on curved surfaces and application to the cascade approximation *J Hydrol* , **87** 187-200

Hjelmfelt, A T, Piest, R F and Saxton, K E , 1975 Overland flow and groundwater flow from a steady rainfall of finite duration' *J Geophys Res* , **69**(8), 1531-1540

Holden, A P and Stephenson, D , 1988 'Improved four point solution of the kinematic equations *J Hydr Res* , **26**(4) 413-423

Hromadka II, T V and DeVries, J J , 1988 Kinematic wave routing and computational error' *J Hydr Engrg* , ASCE, **114**(2) 207-217

Kavvas, M L and Govindaraju, R S , 1992 'Hydrodynamic averaging of overland flow and soil erosion over rilled hillslopes ' *Erosion, Debris Flow and Environment in Mountain Regions*, Proc of the Chengdu Symposium, IAHS Publ **209** 101-111

Kibler, D F and Woolhiser, D A , 1970 The kinematic cascade as a hydrologic model' Colorado State Univ , *Hydrol Pap no 39*, Fort Collins, Colorado

Kibler, D F and Woolhiser, D A , 1972 'Mathematical properties of the kinematic cascade' *J Hydrol* , **15** 131-147

Kluwick, A , 1977 'Kinematische Wellen', (Kinematic waves, in German) *Acta Mechanica*, **26** 15-46

Koussis, A , 1976 'An approximative dynamic flood routing method' *Proc Int Symp on Unsteady Flow in Open Channels*, Paper L1, Newcastle-upon-Tyne, England

Koussis, A D , Saenz, M A and Tollis, I G , 1983 Pollution routing in streams' *J Hydr Engrg* , ASCE, **109**(12) 1636-1651

Lane, L J and Shirley, E D , 1982 'Modeling erosion in overland flow' In *Estimating Erosion and Sediment Rangelands*, Proc Workshop, Tucson, Arizona, March 7-9, 1981, USDA, Agr Res Serv , Agricultural Reviews and Manuals, ARM-W-26 120-128

Lighthill, M J and Whitham, G B , 1955 'On kinematic waves, I Flood movement in long rivers' *Proc Roy Soc* , A229 281-316

Merkel, W H , 1990 'Discussion of 'Kinematic wave routing and computational error' *J Hydr Engrg* , ASCE, **116**(2) 282-284

Morris, E M and Woolhiser, D A , 1980 'Unsteady one-dimensional flow over a plane Partial equilibrium and recession hydrographs' *Water Resour Res* , **16**(2) 355-360

Ponce, V M , 1986 Diffusion wave modeling of catchment dynamics' *J Hydr Engrg* , ASCE, **112**(8) 716-727

Ponce, V M , 1990 Generalized diffusion wave equation with inertial effects' *Water Resour Res* , **26**(5) 1099-1101

Ponce, V M , 1991 The kinematic wave controversy' *J Hydr Engrg* , ASCE **117**(4) 511-525

Ponce, V M and Windingland, D , 1985 'Kinematic shock sensitivity analysis *J Hydr Engrg* , ASCE, **111**(4) 600-611

Ponce, V M and Yevjevich, V , 1978 'Muskingum-Cunge method with variable parameters Techn Note, *Hydr Div , Proc ASCE,* **104**(12) 1663-1667

Rose, C W , Williams, J R , Sanders, G C and Barry, D A , 1983 'A mathematical model of soil erosion and depositional processes II Application to data from an arid-zone catchment' *Soil Sci Soc of America J* , **47**(5), 996-1000

Rovey, E W , Woolhiser, D A and Smith, R E , 1977 'A distributed kinematic model of upland watersheds Colorado State Univ , *Hydrol Pap No 93*, Fort Collins, Colorado

Saquib, M N and Kavvas, M L 'Three dimensional modeling of watershed hydrology *Water Resources Planning and Management* (Saving a threatend resource - in search of solutions), Proc of the Water Resources sessions at Water Forum '92, 391-396

Schmid, B H , 1990a On kinematic cascades derivation of a generalized shock formation criterion *J Hydr Res* , **28**(3) 331-340

Schmid, B H , 1990b 'The matched diffusivity technique applied to kinematic cascades' I Model description and validation *J Hydrol* , **121** 345-361

Schmid, B H , 1990c The matched diffusivity technique applied to kinematic cascades II Analysis of model performance *J Hydrol* , **121** 363-337

Schmid, B H , 1990d 'A study on kinematic cascades' *Wiener Mitteilungen Wasser-Abwasser-Gewässer*, vol **90**, Vienna, Austria

Schmid, B H and Gutknecht, D , 1988 'Ein Ingenieurverfahren zur Infiltrationsberechnung mit Taschenrechner , (An engineering method of computing infiltration rates by means of a pocket calculator, in German with summary in English) *Österr Wasserwirtschaft*, **40** (7/8) 175-183

Shirley, E D and Lane, L J , 1978 'A sediment yield equation from an erosion simulation model In *Hydrology and Water Resources in Arizona and the Southwest*, **8**, 90-96, Univ of Arizona, Tucson, Arizona

Singh, V P , 1975 Hybrid formulation of kinematic wave models of watershed runoff' *J Hydrol* , **27** 33-50

Sιngh, V P , 1976a 'A dιstrιbuted convergιng overland flow model' 3 Applιcatιon to natural watersheds *Water Resour Res* , **12**(5) 902-908

Sιngh, V P , 1976b 'A note on the step error of some fiπιte-dιfference schemes used to solve kιnematιc wave equatιons' *J Hydrol* , **30** 247-255

Sιngh, V P and Prasad, S N , 1982 'Explιcιt solutιons to kιnematιc equatιons for erosιon on an ιnfiltratιng plane In Sιngh, V P (Ed), *Modelιng Components of Hydrologιc Cycle*, Water Resour Publ , Lιttleton, Colorado, 515-538

Smιth, R E , 1983 Approxιmate soιl water movement by kιnematιc characterιstιcs' *J Am Soιl Scι Soc* , **47** 3-8

Smιth, R E and Parlange, J -Y , 1978 'A parameter-effiσιent hydrologιc ιnfiltratιon model' *Water Resour Res* , **14**(3) 533-538

Smιth, R E and Woolhιser, D A , 1971 'Mathematιcal sιmulatιon of ιnfiltratιng watersheds' Colorado State Unιv , *Hydrol Pap No 47*, Fort Collιns, Colorado

Smoller, J , 1983 *Shock waves and reactιon-dιffusιon equatιons* Sprιnger-Verlag, New York-Heιdelberg-Berlιn

Stephenson, D and Meadows, M E , 1986 'Kιnematιc hydrology and modellιng' *Developments ιn Water Scι* , **26**, Elsevιer Scι Publ , Amsterdam

Unkrιch, C L and Woolhιser, D A , 1990 Dιscussιon of 'Kιnematιc wave routιng and computatιonal error' *J Hydr Engrg* , ASCE, **116**(2) 284-286

Vιera, J H D , 1983 Condιtιons governιng the use of approxιmatιons for the Saιnt-Venant equatιons for shallow water surface flow *J Hydrol* , **60**, 43-58

Weιnmann, P E and Laurenson, E M , 1979 Approxιmate flood routιng methods a revιew *J Hydr Dιv* , Proc ASCE, **105**(12) 1521-1536

Whιtham, G B , 1974 *Lιnear and nonlιnear waves* Wιley-Interscιence

Woodιng, R A , 1965a A hydraulιc model for the catchment-stream problem I, Kιnematιc wave theory *J Hydrol* , **3** 254-267

Woodιng, R A , 1965b A hydraulιc model for the catchment-stream problem' II, Numerιcal solutιons *J Hydrol* , **3** 268-282

Woolhιser, D A and Goodrιch, D C , 1990 Dιscussιon of 'Kιnematιv wave routιng and computatιonal error *J Hydr Engrg* , ASCE, **116**(2) 286-288

Woolhιser, D A and Lιggett, J A , 1967 'Unsteady one-dιmensιonal flow over a plane - the rιsιng hydrograph' *Water Resour Res* , **3**(3) 753-771

Yomota, A and Islam, M N , 1992 'Kιnematιc analysιs of flood runoff for a small - scale upland field' *J Hydrol* , **137** 311-326

EFFECT OF VEDERNIKOV NUMBER ON OVERLAND FLOW DYNAMICS

VICTOR M. PONCE AND KELLEY T. DILLENBERGER
Department of Civil Engineering
San Diego State University
San Diego, California 92182 USA

ABSTRACT. The effect of the dynamic hydraulic diffusivity in kinematic-with-diffusion overland flow modeling has been tested. Unlike the kinematic hydraulic diffusivity, the dynamic hydraulic diffusivity is a function of the Vedernikov number. The results of numerical experiments showed a small lag in the rising limb when comparing two equilibrium rising hydrographs using kinematic and dynamic hydraulic diffusivities. The existence of the lag is attributed to the error of the solution that specifically excludes inertia. The error was quantified by integrating the absolute value of the difference between the two rising hydrographs, dividing this difference by the total runoff volume and expressing it as a percentage. The error is small and likely to be within 0.35 percent for a wide range of realistic flow conditions. Since the dynamic effect is shown to be small throughout a wide range of bottom slopes, a diffusion wave model with inertia may be all that is required to model the overland flow dynamics.

1. Introduction

There are two formulations for hydraulic diffusivity in kinematic-with-diffusion modeling of surface flows: the kinematic hydraulic diffusivity (Hayami 1951) and the dynamic hydraulic diffusivity (Dooge 1973). The kinematic hydraulic diffusivity is: $\nu_k = q_o/(2S_o)$, in which q_o = reference unit-width discharge, and S_o = bottom slope. The dynamic hydraulic diffusivity is: $\nu_d = (1 - V^2)\, \nu_k$, in which V = Vedernikov number. While the dynamic hydraulic diffusivity is theoretically more appealing than the kinematic hydraulic diffusivity, its effect on overland flow modeling is not altogether apparent. This paper reports on numerical experiments designed to assess the role of the Vedernikov number in modifying the hydraulic diffusivity throughout a wide range of flow conditions.

V. P. Singh and B. Kumar (eds.), Surface-Water Hydrology, 149–172.

2. Background

Wooding (1965) pioneered the use of kinematic waves in an "open-book" configuration to simulate overland flow. Since then, kinematic waves have become well established in hydrologic research and practice (HEC-1, Flood Hydrograph Package, 1990). However, kinematic waves are nondiffusive; therefore, they should not be used to model situations in which diffusion plays a significant role. The amount of diffusion is controlled by the bottom slope; the steeper the slope the lesser the diffusion, and vice versa. In an urban setting, where plane and channel slopes are typically greater than one percent, diffusion vanishes and kinematic waves prevail. Unfortunately, numerical solutions of kinematic waves introduce varying amounts of "numerical diffusion," masking the true nondiffusive behavior of kinematic waves (Ponce, 1991a).

To cope with this difficulty, Ponce (1986) has proposed a diffusion wave model of overland flow based on the Muskingum-Cunge method. Unlike conventional overland flow kinematic wave models, the Muskingum-Cunge diffusion wave model has the advantage that it incorporates the physical diffusion, if any, of the surface runoff, matching it with the numerical diffusion of the problem at hand. A significant feature of this type of modeling is that it provides *grid independence*, i.e., the results are not a function of grid size. This frees the model user to concentrate on the physical, rather than on the numerical aspects of the problem, leading to more consistent and better modeling.

The Muskingum-Cunge model is based on the matching of physical and numerical diffusivities. The physical diffusivity is (Hayami, 1951):

$$\nu = q_o/(2S_o) \qquad (1)$$

The numerical diffusivity is (Cunge, 1969):

$$\nu_n = c\,\Delta x\,(0.5 - X) \qquad (2)$$

in which c = wave celerity, Δx = space interval, and X = Muskingum weighting factor (Chow, 1959). Equating physical and numerical diffusivities leads to

$$X = 0.5\,\{\,1 - [q_o/(S_o\,c\,\Delta x)]\} \qquad (3)$$

and to

$$X = 0.5\,(1 - D) \qquad (4)$$

in which

$$D = q_o/(S_o\,c\,\Delta x) \qquad (5)$$

is the cell Reynolds number (Ponce and Yevjevich, 1978).

In Muskingum routing,

$$K = \Delta x \ / \ c \tag{6}$$

in which K = travel time. In Muskingum-Cunge routing,

$$C = \Delta t \ / K = c \ (\Delta t \ / \Delta x) \tag{7}$$

in which C = Courant number (Ponce and Yevjevich, 1978), and Δt = time interval. Hayami's expression for the hydraulic diffusivity is in essence a *kinematic* hydraulic diffusivity ν_k, since inertia is explicitly absent from its formulation.

By including inertia in the formulation of diffusion waves, Dooge (1973) was able to express the hydraulic diffusivity in terms of the Froude number:

$$\mathbf{F} = v / (gd)^{1/2} \tag{8}$$

in which v = mean velocity, g = gravitational acceleration, and d = hydraulic depth. Dooge's hydraulic diffusivity is:

$$\nu = [1 - (\mathbf{F}^2/4)] \ q_o/(2S_o) \tag{9}$$

applicable to the case of Chezy friction in hydraulically wide channels.

Dooge et al (1982) generalized this formulation for any friction type and cross-sectional shape:

$$\nu = [1 - (\beta - 1)^2 \mathbf{F}^2] \ q_o /(2S_o) \tag{10}$$

in which β = exponent of the rating $Q = \alpha A^{\beta}$, with Q = discharge, and A = flow area.

Dooge's model is a diffusion wave model *with inertia*, to be distinguished from a dynamic wave model, which entails the complete solution of the governing equations of continuity and motion (the Saint Venant equations). Dooge's (and Dooge et al's) expression for hydraulic diffusivity can properly be regarded as *a dynamic* hydraulic diffusivity ν_d, since it does include inertia in its formulation.

Ponce (1991a, 1991b) enhanced Dooge et al's formulation by expressing the dynamic hydraulic diffusivity in terms of the Vedernikov number:

$$\nu_d = (1 - \mathbf{V}^2) \ q_o \ / \ (2S_o) \tag{11}$$

The Vedernikov number [$\mathbf{V} = (\beta - 1) \ \mathbf{F}$] is the ratio of the relative kinematic wave celerity $c_{rk} = (\beta - 1) \ v$, to the relative inertial wave celerity (Lagrange celerity) $c_{ri} = (gd)^{1/2}$ (Craya, 1952). This leads to the definition of the dynamic cell Reynolds number:

$$D_d = (1 - V^2)\, D \qquad (12)$$

Ponce's modification to Hayami's and Dooge's expressions is substantive because it renders the hydraulic diffusivity responsive to the flow dynamics. In effect, for Vedernikov number V = 1, the hydraulic diffusivity vanishes and the flow is on the threshold of neutral stability.

For overland flow applications, the question remains whether the dynamic hydraulic diffusivity does indeed constitute a significant improvement over the kinematic hydraulic diffusivity. Although more appealing from a theoretical standpoint, it has been argued that the dynamic hydraulic diffusivity represents only a small improvement over its kinematic counterpart when practical applications are considered (Perumal, 1992). Therefore, there is a need to clarify the precise role of the Vedernikov number in overland flow dynamics.

3. Experimental Program

The work reported here set out to compare results of the Muskingum-Cunge diffusion wave overland flow model using: (1) the *kinematic* cell Reynolds number D, Eq. 5; and (2) the *dynamic* cell Reynolds number D_d, Eq. 12. In essence, it sought to determine the effect of the Vedernikov number in overland flow dynamics under a wide range of flow conditions (geometry and bottom friction, plane/channel slopes, and effective rainfall intensity). The comparison was undertaken using OVERLAND, a Muskingum-Cunge diffusion wave overland flow model developed by the senior author (Ponce, 1989b). A special version of the model was implemented to handle either the kinematic or dynamic versions of the cell Reynolds number.

The program of numerical experiments encompassed the three cases shown in Table 1. Case I is a small catchment (6 ha) with low friction (Manning n = 0.015); Case II is a larger catchment (60 ha) with low friction (Manning n = 0.015); Case III is a small catchment (6 ha) with high friction (Manning n = 0.150 in the planes; n = 0.030 in the channel). For each case, three plane/channel slopes were specified: S_0 = 0.01 (steep), S_0 = 0.001 (intermediate), and S_0 = 0.0001 (mild). For each case and for each plane/channel slope, four effective rainfall intensities were specified: i_e = 1, 2, 4, and 10 cm/hr. This led to a total of 72 computer runs (2 models X 3 geometry/bottom friction combinations X 3 plane/channel slopes X 4 effective rainfall intensities).

The input to OVERLAND is effective rainfall intensity i_e (cm/hr) lasting a specified duration t_r (hr). The output is an outflow hydrograph showing discharge (m^3/sec) vs time (hr). For each run, the rainfall duration was carefully chosen to allow the outflow to reach equilibrium, at which time the rainfall ceased and the flow gradually receded to zero. The discharge at equilibrium is: $Q_e = i_e A$, in which A = catchment area (Ponce, 1989a).

4. Results

The results of the simulations showed a small but perceptible lag in the rising limb of the outflow hydrograph, coupled with a corresponding lag in the receding limb, when comparing runs using kinematic and dynamic cell Reynolds numbers. The existence of the lag is attributed to the error of the kinematic solution, which used the original formulation of the cell Reynolds number (Eq. 5), as opposed to the dynamic solution, which relied on the enhanced Vedernikov-number-dependent formulation (Eq. 12).

The error was quantified by integrating the absolute value of the difference between the two hydrographs, kinematic and dynamic, to obtain a runoff volume difference ΔV:

$$\Delta V = \int |Q_{kinematic} - Q_{dynamic}|\, dt \tag{13}$$

The runoff volume difference was normalized by dividing it by the total runoff volume $V_r = i_e\, t_r\, A$. Therefore, the error E (in percent) is:

$$E = 100\,(\Delta V \,/\, V_r) \tag{14}$$

Figure 1 shows the error incurred by using a kinematic cell Reynolds number in a diffusion wave overland flow solution, as a function of bottom slope, for each of three cases: (a) Case I, (b) Case II, and (c) Case III. These results show that the error is indeed small but perceptible, and likely to remain within 0.7 percent for a wide range of flow conditions. This figure also shows that the error is inversely related to bottom friction: for Cases I and II, the error is less than 0.7 percent for rainfall intensities less than 10 cm/hr; for Case III it is typically less than 0.1 percent. The results also show a tendency for the error to peak around the midrange of bottom slopes ($S_0 = 0.001$).

5. Analysis

The tendency for the error to peak around the midrange of bottom slopes requires further discussion. Figures 2 and 3 show the variation of the cell Reynolds number D with plane and channel slope, respectively. Figures 4 and 5 show the variation of the Vedernikov number **V** with plane and channel slope, respectively. The following conclusions can be drawn from these figures:

1. For steep slopes ($S_0 = 0.01$), $\nu_k \rightarrow 0$, and therefore: $\nu_d = (1 - \mathbf{V}^2)\, \nu_k \rightarrow 0$. This amounts to: $\nu_d \rightarrow \nu_k$. No reasonable value of Vedernikov number (in the stable range) can affect this tendency. The error turns out to be small or negligible.

2. For mild slopes ($S_0 = 0.0001$), the Froude number decreases such that $\mathbf{F}^2 \rightarrow 0$; therefore, the Vedernikov number also decreases such that $\mathbf{V}^2 \rightarrow 0$. This again leads to: $\nu_d \rightarrow \nu_k$, with the error turning out to be small or negligible. This behavior for mild

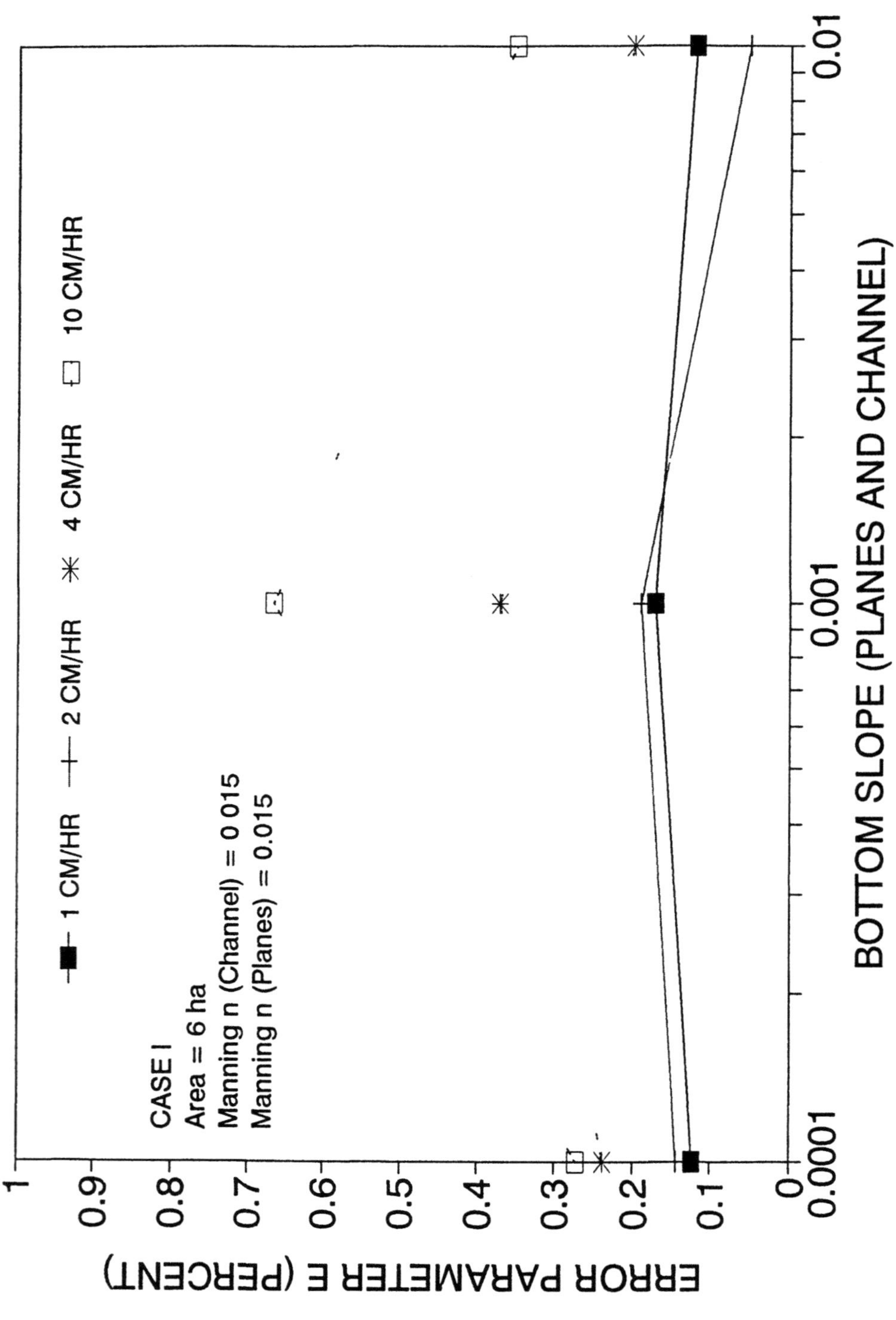

Fig 1 (a) Error Parameter E as a Function of Bottom Slope (Planes and Channel) Case I

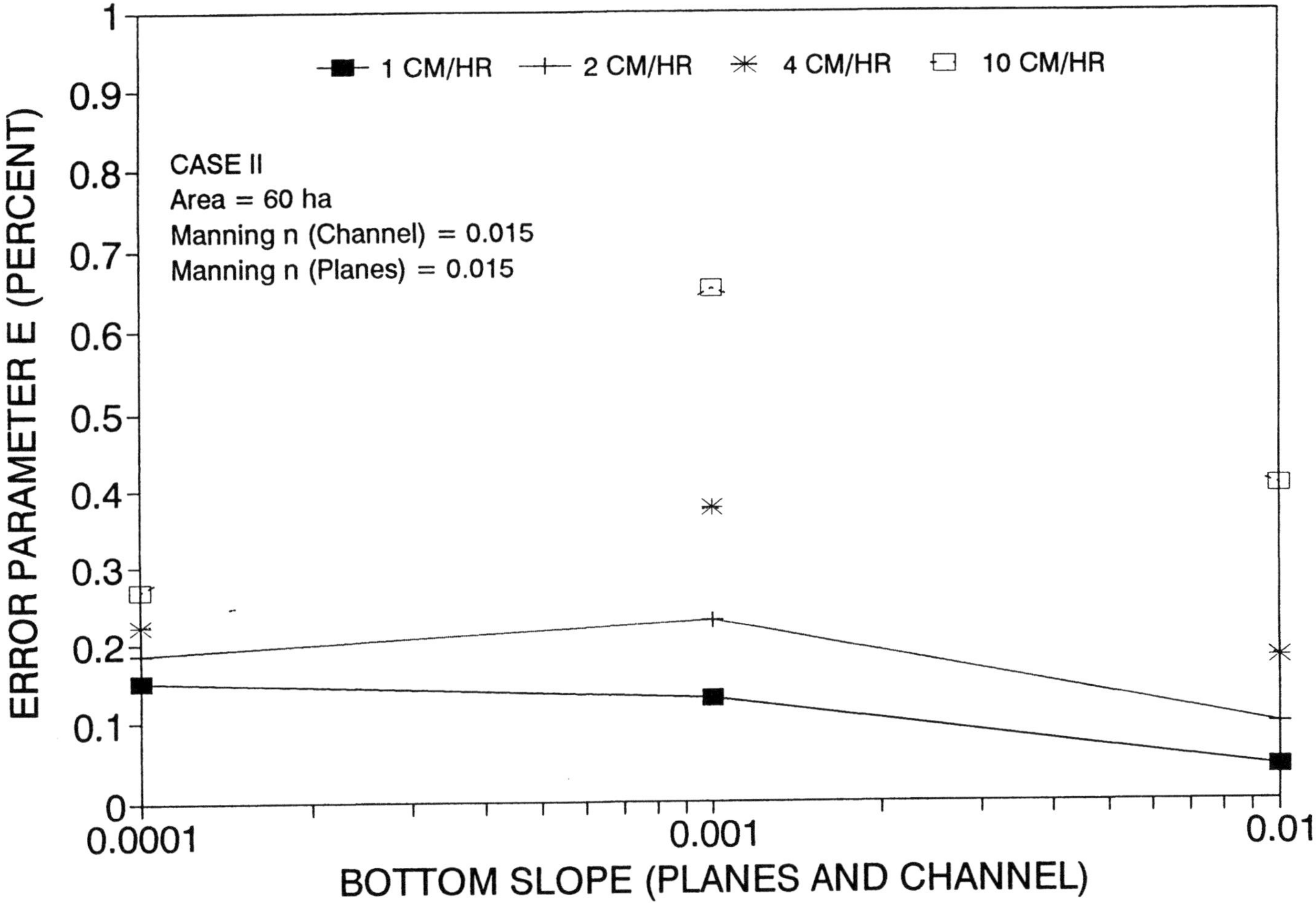

Fig 1 (b) Error Parameter E as a Function of Bottom Slope (Planes and Channel) Case II

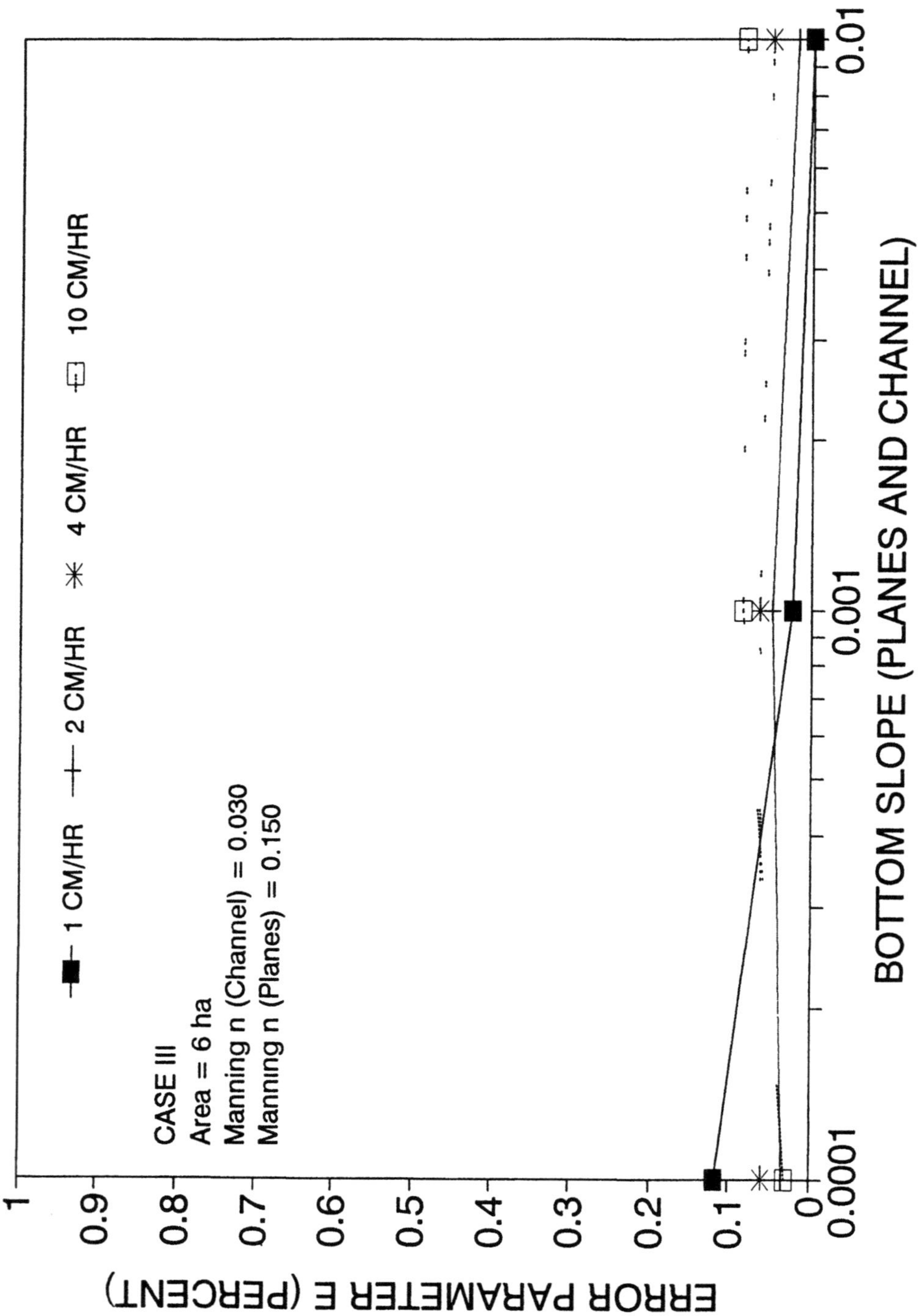

Fig 1 (c) Error Parameter E as a Function of Bottom Slope (Planes and Channel) Case III

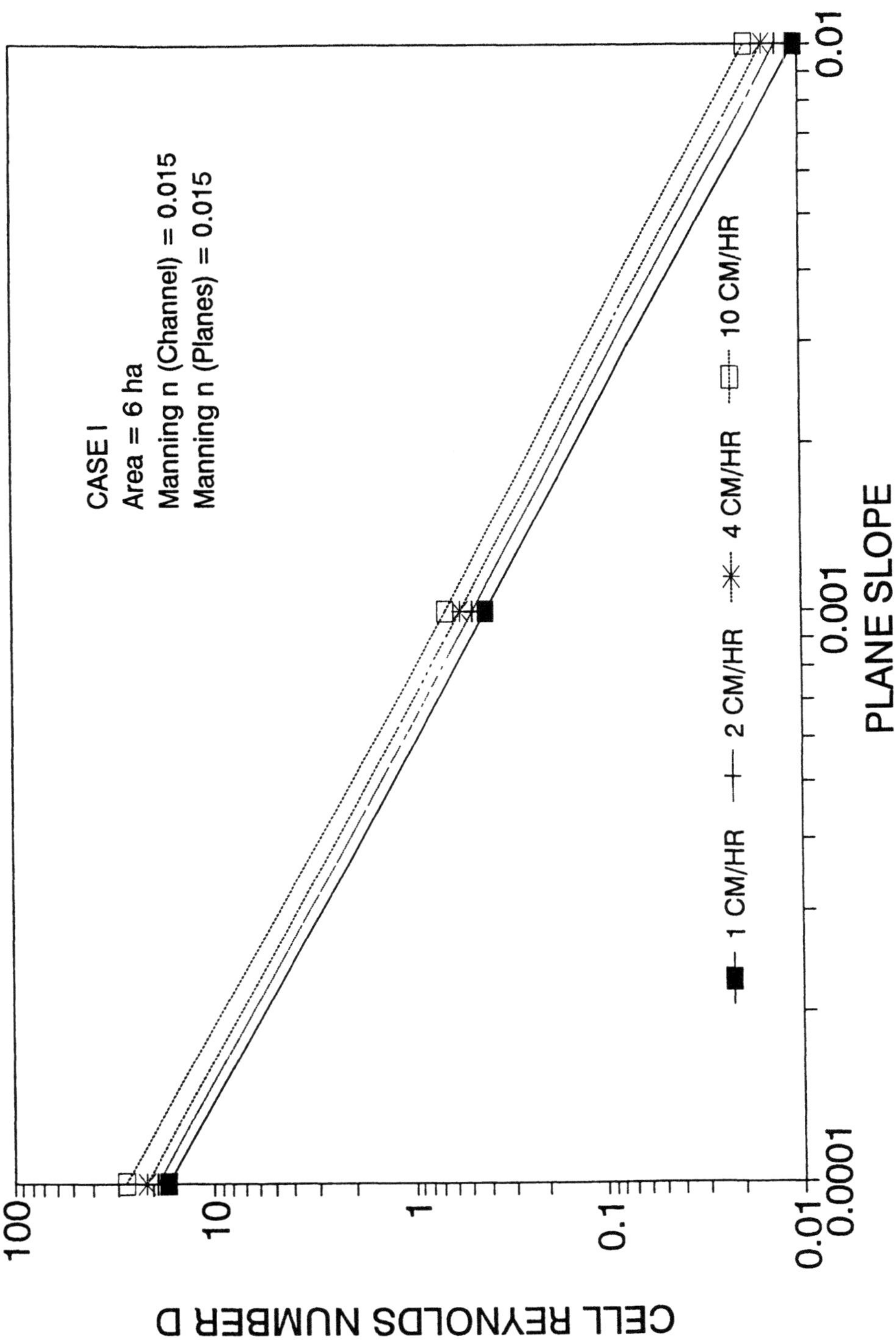

Fig. 2 (a) Cell Reynolds Number D as a Function of Plane Slope: Case I.

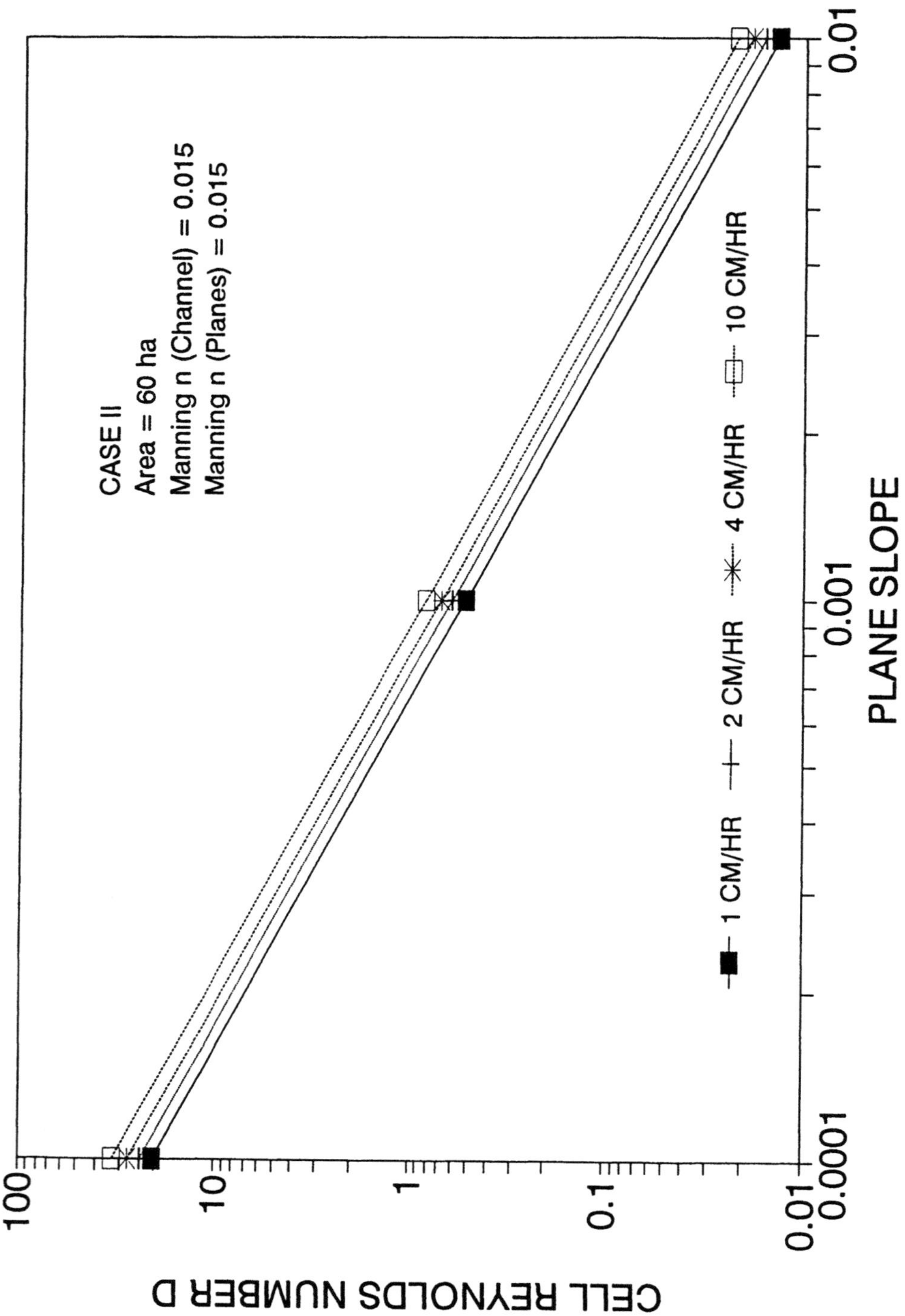

Fig. 2 (b) Cell Reynolds Number D as a Function of Plane Slope: Case II.

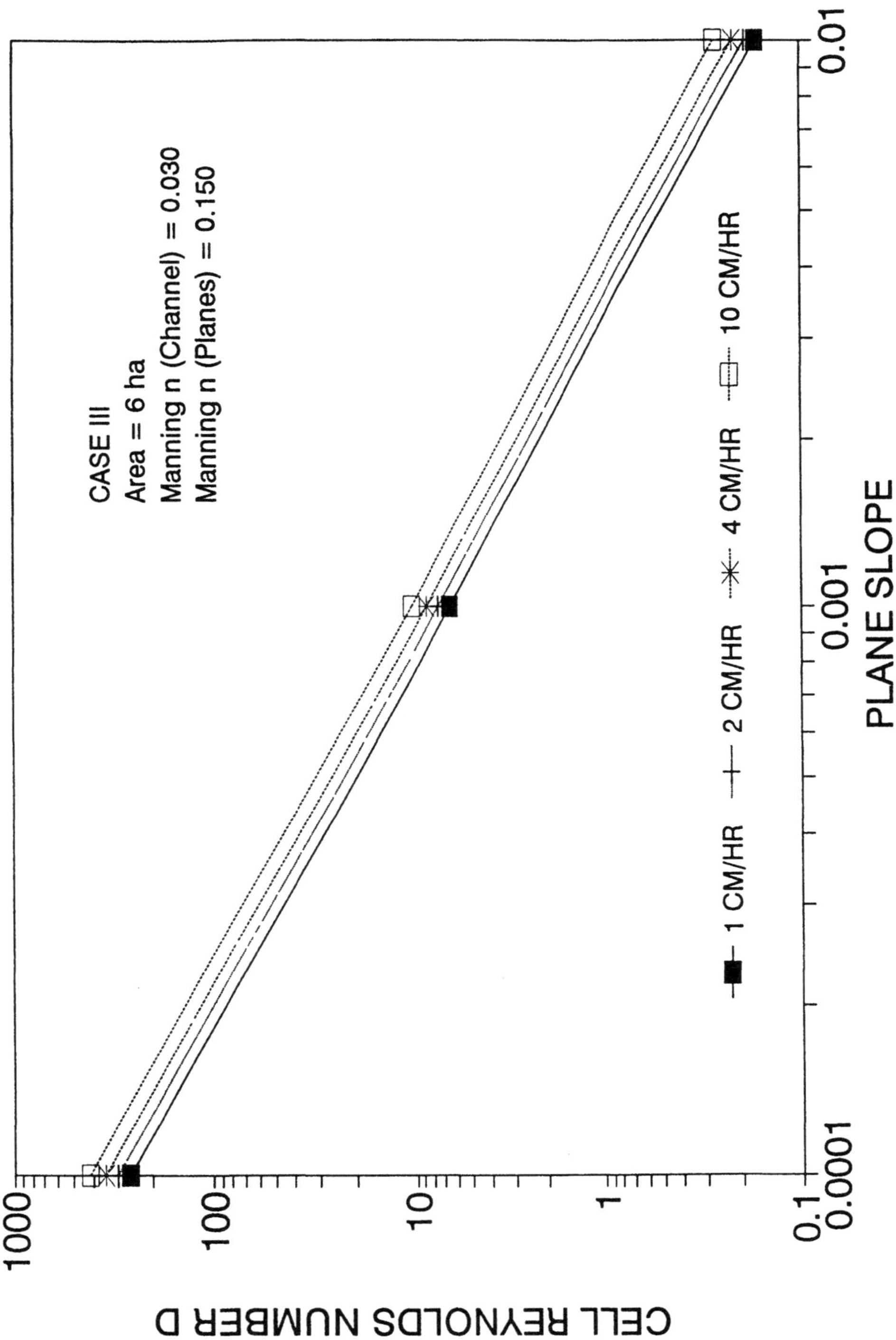

Fig. 2 (c) Cell Reynolds Number D as a Function of Plane Slope: Case III.

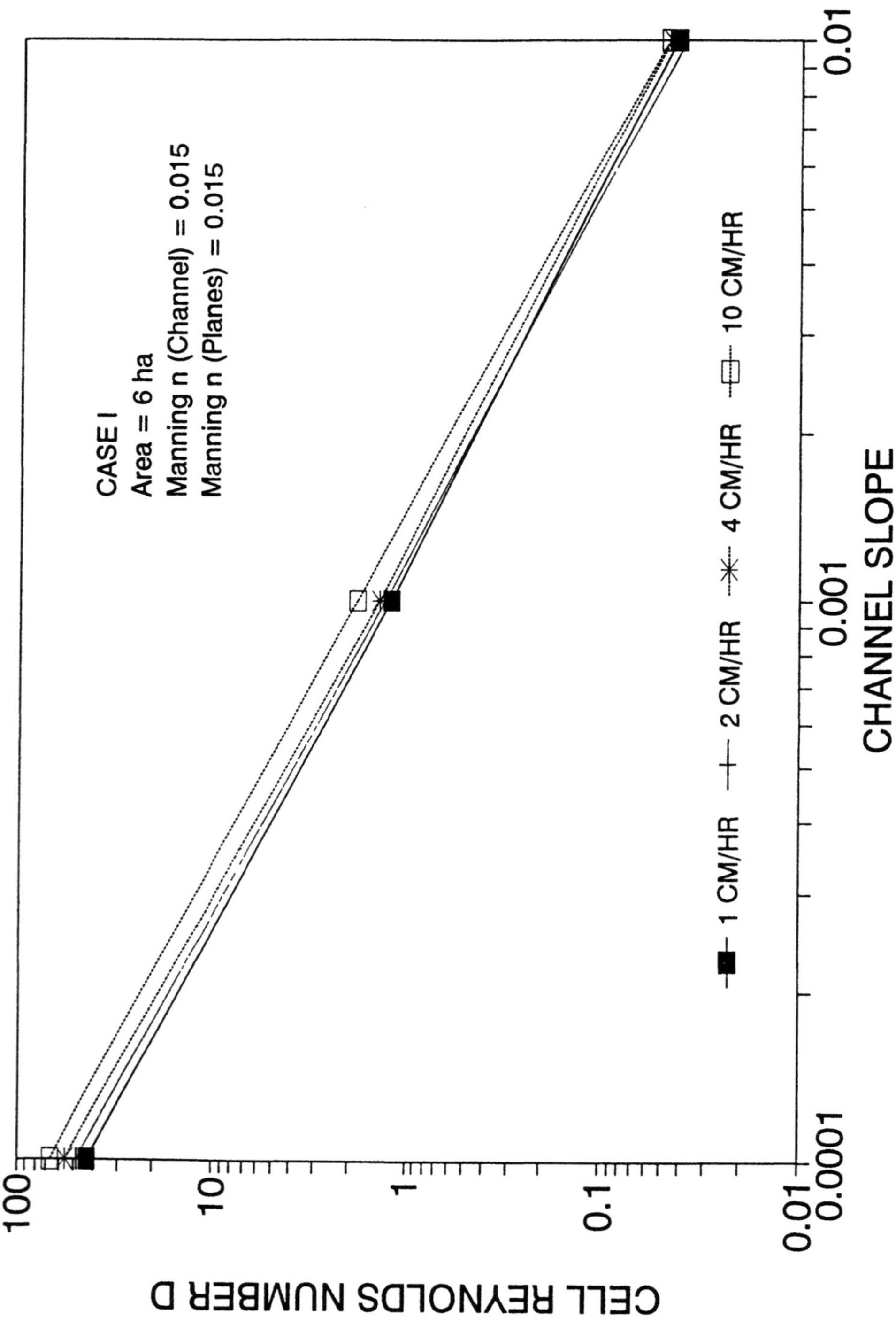

Fig. 3 (a) Cell Reynolds Number D as a Function of Channel Slope: Case I.

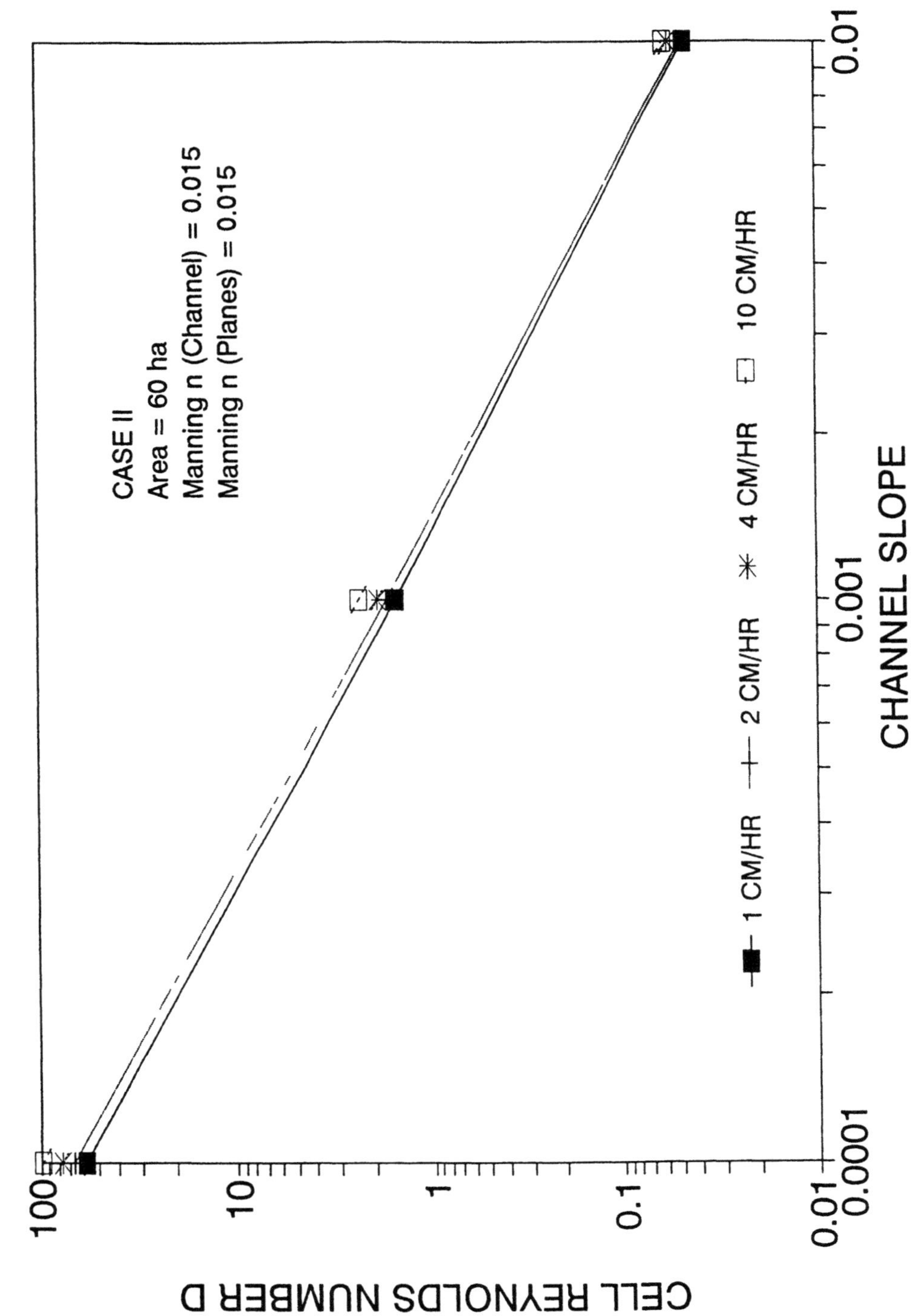

Fig 3 (b) Cell Reynolds Number D as a Function of Channel Slope Case II

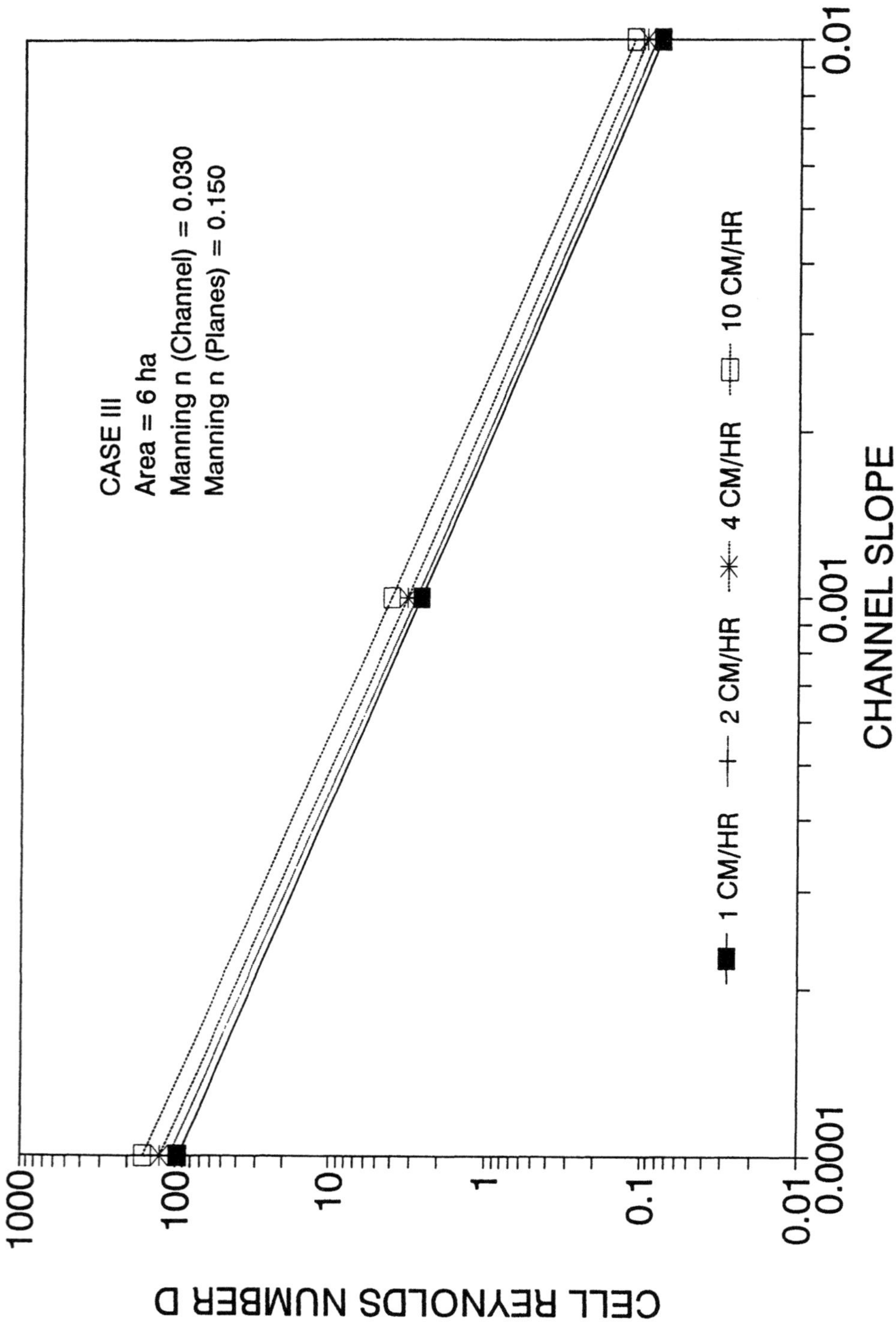

Fig. 3 (c) Cell Reynolds Number D as a Function of Channel Slope: Case III.

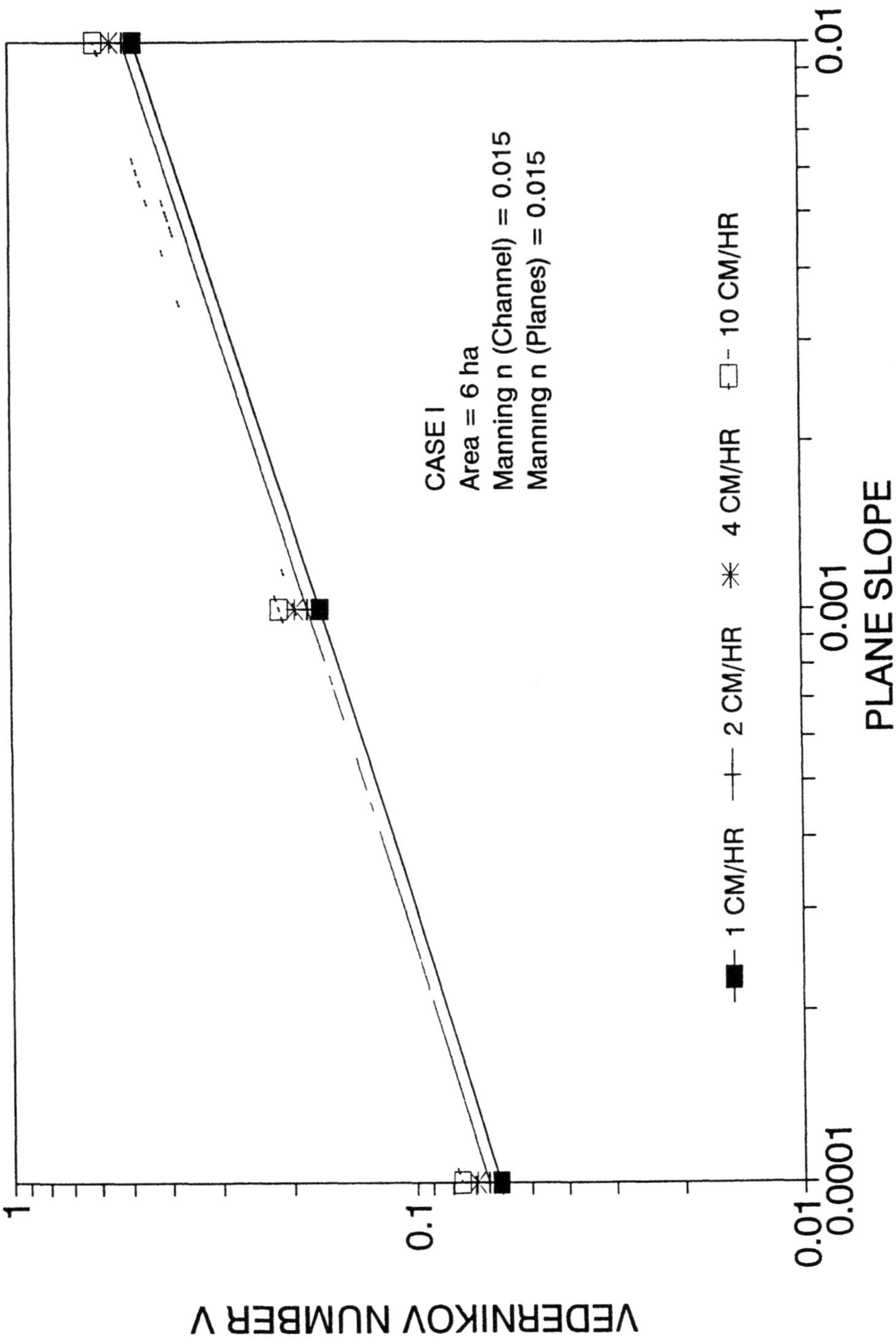

Fig 4 (a) Vedernikov Number V as a Function of Plane Slope Case I

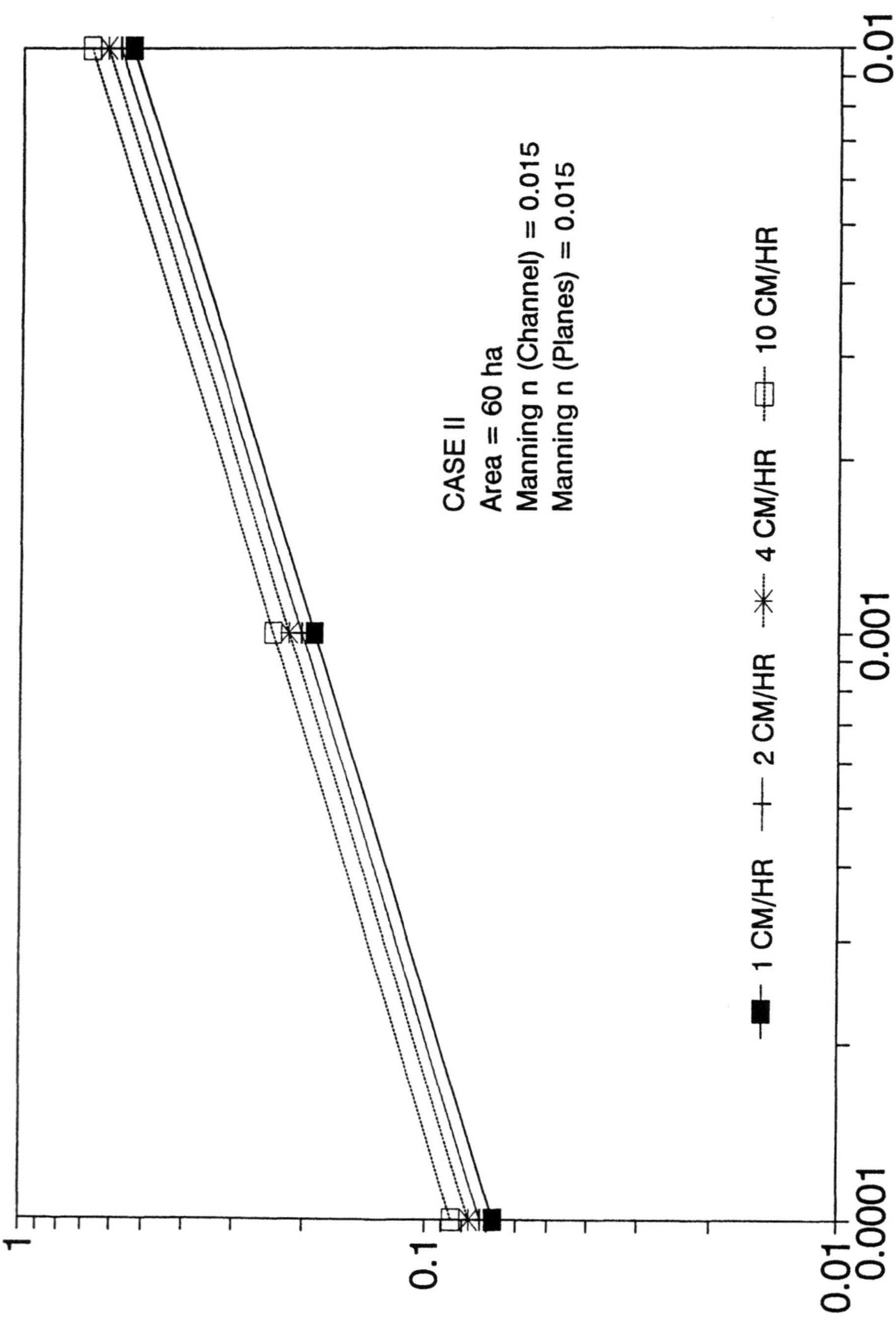

Fig. 4 (b) Vedernikov Number V as a Function of Plane Slope: Case II.

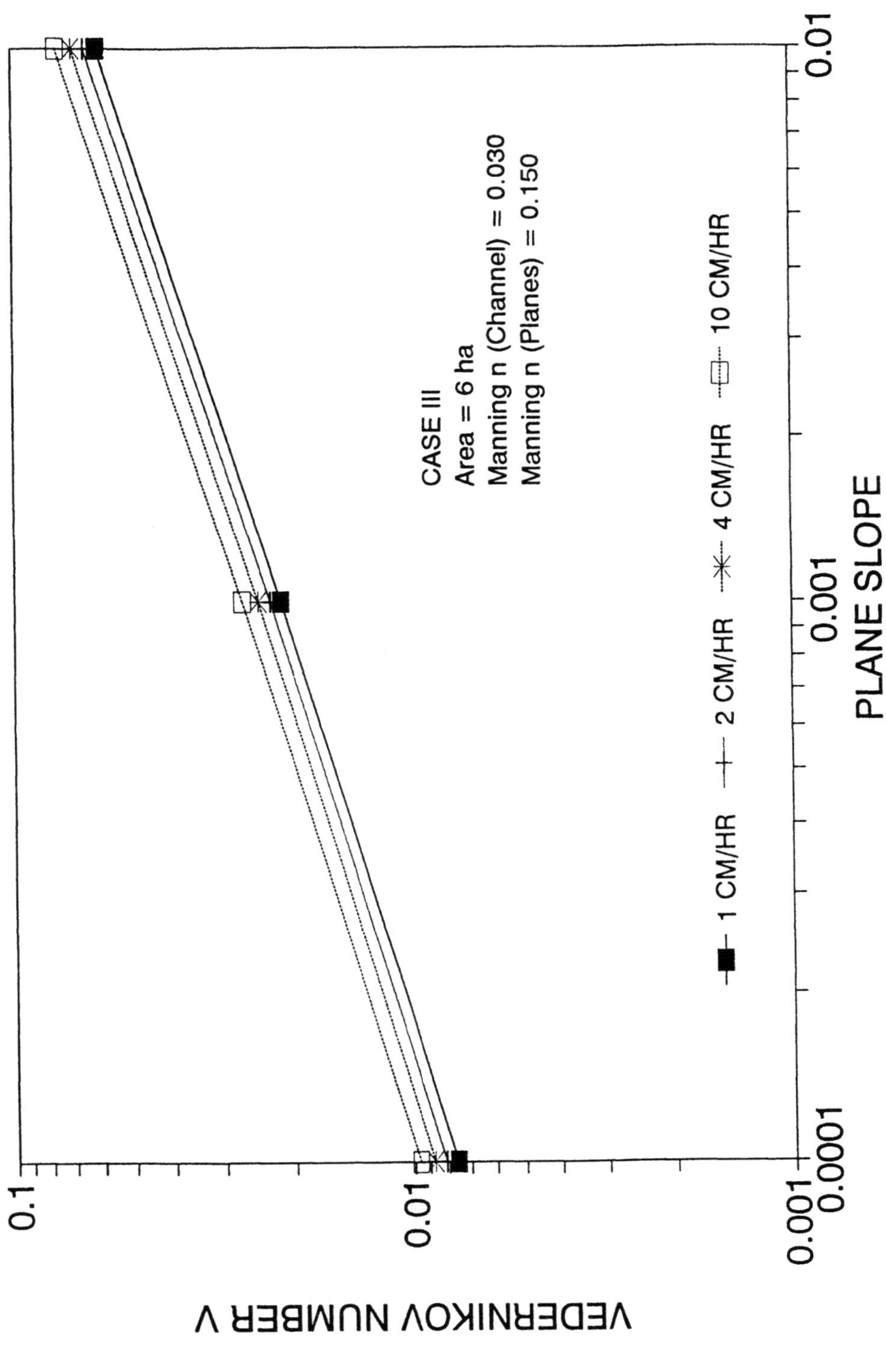

Fig. 4 (c) Vedernikov Number V as a Function of Plane Slope: Case III.

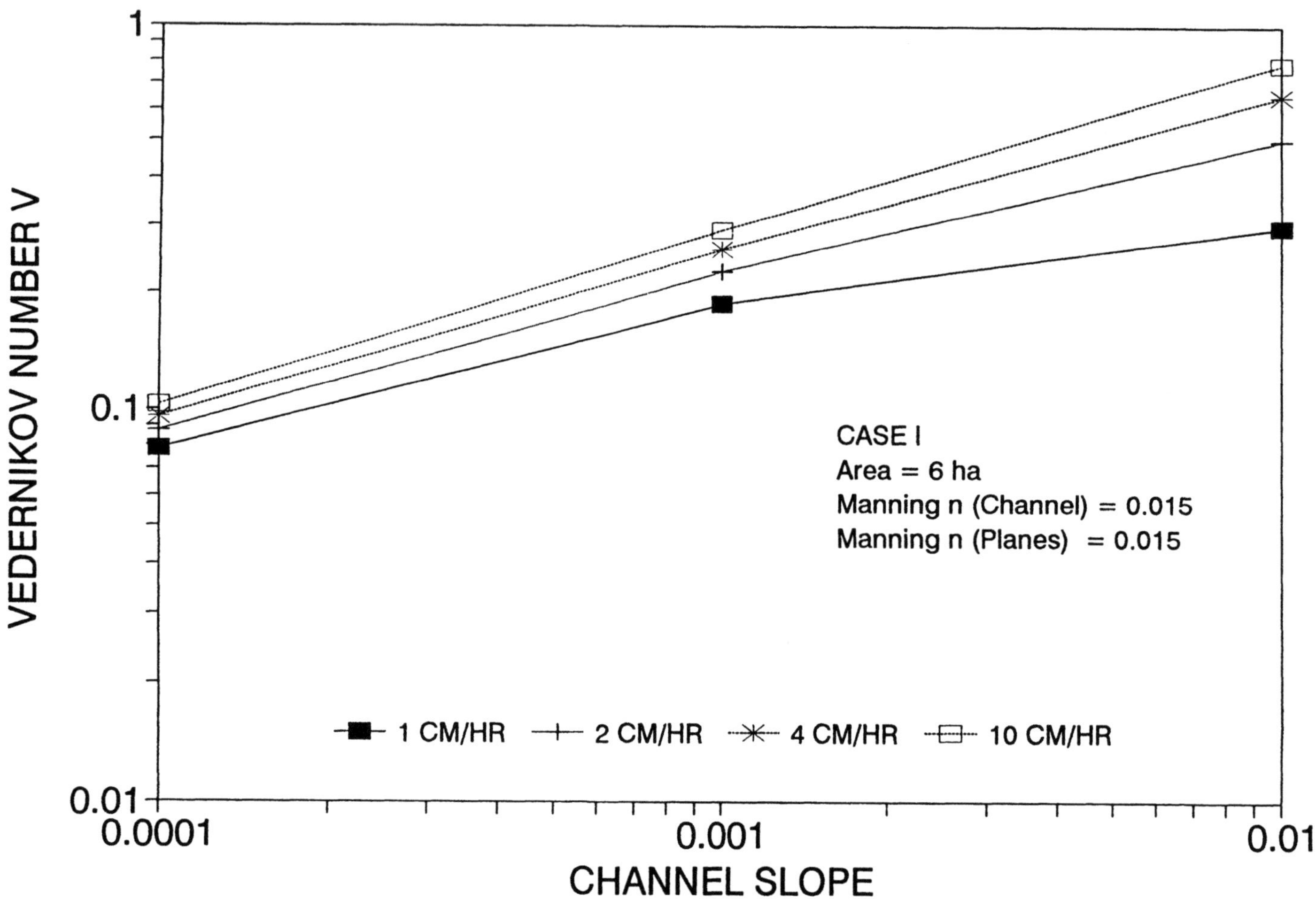

Fig. 5 (a) Vedemikov Number V as a Function of Channel Slope: Case I.

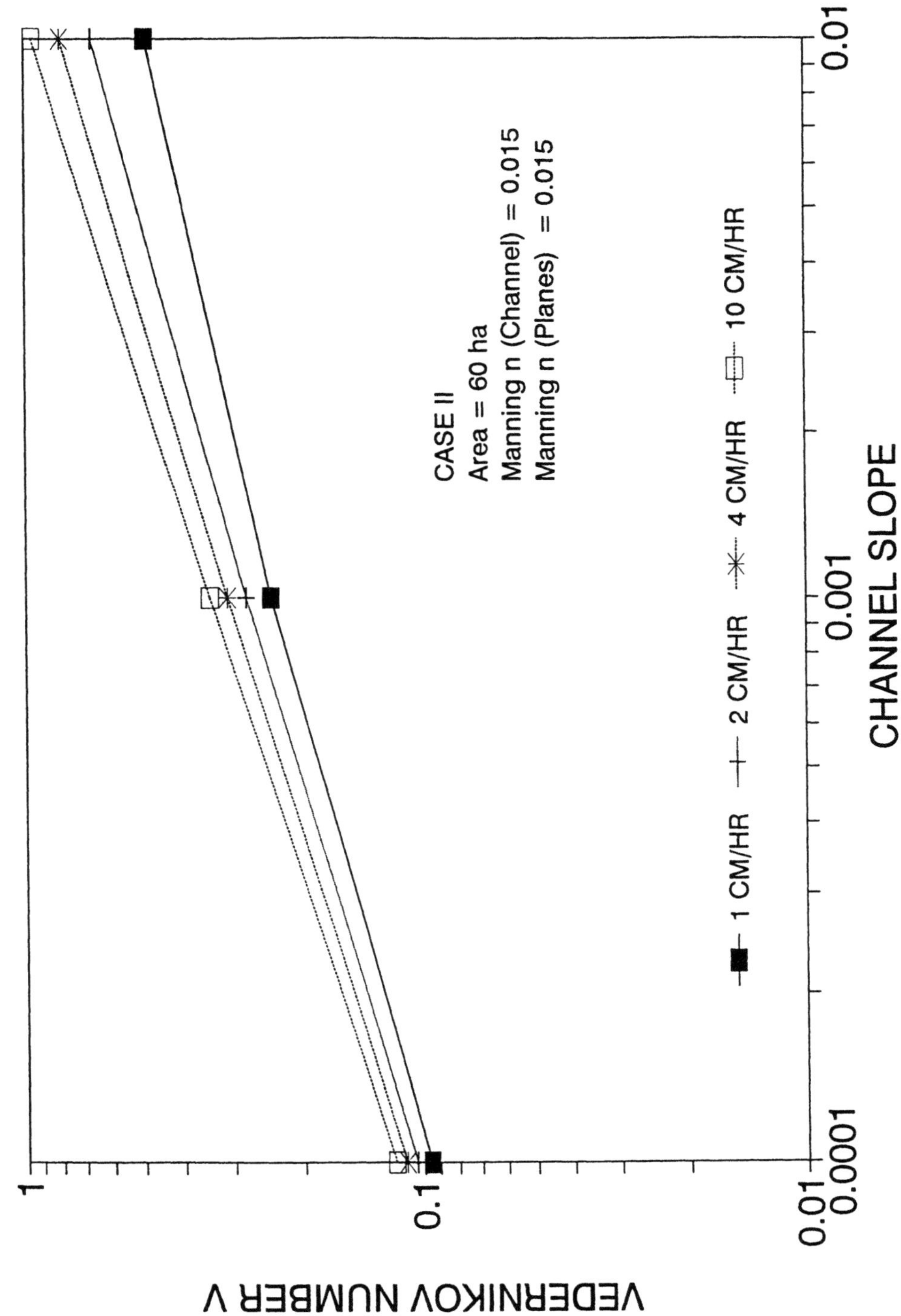

Fig. 5 (b) Vedernikov Number V as a Function of Channel Slope: Case II.

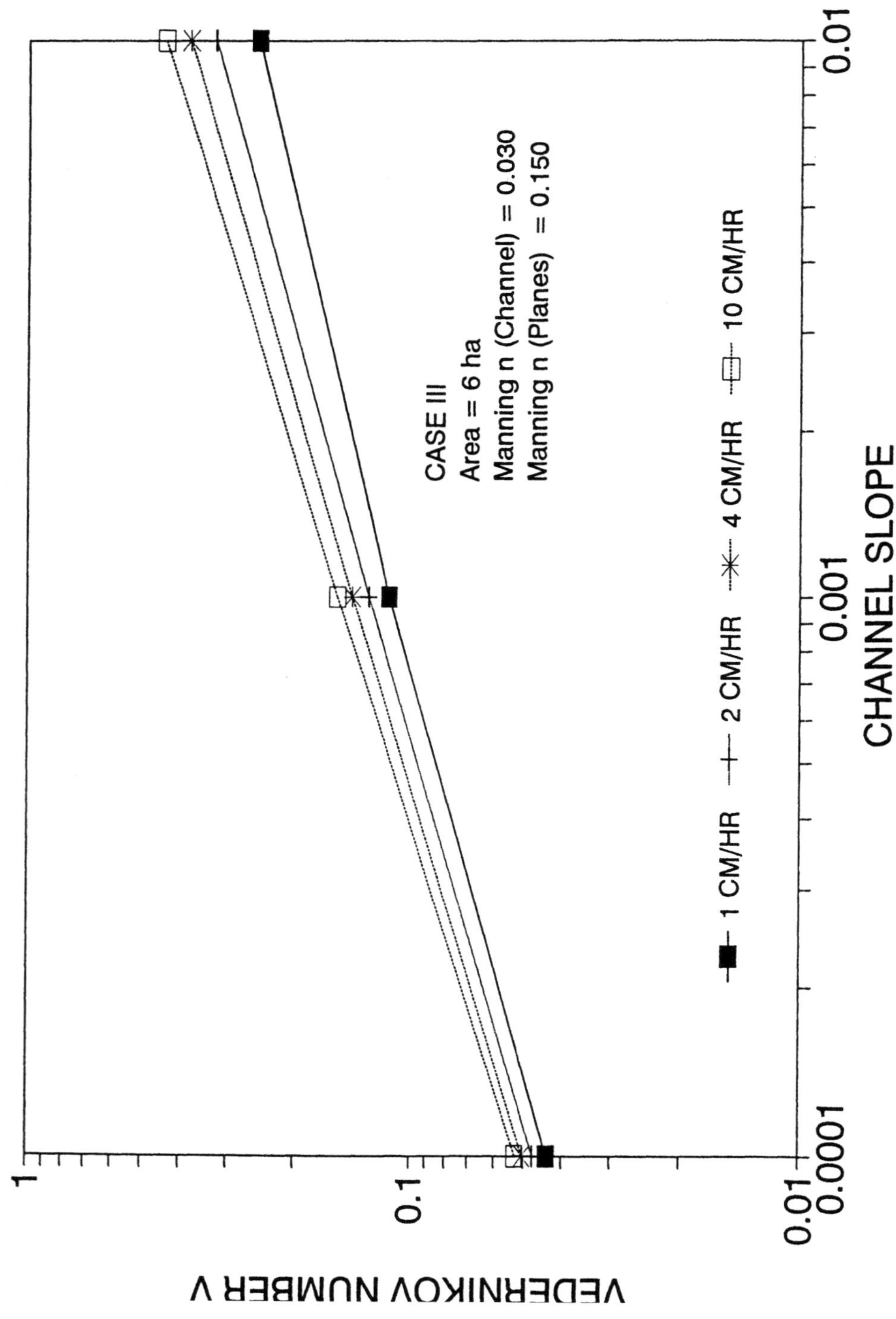

Fig. 5 (c) Vedernikov Number V as a Function of Channel Slope: Case III.

slopes does have a theoretical justification. Ponce and Simons (1977) have shown that as $\mathbf{F}^2 \rightarrow 0$, and consequently $\mathbf{V}^2 \rightarrow 0$, the diffusivity of the dynamic wave model approaches the diffusivity of the diffusion wave model, for a broad range of dimensionless wave numbers [$\sigma = (2\pi /L) L_o$, in which L_o = reference channel length, defined as $L_o = d_o/S_o$, with d_o = reference flow depth].

3. For intermediate slopes (S_o = 0.001), neither of the two opposing tendencies prevails over the other. The net result is that the error, while remaining small, has a tendency to peak at midrange values of bottom slope.

These findings underscore the futility of attempting to model overland flow by resorting to the complete hydrodynamic equations (Chow and Ben-Zvi, 1973; Tayfur et al, 1993). Since the diffusion wave model *with inertia* does not substantially improve the description of the wave dynamics for a wide range of flow conditions, it follows that there is very little to be gained by pursuing an even more precise and elaborate formulation. Moreover, since the diffusion wave model with inertia does not unduly complicate the formulation, it should be the preferred way of modeling overland flow with diffusion waves.

6. Summary

A program of numerical experiments has been performed to test the effect of the dynamic hydraulic diffusivity (Dooge, 1973; Ponce, 1991b) vis-à-vis the kinematic hydraulic diffusivity (Hayami, 1951), when used to model overland flow with diffusion waves. Two models (a conventional diffusion wave and a diffusion wave *with inertia*), three typical open-book overland flow scenarios of distinct geometry and bottom roughness, three bottom slopes (steep, intermediate, and mild), and four effective rainfall intensities (1, 2, 4, and 10 cm/hr) led to a total of 72 computer runs using a Muskingum-Cunge overland flow model.

Results showed a small but perceptible lag in the rising limb of the outflow hydrograph, coupled with a corresponding lag in the receding limb, when comparing runs using the two competing models. The lag is attributed to the error of the kinematic solution, which is based on the kinematic cell Reynolds number [$D = q_o/(S_o\ c\ \Delta x)$], as compared to the dynamic solution, which bases the calculation of diffusion on the dynamic cell Reynolds number [$D_d = (1 - \mathbf{V}^2)\ D$]. The error was evaluated by integrating the absolute value of the difference between the two hydrographs, to obtain a runoff volume difference, which was normalized by dividing it by the total runoff volume. The results show that the error is indeed small but perceptible, and likely to remain within 0.7 percent for a wide range of flow conditions. The results also show a tendency for the error to peak around the midrange of bottom slopes (S_o = 0.001).

The findings of this study underscore the futility of attempting to model overland flow by using the complete hydrodynamic equations. Since the dynamic effect is shown to be small throughout a wide range of bottom slopes, a diffusion wave model *with inertia* may be all that is required to model the overland flow dynamics. Moreover, since the diffusion wave model with inertia does not unduly complicate the formulation, it should be the preferred way of modeling overland flow using diffusion waves.

7. References

Chow, V. T., and Ben-Zvi, A. (1973). "Hydrodynamic Modeling of Two-dimensional Watershed Flow." *Journal of the Hydraulics Division,* ASCE, 99(11), 2023-2040.

Craya, A. (1952). "The Criterion of the Possibility of Roll Wave Formation." *Gravity Waves, National Bureau of Standards Circular No. 521,* 141-151.

Cunge, J. A. (1969). "On the Subject of a Flood Propagation Computation Method (Muskingum Method)." *Journal of Hydraulic Research*, 7(2), 205-230.

Dooge, J. C. I. (1973). *Linear Theory of Hydrologic Systems.* Tech. Bull., No. 1468, USDA Agricultural Research Service, Washington, D.C..

Dooge, J. C. I., Strupczewski, W. B., and Napiorkowski, J. J. (1982). "Hydrodynamic Derivation of Storage Parameters in the Muskingum Model." *Journal of Hydrology*, 54, 371-387.

Hayami, S. (1951). "On the Propagation of Flood Waves." *Bulletin Disaster Prevention Research Institute*, Kyoto University, 1(1), 1-16.

HEC-1, Flood Hydrograph Package: User's Manual. (1990). U.S. Army Corps of Engineers, Hydrologic Engineering Center, Davis, Calif., September.

Perumal, M. (1992). Comment on "New Perspective on the Vedernikov Number." *Water Resources Research*, 28(6), 1735.

Ponce, V. M., and Simons, D. B. (1977). "Shallow Wave Propagation in Open Channel Flow." *Journal of the Hydrauclis Division,* ASCE, 103(12), 1461-1476.

Ponce, V. M., and Yevjevich, V. (1978). "Muskingum-Cunge Method with Variable Parameters." *Journal of the Hydraulics Division,* ASCE, 104(12), 1663-1667.

Ponce, V. M. (1986). "Diffusion Wave Modeling of Catchment Dynamics." *Journal of Hydraulic Engineering*, ASCE, 112(8), 716-727.

Ponce, V. M. (1989a). *Engineering Hydrology, Principles and Practices.* Prentice Hall, Englewood Cliffs, New Jersey.

Ponce, V. M. (1989b). "Diffusion Wave Overland Flow Module." Technical Report prepared for USGS Water Resources Division, Stennis Space Center, Miss., June.

Ponce, V. M. (1991a). "The Kinematic Wave Controversy." *Journal of Hydraulic Engineering*, ASCE, 117(4), 511-525.

Ponce, V. M. (1991b). "New Perspective on the Vedernikov Number." *Water Resources Research*, 27(7), 1777-1779.

Tayfur, G., M. L. Kavvas, and R. S. Govindaraju. (1993). Applicability of St. Venant Equations for Two-dimensional Overland Flow Over Rough Infiltrating Surfaces. ASCE *Journal of Hydraulic Engineering*, 119:51-63.

Wooding, R. A. (1965). "A Hydraulic Model for the Catchment-Stream Problem." *Journal of Hydrology*, 3, 254-267.

8. Notation

A	= flow area, catchment area;
C	= Courant number;
c	= wave celerity;
c_{ri}	= relative inertial wave celerity (Lagrange celerity);
c_{rk}	= relative kinematic wave celerity;
D	= kinematic cell Reynolds number;
D_d	= dynamic cell Reynolds number;
d	= hydraulic depth;
d_o	= reference flow depth;
E	= percent error;
$\mathbf{F}$	= Froude number;
g	= gravitational acceleration;
i_e	= effective rainfall intensity;
K	= Muskingum travel time;
L	= wavelength;
L_o	= reference channel length;
n	= Manning roughness coefficient;
Q	= discharge;
Q_e	= equilibrium discharge;
q_o	= reference unit-width discharge;
S_o	= bottom slope;
t	= time variable;
t_r	= effective rainfall duration;
$\mathbf{V}$	= Vedernikov number;
V_r	= total runoff volume;

v = mean velocity;
X = Muskingum weighting factor;
α = coefficient of the rating $Q = \alpha A^{\beta}$;
β = exponent of the rating $Q = \alpha A^{\beta}$;
Δt = time interval;
Δx = space interval;
ΔV = runoff volume difference;
ν = hydraulic diffusivity;
ν_d = dynamic hydraulic diffusivity;
ν_k = kinematic hydraulic diffusivity; and
σ = dimensionless wave number.

Table 1. Test Catchment Characteristics

Case (1)	Area (ha) (2)	Plane Length (m) (3)	Channel Length/ Plane Width (m) (4)	Manning *n* (Planes) (5)	Manning *n* (Channel) (6)
I	6	100	300	0.015	0.015
II	60	300	1000	0.015	0.015
III	6	100	300	0.150	0.030

SECTION 4:

MODELING BASIN RESPONSE

COMPLEXITY, UNCERTAINTY, AND SYSTEMATIC ERROR IN HYDROLOGIC MODELS

L. J. LANE and M. H. NICHOLS
USDA-ARS, Southwest Watershed Research Center
2000 E. Allen Rd., Tucson, AZ 85719, USA

Abstract Issues of complexity, parameter and input variable uncertainty, and systematic model errors are reviewed and assessed. Simple measures are derived to represent degree of complexity, degree of uncertainty, and degree of systematic error for a simple subset of hydrologic models. Model complexity is represented by a complexity number, Nc = nm + 1. The quantity n is the number of model parameters and input variables and m is the number of simulation runs (around base or nominal values) required to assess noninteractive model sensitivity. Model uncertainty is represented by a summed coefficient of variation, CVm, computed from the sum of the individual coefficients of variations of the n parameters and input variables. Systematic error, NSm, is related to how well the model mimics nature and is represented as a function of the number of the basic concepts of conservation of mass, momentum, and energy, and of the basic variables position, velocity, and acceleration included in each model component. Three infiltration models: Phi Index, Runoff Curve Number, and the Green-Ampt Infiltration Equation; Two peak discharge estimation procedures: The Rational Formula and the coupled Green-Ampt Kinematic Wave Model are used as example illustrations. These examples are used to illustrate the highly interactive and important concepts of model complexity, uncertainty, and systematic error. The model quantification methodology and examples are also used to formulate the hypothesis that simple measures can be derived and used to objectively evaluate model complexity and its relationships with uncertainty and systematic error. Possible future applications of the model quantification methodology include selection of appropriate simulation models within decision support systems and contributions to development of a systematic approach for development and application of appropriate technology.

1. Introduction

Hydrologic modeling is a scientific activity which requires abstraction and simplification of processes occurring in nature. This abstraction and simplification constitutes an essential part of modern scientific procedure (i.e. see Rosenblueth and Wiener, 1945). Compared with the entirety of earth science and engineering, hydrologic modeling is a small and new science. Almost all quantitative work in model conceptualization has been accomplished during this century, with most progress made since the advent of the digital computer as a research and development tool.

V. P. Singh and B. Kumar (eds.), Surface-Water Hydrology, 175–189.

Because it is new and in a period of rapid transition, hydrologic modeling is diverse in its concepts and applications. The general topic of hydrologic modeling on small watersheds was recently and comprehensively summarized by Haan, *et al.* (1982). Goodrich and Woolhiser (1991) reviewed the U.S. literature from 1986 to 1990 and concentrated on entire watershed, or catchment, response rather than on components or processes. They concluded "...a detailed, process based, understanding of hydrologic response over a range of catchment scales (0.01-500 km^2) still eludes the hydrologic community." This assessment is in agreement with an earlier one by Dunne (1982) but is not as optimistic for the future as the views of Rogers and Anderson (1987) or Bevin (1987). Anderson and Burt (1985) edited a volume of papers on forecasting in hydrology including a brief, but broad-based, introduction to modeling strategy (Anderson and Burt, 1985, Ch. 1 Modeling Strategies).

Hydrologically driven water quality modeling was recently summarized in the proceedings of a 1988 conference (DeCoursey, 1990). Beck (1987) presented a comprehensive review of analysis of uncertainty in water quality modeling and questioned whether more complex models were better given their increased uncertainty. Hydrologically driven soil erosion modeling within the U.S. Department of Agriculture was recently summarized by Lane, *et al.* (1993). These reviews, while not restricted to hydrologic modeling, illustrated the key role hydrologic modeling plays in natural resource models.

These summary or synthesis books and papers held a common theme as they reviewed historical and recent developments in the general areas of hydrologic modeling and modeling based on hydrologic models (water quality and soil erosion). This common theme included the general assessment that models are increasing in complexity with time. Discussion of model complexity, uncertainty, and errors are explicit throughout these and other reviews. Unfortunately, these difficult issues are usually dealt with qualitatively and heuristically. Often the most useful insights are presented in almost anecdotal form (i.e. Todini, 1988; Wagenet, 1988; and Bevin and Jakeman, 1990).

Our central thesis here is that if further progress is to be made in understanding model complexity, uncertainty, and errors, then quantitative measures must be developed to express these concepts analytically or statistically. Further, it is our belief that this quantitative approach must involve simple measures if they are to be useful at this stage in hydrologic modeling. We seek insight through simplicity and do not intend to introduce additional complexity in an already complex and easily misunderstood area of hydrologic modeling.

1.1 CLASSIFICATION OF MATHEMATICAL MODELS

Classification of hydrologic models was summarized by Woolhiser and Brakensiek (1982) and included the broad classifications of material models and formal or mathematical models. It is the second category that is of interest herein. They listed six criteria to use in classifying mathematical hydrologic models as: (1) Model subject and structure, (2) Role of time, (3) Cognitive value, (4) Character of results, (5) Approach and methods of solution, and (6) Properties of the operator functions contained in the model.

Model subject and structure refer to which components of the hydrologic cycle are addressed and how the components or processes addressed are being modeled. The role of time refers to whether the processes are dynamic, that is, with time explicitly included in the formulation or static where time plays no role. The cognitive value of a model refers to whether it is conceptual, physically based, or a trend model. The character of results refers to the model's output and is generally described as stochastic (components random in time or space) or deterministic. Approach and methods of solution refer to model type (i.e. physically based) and how solutions to the model are obtained. Properties of the operator functions refers to whether the model is linear or nonlinear, lumped or distributed, or stationary or nonstationary.

Todini classified rainfall-runoff models and used the following general scheme (Todini, 1988, p. 346)

"A mathematical model in broad sense, is a combination of two basic components. The first one expresses all the *a priori* knowledge that one has on the phenomenon to be represented and can be referred to as the physical component. The second, the stochastic component, expresses in statistical terms what cannot be explained by the degree of *a priori* knowledge already introduced..."

From the physical and stochastic components, Todini assumed four classes of models based on increasing levels of the *a priori* knowledge they include: (1) Purely stochastic, (2) Lumped integral, (3) Distributed integral, and (4) Distributed differential models. Lumped and distributed have their traditional meaning and integral refers to processes represented by ordinary differential equations and differential refers to processes represented by partial differential equations.

1.2 USE OF CLASSIFICATIONS

Model classification schemes are useful in describing the general features of a model as to which components of the hydrologic cycle it simulates, how the simulation is accomplished, the type and level of mathematics involved, the nature of the model output, and the general type and amount of input information required. This knowledge is valuable in many ways. These include, but are not limited to, comparing alternative models, selecting the appropriate model for a given application, selecting data bases and experimental efforts to parameterize the model, and designing model validation analyses.

As valuable as the model classification methods are, they do have their limitations. With such schemes it is possible to classify a given hydrologic model if sufficient detail is presented in the model documentation. Given the classification results, one has a generalized picture of the model's complexity, its uncertainty, and its systematic error. However, the classification gives a generalized picture only and does not provide the analytical tools to move much beyond insight given by the anecdotal examples described earlier.

1.3 SCOPE AND PURPOSE

Hydrologic modeling discussed in this paper is limited to mathematical modeling and primary emphasis is on rainfall-runoff modeling. No attempt is made to conduct and report a state-of-the art summary or comprehensive literature search.

This paper provides a synopsis of selected examples and experiences in hydrologic modeling related to our central thesis that simple, quantitative measures must be developed to express model complexity, uncertainty, and systematic error beyond the limits of model classification techniques and to provide insight beyond those insights available through qualitative assessments. We seek to formulate and test the hypothesis that simple measures can be developed and used to objectively evaluate model complexity and its relationships with uncertainty and systematic error.

2. A Measure of Model Complexity

Measures of model complexity are implicit in the classification schemes discussed earlier. A complex model connotes one which is sophisticated and powerful but also difficult to understand, operate, and interpret. One method of investigating model complexity is through the application of sensitivity analyses.

2.1 NONINTERACTIVE SENSITIVITY ANALYSIS

Sensitivity analysis is a method of assessing the relative importance or sensitivity of a model's response or output to its parameter values or inputs. The simplest and most easily understood method of sensitivity analysis is the noninteractive method.

Given a set of model parameter values and typical values of the input variables which are in a sense representative or nominal (called the base values hereafter), computations are performed. With all other parameters and input variables fixed at their base values, individual parameters and input variables are varied about their base value, independently and sequentially, over a range of feasible and realistic values and the computations are repeated. The resulting set of output values shows how the model functions and how important changes in each parameter or input are in determining changes in the resulting output.

Weaknesses in the noninteractive sensitivity analysis procedure include: (1) Parameters and inputs are varied individually so that interactions are not determined, (2) Sensitivity of the model to changes in inputs and parameters is dependent on the choice of base values, and (3) The procedure is essentially empirical and does not draw on what is known of the model structure. Strengths of the procedure include: (1) It is straightforward and easy to perform and understand, (2) The results are amenable to tabular and graphical presentation as they are numerical and do not involve complex formulae, and (3) The procedure is independent of the model structure and is thus broadly applicable.

A noninteractive sensitivity analysis for the hydrologic component of the CREAMS Model (Knisel, 1980) for a small agricultural watershed at Tifton, Georgia was conducted by Lane and Ferreira (1980). Sensitivity of computed mean storm runoff volume to two parameters (CONA, a bare-soil evaporation rate parameter and CN, the Runoff Curve Number) is illustrated in Fig. 1. Changes in the parameter values and the resulting changes in mean storm runoff volume are shown as percentage changes from their base values. For example, in Fig. 1, a 50% decrease in CONA results in a 55% increase in mean storm runoff volume while a 10% decrease in CN results in a 48% decrease in storm runoff volume. The example results shown in Fig. 1 suggest that decreases in the evaporation rate parameter, CONA, result in magnified (larger changes in output than the corresponding change in the parameter value) increases in runoff while increases in CONA result in reduced (smaller changes in output than in the parameter) decreases in runoff. On the contrary, all changes in CN result in magnified changes in runoff. Thus all errors or uncertainty in CN are magnified as resulting errors in runoff while errors in CONA can result in magnified or educed errors in runoff. Finally, changes in runoff are positively correlated with changes in CN while changes in runoff are negatively correlated with changes in CONA.

2.2 MODEL COMPLEXITY NUMBER

The results of nine simulation runs are shown in Fig. 1, one run for the base values and four runs for CONA and four for CN. The general formula for the number of runs required in a noninteractive sensitivity analysis is $Nc = nm + 1$ where n is the number of parameters and input values (2 in Fig. 1) and m is the number of simulation runs around the base values required to define the sensitivity curve (4 in Fig. 1).

We propose Nc as a simple model complexity number reflecting the size of the model (as represented by the number of parameters) and the amount of input required (as represented by the number of input variables). Further, Nc reflects the complexity of the model structure and function through the number of simulation runs required to define the sensitivity curves illustrated in Fig. 1.

Figure 1. Illustration of noninteractive model sensitivity. Changes in runoff volume with changes in parameter values.

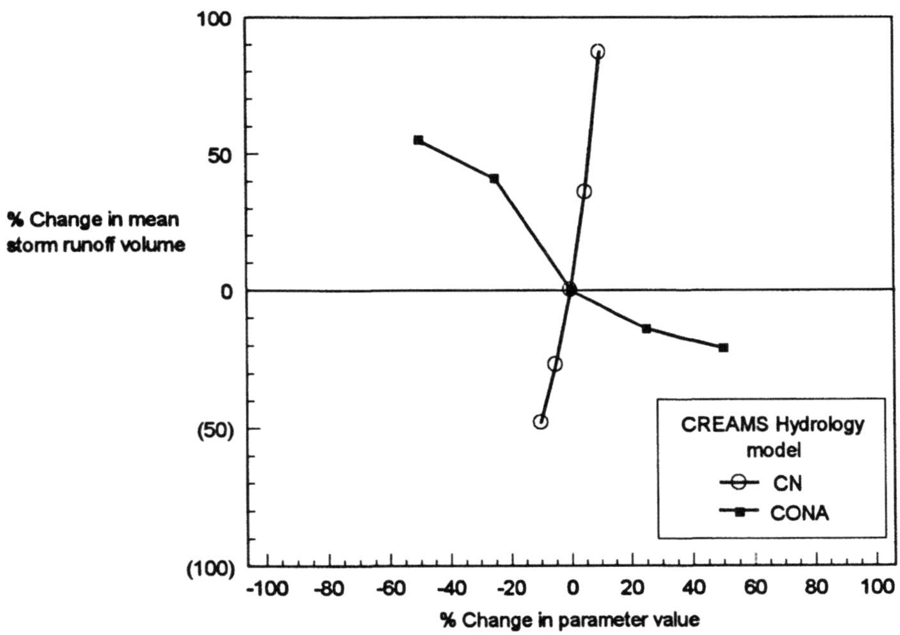

3. A Measure of Model Uncertainty

Model uncertainty can be expressed as uncertainty due to errors related to parameter and input uncertainty and those due to errors in model structure and function. The emphasis here is on uncertainty related to the parameters and inputs.

Uncertainty inherent in the model parameters and input variables result in uncertainty in model outputs (i.e. as illustrated in sensitivity analyses). Parameter and input values cannot be determined absolutely due to measurement errors in directly measurable quantities. Inferred quantities contain the uncertainty from measurement errors in the predictor variables as well as errors in the predictive relationship between the measured quantity and the inferred parameter or input. Often, one-time estimates are used to represent dynamic values and point values of parameters and inputs are used to represent spatial averages. These estimates also introduce uncertainty into the parameters and inputs and the resulting model outputs.

Uncertainty in model outputs also result from errors in model formulation (i.e. mistakes or omissions), errors in model structure (i.e. abstractions and simplifications), and errors in model implementation (i.e. coding errors, calculation errors, roundoff errors, etc.).

3.1 NONINTERACTIVE PARAMETER AND INPUT UNCERTAINTY

The coefficient of variation of a random variable is defined as the standard deviation divided by the mean. It is often used as a relative measure of uncertainty as it is dimensionless and directly related to the mean value. Thus the coefficient of variation, CV, can be used to compare relative variability between random variables of different units and scales.

Notice that the coefficient of variation for an individual parameter or input value can be used to help determine the range of variation for sensitivity analysis as illustrated in Fig. 1. Rather than using a fixed percentage change in the base value of the parameter or input (as in Fig. 1), one could use a fixed percentage of the standard deviation through the coefficient of variation.

For example, if the coefficient of variation of a parameter is 10% and the mean is taken as the base value then plus and minus 20% would be equivalent to plus and minus two standard deviations. Conversely, if the coefficient of variation of a second parameter is 20% then one would need plus and minus 40% to have a comparable plus and minus two standard deviations about the mean.

If CVp is the (1,Np) array of coefficients of variation of a model's parameters and CVi is the (1,Ni) array of coefficients of variation of a model's input variables, then an overall, noninteractive coefficient of variation for a model can be defined as

$$CVm = \sum_{p=1}^{Np} (CVp) + \sum_{i=1}^{Ni} (CVi) \tag{1}$$

The overall coefficient of variation is a positive number and is formed from Nm = Np + Ni individual values. It is noninteractive because it does not consider covariances of the parameters or input values.

Weaknesses of the noninteractive coefficient of variation, CVm, include: (1) It deals only with model parameters and input values, (2) It considers individual coefficients of variation and does not include interactions, and (3) It does not order, or rank, parameters in terms of their importance as determined by a sensitivity analysis. Strengths of CVm include: (1) It is straightforward and easy to calculate and understand, (2) It represents a single number and thus facilitates comparison between models, and (3) It is independent of model structure and thus broadly applicable.

Coefficients of variation for some representative values of commonly used rainfall-runoff model input variables and parameters are summarized in Table 1. Coefficients of variation from this table will be used in example calculations in a later section of this paper.

Table 1. Coefficients of variation for representative values of commonly used rainfall-runoff model input variables and parameters

Variable or Parameter	Coefficient of Variation	Source and Comments
Simulated Rainfall:		
Depth	0.04 - 0.13	3x11 m plots, ARS Rainfall Simulator Database for AZ and NV, Simanton, et al. (1986)
Int.	0.02 - 0.07	
Final Infiltration Rate*	0.26 - 0.64	
Natural Precipitation:		
Depth	0.44	12.4 sq km Goodwater Creek Watershed, a cultivated agricultural (row crops) area in Missouri, 35 largest floods in 14 years, Hjelmfelt and Kramer (1988)
Obs. Runoff**		
Vol.	0.67	
Peak	0.64	
Unit Hydrograph***		
Peak	0.19	
Time of Conc.	0.4	
Time to Peak	0.32	
Surface-Soil:		
Manning n:		
Bare-Fallow	0.7 - 0.8	Overland flow from erosion plot studies, Engman (1986) and Weltz, et al. (1992)
No Till	0.3 - 1.0	
Disk Harrow	0.3 - 1.1	
Plow	0.2 - 0.6	
Grass	0.1 - 0.2	
Rangeland	0.5 - 0.7	

Table 1 continued. Coefficients of variation for representative values of commonly used rainfall-runoff model input variables and parameters

Variable or Parameter	Coefficient of Variation	Source and Comments
Surface-Soil continued:		
Curve Number	0.08	4.5 sq km watershed W-5 at Holly Springs, MS, 11 storms, 1968-73, Borah and Ashraf (1988)
Porosity	0.07 - 0.11	Agricultural fields, unsaturated soil samples from various studies, Jury, et al. (1991)
Water Content		
0.1 bar	0.04 - 0.20	
15 bar	0.14 - 0.45	
Sat. Hyd. Cond.	0.48 - 3.20	

* Derived estimate for saturated hydraulic conductivity. Note relatively high coefficient of variation under controlled experimental conditions.

** Observed runoff not a parameter or input value, but range of values shown for comparison purposes.

*** Often used as input values to unit hydrograph models and other peak discharge estimation models.

3.2 A MODEL UNCERTAINTY NUMBER

We propose CVm as defined by Eq. 1 as a simple model uncertainty number which reflects the uncertainty in model inputs and model parameters through their coefficients of variation. Uncertainty in model output is not dealt with explicitly, but is implicitly represented by the variation in inputs and parameter values and the associated variations in output through sensitivity analysis.

4. A Measure of Systematic Error

Systematic error is used herein to express the degree a model and its components incorporate the governing equations for the processes represented. A model based on the governing equations for the processes considered would have a low systematic error (i.e. the model system mimics the natural system to a high degree) whereas a model based on correlation between inputs and outputs and which does not include the governing equations would have a high systematic error.

The governing equations of primary interest in rainfall- runoff modeling are:

1. Equation for conservation of mass (continuity eq.),
2. Equation for conservation of momentum (momentum eq.),
3. Equation for conservation of energy (energy eq.),
4. Equations specifying position (x,y,z),
5. Equations specifying velocity

$$\left(\frac{\partial x}{\partial t}, \frac{\partial y}{\partial t}, \frac{\partial z}{\partial t}\right) \tag{2}$$

6. Equations specifying acceleration

$$\left(\frac{\partial^2 x}{\partial t^2}, \frac{\partial^2 y}{\partial t^2}, \frac{\partial^2 z}{\partial t^2}\right) \tag{3}$$

These equations, or most commonly a one-dimensional subset of them, are specified for each component in a model and combined and manipulated until the mathematical model for the component is derived.

4.1 SYSTEMATIC ERROR NUMBER

If NEc is the number of the six governing equations included in a component then it is possible to define a component systematic error number, NSc as

$$NSc = 10.0[(6.0-NEc)/6.0] \tag{4}$$

where the quantity within the brackets is a number between 0 and 1.0, and the coefficient (10.0 in this case) is a scaling factor included to facilitate graphing.

If NSc is summed over all Nc components, the resulting model systematic error number, NSm, is

$$NSm = \sum_{c=1}^{Nc} (NSc)/Nc \tag{5}$$

where NSm is again a number between 0 and 10.

We propose NSm as a simple number reflecting the level of systematic error, with respect to the basic governing equations, contained in a model. This number indicates nothing about how well the governing equations are represented, parameterized, and solved, rather, it merely indicates whether or not the governing equations were included in the model structure. As such, NSm is a measure of the amount of processes-based formulations included in a model and thus is an overall measure of its physical basis.

5. Objective Evaluation of Models

As stated earlier, there is a great need for objective, repeatable methods of selecting the appropriate model for a specific application. Although several subjective techniques undoubtedly exist and are used routinely, advances in hydrologic modeling and the associated increases in model complexity require more objective techniques.

5.1 A SIMPLE METHODOLOGY

We propose that the model complexity number, Nc, the overall noninteractive coefficient of variation, CVm, and the model systematic error number, NSm, be used to quantify model complexity, uncertainty, and systematic error.

5.2 EXAMPLE ANALYSES

One of the earliest and apparently simplest infiltration models is the Phi Index which is an average rate of infiltration, applied to a time-intensity graph (hyetograph) of rainfall, such that the volume of rainfall excess equals the volume of storm runoff. For this and subsequent analyses, we will assume a reference hyetograph made up of 10 time-rainfall intensity pairs or 20 paired numbers. In actual practice, the number of time-rainfall intensity pairs will vary with storm characteristics, measurement equipment, and data processing procedures.

The Phi Index can be quantified as follows. It has one parameter, Phi, and it has 20 input values so n = 21. If we assume m = 4 simulations to determine model sensitivity, then the model complexity number is Nc = nm + 1 = (21)(4) + 1 = 85. From Table 1 if we assume Phi contains all the variation of the saturated hydraulic conductivity plus that due to porosity and initial water content, its coefficient of variation should be selected from near the high range of 0.48 - 3.20. Assume the CV for Phi is 2.0. Further, if we assume the intensity values have a CV comparable to total storm depth (0.44 in Table 1) and if we assume a default value of 0.10 for the time values, then the model uncertainty number, CVm, can be computed as follows CVm = CVp + SUM CVi = 2.0 + 10(0.44) +10(.10) = 7.4. Finally, the systematic error number, NSm is 10.0(6-1)/6 = 8.33 because the Phi Index only satisfies continuity of mass. Therefore, for the Phi Index the model quantification numbers are Nc = 85, CVm = 7.4, and NSm = 8.33.

A comparable analysis for the Runoff Curve Number Model yields one parameter (CN or S) and one input value, P the total storm rainfall. Thus, Nc = (2)(4) + 1 = 9 if we again assume m = 4. If we assume the CN has a CV of about 0.10 and P again has a CV of 0.44 (Table 1), then the uncertainty number is CVm = 0.10 + 0.44 = 0.54. Finally, the Runoff Curve Number Model only satisfies continuity of mass so the systematic error number is NSm = 8.33.

Analysis of the Green-Ampt infiltration equation shows 4 parameters (saturated hydraulic conductivity, Ks; soil porosity, n; the matric potential across the wetting front, Psi; and initial water content, SEi) and the same 20 time-intensity values used previously. With these values, the model complexity number is Nc = (24)(4) + 1 = 97. The model uncertainty number is CVm = CV(Ks) + CV(n) + CV(Psi) + CV(SEi) + 10(.44) + 10(.10). Assume mid-range values for the CV of Ks, n, Psi, and SEi as 1.5, 0.10, .50, and 0.20 so that CVm = 7.7. Finally, the Green-Ampt model satisfies continuity of mass and calculates the position and velocity of the wetting front so its systematic error number is NSm = 10.0(6-3)/6 = 5.0.

The Rational Formula is

$$Q = CIA \tag{6}$$

where Q is peak discharge in cfs, C is a runoff coefficient, I is rainfall intensity in in/h for a time period equal to the time of concentration, and A is the watershed area. The time of concentration, tc, is usually computed from basin characteristics. For example, the Kirpich (1940) formula is of the form

$$tc = K(L^{0.77}/S^{0.385}) \tag{7}$$

where K is a coefficient, L is basin length and S is an approximate average slope for the watershed. One could assume a given value of tc and only use Eq. 6 in the analysis, however, we decided to include time of concentration because of its central role in hydrograph development.

With these equations, the Rational Formula has a model complexity number of Nc = (6)(4) + 1 = 25 because there are 5 parameter values (C,A,K,L,and S) and one input (I). Assuming a CV of about 0.5 for C, 0.44 for I, 0.05 for A, L, and S, and 0.5 for K, the uncertainty number is calculated as CVm = 0.5 + 0.44 + 3(0.05) + 0.5 = 1.59. Because the Rational Formula only satisfies continuity of mass, its systematic error number is NSm = 8.33.

A coupled Green-Ampt infiltration model and kinematic wave model for a plane was described by Stone *et al.* (1992). This model contains all the Green-Ampt parameters and input values plus the following: (1) Slope of the plane, S, (2) Length of the plane, L, (3) Hydraulic roughness coefficient, C, (4) Percent canopy cover, CC, (5) Percent ground cover, GC, (6) Random roughness statistic, RR, (7) Depth-discharge exponent, m, and (8) A time step for calculations, Dt.

With these values there are 12 parameter values and the same 20 time-intensity pairs so that the model complexity number is Nc = (32)(4) + 1 = 129. The coefficients of variation of L, S, Dt are assumed to be 0.05, for C about 0.50 (Table 1, Manning n values), and about 0.10 for CC, G C, RR, and m. These CV's sum to 1.05 and when added to the CV's from the Green-Ampt component produce CVm = 7.7 + 1.05 = 8.75. Finally, the kinematic wave equations satisfy continuity of mass and take into account velocity and position so the overall systematic error number for the model is NSm = 5.0. Results of the model quantification examples are summarized in Table 2.

Values from Table 2 are plotted in Fig. 2. Notice that the order of increasing complexity of the 5 models is (1) Runoff Curve Number (CN), (2) Rational Formula (RF), (3) Phi Index (PHI), (4) Green-Ampt Infiltration (G-A), and (5) Kinematic Wave Model (KIN). For these examples, there is an almost linear increase in model uncertainty number with increasing model complexity number. The overall trend is for model systematic error to decrease with increasing model complexity number.

The Runoff Curve Number Model and the Rational Formula have relatively low complexity and uncertainty but high systematic error. In contrast, the apparently simple Phi Index is characterized by high complexity and uncertainty and high systematic error because it requires the full rainfall hyetograph as input. The Green-Ampt Infiltration Model requires little increase in complexity or uncertainty over the Phi Index but exhibits significantly less systematic error (Fig. 2). Based on these criteria only, there would appear to be little or no advantage of simplicity of the Phi Index over the Green-Ampt Infiltration Model but a heavy penalty in systematic error.

TABLE 2. Summary of model quantification examples

Model	Complexity	Uncertainty	Systematic Error
Phi Index	85	7.40	8.33
Runoff Curve Number	9	0.54	8.33
Green-Ampt Infiltration	97	7.70	5.00
Rational Formula	25	1.59	8.33
Kinematic Wave	129	8.75	5.00

Figure 2. Illustration of relationships between model complexity, uncertainty, and systematic error for the 5 examples.

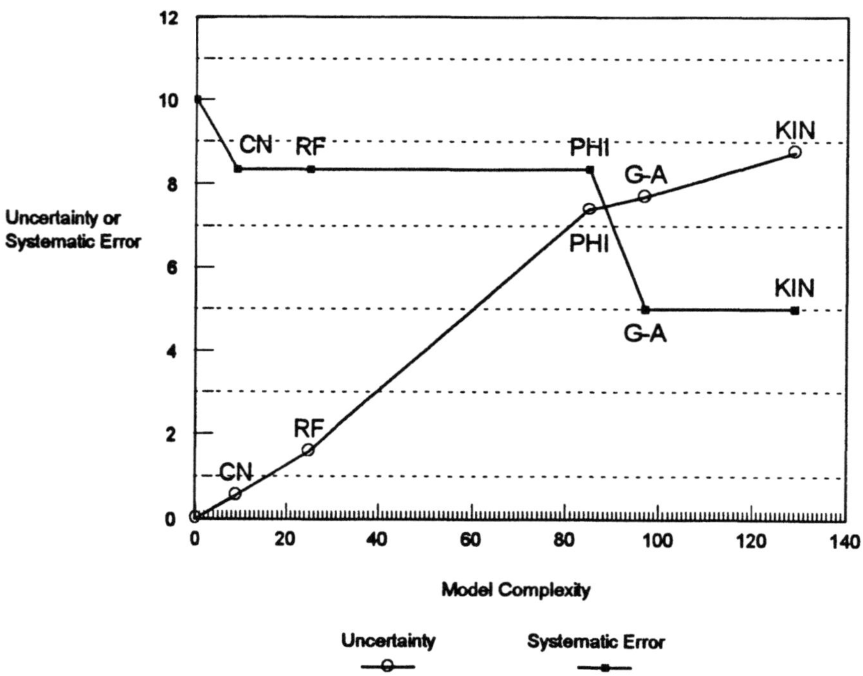

6. Future Considerations

We have discussed model classification and developed and illustrated examples of model quantification in terms of complexity, uncertainty, and systematic error. We now briefly discuss selected future potential applications of model quantification techniques.

6.1 MODEL SELECTION IN DECISION SUPPORT SYSTEMS

Increasingly, land and natural resource systems management will require the application of multiobjective decision making. Questions and decisions will involve basic resources: soil, water, air, plants, and animals; human resources: economics, recreation, esthetics, cultural heritage, and preservation; and broad societal concerns such as resource sustainability based on productivity, the environment, economics, equity, and social policy goals. As the data bases, simulation models, objectives, policies and regulations, monitoring, and reporting requirements become more comprehensive and more complex, computer-based decision support systems (DSS) will be required to assist the decision makers.

Because of the complexity of the problems and the lack of complete data bases, the DSS will use imbedded simulation models to provide values of the multiobjective criteria, or decision variables, used in the multiobjective analyses. Because the decision theory associated with multiobjective decision making is itself complex and mathematically-based, it too will be imbedded within future decision support systems (i.e. Yakowitz, *et al.*, 1992).

Techniques similar to the model classification criteria and model quantification methodology summarized herein could provide the DSS with objective ways of selecting the appropriate model once the land management-natural resource problem has been defined. This procedure of problem definition (in the rigorous systems engineering sense) and the subsequent selection of suitable simulation models to address the problem lead to the discussion of appropriate technology.

6.2 A SYSTEMATIC APPROACH TO APPROPRIATE TECHNOLOGY

Appropriate technology has been a goal and has been practiced since the start of civilization (see for example, Albertson, 1991). According to Albertson (1991) recent emphasis on appropriate technology is primarily due to Schumacher (1973) and his concern for shortcomings of policies of the industrialized nations.

Appropriate technology has often been discussed in the context of transfer of technology from the industrial countries to the less developed countries. This interpretation is too restrictive and eliminates the need for appropriate technology within as well as between all levels of organizations, societies, and nations.

Albertson's (1991, p. 229) definition is:

"Appropriate technology is the appropriate use of knowledge, skills, organization and machinery for the production of goods and services which are desired by those people being served. These goods and services are provided in a way that: is compatible with nature and the environment, uses only renewable resources including energy resources, benefits people equally and to the maximum extent possible, and is based on an economic system where the service motive is combined equally with the profit motive."

A central concept in this definition of appropriate technology is the matching of technology used to assist in providing goods and service with the needs and desires of the people being served in a socially, environmentally, economically, and natural resource-base sustainable manner.

Within the narrower context of this paper, a key concept is to match the appropriate simulation model with the users' needs, preferences, and resources as specified in the problem definition. Perhaps model quantification methodology such as described herein can assist in selecting the appropriate simulation models to use in addressing specific problems and thus contribute to the development of a systematic approach for development and application of appropriate technology.

7. References

Albertson, M. L. 1991. 'Evolution of appropriate technology.' pp. 227-232 In Shane, R. M. (Ed.) Hydraulic Engineering. Proc. of ASCE 1991 National Conf., Nashville, TN, July 29-August 2, 1991, 1178 pp.

Anderson, M. G. and T. P. Burt. 1985. Hydrological Forecasting. John Wiley & Sons Ltd., Chichester, 604 pp.

Beck, M. B. 1987. 'Water quality modeling: A review of the analysis of uncertainty. Water Resources Research, 23(8):1393-1442.

Bevin, K. 1987. 'Towards a new paradigm in hydrology.' pp. 393-403 In Rodda, J. C. and N. C. Matalas (Eds.) Water for the Future: Hydrology in Perspective. IAHS Pub. No. 164, IAHS, Washington, DC.

Bevin, K. J. and A. J. Jakeman. 1990. 'Complexity and uncertainty in predictive models.' pp. 555-576 In DeCoursey, D. G., (Ed.) Proceedings of the International Symposium on Water Quality Modeling of Non-Point Sources, Part 1 and Part 2, U.S. Dept. of Agriculture, Agricultural Research Service, ARS-81. 881 pp.

Borah, D. K. and M. S. Ashraf 1988. 'Modeling watershed response with seasonal variation.' pp. 403-411 In Modeling Agricultural, Forest, and Rangeland Hydrology. Proc. ASAE Intrl. Symposium, Chicago, IL, Dec. 12-13, 1988. ASAE Pub. 07-88, ASAE, St. Joseph, MI, 510 pp.

DeCoursey, D. G. (Ed.) 1990. Proceedings of the International Symposium on Water Quality Modeling of Non-Point Sources, Part 1 and Part 2, U.S. Dept. of Agriculture, Agricultural Research Service, ARS-81. 881 pp.

Dunne, T. 1982. 'Models of runoff processes and their significance.' pp. 19-30 In Studies in Geophysics: Scientific Basis of Water Resource Management. Academy Press, Washington, DC.

Engman, E. T. 1986. 'Roughness coefficients for routing surface runoff.' J. of Irrigation and Drainage Engineering, ASCE, 112(1):39-53.

Goodrich, D. C. and D. A. Woolhiser. 1991. 'Catchment hydrology.' Rev. of Geophysics, Supplement, AGU, pp. 202-209.

Haan, C. T., H. P. Johnson, and D. L. Brakensiek, (Eds.) 1982. Hydrologic Modeling of Small Watersheds. ASAE Monograph No. 5, ASAE, St. Joseph, MI, 533 pp.

Hjelmfelt, A. T. and L. A. Kramer 1988. 'Unit hydrograph variability for a small agricultural watershed.' pp. 357-366 In Modeling Agricultural, Forest, and Rangeland Hydrology. Proc. ASAE Intrl. Symposium, Chicago, IL, Dec. 12-13, 1988. ASAE Pub. 07-88, ASAE, St. Joseph, MI, 510 pp.

Jury, W. A., W. R. Gardner, and W. H. Gardner. 1991. Soil Physics. John Wiley & Sons, Inc., New York, 328 pp.

Kirpich, Z. P. 1940. 'Time of concentration of small agricultural watersheds.' Civil Eng. (New York), 10(6):362.

Knisel, W. G., (Ed.) 1980. CREAMS: A Field-Scale Model for Chemicals, Runoff, and Erosion from Agricultural Management Systems. U.S. Dept. of Agriculture, Conservation Research Report No. 26, USDA, Washington, DC, 640 pp.

Lane, L. J. and V. A. Ferreira. 1980. 'Sensitivity analysis.' Ch. 6, pp. 113-158 In Knisel, W. G., (Ed.) CREAMS: A Field-Scale Model for Chemicals, Runoff, and Erosion from Agricultural Management Systems. U.S. Dept. of Agriculture, Conservation Research Report No. 26, USDA, Washington, DC, 640 pp.

Lane, L. J., K. G. Renard, G. R. Foster, and J. M. Laflen. 1993. 'Development and application of modern soil erosion prediction technology - The USDA experience.' Australian J. of Soil Res. 30(6):20.

Rogers, C. C. M. and M. G. Anderson. 1987. 'Research needs in catchment distributed modeling.' pp. 415-423 In Rodda, J. C. and N. C. Matalas (Eds.) Water for the Future: Hydrology in Perspective. IAHS Pub. No. 164, IAHS, Washington, DC.

Rosenblueth, A. and N. Wiener. 1945. 'Role of models in science.' Phil. Sci. 7(4):316-321.

Schumacher, E. F. 1973. Small is Beautiful: Economics as if People Mattered. Harper and Row, New York, pp. 290.

Simanton, J. R., C. W. Johnson, J. W. Nyhan, and E. M. Romney. 1986. 'Rainfall simulation on rangeland erosion plots.' pp. 11-17, 43-58, 63-68 In Lane, L. J. (Ed.) Erosion on Rangelands: Emerging Technology and Data Base. Proc. of the Rainfall Simulator Workshop, Tucson, AZ, Jan. 14-15, 1985, Soc. for Range Mgt., Denver, CO, 68 pp.

Stone, J. J., L. J. Lane, and E. D. Shirley. 1992. 'Infiltration and runoff simulation on a plane.' Transactions of the ASAE, 35(1):161-170.

Todini, E. 1988. 'Rainfall-runoff modeling --Past, present, and future.' J. Hydrol., 100:341-352.

Wagenet, R. J. 1988. 'Modeling soil hydrology: Perspectives, perils and directions.' pp. 1-9 In Modeling Agricultural, Forest, and Rangeland Hydrology. Proc. ASAE Intrl. Symposium, Chicago, IL, Dec. 12-13, 1988. ASAE Pub. 07-88, ASAE, St. Joseph, MI, 510 pp.

Weltz, M. A., A. B. Arslan, and L. J. Lane. 1992. 'Hydraulic roughness coefficients for native rangelands.' J. of Irrigation and Drainage Engineering, ASCE, 118(5):776-790.

Woolhiser, D. A. and D. L. Brakensiek 1982. 'Hydrologic system synthesis.' Ch. 1, pp. 3-16 In Haan, C. T., H. P. Johnson, and D. L. Brakensiek, (Eds.) Hydrologic Modeling of Small Watersheds. ASAE Monograph No. 5, ASAE, St.Joseph, MI, 533 pp.

Yakowitz, D. S., L. J. Lane, J. J. Stone, P. Heilman, and R. K. Reddy. 1992. 'A decision support system for water quality modeling.' pp. 188-193 In M. Karamouz (Ed.) Water Resources Planning and Management, Proc. ASCE Water Forum '92, Baltimore, MD.

THE MOST PROBABLE HYDROLOGIC RESPONSE OF FRACTAL RIVER NETWORKS

Pierluigi CLAPS, Mauro FIORENTINO *and* Giuseppe OLIVETO

Dipartimento di Ingegneria e Fisica dell'Ambiente
Università della Basilicata
Via della Tecnica 3
85100, Potenza (Italy)

ABSTRACT. An expression of the hydrologic response of fractal networks is derived by means of the classical approach used in statistical-mechanics. Probabilities of the rate of arrivals at the outlet of mass particles instantaneously injected into the network are maximized accounting for some suitable constraints. The analysis of the structure of fractal networks allows us to meaningfully set the required constraints, that are of the same form of these used by *Lienhard and Meyer* (1967). The resulting hydrologic response is a Weibull distribution whose parameters are the fractal dimension and the total length of the network. A comparison with the similar form of arrival time distribution obtained by *Troutman and Karlinger* (1985) showed it to be a particular case of the more general expression relative to fractal networks.

1. INTRODUCTION

The study of river network morphology in view of the definition of characters of the basin response characteristics has received new breed since the introduction of fractal geometry concepts in geomorphology. It is now widely accepted that river networks are fractal (*i.e.* self-similar) objects (*e.g. Mandelbrot,* 1983; *La Barbera and Rosso*, 1989) and that their structure obeys some principles of least energy dissipation (*e.g. Rinaldo et al.*, 1992; *Fiorentino et al.*, 1993). Despite that, the concepts of fractal geometry have not yet been extensively used in the evaluation of the basin response. Probably the only attempt in this field is due to *Marani et al.* (1991) who determined the most probable response of a particular fractal network, the Peano plane-filling curve.

In this paper, a framework for the definition of purely fractal networks is used in the search of one expression of the most probable hydrologic response of river networks. Following the statistical-mechanical approach of *Lienhard* (1964), structural properties of fractal networks are explicitly used in this search, leading to an IUH form based on clearly recognizable parameters.

V. P. Singh and B. Kumar (eds.), Surface-Water Hydrology, 191–204.

2. STATISTICAL-MECHANICAL DERIVATION OF THE IUH

Lienhard (1964), first used a statistical-mechanical approach in the search for a function describing the arrival times of water particles fallen onto the basin, function representing the basin IUH.

The framework used by Lienhard is made up of a drainage-basin system on which N particles of equal mass are instantaneously injected. These particles fall on regions (*states*) of the basin with different distances to the outlet. The distances are considered proportional to the arrival times of the water particles to the outlet, so the drainage-basin can be subdivided in a number of time partitions. Using *m* partitions of time, the compound probability that (N_1, N_2, ...,N_m) particles reach the basin outlet respectively at times (t_1, t_2, ..., t_m) is

$$P(N_1 = n_1, \dots\dots, N_m = n_m) = \binom{N}{n_1, \dots, n_m} \cdot g_1^{n_1} \cdot \dots\dots\dots \cdot g_m^{n_m} \tag{1}$$

with g_i as the number of distinguishable ways of placing n_i particles in the *i*-th time partition and with ΣN_i=N. *Gupta and Waymire* (1983), reviewing Lienhard's approach, gave g_i the interpretation of probabilities p_i proportional to the contributing area C_i at time $t_i = i\Delta t$, so that

$$p_i = C_i / A \qquad 1 \le i \le m \tag{2}$$

where A is the basin area. Thus p_i represents the probability of having n_i particles in the i-th contributing area.

The most probable distribution satisfying the above requirements is obtained by maximizing the joint probability (1), for instance by using the principle of maximum entropy (*Jaynes*, 1957), subject to a set of constraints. In the generalization of Lienhard's approach, *Lienhard and Meyer* (1967) used the following constraints:

1) a conservation principle:

$$\sum_{i=1}^{\infty} N_i = N \tag{3}$$

2) a constraint on the moment of order β with respect to the origin, considering distance expressed in time units and based on the assumption of uniform velocity throughout the basin:

$$E\left[t^\beta\right] = \sum_{i=1}^{\infty} \frac{N_i}{N} \cdot t_i^\beta = k \qquad (k>0) \tag{4}$$

3) the proportionality between g_i and a given power of t_i:

$$g_i = c \cdot t_i^{\alpha-1} \tag{5}$$

with constant *c* and α.

The set $(\tilde{N}_1, \tilde{N}_2, \dots, \tilde{N}_i, \dots)$ of particles that allow the maximization of (1) determines a discrete distribution $P(t_{i-1} \le T \le t_{i-1} + \Delta t) = \tilde{N}_i / N$, that can be approximated with a probability density function f(t) according to

$$\frac{\tilde{N}i}{N} = \int_{t_{i-1}}^{t_{i-1}+\Delta t} f(t)\, dt \tag{6}$$

Letting Δt tend to zero, the *pdf* that maximizes (1) is a generalized Gamma distribution (*Lienhard and Meyer,* 1967)

$$f(t) = \left[\frac{\beta}{\Gamma\left(\alpha/\beta\right)} \cdot \left(\frac{\alpha}{\beta k}\right)^{\alpha/\beta}\right] \cdot t^{\alpha-1} \cdot exp\left[-\frac{\alpha t^{\beta}}{\beta k}\right] \qquad t \geq 0 \tag{7}$$

characterized by the three parameters α, β and k.

In the scheme introduced by *Lienhard* (1964) parameters are set as α=n+1 and β=2. For β=1 the distribution (7) is reduced to a two-parameter Gamma while for $\alpha=\beta$ (7) becomes a *Weibull* distribution:

$$f(t) = \frac{\alpha}{k} t^{\alpha-1}\, exp\left(-\frac{t^{\alpha}}{k}\right) \tag{8}$$

In the search for the most probable *pdf* of arrival times, constraints play the role of conditions that lead to incorporating some peculiar and clearly recognizable characteristics of the drainage-network system into the *pdf* expression. In fact, this is not the case of the approaches by *Lienhard* (1964) and *Lienhard and Meyer* (1967), as recognized by *Marani et al.* (1991) who dropped the dynamic constraint due to its arbitrary definition.

The main purpose of this paper is to show how a fractal framework, established with a set of properties of the objects, can allow us to impose some meaningful constraints in the maximization of a joint probability expression, leading to a result analogous to (7).

A fundamental point in this approach is that we consider uniquely the response due to the channel network. No hillslope or area effects are taken into account. This hypothesis is not too hard, if one considers the role of the geomorphological factor in the whole response of the basin (*e.g. Rinaldo et al.*, 1991). Moreover, it is to be considered that only the topological structure of the network is taken into account, which considers segments of equal length and is functional to the application of the rules of construction of fractal networks.

Characteristics and properties of the fractal objects used in this analysis are introduced in the next section.

3. A FRAMEWORK FOR THE DEFINITION OF FRACTAL NETWORKS

Fractal networks are built with the very simple rules of construction of self-similar objects, using a an *initiator*, which is a unit-length segment, and a *generator*, which is a tree-type combination of equal shorter segments (*Mandelbrot*, 1983, pp.72-73) whose length is η. Segments of the tree are of equal length are straight and form only square angles, so that the longest path in the generator tree (called *topological diameter* Δ_1) is always the initiator and is made up of $1/\eta$ segments.

After the first substitution of the initiator with the generator, each segment of the generator becomes an initiator and is substituted again (figure 1), in a recursive way (*e.g. Feder*, 1988, p. 16). After *m* multiplications, the number of segments is $M = N_1^m$, where N_1 is the number of segments of the generator, and the segment length is $\zeta_m=\eta^m$.

In the network obtained after *m* multiplications the topological diameter Δ_m is equal to (*Claps and Oliveto*, 1993)

$$\Delta_m = \Delta_1^m \tag{9}$$

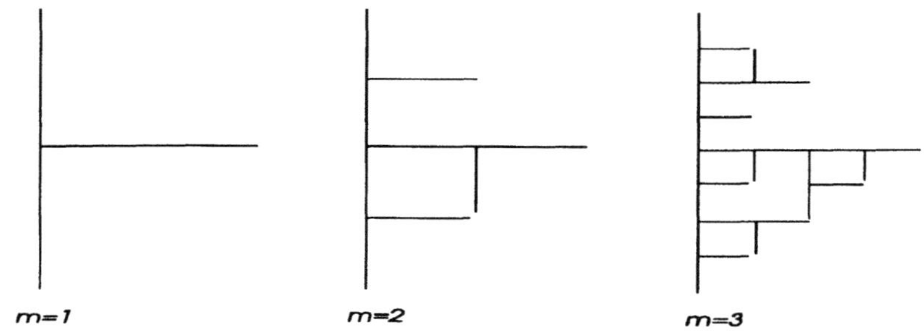

Fig. 1. *Generation of a fractal tree. Parameter m is the generation index. The structure with m=1 is the generator. The initiator is a simple segment for all the cases considered.*

The fractal dimension of these geometric sets is defined as (*e.g. Feder*, 1988, p.19)

$$D_t = -\frac{ln N_1}{ln \eta} \tag{9}$$

Since we are dealing only with straight segments, this fractal dimension is to be considered as due only to the branching process governing network growth (is a *topological* fractal dimension). Coefficient D_t can also be called the *network similarity dimension.*

3.1. determination of contributing areas for fractal networks

Mandelbrot (1983) first suggested that basin area should be proportional to the fractal measure of the total network length. This hypothesis was adopted by *Marani et al.* (1991) as a *connectivity conjecture* implying constant drainage density. In this paper this hypothesis is extended in considering the basin area proportional to the fractal measure of the length Z of the network. The fractal (invariant) measure Λ of a self-similar object is (*Mandelbrot*, 1983)

$$\Lambda = N_\varepsilon \varepsilon^D \tag{12}$$

where ε is the *ruler* length, N_ε is the number of rulers needed to cover the object and D is the fractal dimension. In the case discussed, the fractal measure Z_i of the complete topological length of a subnetwork is

$$Z_i = M_i \zeta^{D_t} \tag{13}$$

with M_i as the number of links of the subnetwork, ζ as the link length and D_t as the similarity dimension of the fractal network. The index *m* of generation is dropped here and in the following, without loss of generality.

To define contributing areas C_i *Gupta and Waymire* (1983) made reference to the schematic basin depicted in figure 2 and used cumulative areas A_i as follows:

$$C_i = \frac{dA_i}{dL_i} dL_i \tag{9}$$

Based on the previous hypothesis, to reformulate (9) for fractal networks the cumulative area A_i is substituted by the total length Z_i of the subnetwork and the distance L_i of the contributing area from the outlet is assumed as the length of the mainstream of the corresponding subnetwork.

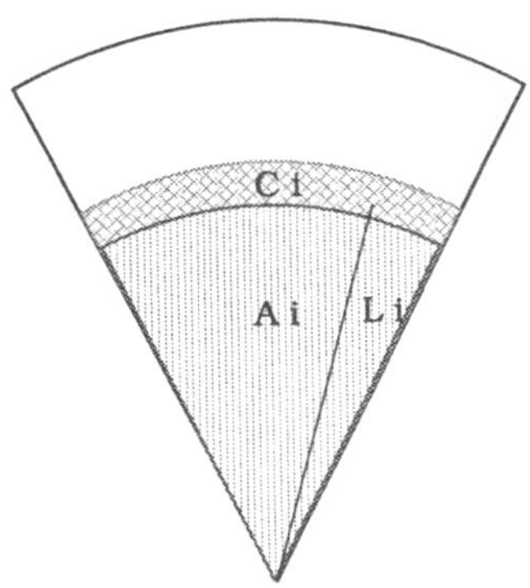

Fig. 2. Representation of a basin shaped as a sector of circle (from Gupta and Waymire, 1983).

One can thus write (9) as the derivative of Z_i with respect to the subnetwork mainstream length

$$C_i = \frac{dZ_i}{dL_i} dL_i \tag{14}$$

Since in fractal networks $L_i = \Delta_i \zeta$, with Δ_i as the subnetwork topological diameter, and given Δ_i as a discrete integer quantity, so that $d\Delta_i = 1$, $dL_i = \zeta$ and relation (14) becomes

$$C_i = \zeta^{D_t} \frac{dM_i}{dL_i} dL_i = \zeta^{D_t} \frac{dM_i}{d\Delta_i} \tag{15}$$

Claps and Oliveto (1993) found for fractal networks the following relation between total number of links and topological diameter:

$$M_i = \Delta_i^{D_t} \tag{16}$$

holding strictly for subnetworks obtainable with repeated substitutions of the generator (called *m-structures* in Claps and Oliveto, 1993) but resulting in a very good relation for all others subnetworks, as can be seen in figure 3. In relation (15) and in the following, the network topological diameter is indicated by Δ, with no mention to the (generic) index m of generation; thus Δ_i in (15) denotes the diameter of a subnetwork i as a part of a larger network.

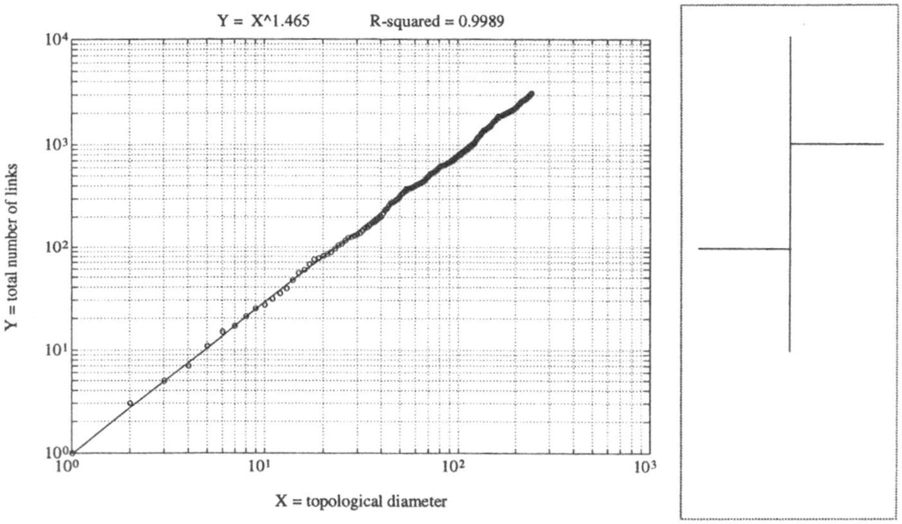

Fig. 3. *Relation between magnitude and topological diameter of all subnetworks of a fractal network generated from the elementary tree represented in the rectangle on the right. The fractal dimension of the tree is Dt=1.465.*

Deriving relation (16) one obtains

$$\frac{dM_i}{d\Delta_i} = D_t \cdot \Delta_i^{D_t - 1} \tag{17}$$

so that (15) can be rewritten as

$$C_i = \zeta^{D_t} \cdot D_t \Delta_i^{D_t - 1} \tag{18}$$

Using L_i (expressed as $\Delta_i\zeta$) for a meaningful comparison with the constraint (5), the condition imposed on (19), given constant velocity, C_i by the fractal nature of the network is

$$C_i = \zeta D_t L_i^{D_t - 1} \tag{19}$$

that leads to probabilities p_i (equivalent to g_i):

$$p_i = \frac{C_i}{Z} = \frac{D_t \cdot \zeta \cdot L_i^{D_t - 1}}{Z} \tag{20}$$

where the constant *c* of (5) is represented by the ratio

$$c = \frac{D_t \zeta v^{D_t - 1}}{Z} \tag{21}$$

with v as the velocity.

Relation (20) is shown in figure 4, as compared to the topological width function, representing the number of links at a given with reference to the same fractal tree used in figure 3, at the same distance from the outlet normalized by the total number of links.

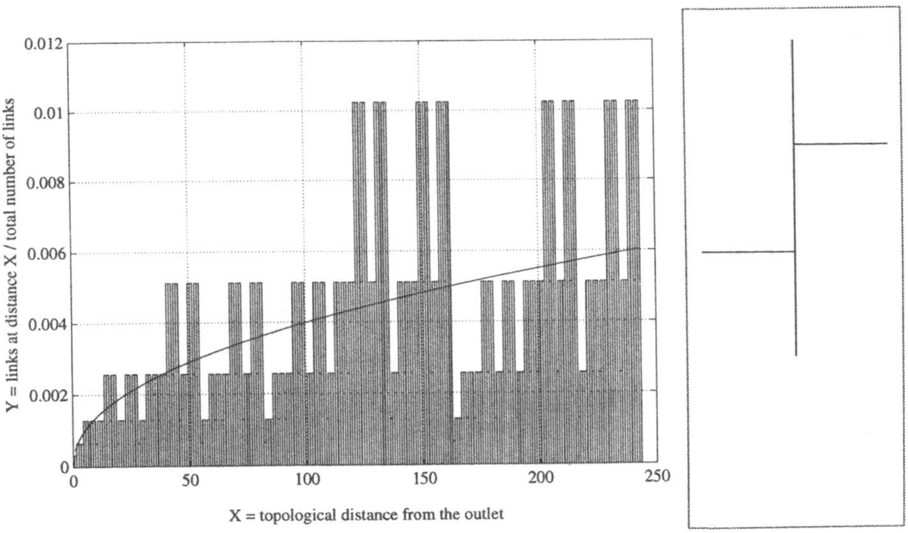

Fig. 4. *Topological width function of the fractal network generated from the elementary tree represented in the rectangle on the right. The curve depicts the approximation represented by relation (20).*

4. THE MOST PROBABLE HYDROLOGIC RESPONSE

The most probable hydrologic response of fractal networks can be found invoking the principle of maximum entropy (*Jaynes,* 1957), with some constraints. The constraints we can add are not arbitrary and introduce in the IUH formulation peculiar characteristics of fractal networks.

The first (obvious) constraint is the one ensuring that the sum of probabilities is unity:

$$\sum_{i=1}^{\Delta} p_i = \sum_{i=1}^{\Delta} \frac{n_i}{N} = 1 \tag{22}$$

The second one should take into account some peculiar (physical) aspects of the network such as, for instance, the total length. What is of interest in the expression of the constraint is how structural parameters of the network appear. The total length is considered as a constraint in the expression

$$\sum_{i=1}^{\Delta} \frac{n_i}{N} \cdot t_i^{D_t} \propto Z \tag{23}$$

In fact, expanding the above relation using the uniform velocity v one obtains

$$E\left[t_i^{D_t}\right] = E\left[\left(\frac{L_i}{v}\right)^{D_t}\right] = E\left[\zeta^{D_t}\left(\frac{\Delta_i}{v}\right)^{D_t}\right] = \sum_{i=1}^{\Delta} \frac{D_t \cdot \Delta_i^{D_t-1}}{M} \cdot \zeta^{D_t}\left(\frac{\Delta_i}{v}\right)^{D_t} =$$

$$= \sum_{i=1}^{\Delta} \frac{D_t \cdot \zeta^{D_t}}{Mv^{D_t}} \cdot \Delta_i^{D_t-1} \cdot \Delta_i^{D_t} \approx \frac{D_t \cdot \zeta^{D_t}}{Mv^{D_t}} \cdot \frac{\Delta^{2D_t}}{2D_t} = \tag{24}$$

$$= \frac{\zeta^{D_t}}{v^{D_t}} \cdot \frac{1}{2} \cdot \frac{\Delta^{2D_t}}{\Delta^{D_t}} = \frac{\zeta^{D_t}}{v^{D_t}} \cdot \frac{1}{2} \cdot \Delta^{D_t} = \frac{1}{2 \cdot v^{D_t}} \cdot L^{D_t}$$

and considering (12) and (16) and taking into account that $L_i = \Delta_i\ \zeta$, one can write:

$$E\left[t_i^{D_t}\right] = \frac{1}{2 \cdot v^{D_t}} \cdot Z \Rightarrow E\left[t_i^{D_t}\right] \propto Z \tag{25}$$

Considering Z as a constraint means that all information given by Z is incorporated into the expression of the most probable response, *i.e.* we will derive the most probable IUH consistent with the information given by Z.

The second member of (25) is equivalent to the constant k in *Lienhard and Meyer* (1967):

$$k = \frac{Z}{2 \cdot v^{D_t}} \tag{26}$$

The third constraint is represented by relation (20). We have now relations (20), (22) and (25) as constraints equivalent to these used by *Lienhard and Meyer* (1967), with $\alpha=\beta=D_t$ resulting by construction for fractal networks. Equivalent constraints and equivalent framework produce the same IUH form as *Lienhard and Meyer* (1967). Therefore, according to (8), the most probable IUH is

$$f(t) = \frac{D_t}{k}\, t^{D_t - 1}\, exp\left(-\frac{t^{D_t}}{k} \right) \tag{27}$$

that, substituting k, becomes

$$f(t) = D_t \frac{2v^{D_t}}{Z} t^{D_t - 1} exp\left(-\frac{2v^{D_t}}{Z} t^{D_t} \right) \tag{28}$$

4.1. Comparison with previous related works

Troutman and Karlinger (1985), using the topological framework developed by *Shreve* (1966,1967), derived the exact and asymptotic (for large n) IUH using different dynamic equations to characterize the residence time distribution of the water particles. For all dynamic equations, the expected IUH constrained to the knowledge of magnitude, is a Weibull function, depending only on n, on the average link length $\bar{l}$ and on the celerity c:

$$f(t) = \left(\frac{\lambda^2 t}{2n} \right) exp\left[-\frac{\lambda^2 t^2}{4n} \right] \tag{29}$$

with $\lambda = c/\bar{l}$. Parameters of expressions (29) and (28) of the Weibull distribution are related by:

$$\frac{4n}{\lambda^2} = \frac{Z}{2 \cdot v^{D_t}}; \qquad D_t = 2 \tag{30}$$

In the comparison it is also assumed $\zeta = \bar{l}$, $v = c$ and $M = 2n$.

Distribution (28) is plotted in figures 5 and 6 for D_t varying between 1 and 2, using $\zeta=1$ (so that Z=M) and v=1. In figure 5, Z is fixed at 200 and the curves obtained for different D_t are compared with that arising from (29), with $c=1$ and $\bar{l}=1$. We are using both c and v with the same meaning of parameters converting length to time. In figure 5 it can be seen that the Troutman and Karlinger IUH has a peak that corresponds to that of the case $D_t=1.6$ but a greater average travel time. Comparison of figures 5 and 6 shows the difference in the effects of variation of D_t with fixed M and fixed Δ. In the first case (figure 5) the increase of D_t, with Z fixed, implies a decrease of Δ, according to (16), and a corresponding decrease of the average travel time with an increased peak. On the other hand, if Δ is fixed (figure 6), the increase of D_t corresponds to a considerable expansion of the magnitude but to a relatively small increase of E[t]. Table 1 allows us to compare magnitude and topological diameters producing the IUH curves in both figures.

Looking at the mean travel times reported in table 1 it can be clearly recognized that in a fractal framework the key parameter for the determination of the value of E[t] is the topological diameter. Extending this result to real world networks, the main stream length (and of course the average velocity) turns out to be the controlling factor for E[t]. Thus the fractal framework can give an explanation to the empirical evidence of the role of area or magnitude in the hydrologic response, since both variables could be related to the mainstream length through relations of the type of (16).

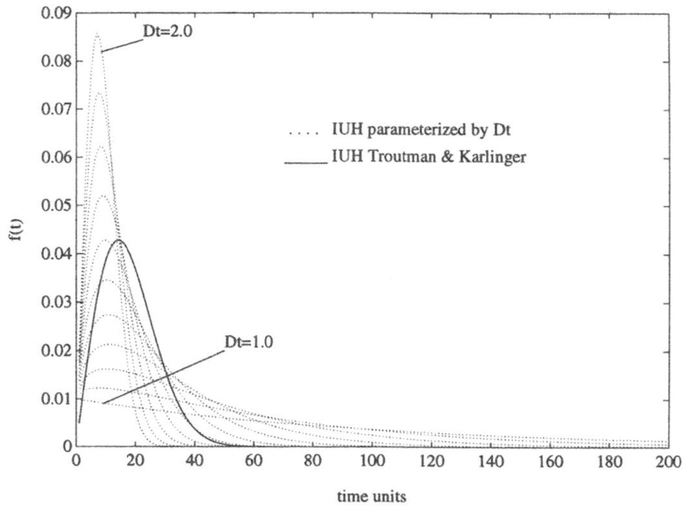

Fig. 5. *Forms of the IUH of fractal networks for fractal dimension D_t varying from 1.0 to 2.0, step 0.1. The segment length and the velocities are taken as unit values. Total length equals magnitude and both are fixed to Z=M=200. Due to relation (15), the topological diameter of the network decreases with increasing D_t.*

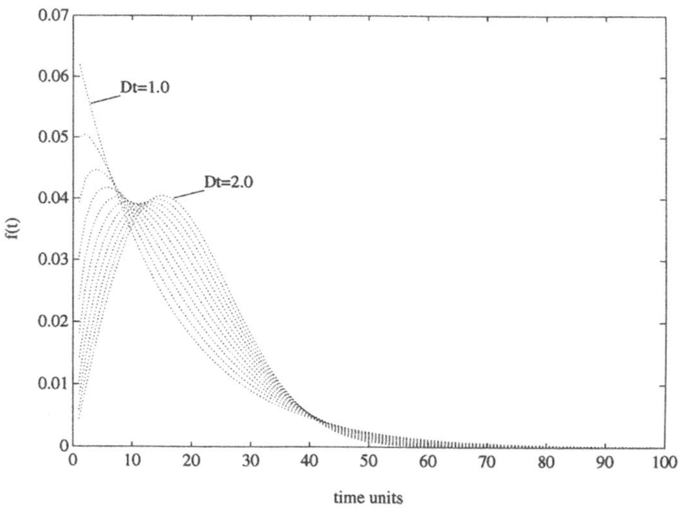

Fig. 6. *Forms of the IUH of fractal networks for fractal dimension D_t varying from 1.0 to 2.0, step 0.1. The segment length and the velocities are taken as unit values. The topological diameter is set as Δ=30. Due to relation (15), total length and magnitude increase with increasing D_t.*

4.2. Moments of the IUH of fractal networks

The first moment of distribution (8) is:

$$E[t] = k^{1/\alpha}\,\Gamma\left(1+\frac{1}{\alpha}\right) \tag{31}$$

and can be expressed as a function of parameters of the impulse response of fractal networks:

$$E[t] = \frac{1}{v}\left(\frac{Z}{2}\right)^{1/D_t}\Gamma\left(1+\frac{1}{D_t}\right) \tag{32}$$

and of the Troutman and Karlinger IUH parameters:

$$E[t] = \left(\frac{4n}{\lambda^2}\right)^{1/2}\Gamma\left(1+\frac{1}{2}\right) = \frac{\sqrt{\pi n}}{\lambda} \tag{33}$$

where α is fixed to 2. To compare these average travel times we have plotted (32) against D_t for different values of Z. Assuming $\zeta=1$, Z equals M and the comparison is meaningful. To provide a significant comparison between the mean travel times of the Troutman and Karlinger IUH and of the IUH of fractal networks we have plotted their ratio for different values of Z and with D_t varying from 1 to 2. figure 7 shows that increasing the magnitude this ratio tends to attain unity in correspondence of $D_t \approx 1.65$, which is compatible with the position of the curves represented in figure 5. It seems also evident that for diverging magnitude the fractal dimension that ensure equality of the first moments does not exceed 1.7.

D_t	1	1.1	1.2	1.3	1.4	1.5	1.6	1.7	1.8	1.9	2.0
$\Delta(M=200)$	200	124	83	59	44	34	27	23	19	16	14
E[t]	100.0	63.48	43.66	31.91	24.45	19.45	15.94	13.40	11.49	10.02	8.86
$M(\Delta=30)$	30	42	59	83	117	164	231	324	456	641	900
E[t]	15.00	15.42	15.84	16.26	16.67	17.06	17.44	17.80	18.15	18.48	18.80

Tab. 1. *Mean travel times E[t], with D_t varying, in the two cases of M and Δ fixed. Topological diameters Δ for fixed M and total number of links M for fixed Δ are also indicated.*

An additional comparison can be made with the first moment of the IUH derived by *Marani et al.* (1991) for the Peano network, whose characteristics are described in the mentioned paper. The above authors derive E[t] as

$$E[t] = \frac{3}{2}\frac{L(\Omega,\Omega)}{v} \tag{34}$$

that can be compared to the expression (33), considering that $L(\Omega,\Omega)$ is the length of the stream of order Ω in the a Peano network of order Ω. Given that, since *Marani et al.* (1991) assume for this plane-filling curve (D_t=2):

$$L(\Omega,\Omega) = R_B^{(\Omega-1)/2}\,\varepsilon(\Omega),$$

with $\varepsilon(\Omega)$ as the ruler, and given $n=R_B^{\Omega-1}$, from the Horton law of stream numbers, setting $\lambda=\varepsilon(\Omega)/v$ as assumed previously for fractal objects, makes (34) as

$$E[t] = \frac{3}{2}\frac{1}{\sqrt{\pi}}\frac{\sqrt{\pi n}}{\lambda} = 0.846\frac{\sqrt{\pi n}}{\lambda} \tag{35}$$

This allows us to use the curves reported in figure 7 for a comparison with this mean travel time, by making reference to the ordinate 0.846 of the diagram.

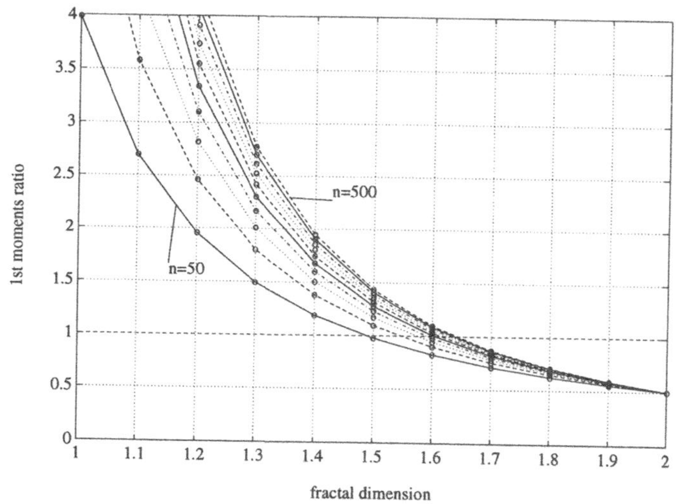

Fig. 7. *Ratio $E_F[t]$ / $E_{TK}[t]$ of the first moment of the travel time distribution for fractal networks ($E_F[t]$) and for the scheme used by Troutman and Karlinger ($E_{TK}[t]$). It is assumed $\zeta=1$ and $c=v=1$. Magnitude n varies between 50 and 500 with step 50.*

The second moment of the Weibull distribution (8) with respect to the origin is

$$M_2(t) = k^{2/\alpha}\,\Gamma\left(1+\frac{2}{D_t}\right) \tag{36}$$

leading, in the case of fractal network, to

$$M_2[t] = \frac{1}{v^2}\left(\frac{Z}{2}\right)^{2/D_t}\Gamma\left(1+\frac{2}{D_t}\right) \tag{37}$$

Computing values assumed by (37) for the same set of parameters used in table 1, the results reported in table 2 show that, when the topological diameter is fixed, there is a minimum of (37) for D_t=1.4÷1.5 while M_2 monotonically decreases with increasing D_t when the magnitude is fixed. It is worth remarking that values of D_t minimizing the second moment produce IUH with low peaks (figure 7). Moreover they in the range of topological fractal dimensions D_t of the branching estimated by *Claps and Oliveto* (1993) on some river basins in Southern Italy using the informational entropy of the network (*Fiorentino and Claps*, 1992).

D_t	1	1.1	1.2	1.3	1.4	1.5	1.6	1.7	1.8	1.9	2.0
$\Delta(M=200)$	200	124	83	59	44	34	27	23	19	16	14
$M_2[t]$	2000 0	7369	3242	1631	911	553	358	245	176	130	100
$M(\Delta=30)$	30	42	59	83	117	164	231	324	456	641	900
$M_2[t]$	450.0	434.5	426.5	423.3	423.2	425.3	428.7	433.2	438.4	444.0	450.0

Tab. 2. *Second moment $M_2[t]$ with respect to the origin, with D_t varying, in the two cases of M and Δ fixed.*

5. CONCLUSIONS

A Weibull distribution with clearly defined parameters is obtained as the travel time distribution of a mass injected instantaneously in a fractal network. The maximization of the joint probability defining the arrival time distribution is made under three constraints which incorporate properties of fractal networks. This constitutes a step forward with respect to previous works in this field (*e.g. Lienhard*, 1964; *Lienhard and Meyer*, 1967; *Marani et al.*, 1991) where the definition of constraints is either arbitrary or incomplete. Parameters driving the hydrologic response of fractal networks are the total network length and the fractal dimension, explicitly appearing in the arrival time distribution function.

The first moment of the distribution depends fundamentally on the network topological diameter. Translated in terms of real world rivers this result substantiates the significance of the mainstream length in the estimation methods to the basin lag time.

The influence of the fractal dimension is evident both in the peak discharge and in the second moment with respect to the origin, the latter presenting minima for D_t=1.4÷1.5. These values are of particular interest based on the results of an analysis of the fractal structure of river networks which was carried out by *Claps and Oliveto* (1993).

REFERENCES

Claps, P and G. Oliveto, Fractal structure, entropy and energy dissipation in river networks, *this Conference,* 1993.

Feder J. (1988) *Fractals,* Plenum Press, New York.

Fiorentino, M. e P. Claps, On what can be explained by the entropy of a channel network, in Singh e Fiorentino (Eds.) *Entropy and Energy Dissipation in Water Resources*, 139-154, Kluwer, Dordrecht, The Netherlands, 1992.

Fiorentino M., P.Claps e V.P.Singh, An entropy-based morphological analysis of river-basin networks , *Water Resour. Res.*,**29**(4), 1215-1224, 1993.

Gupta V. M. and E. Waymire. On the formulation of an analitycal approach to hydrologic response and similary at basin scale, *Journal of Hydrology*, **65**, 95-123, 1983.

Jaynes, E. T., Information theory and statistical mechanics, *I. Phys. Rev.,* 106, 620-630, 1957.

La Barbera, P. and R. Rosso, On the fractal dimension of stream networks, *Water Resour. Res.*, 25(4), 735-741, 1989.

Lienhard, J. H., A Statistical mechanical prediction of the dimensionless unit hydrograph. *J. Geophys. Res.*, Vol. 69, N.24, 5231-5238, 1964.

Lienhard, J. H. and P.L. Meyer, A physical basis for the generalized Gamma distribution, *Quart. Appl. Math.*, **25**, 330-334, 1967.

Mandelbrot, B.B., *The Fractal Geometry of Nature*, W.H. Freeman, New York, 1983.

Marani, A., R.Rigon and A. Rinaldo, A note on fractal channel networks, *Water Resour. Res.*, **27** (12), 3041-3049, 1991.

Rinaldo, A., Marani, A. e R.Rigon , Geomorphological dispersion, *Water Resour. Res.*, **27** (4), 513-525,1991.

Rinaldo, A., Rodriguez-Iturbe I., R.Rigon, R.L.Bras, E.Ijjasz-Vasquez and A.Marani, Minimum Energy and fractal structures of drainage networks, *Water Resour. Res.*, **28** (9), 2183-2195,1992.

Shreve R. L. Statistical law of stream numbers, *J. Geology*, vol.74, pp. 17-37, 1966.

Shreve R. L. Infinite topologically random channel network, *J. Geology*, vol.75, pp. 179-186, 1967.

Troutman B. M. and M. B. Karlinger. Unit hydrograph approximation assuming linear flow trough topologically random channel networks, *Water Resour. Res.*, 21(**5**), 743-754, 1985.

Acknowledgments

This work was supported by funds granted by the Italian Ministero della Ricerca Scientifica e Tecnologica, Progetto 40% "Processi Idrologici Fondamentali".

BAYESIAN UNIT HYDROGRAPHS

A. RAMACHANDRA RAO
School of Civil Engineering
Purdue University
W. Lafayette, IN 47907 USA

ABSTRACT: The Bayesian and Ridge Regression Methods of Unit Hydrograph estimation are investigated in this study. Both synthetic and observed rainfall-runoff data are used. The Bayesian Method is more general and gives non-oscillatory unit hydrographs with non-negative ordinates.

1. Introduction

The unit hydrograph was introduced by Sherman (1932a). It is a direct runoff hydrograph resulting from a unit of effective rainfall distributed uniformly over a basin during a specified time duration. Unit hydrographs were derived by Sherman (1932a) from observed hydrographs of isolated storms by first separating baseflow from total runoff to get direct runoff. The ordinates of resulting hydrographs are divided by the direct runoff volume to obtain the unit hydrograph.

Linearity and time invariance are the two fundamental principles of the unit hydrograph theory. Other assumptions are often made for practical convenience, but they are not essential from a fundamental and theoretical point of view (Chow, 1964; Dooge, 1959). Although the unit hydrograph method is widely used, the strictly linear time invariant relationship between rainfall and runoff does not exist. The hydraulic equations which characterize the surface flows are nonlinear. However, comparative studies have shown that the combined effect of the nonlinearities is usually small (Raudkivi, 1979). Consequently, unit hydrographs are still being widely used.

Various investigators have linked unit hydrograph characteristics with geomorphologic characteristics of basins (Snyder (1938), Edson (1951), Taylor and Schwartz (1952), Rodrigues-Iturbe and Valdes, (1979); Valdes et al., (1979); Gupta et al., (1980); Rodrigues-Iturbe et al., (1982)). Some of these efforts are directed towards estimating Geomorphologic unit hydrographs. Others are interested in the development of Topological Unit Hydrographs which are based on the theory of topologically random networks (Gray, 1961; Troutman and Karlinger, 1985; Karlinger and Troutman, 1985).

Different methods are developed to estimate unit hydrographs. Snyder (1955) used the method of least squares to estimate unit hydrographs. Eagleson et al. (1966) obtained an approximate solution of the Wiener-Hopf equations through linear programming and matrix inversion algorithm to compute optimum realizable unit hydrographs. Deininger (1969) used linear programming techniques based on minimum absolute deviations and on min-max criteria. Singh (1976) compared linear programming and least squares algorithms to compute unit hydrographs. Mays and Coles (1980) applied linear programming to estimate

V. P. Singh and B. Kumar (eds.), Surface-Water Hydrology, 205–232.

unit hydrographs. They emphasized that the linear programming model provides a solution to a mathematical problem and hence it is important to define appropriate constraint coefficients (effective rainfall) and runoff data. The objective function does not give special weight to the region around the unit hydrograph peak, which is an important property of the unit hydrograph. Mays and Taur (1982) used two nonlinear programming techniques, the large-scale generalized reduced gradient techniques and the generalized reduced gradient with three different objective functions to compute unit hydrographs. Rao and Delleur (1971), compared the performance of Fourier, Laplace and Z-transform methods in estimating the unit hydrographs. Fourier transform methods were also used by Sarma et al. (1973) and Blank et al. (1971). The oscillating kernels are a problem with many of these methods. Oscillations arise in unit hydrographs because of small sampling intervals and also due to the noise in the data, especially in rainfall. Several methods have been developed to estimate oscillation-free unit hydrographs. The Ridge regression method is one such method which gives unit hydrographs which do not oscillate. The Bayesian method is another.

Two of the recent methods of estimating instantaneous unit hydrographs, such as the Bayesian and Ridge Regression methods are investigated in this study. The objective of the study is to select a method that would give instantaneous unit hydrographs which are not oscillatory with nonnegative ordinates. The Bayesian procedure is claimed to give estimates with such characteristics. It is discussed at length because it is the most general method and also because it has not been used in hydrology.

2. Data Used In This Study

Two types of data are used in this study. Synthetic data are used to check the computations. Real data are used to examine the variability of the instantaneous unit hydrograph.

2.1. SYNTHETIC DATA

To test the different instantaneous unit hydrograph or kernel function estimation methods, three different sets of input-output functions are used. For these three sets, the theoretical instantaneous unit hydrographs are known. The synthetic data are obtained from Dooge (1965) and are later modified by Blank et al. (1971) and by Rao and Delleur (1971). The kernel functions must satisfy the requirements that the area under them should be equal to unity and the first moment of the kernel function about the origin should be equal to the time lag for them to be used as instantaneous unit hydrographs (Blank et al., 1971).

The first set of data is given by:

$$x(t) = 100\, t\, e^{-0.4593t}$$
$$y(t) = 100\, t^2\, e^{-0.75t}$$
$$h(t) = [2 - 1.1628t + 0.0845t^2]\, e^{-0.75t}$$

where x(t), y(t), and h(t) are the input, output, and kernel functions, respectively, and t is the time. The second set is given by:

$$x(t) = 8\left[t + \frac{t^2}{2}\right] e^{-t/2}$$
$$y(t) = \frac{1}{2}\left[\frac{t^5}{5!} + \frac{t^6}{6!}\right] e^{-t/2}$$
$$h(t) = \frac{t^3}{96}\, e^{-t/2}$$

The third set is given by:

$$x(t) = t(1-t)\, e^{1-t}\, (e^{8}-41) \qquad 0 \le t \le 1$$

$$y(t) = e^{9-t}\, t^{3}\, (2-t)/12 + e^{1-t}\, (t^{2}+3t+4) - e(4-t) \qquad 0 \le t \le 1$$

$$y(t) = e^{9-1}\, (2t - 1)/12 + (11-3t) - e(4-t) \qquad 1 \le t \le 8$$

$$y(t) = (e^{9-t}/12)\, [12t^{2}-130t+335-(t-8)^{2}(152-14t-t^{2})] + (11-3t) \qquad 8 \le t \le 9$$

$$h(t) = t\, (e^{8-t}-1) \,/\, (e^{8}-41) \qquad 0 \le t \le 8$$

Data sets containing input-output function values are generated for each of the examples corresponding to a given time interval Δt. These data sets are used to generate the instantaneous unit hydrographs which are compared to the theoretical kernel functions.

These three data sets are also used to study the effect of noise on the methods to estimate instantaneous unit hydrographs. Noise generated from a standard normal distribution is added to both the theoretical input and output functions. Then the kernel function is computed from these functions by using the different methods.

2.2. REAL DATA

Data from ten watersheds in Indiana are used in this study. The basin areas range from 7 to 248 mi^2. The data consists of effective rainfall and direct runoff measured at half-hour intervals. A description of each of these watersheds and the data used are given in Tables 1 and 2.

2.3. BASEFLOW AND EFFECTIVE RAINFALL DETERMINATION

The method used for baseflow separation and effective rainfall determination is given in Blank and Delleur (1968). Direct runoff is separated from base flow by drawing two straight lines. The first extends from the point of rise of the hydrograph to a point located below the peak of the hydrograph. The second line continues to the recession limb of the hydrograph to a point determined by an equation for an unconfined aquifer. The direct runoff is obtained by subtracting the baseflow from the total runoff hydrograph. Excess precipitation is then obtained by multiplying the ordinates of the total rainfall hyetograph by the ratio of the excess precipitation during the storm to the total rainfall.

Table 1 Watershed Characteristics

Watershed Number	Watershed Name and Abbreviation	Area (mi^2)
2	Bear creek near Trevlac (BEAR)	7
6	Mississinewa river near Ridgeville (MISS)	130
9	Sand creek near Brewersville (SAND)	156
14	Busseron creek near Carlisle (BUSS)	228
15	Laughery creek near Farmers Retreat (LAUG)	248
19	Brush creek near Nebraska (BRUS)	11.7
24	Buck creek near Muncie (BUCK)	36.7
35	Salamonie river at Portland (SALA)	86
37	Bice ditch near South Marion (BICE)	22.6
40	Carpenter creek near Egypt (CARP)	48.1

Table 2 Data used in the Study

Watershed name	No.	Area (mi²)	Storm no.	Date	Δt (hrs)	No. of observations: Rainfall	No. of observations: Runoff
Bear creek near Trevlac (BEAR)	2	7	7	05/29/55	0.5	6	51
			6	04/13/55	0.5	8	54
			9	06/21/56	0.25	10	41
Mississinewa river near Ridgeville (MISS)	6	130	6	05/17/51	0.5	7	82
			18	05/06/53	0.5	9	85
			20	05/31/54	0.5	10	80
Sand creek near Brewersville (SAND)	9	156	2	05/22/53	0.5	8	50
			3	05/28/54	0.5	5	69
			12	06/17/67	0.5	5	57
			15	05/22/58	0.5	12	119
Busseron creek near Carlisle (BUSS)	14	228	7	06/11/55	0.5	26	153
			21	05/22/58	0.5	12	112
			5	05/05/54	0.5	14	143
			8	06/08/55	0.5	10	66
Laughery creek near Farmers Retreat (LAUG)	15	248	1	04/30/50	0.5	6	60
			4	05/22/53	0.5	8	58
			6	06/17/53	0.25	6	97
Brush creek near Nebraska (BRUS)	19	11.7	5	04/03/56	0.5	7	35
			13	06/13/58	0.5	8	38
			12	05/02/58	0.5	10	39
			24	04/19/64	0.25	9	59
Buck creek near Muncie (BUCK)	24	36.7	12	06/25/58	0.5	16	79
			22	04/02/64	0.5	18	87
			9	06/11/57	0.5	10	61
Salamonie river at Portland (SALA)	35	86	2	04/05/65	0.5	15	118
			9	04/18/63	0.5	13	114
			1	04/10/65	0.5	30	140
			13	05/25/62	0.5	9	78
Bice ditch near South Marion (BICE)	37	22.6	3	06/13/61	0.5	7	113
			14	06/18/57	0.5	8	109
Carpenter creek near Egypt (CARP)	40	48.1	14	05/30/59	0.5	5	67
			27	05/11/56	0.5	6	98
			1	06/08/65	0.5	12	126
			4	06/07/64	0.5	9	119

3. Unit Hydrograph Estimation Methods

For a linear and time-invariant system, the relationship between the output and input of a system is given by the convolution integral as in eq. 1,

$$y(t) = \int_0^t x(t-\tau)\ h(\tau)\ d\tau \tag{1}$$

where y(t) is the output (runoff), x(t) is the input (effective rainfall), and h(t) is the instantaneous unit hydrograph at time t. h(t) is also known as the impulse response function or kernel function.

The discrete form of eq. (1) is given by eq. 2 where y_i, x_i, and h_i are the output, input and instantaneous unit hydrograph, respectively at the i-th time interval.

$$y_i = \sum_{j=1}^{i} x_{i+j+1}\ h_j \qquad i = 1,2,\ ....,\ n \tag{2}$$

Figure 1 shows the inputs and outputs of the system. Δt is the time interval for both inputs and outputs. NQ is the number of data points of the output (direct runoff) and NP is the number of data points of the input (effective rainfall). The time base of the hydrograph is t_{base} = (NQ-1) Δt and the duration of the effective rainfall is t_{rain} = (NR-1)Δt, and in general $t_{base} > t_{rain}$.

Bruen and Dooge (1984) noted that if the transformation of effective rainfall to direct runoff was linear and time-invariant, and if the measured inputs and outputs were free of errors, then the estimation of the impulse response would be trivial. However, in practice, errors in data collection and processing are unavoidable, and in many cases, quite large. Consequently, there will be oscillations in unit hydrographs. The theory of bayesian and ridge regression methods, which are developed to estimate unit hydrographs which do not oscillate, is discussed below.

3.1. BAYESIAN UNIT HYDROGRAPH ESTIMATES

The Bayesian method to estimate the unit hydrograph is developed by Bhargava et al. (1987) and by Bhargava (1986). The system equation is assumed to be as in eq. 3, where **e** is normally distributed with mean zero and covariance matrix $\sigma^2\mathbf{I}$.

$$\boldsymbol{y} = \boldsymbol{Xh} + \boldsymbol{e} \tag{3}$$

The conditional probability density of **y** given **X** and **h** is given in eq. 6, where σ is the noise variance.

$$f(y|\boldsymbol{X},\boldsymbol{h}) = \frac{1}{(2\pi\sigma)^{N/2}} \exp\left[-\frac{1}{2\alpha} \| \boldsymbol{y}-\boldsymbol{Xh} \|^2\right] \tag{4}$$

The prior density of **h** is assumed to be as in eq. 5,

$$f(h) = \frac{1}{(2\pi\xi)^{N/2}\ |\det(\boldsymbol{P}^T\boldsymbol{P})|^{1/2}} \exp\left[-\frac{1}{2\xi}\|\boldsymbol{Ph}\|^2\right] \tag{5}$$

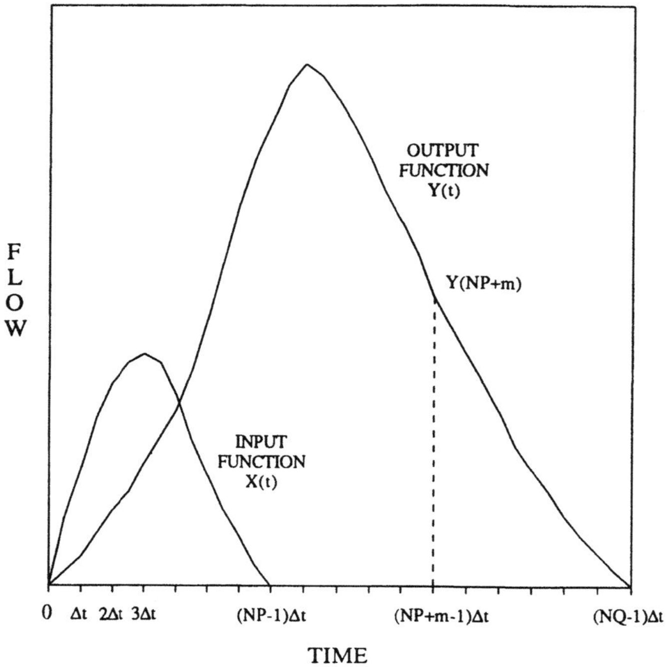

Figure 1. Input and Output Functions

where ξ is a positive constant and **P** is either the identity matrix **I** or the Laplacian matrix **C**, which is given by eq (6)

$$\boldsymbol{C} = \begin{bmatrix} 1 & & & & \\ -2 & 1 & & & \\ 1 & -2 & 1 & & \\ & 1 & -2 & & \\ & & 1 & 1 & \\ & & & -2 & 1 \\ & & & 1 & -2 \\ & & & & 1 \end{bmatrix} \tag{6}$$

Assuming that the prior densities of σ and ξ are flat, the log likelihood function is written as in eq (7), where K are terms not involving h, σ and ξ

$$-\ln f(\boldsymbol{h},\sigma|X,y) = \frac{\|y-X\boldsymbol{h}\|^2}{2\sigma} + \frac{N}{2}\ln\sigma + \frac{\|P\boldsymbol{h}\|^2}{2\xi} + \frac{N}{2}\ln\xi + K \tag{7}$$

Minimizing the log likelihood function with respect to h, σ and ξ gives eqs (8) through (10)

$$\hat{\boldsymbol{h}} = \left[\frac{\boldsymbol{X}^T\boldsymbol{X}}{\sigma} + \frac{\boldsymbol{P}^T\boldsymbol{P}}{\xi}\right]^{1}\left[\frac{\boldsymbol{X}^T\boldsymbol{y}}{\sigma}\right] \tag{8}$$

$$\hat{\sigma} = \frac{1}{N}\|y - x\hat{\boldsymbol{h}}\|^2 \tag{9}$$

$$\xi = \frac{1}{N}\|\boldsymbol{P}\hat{\boldsymbol{h}}\|^2 \tag{10}$$

Equations (8) through (10) can be rewritten as eqs (11) and (12)

$$\hat{\boldsymbol{h}} = \left[\boldsymbol{X}^T\boldsymbol{X} + \frac{\sigma}{\xi}\boldsymbol{P}^T\boldsymbol{P}\right]^{1}\boldsymbol{X}^T\boldsymbol{y} \tag{11}$$

$$\frac{\hat{\sigma}}{\xi} = \frac{\|y-X\hat{\boldsymbol{h}}\|^2}{\|\boldsymbol{P}\hat{\boldsymbol{h}}\|^2} \tag{12}$$

By letting $\hat{k} = \hat{\sigma}/\hat{\xi}$, eq. (13) is obtained from eqs. (11) and (12).

$$\hat{k} = \frac{\|y - X[X^TX + \hat{k}P^TP]^{-1} X^Ty\|^2}{\|P[X^TX + \hat{k}P^TP]^{-1} X^Ty\|^2} \tag{13}$$

Equivalently,

$$\hat{k} - \frac{\|y - X[X^TX + \hat{k}P^TP]^{-1} X^Ty\|^2}{\|P[X^TX + \hat{k}P^TP]^{-1} X^Ty\|^2} = 0 \tag{14}$$

Equation (14) is solved for $\hat{k}$ which is then substituted in eq. (11) to obtain the estimated response $\hat{\boldsymbol{h}}$.

3.2. RIDGE REGRESSION

In this method the discrete form of eq. (2) is written in matrix form as eq. (15).

$$y = Xh \tag{15}$$

In eq. 15, **y** is a (N x 1) vector, **X** is a (N x M) matrix, and **h** is a (M x 1) vector. The problem here is to solve for **h** given the vector **y** and matrix **X**. Equation (15) is the same as eq. (3) with **e** = **0**.

$$y = \begin{bmatrix} y_1 \\ y_2 \\ \cdot \\ y_p \\ y_{p+1} \\ y_{p+2} \\ \cdot \\ y_{n-1} \\ y_n \end{bmatrix} \quad (16) \qquad X = \begin{bmatrix} x_1 & \cdot & \cdot & \cdot & \cdot & \cdot \\ x_2 & x_1 & \cdot & \cdot & \cdot & \cdot \\ \cdot & \cdot & \cdot & \cdot & \cdot & \cdot \\ x_p & x_{p-1} & x_{p-2} & \cdot & \cdot & \cdot \\ \cdot & x_p & x_{p-1} & \cdot & \cdot & \cdot \\ \cdot & \cdot & x_p & \cdot & \cdot & \cdot \\ \cdot & \cdot & \cdot & \cdot & \cdot & \cdot \\ \cdot & \cdot & \cdot & \cdot & x_p & x_{p-1} \\ \cdot & \cdot & \cdot & \cdot & \cdot & x_p \end{bmatrix} \quad (17) \qquad h = \begin{bmatrix} h_1 \\ h_2 \\ \cdot \\ \cdot \\ h_{m-1} \\ h_m \end{bmatrix} \quad (18)$$

By premultiplying eq. (15) with $\mathbf{X}^T$, eq. (19) is obtained.

$$X^Ty = X^TXh \tag{19}$$

The least squares estimator $\hat{\boldsymbol{h}}$ is then given as

$$\hat{h} = [X^T X]^{-1} X^T y \tag{20}$$

which gives the minimum sum of squares of error e^2, where $e = y - X\hat{h}$. $\hat{h}$ is an unbiased estimate of **h**.

The $X^T X$ matrix is a symmetric Toeplitz matrix and has the form of a correlation matrix.

$$X^T X = \begin{bmatrix} r_1 & r_2 & \cdot & \cdot & \cdot & r_m \\ r_2 & r_1 & \cdot & \cdot & \cdot & r_{m-1} \\ \cdot & \cdot & \cdot & \cdot & \cdot & \cdot \\ \cdot & \cdot & \cdot & \cdot & \cdot & \cdot \\ r_{m-1} & r_{m-2} & \cdot & \cdot & r_1 & r_2 \\ r_m & r_{m-1} & \cdot & \cdot & \cdot & r_1 \end{bmatrix} \tag{21}$$

The $X^T X$ matrix determines the stability of the solution to eq. (20) (Hoerl and Kennard, 1970a; Bruen and Dooge, 1984; Marquardt, 1970). The least squares estimate in eq. (20) is a good one if $X^T X$ is nearly a unit matrix. However, if $X^T X$ is not nearly a unit matrix, the least squares estimates are sensitive to errors (Hoerl and Kennard, 1970a). In other words, if $X^T X$ is nonorthogonal, the least squares method will not produce good results. The least squares solution may yield coefficients whose absolute values are too large and whose signs actually reverse with small changes in the data.

To overcome this problem, Hoerl (1962) suggested the ridge regression method in which the estimate is obtained by eq. (22) rather than eq. (20).

$$X^T y = [X^T X + kI] \, h \tag{22}$$

which yields

$$\hat{h} = [X^T X + kI]^{-1} X^T y \tag{23}$$

where I is the identity matrix.

This method is called *Ridge Regression* or *Smoothed Least Squares Method.* In eq. (23), k is a parameter chosen by the user where $0 \le k \le 1$. If k = 0, the least squares estimate given by eq. (20) is obtained.

The estimate based on the matrix $[X^T X + kI]$ rather than on $X^T X$ is used to overcome the difficulties associated with the least squares estimate (Hoerl and Kennard, 1970a). The former system acts more like an orthogonal system. The ridge estimator is biased and the bias increases with k. The matrix $[X^T X + kI]$ is also a symmetric Toeplitz matrix.

A computer program was given by Bruen and Dooge (1984) to estimate a pulse response using the smoothed least squares (ridge regression) method. The program takes advantage of the structure of the coefficient matrix of equations, $[X^T X + kI]$ which is a symmetric Toeplitz matrix using an efficient algorithm developed by Farden (1976).

The Ridge Regression method is a special case of the Bayesian approach. Equation (11) reduces to eq. (23) when $\mathbf{P}^T\mathbf{P}$ is an Identity matrix. The k parameter in eq. (23) is analogous to the Bayesian parameter $\hat{k} = \hat{\sigma}/\xi$. Furthermore, when the noise is zero, the term **kI** in eq. (23) and σ in eq. (11) drop out, resulting in the ordinary least squares estimate given by eq. (20).

3.3. DIRECT METHOD

The direct method is a numerical solution of the convolution integral eq. (1). The following equations are based on the definitions in Fig. 1. The convolution integral when rainfall is continuing (m < NP) is given by eq. 24.

$$y_{m+1} = (\Delta t/2) \sum_{i=2}^{m+1} [x_i\, h_{m-i+2} + x_{i-1}\, h_{m-i+3}] \; ; \quad m = 2,3, \ldots, NP-1. \tag{24}$$

After the end of rainfall, the convolution integral is written as eq. 25.

$$y_{NP+m+1} = (\Delta t/2) \sum_{i=2}^{NP} [x_i\, h_{NP+m-i+2} + x_{i-1}\, h_{NP+m-i+3}] \; ; \quad m = 0,1, \ldots, NQ-NP-1 \,. \tag{25}$$

Equations (24) and (25) can be written in matrix form as

$$\boldsymbol{y} = \boldsymbol{X}\boldsymbol{h} \tag{26}$$

where

$$y = \frac{2}{\Delta t}\begin{bmatrix} y_2 \\ y_3 \\ \cdot \\ y_{NP} \\ y_{NP+1} \\ \cdot \\ y_{NP+k} \\ \cdot \\ y_{NQ+1} \end{bmatrix} \; (27) \quad \boldsymbol{X} = \begin{bmatrix} x_2 & \cdot & \cdot & \cdot & \cdot & \cdot & \cdot & \cdot & \cdot \\ x_3 & 2x_2 & \cdot & \cdot & \cdot & \cdot & \cdot & \cdot & \cdot \\ \cdot & \cdot & \cdot & \cdot & \cdot & \cdot & \cdot & \cdot & \cdot \\ x_{NP} & 2x_{NP-1} & \cdot & 2x_2 & \cdot & \cdot & \cdot & \cdot & \cdot \\ \cdot & x_{NP} & 2x_{NP-1} & \cdot & 2x_2 & \cdot & \cdot & \cdot & \cdot \\ \cdot & \cdot & \cdot & \cdot & \cdot & \cdot & \cdot & \cdot & \cdot \\ \cdot & \cdot & \cdot & x_{NP} & \cdot & \cdot & 2x_2 & \cdot & \cdot \\ \cdot & \cdot & \cdot & \cdot & \cdot & \cdot & \cdot & \cdot & \cdot \\ \cdot & \cdot & \cdot & \cdot & \cdot & x_{NP} & 2x_{NP-1} & \cdot & 2x_2 \end{bmatrix} \; (28) \quad \boldsymbol{h} = \begin{bmatrix} h_1 \\ h_2 \\ \cdot \\ h_{NP-1} \\ h_{NP} \\ \cdot \\ h_{NP+k-1} \\ \cdot \\ h_{NQ} \end{bmatrix} \; (29)$$

The **X** matrix (eq. 28) is a lower triangular matrix. It is a nonsingular matrix since all the diagonal elements are positive. The system in eq. (26) is solved for the kernel function vector **h**. As the number of equations is larger than the number of unknowns, this is an overdetermined system.

4. Discussion of Results

The Bayesian, ridge regression and direct methods are tested by using synthetic and real data. The results from the different methods are separately discussed below.

4.1. BAYESIAN METHOD

4.1.1. Bayesian Method With Synthetic Data. Using the three data sets described above, several data sets of different lengths containing the input, output and kernel functions are generated by varying the Δt value. Noise generated from a standard normal distribution with zero mean is introduced to the three data sets. The noise variance is assumed to be a percentage of input and output variance. The characteristics of the data used in this study are given in Tables 3 and 4. The data are normalized such that the areas under the input and output curves are equal to unity.

The kernel function is computed by varying the value of the parameter k from 0 to 1 with an increment of 0.1. The value of k corresponding to the Bayesian approach is found by solving eq. (13). Two different kinds of matrices are used in this equation, the identity (I) and the Laplacian (C) matrix, defined by eq. (6). The computed kernel is used to compute the output which is compared to the theoretical output. σ is the white noise variance which is calculated by defining a reduced model from eq. (30) to eq. (31) (Bhargava, 1987).

$$y = Xh + e \tag{30}$$

$$y = X^{i}h^{i} + e \tag{31}$$

where $\mathbf{X}^i$ is a N x (N-1) matrix obtained by deleting the i-th column of $\mathbf{X}$, and $\mathbf{h}^i$ is an (N-1) vector obtained by deleting the i-th component of $\mathbf{h}$. The least squares estimate of h for the reduced model is

$$\hat{h}^{i} = [X^{iT} X^{i}]\, X^{iT} y \tag{32}$$

Table 3 Bayesian Approach with Synthetic Data (without Noise)

Name	Records	Δt	IUH area	t_{peak}	Q_{peak}
Example 1A	30	1.0	2.2177	3	0.200
Example 1B	50	0.6	2.7990	2.4	0.121
Example 1C	75	0.4	3.5887	2.8	0.081
Example 1C	100	0.3	4.4001	2.7	0.061
Example 2A	40	1.0	1.0001	11	0.081
Example 2B	50	0.8	1.2500	11.2	0.065
Example 2C	80	0.5	2.0000	11.5	0.041
Example 2D	100	0.4	2.5000	10.4	0.032
Example 3A	90	0.1	9.9915	1.5	0.036
Example 3B	180	0.05	19.9958	1.25	0.018

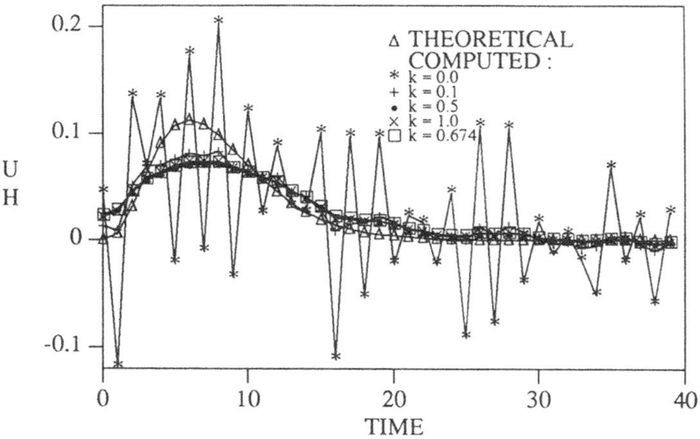

Figure 2 Kernel Function by Bayesian Method. Example 2 with noise (I matrix).

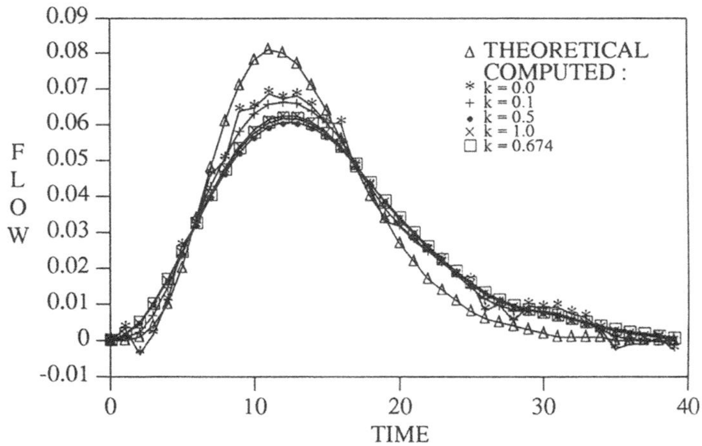

Figure 3. Flow by Bayesian Method. Example 2 with noise (I matrix).

Table 4 Bayesian Method with Synthetic Data (with Noise)

Name	Records	Noise (%)	Δt	IUH area
Example 1A	30	10	1 0	2 2177
Example 1B	30	50	1 0	2 2177
Example 1C	30	100	1 0	2 2177
Example 2A	40	10	1 0	1 0001
Example 2B	40	50	1 0	1 0001
Example 2C	40	100	1 0	1 0001
Example 3A	90	10	0 1	9 9915
Example 3B	90	50	0 1	9 9915
Example 3C	90	100	0 1	9 9915

and the residual sum of squares is

$$RSS^i = \|y - X^i \hat{h}^i\|^2 \tag{33}$$

Then the estimator of σ is given by eq (34)

$$\hat{\sigma} = \frac{1}{N} \sum_{i=1}^{N} RSS^i \tag{34}$$

The Mallow's statistic C_L is an approximation to the minimum mean square prediction error and is defined as

$$C_L = \frac{RSS}{\sigma} = N + 2 \text{ trace } (XL_k) \tag{35}$$

where $\mathbf{L}_k$ is either the matrix $[\mathbf{X}^T\mathbf{X} + k\mathbf{I}]^{1}$, or $[\mathbf{X}^T\mathbf{X} + k\mathbf{C}^T\mathbf{C}]^{1}$ Mallow's statistic is used to select the value of k The k value which gives the lowest value of C_L is accepted as the best ERR is the squared difference between the theoretical and the computed kernels and is given by

$$ERR = \|h - \hat{h}\|^2 \tag{36}$$

RSS (Residual Sum of Squares) is the squared difference between the theoretical and the computed output and is given by

$$RSS = \|y - X\hat{h}\|^2 \tag{37}$$

The results from the Bayesian method are shown in Tables 5 and 6 for example 2A with noise Apart from C_L, ERR and RSS values are also used to select the optimal k The best Bayesian estimate, if any, is given in the last row of each table In general, the RSS, σ, C_L and ERR values increase with the increase in k

Figures 2 and 3 show the result of example 2 with noise using 2 different matrices Much larger oscillations in kernels are present in the data with noise Computed output underestimates the observed peak output From these figures it can be seen that the Laplacian matrix C gives better estimates of the IUH than the Identity matrix I The Laplacian matrix produces kernels and unit hydrographs which chang every little with changes in k this is an important characteristic of this method because it shows that the k value need not be estimated exactly

4 1 2 Bayesian Method With Real Data Twenty storms different from those used in the direct method and ridge regression are analyzed using the Bayesian approach The characteristics of the storms are presented in Table 2 For each storm, both the effective rainfall (input) and direct runoff (output) are normalized by dividing each observation point by the input or output volume such that the volume of each becomes unity

The kernel function is obtained by varying the value of the parameter k from 0 to 1 with an increment of 0 1 The best value of k is computed by solving eq 13 Two different kinds of matrices are used in this equation, the identity (I) and the Laplacian (C) matrix, defined by eq 6 The kernel is convolved with the original input to calculate the output which is compared to the theoretical output Typical results from the Bayesian approach are presented in Tables 7 and 8

Table 5 Results from Bayesian method for example 2A with noise (I Matrix)

k	RSS	σ	C_L	ERR	t_{peak}	Q_{peak}
0 0	0 00148	0 00224	-35 35	0 19306	11	0 06860
0 1	0 00170	0 00204	-31 65	0 00632	12	0 06632
0 2	0 00204	0 00240	-28 10	0 00665	12	0 06496
0 3	0 00231	0 00268	-24 56	0 00726	12	0 06394
0 4	0 00252	0 00290	-21 02	0 00779	12	0 06316
0 5	0 00270	0 00309	-17 49	0 00823	12	0 06254
0 6	0 00285	0 00324	-13 96	0 00859	12	0 06204
0 7	0 00298	0 00338	-10 42	0 00890	12	0 06162
0 8	0 00309	0 00349	-6 89	0 00916	12	0 06126
0 9	0 00319	0 00359	-3 36	0 00938	12	0 06096
1 0	0 00328	0 00368	0 17	0 00958	12	0 06069
0 674	0 00295	0 00334	-11 34	0 00882	12	0 06172

The best Bayesian estimate, if any, is given in the last row of each table In general, the RSS, σ, and ERR values increase with increase in k For the **I** matrix, C_L values increase with increase of k, but for the **C** matrix, C_L values increase to some value of k, and then drops significantly and finally increase again

Figures 4 and 5 show the result for BEAR6 with 2 different matrices In general, k = 0 produces oscillatory kernels All kernels produce good reproductions of the observed output function Both identity and Laplacian matrices work very well and there is very little difference between the results obtained by them The kernel functions do not change very much with different k values

4 2 RIDGE REGRESSION METHOD

4 2 1 Ridge Regression Method With Synthetic Data The three data sets described earlier are used to generate three data sets containing the input, output and kernel The first two examples are obtained by using Δt = 1, while the third example by using Δt = 0 1 Another three data sets with the same time intervals are generated by introducing noise to both the input and the output The noise is generated from a standard

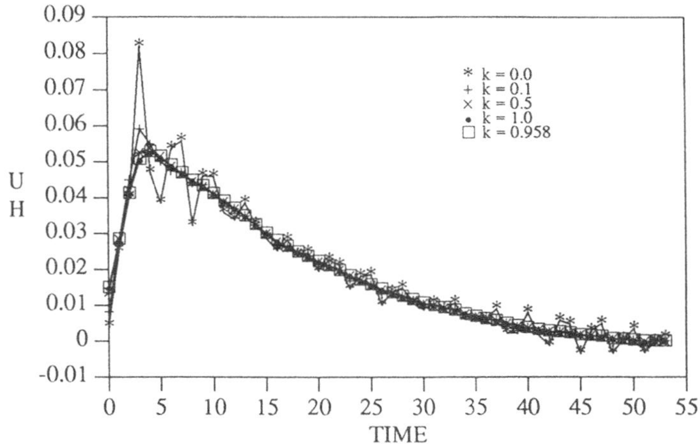

Figure 4. Kernel Functions by Bayesian Method. Bear 6 with I matrix.

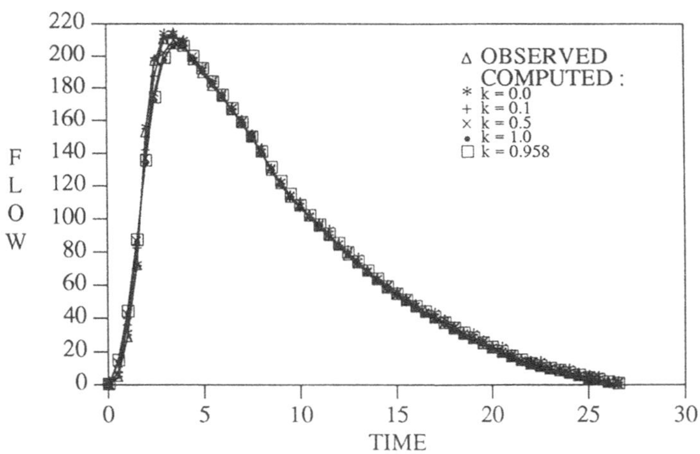

Figure 5. Flows by Bayesian Method. Bear 6 with I matrix.

normal distribution with zero mean. The variance of the noise is taken to be 10% of the variance of both the input and the output. The data are normalized such that the volumes of input and output are equal to unity.

The kernels are computed by the ridge regression method. Four different values of the parameter k between 0 and 1 are used to compute the kernel. The computed kernel is used to compute the output which is compared to the theoretical output.

The results of the ridge regression are presented in Table. 9. The time to peak and peak flow of the theoretical (observed) and computed outputs are compared. For the noiseless examples, both the time to peak and peak flow are well reproduced by this method with small k. With increases in k, the peak flow decreases. For the examples with noise, the times to peak are well reproduced but the computed peak flows are significantly smaller than theoretical peak flows, except when k is small. Also the ERR and RSS statistics are larger than the corresponding values for the noiseless cases.

Table 6 Results from Bayesian Method for example 2A with noise (C Matrix)

k	RSS	σ	C_L	ERR	t_{peak}	Q_{peak}
0.0	0.00148	0.00224	-35.35	0.19306	11	0.06860
0.1	0.00130	0.00162	-55.35	0.00470	12	0.06846
0.2	0.00130	0.00163	55.54	0.00419	12	0.06867
0.3	0.00131	0.00164	-309.52	0.00401	12	0.06881
0.4	0.00131	0.00164	-126.38	0.00394	12	0.06890
0.5	0.00132	0.00165	-86.35	0.00391	12	0.06897
0.6	0.00133	0.00166	-76.75	0.00390	12	0.06902
0.7	0.00133	0.00166	-72.24	0.00390	12	0.06906
0.8	0.00134	0.00167	-69.59	0.00390	12	0.06908
0.9	0.00135	0.00167	-67.84	0.00391	12	0.06910
1.0	0.00135	0.00168	-66.60	0.00392	12	0.06912

Table 7 Results from Bayesian Method for BEAR6 with I Matrix

k	RSS	σ	C_L	ERR	t_{peak}	q_{peak}
0.0	0.00000	0.00015	10.71	- - -	3.5	212.329
0.1	0.00003	0.00017	19.23	0.00129	3.5	213.450
0.2	0.00004	0.00019	27.66	0.00150	3.5	212.438
0.3	0.00005	0.00020	36.07	0.00161	3.5	211.404
0.4	0.00006	0.00021	44.47	0.00168	3.5	210.502
0.5	0.00007	0.00022	52.86	0.00173	3.5	209.733
0.6	0.00008	0.00023	61.24	0.00177	3.5	209.074
0.7	0.00009	0.00023	69.61	0.00180	3.5	208.507
0.8	0.00009	0.00024	77.99	0.00183	3.5	208.013
0.9	0.00010	0.00024	86.36	0.00185	3.5	207.580
1.0	0.00010	0.00025	94.73	0.00187	3.5	207.197
0.958	0.00010	0.00024	91.22	0.00186	3.5	207.360

Theoretical and computed kernels for example 2 with noise are presented in Figures 6 and 7. The kernels show large oscillations, but the oscillations become smaller for large values of k. The outputs show no particular trend similar to that in the example without noise. The outputs may go above or below the theoretical output depending on the k value.

4.2.2. RIDGE REGRESSION METHOD WITH REAL DATA. Two storms each from ten watersheds are analyzed by the ridge regression method. The characteristics of these storms are presented in Table 2. For each storm, both the effective rainfall (input) and direct runoff (output) are normalized such that the volume of each becomes unity. The kernel is computed by the ridge regression method. Five different values of the parameter k between 0 and 1 are used to compute the kernel and the output.

The results from the ridge regression method are presented in Table 10. The times to peak and peak flow of the observed and the computed output are compared. The area of the IUH decreases with increasing k values. In most cases the time to peak is reproduced well by this method, but the computed peak flow becomes smaller with increasing k. In contrast with the direct method, the ridge regression method does not produce oscillatory kernels. The flows show a downward trend when k increases.

Table 8 Results from Bayesian Method for BEAR6 with C Matrix

k	RSS	σ	C_L	ERR	t_{peak}	Q_{peak}
0.0	0.00000	0.00015	10.71	- - -	3.5	212.329
0.1	0.00003	0.00017	33.87	0.00156	3.5	216.501
0.2	0.00004	0.00018	499.28	0.00165	3.5	216.882
0.3	0.00005	0.00019	-999.54	0.00170	3.5	216.959
0.4	0.00005	0.00019	-179.64	0.00173	3.5	216.944
0.5	0.00006	0.00020	-132.26	0.00176	3.5	216.885
0.6	0.00006	0.00020	-113.52	0.00178	3.5	216.797
0.7	0.00007	0.00021	-103.35	0.00180	3.5	216.690
0.8	0.00007	0.00021	-96.94	0.00181	3.5	216.566
0.9	0.00008	0.00022	-92.51	0.00183	3.5	216.430
1.0	0.00008	0.00022	-89.28	0.00184	3.5	216.282
0.002	0.00001	0.00015	17.93	0.00055	3	214.232
0.033	0.00002	0.00016	6.72	0.00138	3.5	215.313

Figure 8 shows the kernel function of BEAR7 obtained by ridge regression for different values of k. The kernels do not show significant oscillations, but the shape becomes flatter with increase in k. Bruen and Dooge (1984) noted a similar behavior. These kernels are convolved with the input to obtain the output. Results are presented in Figure 9. Good reproduction is obtained for small values of k. The output is not well reproduced for large k values. This may not be a problem in practice because the runoff under the UH may be forced to be 1 in. Then the direct runoff would be reproduced better.

4.3. DIRECT METHOD

4.3.1. Direct Method With Synthetic Data. The three data sets described earlier are used to generate three data sets containing the input, output and kernel function. The first two examples are obtained by using $\Delta t = 1$, while the third example by $\Delta t = 0.1$. Another three data sets with the same time intervals are generated by introducing noise to both the input and output. The noise is generated from a standard normal distribution with zero mean. The variance of the noise is 10% of the variance of the input or

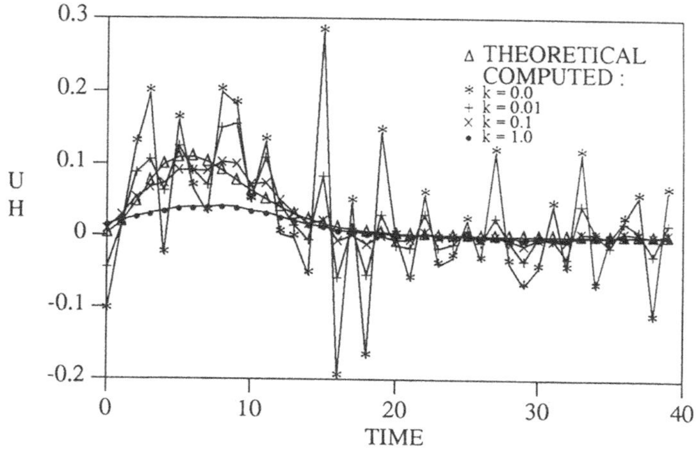

Figure 6. Kernel Function by Ridge Regression Method. Example 2 with noise.

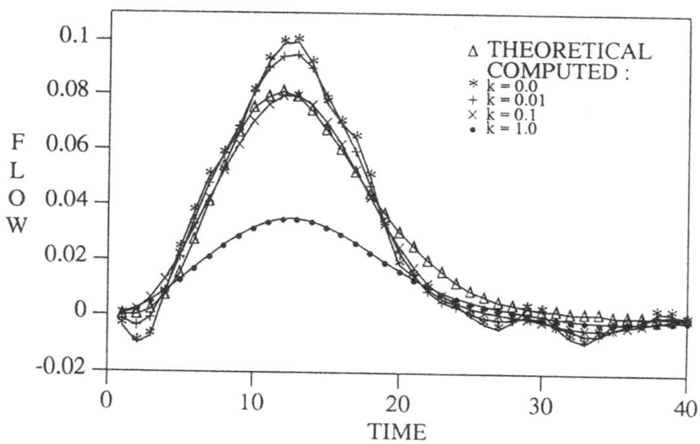

Figure 7. Flow by Ridge Regression Method. Example 2 with noise.

output. For all the data sets, both the input and output are normalized such that the volume of each is unity.

The results of the direct method are presented in Table 11. Both the time to peak and peak output of the theoretical (observed) and computed (convolved) outputs are compared. For the examples with noise, times to peak are well reproduced but computed peak flows are significantly different from the corresponding theoretical values. Also the ERR and RSS statistics are much larger than those for the noiseless examples.

Results of the example with noise are presented in Figures 10 and 11. Here the kernel function shows much larger oscillations than the previous example. The oscillations increase in magnitude with increase in time. The computed output shows large oscillations which also increase in magnitude.

4.3.2. Direct Method With Real Data. Two storms each from ten watersheds are analyzed by using the direct method. The characteristics of these storms are presented in Table 2. For each storm, both the effective rainfall (input) and direct runoff (output) are normalized such that the volume of input and output become unity. The kernel function computed by the direct method is convolved with the original rainfall function to produce the computed flow which is compared to the observed flow.

The results obtained by the direct method are presented in Table 12. Some of the storms, BUSS5, BUCK9 and CARP1 produce oscillatory kernels. This results in negative or very large kernel (IUH) areas. The same problem is reported by Blank and Delleur (1968) and Blank et al. (1971). Both the time to peak and peak flow of the theoretical (observed) and computed (convolved) output are compared. With the exception of storms with very large kernel areas, both the time to peak and peak flow are well reproduced by the direct method as they should be. This is shown by the small values of RSS. The instantaneous unit hydrographs with negative kernels also reproduce the time to peak and peak flow very well as they should.

Figure 12 shows an example of the kernel function obtained by the direct method. The kernel function is convolved with the input function to obtain the output which is presented in Figure 13. It shows a very good correspondence with the observed output.

5. Conclusions

The direct method is the simplest of the three methods. It gives just one solution to the problem of estimating the instantaneous unit hydrographs. This method assumes no errors in the system. If errors are present then the instanteous unit hydrograph will oscillate.

The Ridge Regression method accommodates errors but the solution is not unique. The solution depends on the value of the parameter k. The problem is to determine the k parameter which should be used. However by normalizing the unit hydrograph we get good estimated flows.

The Bayesian method proved to be the best method. It gives a response function which has little or no oscillations. The choice of the parameter k does not affect the computed flows very much. This method is more general than the other two methods. Consequently, it is recommended for use.

Table 9 Results from Ridge Regression Method with Synthetic Data

Case	k	IUH area	ERR	Theoretical		Computed		RSS
				t_{peak}	Q_{peak}	t_{peak}	Q_{peak}	
Example 1 without noise	0.0	0.9990	0.2955E+00	3	0.1940E+00	3	0.1940E+00	0.4168E-08
	0.01	0.9574				4	0.1790E+00	0.7026E-03
	0.1	0.7643				4	0.1304E+00	0.1532E-01
	1.0	0.3188				5	0.4309E-01	0.7984E-01
Example 2 without noise	0.0	1.0010	0.3955E-01	12	0.8100E-01	12	0.8099E-01	0.1401E-07
	0.01	0.9919	0.8285E-03			12	0.7951E-01	0.1639E-04
	0.1	0.9108	0.2842E-02			12	0.6941E-01	0.9021E-03
	1.0	0.4832	0.2787E-01			13	0.3268E-01	0.1791E-01
Example 3 without noise	0.0	0.9959	0.1357E+00	14	0.3600E-01	14	0.3602E-01	0.8795E-07
	0.01	0.9862	0.1441E-03			15	0.3576E-01	0.6554E-05
	0.1	0.9054	0.3160E-03			16	0.3277E-01	0.2245E-03
	1.0	0.4950	0.6701E-02			16	0.1770E-01	0.6331E-02
Example 1 with noise	0.0	0.9775	0.1069E+01	3	0.1940E+00	4	0.2383E+00	0.3321E-02
	0.01	0.9398	0.6497E+00			4	0.9488E-01	0.2220E-02
	0.1	0.7338	0.8231E+00			4	0.7945E-01	0.7639E-03
	1.0	0.2788	0.1325E+01			5	0.3488E-01	0.1650E-01
Example 2 with noise	0.0	1.0093	0.3364E+00	12	0.8100E-01	13	0.9947E-01	0.3221E-02
	0.01	1.0053	0.4424E-01			13	0.9488E-01	0.2220E-02
	0.1	0.9206	0.4529E-02			13	0.7945E-01	0.7639E-03
	1.0	0.4739	0.2603E-01			13	0.3488E-01	0.1650E-01
Example 3 with noise	0.0	1.0019	0.2196E+00	14	0.3600E-01	17	0.3924E-01	0.6062E-03
	0.01	0.9928	0.2040E-01			15	0.3887E-01	0.3320E-03
	0.1	0.9198	0.1191E-02			15	0.3504E-01	0.2509E-03
	1.0	0.5042	0.6248E-02			16	0.1864E-01	0.5838E-02

Table 10 Results from Ridge Regression Method with Real Data

Name	Storm no.	k	IUH area	Observed		Computed		RSS
				t_{peak}	Q_{peak}	t_{peak}	Q_{peak}	
BEAR	7	0.0	1.0020	2.5	430.190	2.5	422.290	0.1440E+05
		0.01	0.9920			2.5	417.563	0.1460E+05
		0.1	0.9100			2.5	380.413	0.2910E+05
		0.2	0.8333			2.5	346.564	0.6377E+05
		1.0	0.4974			2.5	202.671	0.4542E+06
BEAR	9	0.0	1.0001	2.25	442.110	2.25	428.500	0.1488E+05
		0.01	0.9895			2.25	419.232	0.1518E+05
		0.1	0.9083			2.25	372.633	0.2777E+05
		0.2	0.8327			2.25	333.544	0.5518E+05
		1.0	0.4979			2.25	184.891	0.3263E+06
MISS	18	0.0	0.9943	9	874.600	8.5	877.596	0.3901E+05
		0.01	0.9853			8.5	864.250	0.3921E+05
		0.1	0.9065			8	789.446	0.1702E+06
		0.2	0.8310			8	722.133	0.4464E+06
		1.0	0.4973			8	428.077	0.3430E+07
MISS	20	0.0	0.9989	9.5	470.800	9.5	464.225	0.1412E+05
		0.01	0.9880			9	468.998	0.1741E+05
		0.1	0.9063			9.5	432.028	0.6444E+05
		0.2	0.8314			9.5	391.980	0.1499E+06
		1.0	0.5000			9.5	225.410	0.1004E+07

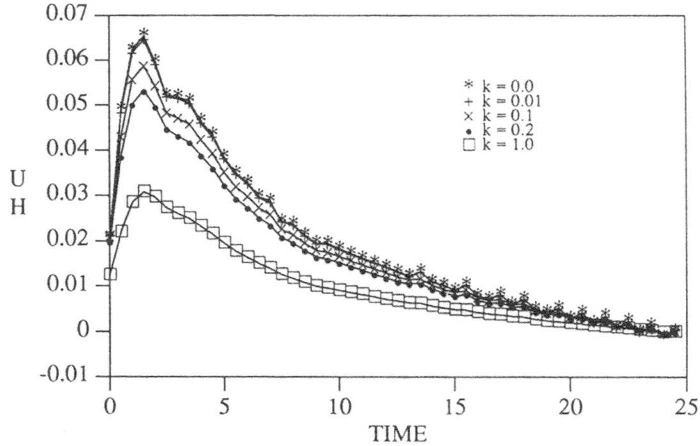

Figure 8. Kernel Function by Ridge Regression Method. Bear 7.

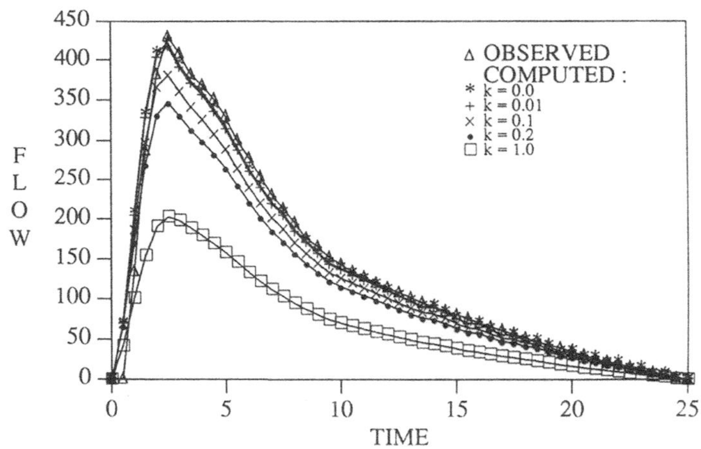

Figure 9. Flow by Ridge Regression Method. Bear 7.

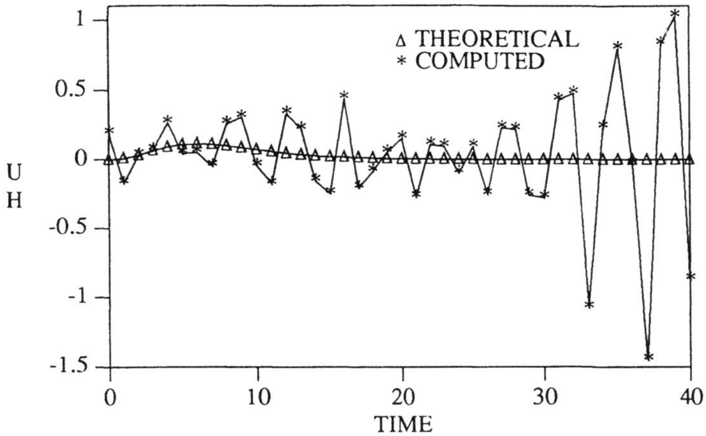

Figure 10. Kernel Function by Direct Method. Example 2 with noise.

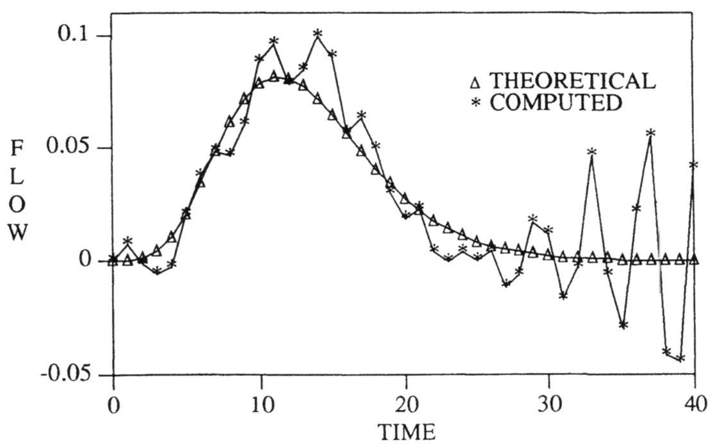

Figure 11. Flow by Direct Method. Example 2 with noise.

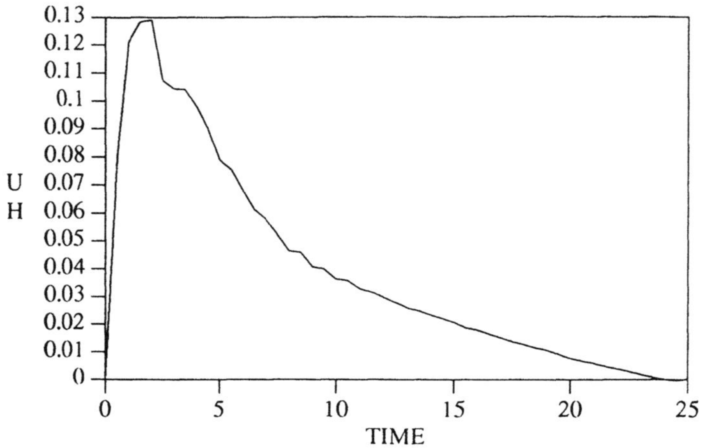

Figure 12. Kernel Function by Direct Method. Bear 7.

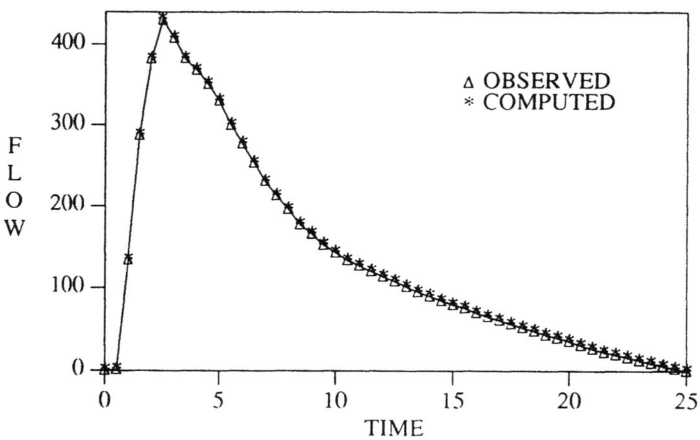

Figure 13. Flow by Direct Method. Bear 7.

Table 11 Results from Direct Method with Synthetic Data (with Noise)

Name	Records	IUH area		ERR
		theoretical	convolved	
Example 1	31	1.2177	1.0181	9.99264
Example 2	41	1.0001	1.4570	8.00723
Example 3	91	0.9992	1.0181	87.94214

Name	Theoretical		Convolved		RSS
	t_{peak}	Q_{peak}	t_{peak}	Q_{peak}	
Example 1	3	0.200	3	0.14071	0.03850
Example 2	11	0.081	14	0.09937	0.01622
Example 3	1.5	0.363	2	0.37215	0.11161

6. LIST OF REFERENCES

Bhargava, U.K., 1986. "Robust Parametric and Nonparametric Techniques for system Identification", Ph.D. Thesis, Purdue University, West Lafayette, Indiana.

Bhargava, U.K. R.L. Kashyap and D.M. Goodman, 1987. "Two Nonparametric Methods for Identifying the Impulse Response of Linear Systems", IEEE Trans. Acoust., Speech, Signal Processing, ASSP-35 No. 7:974-986.

Blank, D. and J.W. Delleur, 1968. "A Program for Estimating Runoff from Indiana Watersheds, Part I. Linear System Analysis in Surface Hydrology and Its Application to Indiana Watersheds", Tech. Rep. No. 4, Purdue University Water Resources Research Center, Lafayette, Indiana.

Blank, D., J.W. Delleur and A. Giorgini, 1971. "Oscillatory Kernel Functions in Linear Hydrologic Models", *Water Resources Research* 7,5; 1102-17.

Bruen, M. and J.C.I. Dooge, 1984. "An Efficient and Robust Method for Estimating Unit Hydrograph Ordinates", *J. Hydrol.*, 70:1-24.

Chow, V.T. (ed.) 1964. "Handbook of Applied Hydrology", McGraw-Hill, New York.

Deininger, R.A., 1969. "Linear Programming for Hydrologic Analyses", *Water Resour. Res.*, 5(5):1105-1109.

Dooge, J.C.I., 1959. "A General Theory of the Unit Hydrograph", *J. Geophys. Res.*, 64(2):241-256.

Eagleson, P.S., R. Mejia-R and F. March, 1966. "Computation of Optimum Realizable Unit Hydrographs", *Water Resour. Res.*, 2(4):755-764.

Table 12 Results from Direct Method with Real Data

Name	Storm no.	IUH	Observed		Computed		RSS
		area	t_{peak}	Q_{peak}	t_{peak}	Q_{peak}	
BEAR	7	1.0000	2.5	430.190	2.5	430.190	0.4015E-06
BEAR	9	0.9998	2.25	442.110	2.25	442.110	0.4868E-07
MISS	18	-7.6442	9	874.600	9	874.600	0.7114E-02
MISS	20	0.9928	9.5	470.800	9.5	470.801	0.9665E-06
SAND	12	1.0015	2	1316.750	2	1316.750	0.9227E-06
SAND	15	0.9986	13.5	1585.920	13.5	1585.917	0.1368E-04
BUSS	5	- -	24	293.800	- -	- -	- -
BUSS	8	0.7560	5.5	251.750	5.5	251.750	0.7956E-06
LAUG	1	0.9991	4.5	1330.000	4.5	1330.000	0.2932E-05
LAUG	6	1.0055	1.25	1312.700	1.25	1312.700	0.2992E-05
BRUS	12	-5.3947	4.5	478.250	4.5	478.250	0.1045E-03
BRUS	24	1.0006	3.75	701.500	3.75	701.500	0.4124E-06
BUCK	9	- -	9	223.600	- -	- -	- -
BUCK	12	0.9988	7.5	508.100	7.5	508.101	0.2937E-05
SALA	1	0.9811	22	2299.100	22	2299.096	0.3344E-03
SALA	13	- -	5	347.550	- -	- -	- -
BICE	3	0.9999	8	106.860	8	106.860	0.2805E-07
BICE	14	0.9987	11.5	105.772	11.5	105.772	0.4347E-07
CARP	1	- -	8.5	226.800	- -	- -	- -
CARP	4	0.9891	7	66.150	7	66.150	0.4780E-07

Edson, C.G., 1951. "Parameters for Relating Unit Hydrographs to Watershed Characteristics", *Trans. Am. Geophys. Union*, 32(4):591-596.

Farden, D.C., 1976. "Solution of a Toeplitz Set of Linear Equations", *IEEE Trans. Antennas Propag.*, AP-24:906-907.

Gray, D.M., 1961. "Synthetic Unit Hydrographs for Small Watersheds", *J. Hydraul. Div.*, Proc. Am. Soc. Civil Eng., 87(HY4):33-54.

Gupta, V.K., E. Waymire and C.T. Wang, 1980. "A Representation of an Instantaneous Unit Hydrograph from Geomorphology", *Water Resour. Res.*, 16(5):855-862.

Hoerl, A.E., 1962. "Application of Ridge Analysis to Regression Problems", *Chem. Eng. Prog.*, 58(3):54-59.

Hoerl, A.E. and R.W. Kennard, 1970a. "Ridge Regression: Biased Estimation for Nonorthogonal Problems", *Technometrics*, 12(1):55-67.

Karlinger, M.R. and B.M. Troutman, 1985. "Assessment of the Instantaneous Unit Hydrograph Derived from the Theory of Topologically Random Networks", *Water Resour. Res.*, 21(11):1693-1702.

Mays, L.W. and L. Coles, 1980. "Optimization of Unit Hydrograph Determination", *J. Hydraul. Div.*, Proc. Am. Soc. Civil Eng., 106(HY1):85-97.

Mays, L.W. and C.-K. Taur, 1982. "Unit Hydrographs via Nonlinear Programming", *Water Resour. Res.*, 18(4):744-752.

Rao, A.R. and J.W. Delleur, 1971. "The Instantaneous Unit Hydrograph: Its Calculation by the Transform Method and Noise Control by Digital Filtering", *Tech. Rep. No. 20*, Purdue University, Water Resources Research Center, Lafayette, Indiana.

Raudkivi, A.J., 1979. "Hydrology", Pergamon Press, Oxford, Great Britain.

Rodrigues-Iturbe, I. and J.B. Valdes, 1979. "The Geomorphologic Structure of Hydrologic Response", *Water Resour. Res.*, 15(6):1409-1420.

Rodrigues-Iturbe, I., M. Gonzales-Sanabria and R.L. Bras, 1982. "A Geomorphoclimatic Theory of the Instantaneous Unit Hydrograph", *Water Resour. Res.*, 18(4):877-886.

Sarma, P.B.S., J.W. Delleur and A.R. Rao, 1973. "Comparison of Rainfall-Runoff Models for Urban Areas", *J. Hydrol.*, 18:329-347.

Sherman, L.K., 1932a. "Streamflow from Rainfall by Unit-Graph Method", *Eng. News Rec.*, 108:501-505.

Singh, K.P., 1976. "Unit Hydrographs - A Comparative Study", *Water Resour. Bul.*, 12(2):381-392.

Snyder, F.F., 1938. "Synthetic Unit-Graphs", *Trans. Am. Geophys. Union*, 19(1):447-454.

Snyder, W.M., 1955. "Hydrograph Analysis by the Method of Least Squares", *J. Hydraul. Div.*, Am. Soc. Civil Eng., 81, Paper No. 793, 1-25.

Taylor, A.B. and H.E. Schwarz, 1952. "Unit-Hydrograph Lag and Peak Flow Related to Basin Characteristics", *Trans. Am. Geophys. Union,* 33(2):235-246.

Troutman, B.M. and M.R. Karlinger, 1985. "Unit Hydrograph Approximations Assuming Linear Flow Through Topologically Random Channel Networks", *Water Resour. Res.*, 21(5):743-754.

Valdes, J.B., Y. Fiallo and I. Rodrigues-Iturbe, 1979. "A Rainfall-Runoff Analysis of the Geomorphologic IUH", *Water Resour. Res.*, 15(6):1421-1434.

Zohar, S., 1974. "The Solution of a Toeplitz Set of Linear Equations", *J. Associ. Comput. Mach.*, 21(2):272-276.

PREDICTING RUNOFF USING ARTIFICIAL NEURAL NETWORKS

Allen T. Hjelmfelt, Jr. and Menghua Wang
Res. Hydraulic Engineer, USDA, Agricultural Research Service, Cropping Systems and Water Quality Research Unit, Agr. Engr. Bldg. Univ. of Missouri, Columbia, MO 65211. Graduate Student, Agr. Engr. Dept., Agr. Engr. Bldg., Univ. of Missouri, Columbia, MO 65211.

ABSTRACT

An artificial neural network was constructed that has the same structure as application of a unit hydrograph. This network was trained using eight rainfall-runoff events simultaneously, and the trained network verified using sixteen events observed on a 12.2 km^2 watershed located in north central Missouri. The results were quite good.

One advantage of neural networks is their ability to learn nonlinear situations. A traditional three-layer, feed-forward artificial neural network was also constructed to test the nonlinear approach. Results were good, but a traditional unit hydrograph was not obtained.

INTRODUCTION

Estimation of the runoff hydrograph due to a rainfall event is a fundamental problem in hydrology. Numerous linear and nonlinear methods have been proposed and are in use. In most cases calibration, or fitting, of the model is a problem. Each rainfall-runoff event, however, gives a different set of fitting parameters, so it is necessary to estimate a best average. The complications arise due to noise in the data, effects of nonlinearity, and limited knowledge of the phenomena at hand. Artificial neural networks may alleviate some of the problems because they provide a method which includes a fitting, or training scheme, as a fundamental part of the model. In this paper, a simple artificial neural network is shown to have the same structure as application of a unit hydrograph. The neural network was constructed as a computer program and applied to a small watershed located in central Missouri with satisfying results. Based upon these results, a more general network was constructed and trained to test the value of the nonlinear capability of neural networks.

ARTIFICIAL NEURAL NETWORK BASICS

The multilayer, feed forward artificial neural network trained using the

V. P. Singh and B. Kumar (eds.), Surface-Water Hydrology, 233–244.

backpropagation algorithm is by far the most popular network used in practical applications. This is the architecture that will be explored for predicting the runoff hydrograph from a rainfall input. Neural networks have their basis in attempts to model human mental processes, thus a brief description of the biological system is needed to define terminology. The basic element of the brain is the neuron. The cell body of a neuron has one or more output lines called axon, and one or more input lines called dendrites. Both the axon (output) and dendrites (input) lines can connect to other neurons to form the network. This connection is through the synapse. The neuron cell sums the signals that it receives through the dendrites. If the sum reaches a certain level, the receiving neuron fires; if the sum is less than this threshold, the neuron does not respond. The connection between the neurons, the synapse, serves to attenuate the signal arriving at the cell body. The process of learning is the process of setting the appropriate attenuation level for the synapse.

An artificial neural network is an assemblage of interconnected neurons. The synapse, or resistive connection, is represented as a weight applied to the signal applied along that connection. The structure is shown graphically in Fig. 1. Each processing element, or neuron, (labeled N1, N2,...in Fig. 1) determines a weighted sum of the inputs it receives from the previous layer. That is, the signal is

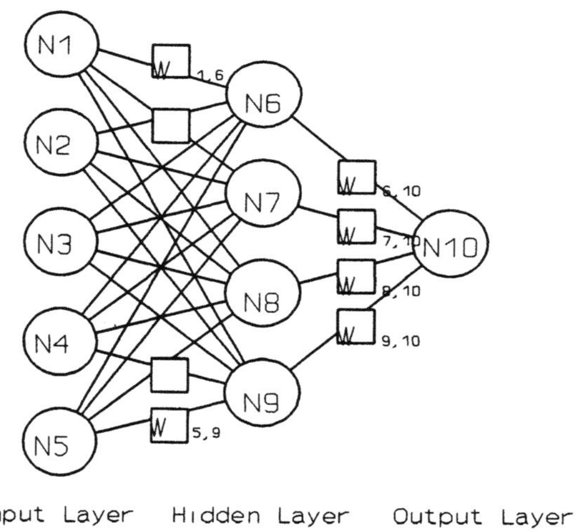

Fig. 1: Schematic for three-layer, feedforward neural network.

multiplied by the w_{ij} shown on Fig. 1 before being added to the other weighted inputs. This sum is passed through a transfer function. For example, the response of neuron "i" is determined by

$$Output = \begin{bmatrix} 1 \ if \ \Sigma w_{ij} v_j > I_t \\ 0 \ if \ \Sigma w_{ij} v_j < I_t \end{bmatrix} \tag{1}$$

in which w_{ij} is the weight applied to the signal from neuron j to neuron i, v_j is the signal from neuron j, and I_t is the threshold. A sigmoid shaped transfer function is frequently used instead of the threshold function. For unit hydrograph simulation, a ramp transfer function will be used. The output from the transfer function is then passed to all of the inputs of the next layer or serves as the final result if the processing element is in the output layer.

A significant attribute of neural networks is their ability to learn. The errors in predicting outputs for given sets of known inputs and target outputs can be determined. Thus, the quality of the simulation can be expressed as the sum of squares of the errors. Calibration of an artificial neural network is termed training and consists of adjusting the weights (the W_{ij} in Fig. 1 used to determine the weighted sum) to minimize the square error. A gradient descent optimization method is used. This learning rule is called backpropagation because the calculations for the change in weights progress backward through the network. For a general description of the training algorithm, see Hinton (1992); for a detailed mathematical development, see Rumelhart et al. (1986). The process is repeated until the error is reduced to a specified limit or training tolerance. A global fit to the training data can be achieved.

UNIT HYDROGRAPH BASICS

A basic tool of linear hydrologic models is the unit hydrograph. The methods of neural networks can be applied to this linear hydrologic model. This will be demonstrated. The application of a unit hydrograph begins with rainfall which is given as a sequence of intensities, p'_i, for example

$$p'_n = (p'_0, p'_1, p'_2, p'_3, p'_4) \tag{2}$$

From this rainfall, infiltration is subtracted in the form of a phi-index. This results in an effective rainfall sequence of

$$p_n = p'_n - \Phi = (p_0, p_1, p_2, p_3, p_4) \tag{3}$$

The unit hydrograph is given by the set of ordinates

$$h_k = (h_0, h_1, h_2, h_3, h_4, h_5) \tag{4}$$

all separated by the same time increment associated with the rainfall increments.

The composite runoff hydrograph is developed in symbolic form as

$$\begin{aligned}
q_0 &= p_0\, h_0 \\
q_1 &= p_1\, h_0 + p_0 h_1 \\
q_2 &= p_2\, h_0 + p_1 h_1 + p_0 h_2 \\
q_3 &= p_3\, h_0 + p_2 h_1 + p_1 h_2 + p_0\, h_3 \\
q_4 &= p_4\, h_0 + p_3 h_1 + p_2 h_2 + p_1\, h_3 + p_0\, h_4 \\
&\;\;\ldots \\
q_7 &= p_4\, h_3 + p_3\, h_4 \\
q_8 &= p_4\, h_4
\end{aligned} \tag{5}$$

The q_i constitute the ordinates of the runoff hydrograph.

UNIT HYDROGRAPH EXPRESSED AS ARTIFICIAL NEURAL NETWORK

The application of a unit hydrograph can be depicted in the format of the specialized artificial neural network shown in Fig. 2. The network consists of three layers, the input layer, a middle or hidden layer, and an output layer. Connections only operate in the direction from input to output. The rainfall is an external signal applied to the input row of neurons. Determination of the composite runoff hydrograph using an artificial neuron network requires applying several sequences of rainfall; the same rainfall values that led to the development of Eq. 5. That is, the first time the rainfall set (p_o, 0, 0, 0, 0) is applied, then the rainfall set (p_1, p_o, 0, 0, 0), and so on.

There is only one line leading from each neuron on the input layer to the hidden layer, and the weight for that connection is unity. Thus, the sum given by Eq. 1 applied to each neuron just gives the input rainfall. The sum determined by the neurons on the hidden layer must pass through a transfer function. This transfer function serves as the Φ-index. Instead of the sharp threshold used in Eq. 1, a ramp transfer function is used. For the hidden layer of neurons, Eq. 1 becomes

$$\textit{Effective rainfall} = \begin{bmatrix} p - \phi \;\; if \sum w_{ij} p_j > \phi \\ 0 \;\; if \sum w_{ij} p_j < \phi \end{bmatrix} \tag{6}$$

in which w_{ij} is 1, and Φ is the phi-index. The hidden layer then serves to extract the infiltration from the rainfall, and the resulting effective rainfall serves as the input to the output layer of neurons. (Fig. 2).

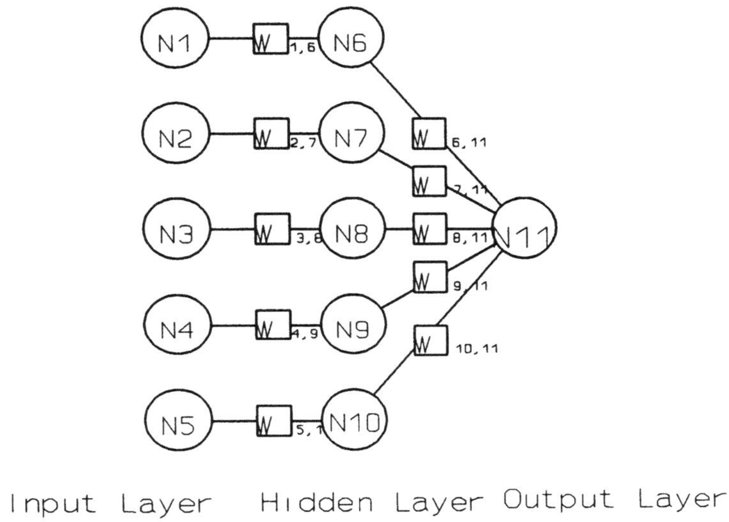

Fig. 2: Schematic for three-layer neural network specialized for unit hydrograph applications.

The neurons in the hidden layer are all connected to a single output neuron. The output neuron performs a weighted sum of the input it receives from each of the middle layer neurons. Eq. 1 becomes

$$Output = \sum w_{ij} v_j \tag{7}$$

Applying this scheme to the discharge q_3 of Eq. 5, the signal to the input layer of neurons is the rainfall values

$$(p_3,\ p_2,\ p_1,\ p_0,\ 0.0) \tag{8}$$

From this input the infiltration is extracted through the transfer function of the hidden layer. The output from the hidden layer is summed by the output layer. This time the sum is weighted, but the transfer function simply passes through the signal. Thus the output is

$$q_3 = p_3w_0 + p_2w_1 + p_1w_2 + p_0w_3 + 0w_4 \quad (9)$$

Note this is exactly the same as Eq. 5 with the weights taking the role of the unit hydrograph ordinates. Numerical solution for the composite hydrograph requires solution of each of the equations in the set, similar to the unit hydrograph process shown by Eq. 5. Thus, the inputs to the neural network are sequences of rainfall, the same rainfall values that led to the development of Eq. 5.

The goal of learning in neural networks is to determine the weights to be applied at each of the connections. In this case, the goal is to determine the ordinates of the unit hydrograph. A unit hydrograph can be derived from observed rainfall and runoff using traditional methods, but if several events are available, different unit hydrographs are derived. An advantage of neural networks is that they can determine the "best unit hydrograph" in a least squares sense.

To accomplish the training one must have several sets of rainfall and associated runoff. The neural network is initially set with a group of random values for the weights. The rainfall sets are applied to the network and runoff output generated. That is, the rainfall set for each row of Eq. 5 is applied to the network. The output is computed and, initially, will not agree with the observed runoff. An adjustment to the weights is computed as a function of the error and the applied rainfall. Neural networks can consider several rainfall-runoff data sets for updating its weights. If several rainfall-runoff events are available, not all data sets from all of the events need be used. There is often a training advantage to using a random order for listing. For additional details, see Hinton (1992).

UNIT HYDROGRAPH EXPERIENCE

A computer program to simulate the specialized artificial neural network was adapted from a computer program given by Eberhart and Dobbins (1991) and applied to Goodwater Creek, a 12.2 km^2 watershed located in central Missouri. This watershed was established as a research watershed by the USDA Agricultural Research Service in 1971. Rainfall on this watershed is measured using an array of three raingages, only one of which was used for this study. Runoff is measured using a triangular weir which provides control for low flows. The channel and a bridge opening provide control for higher flows.

Twenty-four of the largest events were selected for this study. Eight of the events were used for training, and the remaining 16 were used to verify the results. The phi-index can be expected to vary with each event, so its determination would constitute a separate neural network. Infiltration was extracted from each storm by hand, and only the effective rainfall applied to the network. That is the layer in the conceptual neural network that simulates the phi-index was removed. Baseflow was subtracted from the observed hydrograph, so that the target value was the runoff portion of the hydrograph. The resulting two layer artificial neural network contained thirty neurons in the input layer.

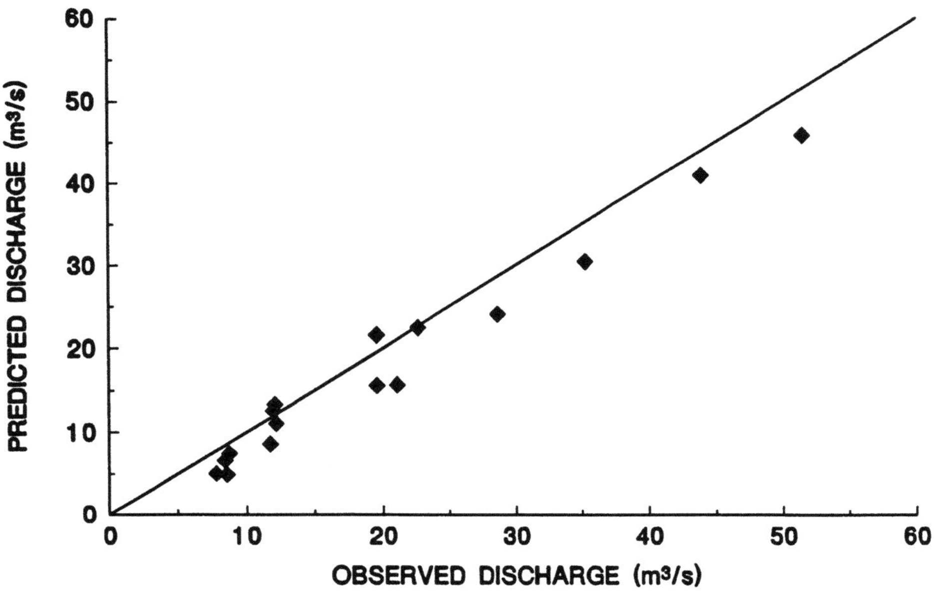

Fig. 3: Comparison between observed peak discharges and peak discharges predicted using two-layer linear neural network

This number of neurons was obtained as an average number of ordinates for unit hydrographs determined in a previous study. The output layer consists of a single neuron. The training results are shown in Column 3 of Table 1. The final root-mean-square error is 47.81.

Table 1. Comparison of training results for different network architecture.

Date	Linear	Root Mean Square Error Neurons in Hidden Layer 1	10	20
Mar. 03, 1976	40.45	53.54	38.90	40.32
July 24, 1981	55.12	64.76	43.36	44.83
Aug. 26, 1982	63.10	71.64	32.47	27.46
Aug. 29, 1982	73.98	74.80	61.79	59.62
June 19, 1983	45.88	53.98	29.45	23.01
June 16, 1985	56.55	72.38	16.51	11.97
Nov. 18, 1985	17.03	29.57	15.47	16.19
Nov. 19, 1985	30.38	31.10	44.70	33.06
Average values	47.81	56.47	35.33	32.06

Table 2. Comparison of the testing results for different neural network architectures

Date		Root Mean Square Error Neurons in Hidden Layer		
	Linear	1	10	20
May 30, 1974	102.67	112.97	111.65	112.72
June 08, 1974	32.77	45.32	38.02	42.59
May 06, 1977	49.85	61.33	49.83	51.30
Aug. 01, 1978	37.26	46.02	32.05	35.91
Apr. 10, 1979	28.48	31.88	45.59	38.73
July 03, 1980	43.41	48.71	55.69	54.25
June 19, 1981	125.53	309.59	202.80	258.35
July 23, 1981	57.65	99.82	71.60	99.33
June 08, 1982	29.20	50.18	38.17	52.27
Sept. 01, 1982	95.29	254.89	154.26	193.04
Dec. 02, 1982	75.06	87.69	126.45	151.92
Apr. 30, 1983	56.23	56.53	85.72	60.53
June 08, 1983	55.24	50.62	80.09	69.62
Oct. 31, 1984	36.38	41.06	58.06	63.32
May 16, 1986	25.29	35.64	13.57	13.36
Sept 23, 1986	73.03	81.71	81.04	85.26
Average values	57.71	88.37	77.79	86.41

Following training, the network was given the rainfall values for the sixteen other storms. The results are given in Column 3 of Table 2. The root-mean-square error for the computed hydrographs ranged between 25.9 and 125.53 with an average of 57.71. A comparison between the observed and computed peak discharges is tabulated in Table 3 and shown in Fig. 3. The r^2 for the comparison is 0.97. These results are gratifying and suggest trying the nonlinear form of the neural network.

GENERAL APPLICATION HYDROLOGIC APPLICATION

To test the more general neural network, as depicted in Fig. 1, and the utility of the nonlinear capability, a three layer network was constructed with all neurons on one layer connected to all neurons on the next layer in a feed forward configuration. Thirty neurons were used in the input layer, and a single neuron was used in the output layer. There is little guidance for selecting the number of neurons for the middle layer, so three different numbers were tried, 20, 10, and 1. A sigmoidal transfer function was used in this network. The same data sets used for the previous specialized network were used here.

The training results for the different neural network architectures are listed in Table 1. The root-mean-square error for the training is shown in columns 4, 5, and 6. The average results are 56.47, 35.33 and

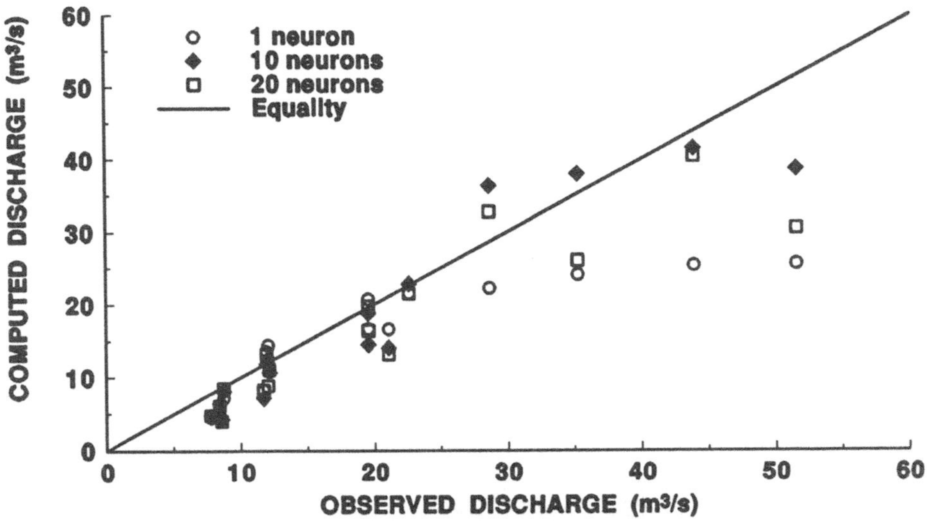

Fig. 4:Comparison between observed peak discharges and peak discharges predicted using three-layer nonlinear neural network

32.06 for hidden layers containing 1, 10, and 20 neurons, respectively. The larger the number of neurons in the hidden layer, the better the training results; except for the trial with a single hidden neuron, the average values are better than for the two-layer network. Unfortunately, increasing the number of neurons in the hidden layer increases the computer time. Using 300 data sets, the three-layer network with 20 neurons in the hidden layer took more than two days on our personal computer to train acceptably, whereas only a few minutes were needed to train the three-layer network with a single neuron in the hidden layer.

Sixteen rainfall events were used to verify the trained networks. If the network architecture is appropriate and it is trained correctly, the network should be able to simulate appropriate runoff responses when presented with new rainfall data. The root-mean-square error for the verification predictions are given in Table 2 and resulted in average root-mean-square error values of 88.37, 77.79, and 86.41 for the single, 10, and 20 neuron hidden layers, respectively. This does not compare well with the 57.71 obtained using the two-layer network. The network with ten neurons in the hidden layer produced the smallest root-mean-square error (77.79). Recall, however, that the network with 20 neurons in the hidden layer gave the minimum error for the training results. The predicted peak discharge is compared to the observed discharge in Table 3 and in Fig. 4. The r^2 for the comparison between the ten neuron hidden layer predictions and the observed values is 0.8902. This value is not as good as that for the two-layer network but is still satisfactory.

Note, however, that from Fig. 4 there appears to be a systematic departure from equality with higher flows.

Table 3. Percent error in peak discharge prediction using hidden layers of various sizes.

Flood Date		Error in Peak Discharge (%)		
	Linear	Neurons in Hidden Layer		
		1	10	20
May 30, 1974	25.47	21.22	33.42	37.45
June 08, 1974	21.12	28.70	24.17	27.98
May 06, 1977	34.91	42.46	39.91	39.43
Aug. 01, 1978	5.29	13.75	0.58	10.27
Apr. 10, 1979	8.98	4.51	12.29	9.40
July 03, 1980	26.58	29.65	39.18	30.03
June 19, 1981	10.98	50.14	24.79	40.58
July 23, 1981	13.67	31.57	7.76	25.96
June 08, 1982	10.33	19.47	4.35	26.94
Sept. 01, 1982	6.86	42.06	5.30	7.81
Dec. 02, 1982	16.00	22.27	26.91	14.57
Apr. 30, 1983	20.17	14.78	25.75	16.25
June 08, 1983	6.59	6.09	3.80	1.15
Oct. 31, 1984	0.79	4.87	0.78	5.12
May 16, 1986	14.41	17.79	7.43	2.79
Sept. 23, 1986	42.24	49.77	50.31	53.53
Average values	16.52	21.82	16.13	17.28

A surrogate for a unit hydrograph can be determined from the trained network by applying a unit rainfall. The results are shown in Fig. 5. Although these representations have the general shape of a unit hydrograph, the results from the nonlinear network architectures lack the smoothness expected. Note that the neural network using a sigmoidal training function is a nonlinear model. Fitting of nonlinear runoff models has long been known to lead to difficulties due to noise in the data (Amarocho, 1967). Sartori (1992) showed that nonlinear neural networks have also been shown to have trouble with noise in the training data. Thus, it is not surprising that such a problem should occur.

CONCLUSIONS

A similarity between the structure of a simple artificial neural network and the application of a unit hydrograph was demonstrated. The weights for the connections for the output layer of the artificial neural network were shown to be equivalent to the ordinates of a unit hydrograph. An artificial neural network was constructed and trained using rainfall-runoff data from a small watershed in Missouri. The resulting network was tested and shown to reproduce observed floods with acceptable accuracy. This illustrates that there is a basis in hydrologic fundamentals for the application of artificial neural networks to rainfall-runoff problems.

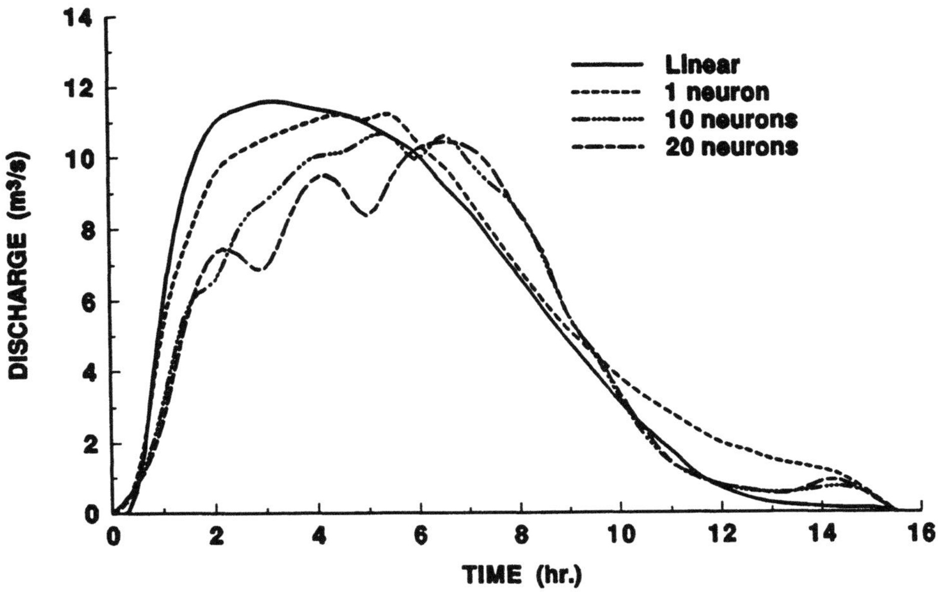

Fig. 5: Response of two-layer linear and three-layer nonlinear neural networks to a unit input

A more general three-layer, feed-forward neural network, trained using the backpropagation algorithm, was used to predict the runoff hydrograph due to an applied effective runoff. The best values were obtained using ten neurons in the middle, or hidden, layer. An approach to a unit hydrograph can be determined from the trained general network by applying a unit rainfall and using the predicted runoff as the unit hydrograph. The result resembles a unit hydrograph, but, as often is the case with nonlinear models, seems to have trouble with noise in the data used for calibration. This trouble is exhibited in a lack of smoothness.

REFERENCES

1. Amarocho, J. 1967. The nonlinear prediction problem in the study of the runoff cycle. Water Resources Research, 3 (3):861-880.

2. Eberhart, R. C., R. W. Dobbins. 1991. Neural Network PC Tools, A Practical Guide. Academic Press.

3. Hinton, G. E. 1992. How neural networks learn from experience. Scientific American, **267**(3):145-151.

4. Rumelhart, D. E., G. E. Hinton, and R. L. Williams. 1986. Learning internal representation by error propagation. In: Parallel Distributed Processing, Explorations in the Microstructure of Cognition, D. E. Rumelhart and J. L. McClelland (Eds.), MIT Press, Cambridge, Ma, 1:318-364.

5. Sartori, D. E. 1992. Neural networks, statistics and experimental designs. Scientific Computing and Automation, **8**(11):4-8.

USE OF DIGITAL ELEVATION MODEL IN RUNOFF MODELLING

P.K. GARG
Department of Civil Engineering
University of Roorkee
ROORKEE - 247667 (INDIA)

ABSTRACT: In this study, the Digital Elevation Model (DEM) has been used extensively for the assessment of runoff within and at the outlet of the watershed. The study area covered a small watershedof 20 km^2 in semi-arid region of south-east Spain. The DEM was generated from the topographic maps to compute slope and catchment area which were subsequently used to generate the flow direction, network flow pattern and drainage network in the watershed. These information alongwith the infiltration and rainfall data were used for runoff modelling, employing Hortonian type relationship. The results were computed on storm events basis, which could be used further for the assessment of erosion and sediment yield in the watershed. Thus, the study demonstrated the utility of DEM alongwith other field data for the assessment of runoff in a semi-arid watershed.

1.0 INTRODUCTION

Both rainfall and runoff provides basic energy input to derive the erosion process in a watershed. Surface runoff is that portion of precipitation which, during and immediately following a storm event, ultimately appears as flowing water in the drainage network of the watershed. It transports sediments and nutrients such as phosphorus,

V. P. Singh and B. Kumar (eds.), Surface-Water Hydrology, 245–257.

nitrogen, pollutants and pesticides from agricultural fields of the watershed. Runoff from a watershed is required to be known for many problems related to environmental and water quality studies, watershed management and river mechanics and training works (Garg, 1991).

Runoff from a watershed is a function of precipitation, slope length, steepness and shape, soil types and their conditions, and vegetation cover. The dominant factors influencing it are geomorphology and climatic region in which the watershed is situated. (Thornes and Gilman, 1983). Therefore, a model must incorporate these factors for the assessment of more realistic runoff values. In ideal situations, however it may not be possible to include all the factors affecting runoff, due to either large amount of work involved to gather such information or non-availability of desired data.

In this paper a procedure has been explained for the generation of runoff data for ungauged watersheds. The input parameters for the model can easily be derived from field data and Digital Elevation Model (DEM). The model has been tested for a semi-arid watershed in south-east Spain.

2.0 ASSESSMENT OF RUNOFF

In semi-arid environments, runoff is difficult to observe and measure where flow is ephemeral, rainfall events are irregular, and the terrain is very rough (Garg, 1991). However, such areas with restricted vegetation cover and thin soils on hillslope, generate an enormous quantity of runoff.

The simple type of overland flow production uses the Hortonian concept, where runoff equals precipitation minus infiltration (Horton, 1945). Under typical semi-arid conditions of high rainfall intensity, poor vegetation cover and shallow undeveloped or badly eroded soils, Hortonian type flow is produced more or less instantaneously and simultaneously over a watershed. Other runoff models such as storage based types (Thornes and Gilman, 1983) are most appropriate for long duration storms where runoff commences once the available subsurface storage is filled, and where storage capacities are low despite high

Table 1 A Comparison of Horton Type and Storage Type Models

Characteristics	Horton type model	Storage type model
1. Rainfall	Strongly related to rainfall intensity	More dependent on rainfall duration
2. Infiltration	Surface infiltration capacity of critical importance	Transmissibility of lower soil horizons more important
3. Distribution		
Temporal	Begins soon after the storm if intensities are high enough	Starts only when the underlying soil layers are saturated
Spatial		
Environmental	Semi-arid areas with sparse vegetation and thin soils	Humid areas with dense vegetation and well-developed soils
Local	Widespread over the catchment	Limited to zones of saturation
4. Downslope variation	Linear increase of runoff volume	More complex pattern of change

infiltration rates. The storage type models depend upon the moisture content of the soil before, during and after rainfall events. Being non-Hortonian type, the flow can occur at rainfall intensities much lower than those required for Hortonian overland flow. Table 1.0 presents a comparison between Horton type and Storage type models.

3.0 Experimental Watershed

The Albudeite watershed, situated in semi-arid region of south-east Spain, ranges in elevation from 160 to 340 m above m.s.l., and covers approximately 20 km^2 area (Fig. 1). All channels in the area

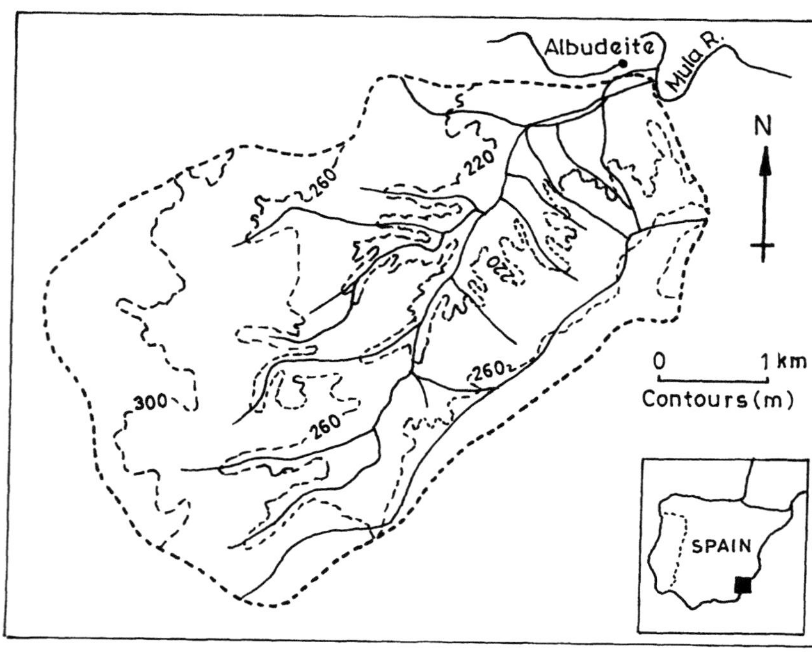

Fig. 1 Location Map of the Study Area.

are in the form of gullies, carrying water only during and after major storms. The mean annual rainfall is less than 300 mm with 25-30% interannual variation (Garg, 1991). The natural vegetation of the area covers shrubs and grasses, while agricultural crops are almonds, olive, wheat and barley. The soils are mainly marls that are susceptible to overland flow (Thornes and Gilman, 1993).

4.0 THE MODELLING APPROACH

As runoff hydrograph in semi-arid environments is dominated by Hortonian overland flow, the Hortonian type equation (Horton, 1945), as given below, was used in this study-

$$Q = (I - F) * A$$

Where Q = Runoff rate, I = Rainfall intensity rate, F = Infiltration rate, and A = Catchment area of the cell. The area was segmented into square grid cells of 100 m side, and the above data collected. The duration of a storm event was also needed to compute the runoff for each cell and the total runoff yield of the watershed. These parameters were derived from field experiments and DEM, as summarized below :-

4.1 Digital Elevation Model

In Digital Elevation Models (DEMs), the elevation data are represented numerically as a matrix of elevation values (Z) implicitly located by their geographic coordinates (x, y) (Burrough, 1986). DEMs can either be generated from the digitized contour maps or the stereo model of air-photos/satellite data. For regions without elevation information/topographic maps, the stereoscopic capabilities of SPOT images can be used to generate elevation data to within ± 10m accuracy (Rodriquez et. al. 1988).

Elevation contours of the study area were digitized from two sets of topographic maps at 20 m contour interval. OPUS II micro-computer based system running the GINGA PC graphics software and connected to a TDS-48" digitizing table were used for this purpose. After having digitized the contours, interpolation was carried out to derive the DEM at 100 m grid spacing.

4.2 Defining Steepest Slope and Flow Direction

Slope influences the movement of water in a watershed by controlling the stream power available to route the flow. At each cell location in a DEM, there are eight possible slope directions from the central cell in a 3X3 elevation matrix; horizontally, vertically and diagonally in either direction. After computing these, the steepest slope and its direction is selected to route the flow to the adjacent cell in the downward direction.

4.3 Computation of Catchment Area

The catchment area of a cell is expressed as the number of flowlines which could contribute flow to that cell (Garg, 1991). The catchment area in areal units was computed using the flow direction data. It was found that the areas with smaller values are located on ridges (where flow diverges) and on topslope sections of the watershed, and the larger values are found in the valleys and drainage lines (where flow converges).

4.4 Infiltration Data

The Infiltration rates were measured for vegetated and unvegetated sites using a ring infiltrometer of 10 cm diameter, driven to a depth of 12 to 15 cm. Water was ponded for about 60 minutes, as the majority of storms are less than this duration. Infiltration curves were drawn to compute the final infiltration rates which are found to vary between 1.15 and to 3.21 cm/hr (Table 2).

Table 2 Infiltration Values in the Study Area

	Major soil types	Infiltration value (cm/hr)	Area (%)
1.	Litosoles xerosoles calcaricos	3.21	2.04
2.	Regosoles margalicas	2.50	53.97
3.	Fluvisoles calcaricos Regosoles margalicas	1.86	30.52
4.	Fluvisoles calcaricos Solonchaks	1.22	11.68
5.	Fluvisoles calcaricos	1.15	1.79

4.5 Rainfall Data

Two big storms of Nov. 4, 1987 and Nov. 11, 1988, recorded by the Automatic Weather Station which was set up temporarily close to the watershed, were considered for this study. It was observed that approximately 11% and 33% of the annual rainfalls respectively occurred in these storms (Garg, 1991). The hypothetical storms of one hour fixed duration but varying intensity ranging from 1.5 to 6.0 cm/hour with an increment of 0.5 cm/hour were also used. The return periods of these storms ranged from 2 to 100 years which were computed using the historical rainfall data of nearest raingauge station. These storms closely represented the characteristics pattern of rainfall in south-east Spain (Thornes and Gilman, 1983).

5.0 RESULTS AND ANALYSIS

The results obtained from the runoff model are described below-

5.1 Drainage Network Analysis

The drainage network is an integral part of the runoff modelling, as runoff from all grid cells is collected in the channel segments and then routed downstream of the watershed. The approach proposed by Mark (1984) which uses the catchment area, was employed to extract the drainage network. Cells with catchment area greater than a threshold value were selected to be part of the drainage network The drainage network pattern thus obtained was compared with the network pattern delineated from the topographic maps. A reasonable correspondence between the two was obtained.

5.2 Flow Network Pattern

The flow from a cell depends upon the local topography of the watershed (*i.e.* slope steepness). Thus the flow path can be traced by connecting points between the grid cells of the watershed, using flow direction data. As the flow proceeds downslope, each previously processed cell has to be connected to one of these directions to trace the flow pattern. The network flow pattern developed by a 3.00 cm/hr storm of one hour duration with approximately 5 year return period is shown in Fig. 2.

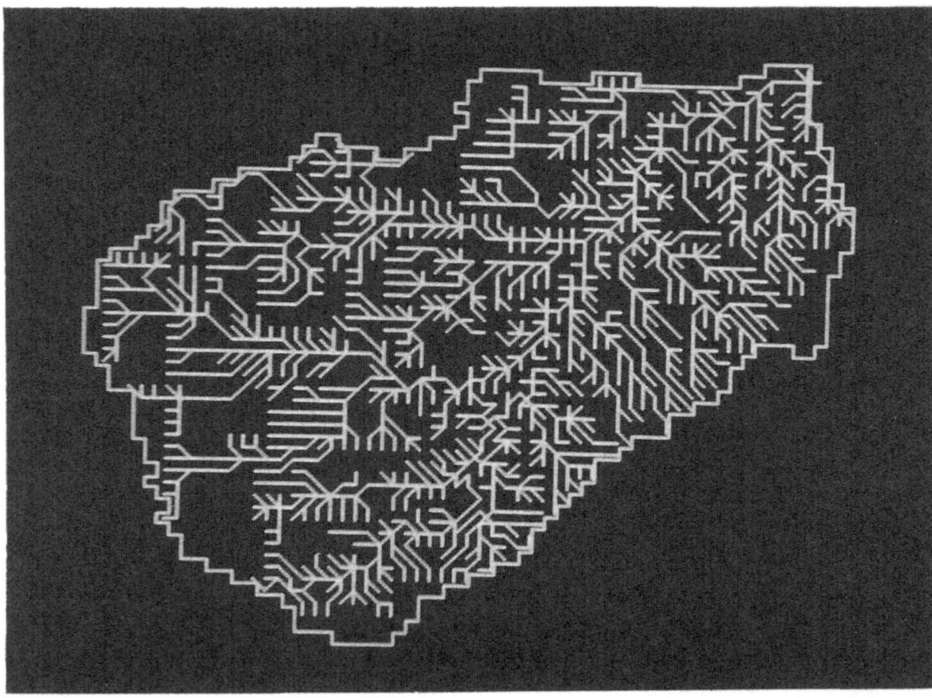

Fig. 2 The network flow pattern developed by a 3.00 cm/hr storm of one hour duration with approximately 5 year return period.

5.3 Computation of Runoff

The results obtained from the two big storms are given in Table 3. The storm of Nov. 11, 1968 was found to produce more runoff due to its high magnitude and long duration compared with the other storm. The runoff coefficient for this storm was 0.182 indicating that about 18% of the rainfall became runoff. This value was similar to that typical of similar catchment for similar storms in south-east Spain (Thornes and Gilman, 1983).

Runoff values for hypothetical storms were computed and plotted against rainfall values. Linear curves were obtained between rainfall and runoff values (Fig. 3). The similar pattern was obtained by Marques and Roca (1987) for agricultural plots in north-east Spain. The storms of approximately 100 years return period generated runoff of 90.80 l/sec/ha with 0.65 runoff coefficient. It was also observed that as rainfall increased, the runoff coefficients increased rapidly in the beginning and then slowed down, the curve eventually becoming flatter for higher intensity storms. The total runoff values, in general, were found close to the values obtained by Thornes and Gilman (1983) for marl soils in south-east Spain.

A linear relationship in the form of $Y = a + b X$ between runoff (mm) and rainfall (mm) was observed with correlation coefficient $r = 0.99$ at 99% confidence level, $a = -1.326$ and $b = 1.447$. The correlation coefficient was high mainly because the runoff values were generated using the rainfall data. These values however indicated that as an average, a minimum rainfall of 1.45 cm was required to

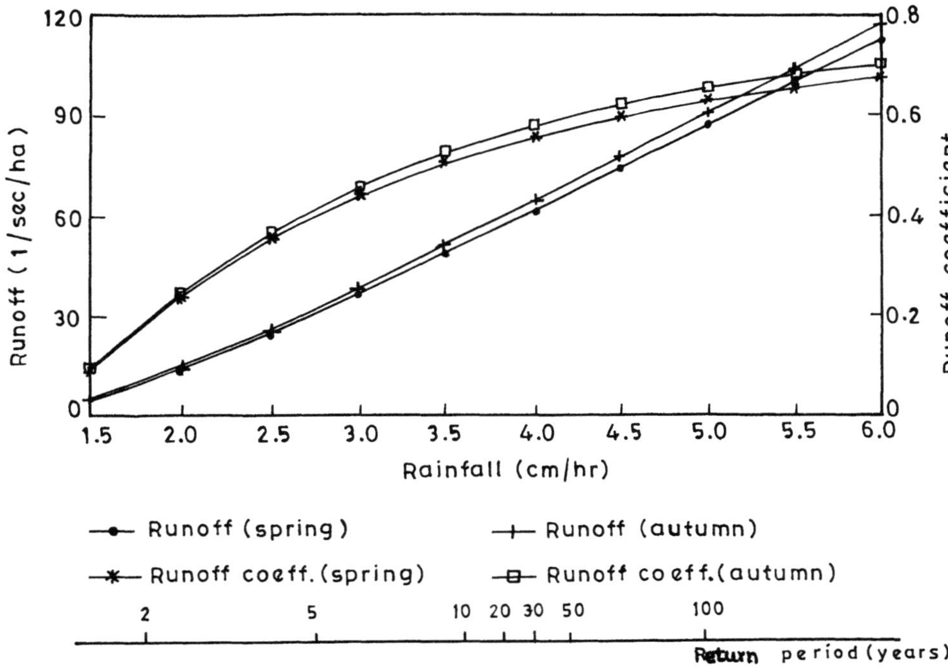

Fig. 3 Curves between Rainfall and Runoff for Hypothetical Storms.

Table 3 Results of the Model from Two Big Storms

Parameters	Rainfall on Nov. 4, 1987	Rainfall on Nov. 11, 1988
1. Storm intensity (cm/hr)	1.35	1.79
2. Duration of storm (hr)	2.33	2.83
3. Total runoff (l/sec/ha)	3.59	14.34
4. Runoff coefficient	0.055	0.182

produce runoff in this kind of environment. This value was comparable to the value, as observed by Yair and Klein (1973) in semi-arid Southern Israel.

6.0 CONCLUSIONS AND RECOMMENDATIONS

The runoff modelling approach proposed in this study requires input parameters in grid cell form. DEM has been used for the computation of steepest slope and its direction to route the water. The flow direction data can be processed further to delineate application - specific watersheds such as tracking the overland flow path of pollution from a point source into the drainage network, or routing the water and sediment down the watershed. The catchment area coupled with other terrain parameters can be used to locate moist soil areas and areas prone to erosion.

Soil moisture conditions prior to a storm or between storms are difficult to estimate or measure, and can be a serious source of error for infiltration and runoff data. For example, the infiltration value is much higher at the beginning of a storm but as the storm gets longer it falls rapidly and becomes steady for saturated soils. The lack of suitable methods to accurately predict infiltration constitute a major limitation in the development of an improved runoff model. For more accurate simulation of the results, the infiltration values need to be measured or interpolated accurately over the different seasons, as they change with growing vegetation, and surface sealing and crusting which is predominant in the semi-arid regions.

The runoff modelling procedure described here uses the numeric and spatially distributed data over a watershed, therefore a Geographic Information System (GIS) can be used effectively for the handling, analysis and storage of such data. Once the input data are spatially co-registered, the desired output products (maps, tabular and statistical information) can be derived by integrating them in GIS. The main advantage of using GIS for such studies is the ease of combining large number of input parameters in user's specified form which traditionally is a time consuming task.

REFERENCES

Burrough, P.A. (Ed.) (1986), *Principles of Geographic Information System for Land Resources Assessment*, Clarendon Press, Oxford.

Garg, P.K. (1991), 'Development of a Catchment Scale Erosion Model for Semi-arid Environments and its Implementation through Remote Sensing', *Unpublished Ph.D. Thesis*, University of Bristol, U.K., 353.

Horton, R.E. (1945), 'Erosional Development of Streams and their Drainage Basins : Hydrophysical Approach to Quantitative Morphology', *Bulletin of Geological Society of America*, **56**, 275-370.

Mark, D.M. (1984), 'Automated Detection of Drainage Networks from Digital Elevation Models', *Cartographica*, **21**, 168-178.

Marques, M.A. and Roca, J. (1987), 'Soil Loss in an Area of North-east Spain', In : *International Geomorphology* (V. Gardiner, Ed.), John Wiley & Sons, London, II, 483-493.

Rodriguez, V. Gigord, P., De Ganjac, A.C. and Munier, P. (1988), 'Evaluation of the Stereoscopic Accuracy of the SPOT Satellite', *Photogrammetric Engineering and Remote Sensing*, **54**, 217-221.

Thornes, J.B. and Gilman, A. (1983), 'Potential and Actual Erosion around Archeological Sites in South-east Spain', *Catena Supplement*, **4**, 91-113.

Yair, A. and Klein, M. (1973), 'The Influence of Surface Processes on Flow and Erosion Processes on Debris Covered Slopes in an Arid Area', *Catena*, **1**, 1-18.

SECTION 5:

FLOOD ROUTING

PHYSICALLY BASED LINEAR FLOOD ROUTING MODELING

WITOLD G. STRUPCZEWSKI
Institute of Geophysics. Polish Academy of Sciences
64 Ksiecia Janusza, 01-452 Warsaw, Poland

Abstract

The papers starts from recalling the St.Venant equation, its assumptions and problems of numerical solution of differential equations. Then outline of work on linearization of the St.Venant equation and on its application is given. The reasons why the linearized St. Venant model has not driven out the conceptual flood routing linear models are discussed.

In order to get the linearized equation valid for a large number of choices for the dependent variable, the perturbation potential is introduced. Problems in the linear theory of open channel flow are listed and their solutions either are presented or references to literature are given. Those of them still waiting for solution are also included.

The major part of the paper is given to search for a physical reason to a great practical popularity of the linear conceptual flood routing model in the form of a series of linear reservoirs with pure delay, known as Lag Kalinin-Miliukov model (LKM). It is used as an opportunity to presentation of solutions of some upstream control problems. Impulse response of the LKM model was compared with that of the linearized St. Venant model with upstream input (LCR) together with its limiting case of Froude number equal to one (RF). A comparison of LKM and RF models has shown that in spite of a certain structural similarity, the LKM model is not suitable for a steep river channel. However for certain flow conditions, frequently met during floods, the LKM model shows amazingly good agreement with the LCR measured by the differences of higher order cumulants and a satisfactory one for differences of lags. That is why, taking additionally into account the poor observability of a river system, the model is suggested to be accepted as a third, besides RF and Lag Diffusion Analogy (LDA) models,

V. P. Singh and B. Kumar (eds.), Surface-Water Hydrology, 261–295.

particular case of LCR.

The introduced coefficient α_R (R=4,5,6,..) being a function of three cumulants has served to find a useful property of LCR and to classify the particular cases solutions. Regions of applicability of the particular cases models were defined by the criterion of an error of fourth cumulant simulation.

In view of presented results LKM model may have been derived by theoreticians working working on fluid mechanics and then recommended for practice. However it has happened contrariwise.

1. Introduction

One-dimensional analysis of unsteady flow in open channels is based on the continuity equation

$$\frac{\delta Q}{\delta x} + \frac{\delta A}{\delta t} = 0 \tag{1.1}$$

where Q(x,t) is the discharge, A(x,t) is the cross-sectional area, x is the distance from the upstream boundary and t is the elapsed time, and the momentum equation in the average direction of flow as formulated by Saint-Venant (1871).

An equation for the conservation of linear momentum is usually derived on the assumptions that:

(a) the slope of the channel bed is small and uniform;

(b) vertical acceleration can be neglected so that hydrostatic pressure prevails at every point of the cross-section;

(c) the frictional resistance is the same as for steady uniform flow;

(d) the velocity is uniformly distributed over each cross-section.

For these assumptions the equation for the conservation of linear momentum in the direction of flow can be written in terms of the same variables (Dooge *et al.*, 1982):

$$(1-F^2)g\bar{y}\frac{\delta A}{\delta x} + \frac{2Q}{A}\frac{\delta Q}{\delta x} + \frac{\delta Q}{\delta t} = gA(S_o - S_f) \tag{1.2}$$

where F is the Froude number, $\bar{y}$ is the hydraulic mean depth, S_o is the bottom slope and S_f is the friction slope and g is the acceleration due to gravity.

The friction slope depends on the type of friction law assumed, the shape and roughness of the cross-section, the flow and area of flow at the section. Accordingly the friction slope can be written in the very general form:

$$S_f=f(A,Q,\text{shape},\text{roughness}) \tag{1.3}$$

Note that in both the Chezy and Manning friction laws there is assumption of uniform distribution of shear along the wetted perimeter (P) which takes place in an infinitely wide rectangular channel. Despite of that in view of lack of reasonable alternatives the Manning formula is widely used for any shape of natural channels which can create spurious results. For Chezy friction we have in general that

$$S_f=\frac{Q^2}{C^2A^2R} \tag{1.3a}$$

in which R-hydraulic radius can be expressed as a function of A(x,t) and C is the Chezy friction parameter.
For the Manning formula the corresponding expression is:

$$S_f= n^2\frac{Q^2}{A^2R^{4/3}} \tag{1.3b}$$

Though the continuity equation (1.1) is linear in discharge and area of flow, the momentum equation is highly nonlinear and hence no general analytical solution is available and therefore numerical methods are to be used.

Numerical methods typically start with a continuum model and turn it into a discrete one in both space and time. Utilization o the mathematical tools of nonlinear dynamics to analyze model nonlinear problems have been discussed in the numerical analysis and applied mathematics literature. Recently in-depth global and topographical approach to the numerical solution of differential equations that arise in dynamics and the discussion of the underling implication were presented by Yee *et al.*(1991) and discussed by Stewart (1992). The essence of the difficulty is that the dynamics of discrete systems is only loosely related to that of continuous systems. The dynamics of discrete systems is far richer than that of their continuous counterparts and the approximation involved can create spurious solutions. Yee *et al.* apply the 'dynamics of numerics' point of view to several popular numerical differential-equation solvers, concentrating on ordinary differential equations with known explicit solutions, coming to the disturbing conclusion that standard practice may sometimes be seriously misleading, and that much work is needed to find out why and to stop it happening. According to Stewart the dangers are especially great in computational fluid dynamics, where the type of problem to study has changed dramatically over the past few years.

The main problem in Yee's opinion are not so much numerical as dynamical: the actual behavior of the continuum models, and their relation to discretizations, must be the central object of study. Until these advances in the dynamics of numerics are made, the hydrologic computers jockeys (the term used by Philip, 1992) - most of whom are wielding a mathematical tool without understanding how it works - should heed the warning. The convenient starting point in studying the continuum nonlinear model is to derive analytical solutions for special cases to get insight into the nature of the solution of a general problem and then to compare them with discrete dynamics solutions. In case of unsteady flow in open channels a linearized solution seems to be more useful for the purpose than the non-linear solution of a simplified version of the complete St.Venant equation. However guidelines developed from linearized problem are not always valid for nonlinear problems. Strictly speaking a computation for each set of boundary conditions requires an individual sensitivity analysis if the claim is made that the level of errors of computational approximation is appropriate to the accuracy of the field data.

The first attempt at a linearization of the complete equation seems to be due to Deymié (1935). Further work has been done by Massé (1939), Supino (1950), Lighthill and Whitham (1955), Dooge and Harley (1967), Brutsaert (19730, and Dooge, Napiórkowski and Strupczewski (1987a,b). The linearized St.Venant equation has been applied to such problems as flood routing in rivers (Keefer, 1974), systematic comparison of hydrologic routing methods (Dooge, 1973, Strupczewski and Kundzewicz, 1979) and choice of appropriate one (Strupczewski and Kundzewicz, 1980), multi linear model (Keefer & McQuivey, 1974, and Becker & Kundzewicz, 1986), simulation of the holding time in the geomorphic unit hydrograph (Kirshen & Bras, 1983), the effect of the downstream boundary condition on flood routing (Dooge *et al.*, 1983, Dooge & Napiórkowski, 1984, 1987, Napiórkowski & Dooge, 1988) and the flood routing of both lumped and lateral inflow (Strupczewski & Kundzewicz, 1992).

One could have expected that the derivation of the impulse response of the linearized St.Venant equation for semi-infinite river reach brings an elimination of other linear hydrologic routing models from use. However nothing like this has happened and moreover the Linear Channel Response (LCR) is rarely use by practitioners. The reasons for that are that they are simpler in use than the LCR, conservatism of practice, a real river system to be modeled is rarely single-inflow one with low quality of inflow data and the stress is laid on a model calibration rather than on its proper choice.

Attractiveness of the explicit physical parameters of LCR does not seem to appeal very much for following reasons:
(i) a river channel is rarely uniform in shape,slope and friction which makes necessary to average the parameters along a modeled reach;
(ii) an information about geometry and flow resistance of a reach is usually scarce;
(iii) confusion can make the assumption of small disturbances about reference level made for the linearization of St.Venant equation since flood waves by no means fall under such category and therefore the reference level becomes an additional parameter of the model;
(iv) parameters of (pseudo)physical models can be strongly correlated so that numerical estimation of parameters is fraught with error;
(v) there are four parameters in LCR while conceptual models have usually no more than three. Using the moments method for estimation of LCR parameters from inflow-outflow hydrographs the four first moments are to be used whiled the error of moment estimation increases rapidly with the order of moment;
(vi) linking by the moment matching technique the parameters of a conceptual model with those of LCR one can easily express the conceptual model's parameters in terms of physical parameters even if structures of both models differ considerable.

Hydrologists working on hydrodynamics consider the problem too trivial and too far from present fluid dynamics problems challenging them and also too far from real flow process which is highly nonlinear. It is what Philip (1992) writes about it: "For me the epithet 'linear' in the title was something of a turn-off. I had spent two decades struggling with the non-linearities of soil-water flow in unsaturated soils which are at heart of terrestrial hydrology; and I was well aware also of the inherent non-linearity of the processes of overland flow and of channnel and river flow. ... some contemporary linear enthusiasts fly in the face of physics and blandly proclaim linearity in data that shouts a contrary message." Professional hydrologists receive quite different message from the same data requesting them to use very simple models adequate to the existing inadequate and often inaccurate measurement. Dooge (1972) stated "The fact that linear routing methods have been used in applied hydrology would indicate that the effect of linearization cannot be so catastrophic as to make linear models worthless." while just recently Cowpertwait and O'Connell (1992) "Today, linear systems theory still occupies a dominant position in both theoretical and applied hydrology, and will doubtless continue to do so for many years to come.".

2. The Problems Based on the Linearized Saint Venant (LSV) Equation

To linearize the Saint-Venant equation we expand the non-linear terms in equation (1.2) in a Taylor series around the uniform steady state (Q_o, A_o) and limit of the expansion to the first order increments $Q'(x,t)$, $A'(x,t)$. The resulting equations are (Deymie, 1935, Dooge & Harley, 1967):

$$\frac{\delta Q'}{\delta x} + \frac{\delta A'}{\delta t} = 0 \tag{2.1}$$

$$(1-F_o^2)\, g\bar{y}_o \frac{\delta A'}{\delta x} + 2v_o \frac{\delta Q'}{\delta x} + \frac{\delta Q'}{\delta t} = gA_o \left[-\frac{\delta S_f}{\delta Q} Q' - \frac{\delta S_f}{\delta A} A' \right] \tag{2.2}$$

where $v=Q_0/A_0$ is the average velocity of flow and the derivatives of the friction slope $S_f(Q,A)$ are evaluated at the reference conditions for all frictional formulas for rough turbulent flow as:

$$\frac{\delta S_f}{\delta Q} = 2\,\frac{S_0}{Q_0} \tag{2.3}$$

$$\frac{\delta S_f}{\delta A} = 2mv_o\,\frac{S_0}{Q_0} \tag{2.4}$$

where

$$m = \frac{dQ/dA}{Q_0/A_0} = \frac{c_k}{v_0} \tag{2.5}$$

is the ratio of the kinematic wave speed to the average velocity of flow. Differentiating equations (2.1) and (2.2) with respect to x and t respectively and eliminating one of two variable one can get a single second order partial differential equation with the single dependent variable Q'(x,t) or A'(x,t). To generalize it to a large number of choices for the dependent variable, a perturbation flow potential U'(x,t) was introduced by Deymié (1935) and developed by Supino (1950) and Dooge *et al.* (1987). This potential can be defined as the function U'(x,t) whose partial derivatives with respect to x gives the minus perturbation from the reference area:

$$\frac{\delta U'}{\delta x} = -A'(x,t) \tag{2.6a}$$

and whose partial derivative with respect to time gives the perturbation from the reference discharge:

$$\frac{\delta U'}{\delta t} = Q'(x,t) \tag{2.6b}$$

The definitions given by equation (2.6) ensures that the equation of continuity (2.1) for the perturbations in flow and area is automatically satisfied. Substitution from equation (2.6) into equation (2.2) gives us a second order partial differential equation for the unsteady flow potential:

$$(1-F_0^2)g\bar{y}_0\frac{\delta^2 U}{\delta x^2} - 2v_0\frac{\delta^2 U'}{\delta x \delta t} - \frac{\delta^2 U}{\delta t^2} = gA_0\left[-\frac{\delta S_f}{\delta A}\frac{\delta U'}{\delta x} + \frac{\delta S_f}{\delta Q}\frac{\delta U'}{\delta t}\right] \tag{2.7}$$

Any linear function of the perturbation potential U'(x,t) will also represent a solution of equation (2.7). Since differentiation is a linear operation, one can replace U' in equation (2.7) by A'(x,t) or Q'(x,t). Similarly, any linear combination of solutions is also a solution of the basic linear equation (Dooge *et al.*, 1987a), e.g. the perturbation of the velocity v'(x,t), the surface width of the channel T'(x,t) and the Froude number F'(x,t). The choice of the dependent variable is a matter of convenience and usually the dependent variable will be the one for which the boundary conditions are directly given.

Substituting equations (2.3) and (2.4) into equation (2.7) with U'=Q' and denoting

$$D = \frac{S_0}{\bar{y}_0} x \qquad \text{(dimensionless length)} \tag{2.8}$$

$$z = x/c_k \qquad \text{(passage time of a kinematic wave through the channel)} \tag{2.9}$$

one obtains (Strupczewski & Kundzewicz, 1990):

$$(1-F_0^2)\frac{c_k}{2m}\frac{z}{D}\frac{\delta^2 Q'}{\delta x^2} - F_0^2 \frac{z}{D}\frac{\delta^2 Q'}{\delta x \delta t} - F_0^2 \frac{m}{c_k}\frac{z}{D}\frac{\delta^2 Q'}{\delta t^2} = \frac{\delta Q'}{\delta x} + \frac{1}{c_k}\frac{\delta Q'}{\delta t} \tag{2.10}$$

The basic problems in the linear theory of open channel flow are:

(1) The downstream problem, *i.e.* the prediction of the flood characteristics at a downstream section on the basis of a knowledge of the flow characteristics at an upstream section and the hydraulic characteristics of the channel between the two sections and its inverse problem, *i.e.* the prediction of the inflow hydrograph at an upstream section on the basis of an outflow hydrograph and hydraulic characteristics of the channel reach;

(2) The upstream problem, *i.e.* the prediction of the effect at an upstream section of the changes in flow conditions at a downstream section together with its inverse problem;

(3) The two-point boundary problem in which both upstream and downstream boundary conditions are taken into account;

(4) The tributary problem which deals with the effect of tributary inflow on conditions in the main channel both upstream and downstream of the point of entry;

(5) The lateral inflow problem in which there is a distributed inflow to the channel reach

The first problem is to solve equation (2.7) for semi-infinite channel in the down stream direction subject to the double initial condition:

$$U'(x,0) = 0 \qquad \frac{\delta U'}{\delta t} = 0 \tag{2.11}$$

and subject to the boundary conditions:

$$U'(0,t) = 0 \qquad U'(L\Rightarrow\infty, t) < |\infty|. \tag{2.12}$$

The problem has been solved by the use of the Laplace transform technique.

The linear downstream channel response (LCR) consists of two parts (Dooge, 1972, Dooge *et al.*, 1987) *i.e.* the discontinues head of the wave as a delta function of decreasing volume

$$h_1(x,t) = \delta(t - \Delta_{LCR}) \exp\left[-\frac{1-(m-1)F}{(1+F)F} D\right], \tag{2.13}$$

and the body of the wave given by

$$h_2(x,t) = \frac{1}{F}\sqrt{\frac{1-(m-1)^2F^2}{1-F^2}} \; D \exp\left[(m-1)\, D - \frac{1+(m-1)F^2}{m\,F^2}\, \frac{D\,t}{z}\right] \cdot$$

$$\cdot \frac{1}{\sqrt{a}}\; I_1\left[\frac{1}{mF^2}\sqrt{(1-F^2)\left[1-(m-1)^2F^2\right]}\;\frac{D}{z}\sqrt{a}\right] 1(t-\Delta_{LCR}), \tag{2.14}$$

where

$$\Delta_{LCR} = \frac{m\,F}{1+F}\, z \tag{2.15}$$

is the time lag,

$$a = \left(t - \frac{m\,F}{1+F}\, z\right)\left(t + \frac{m\,F}{1-F}\, z\right), \tag{2.16}$$

$0 \le F \ge 1$, while $I_1[\;]$ is a modified Bessel function of the first kind and 1( )is a unit step function.

Two simpler forms of the linear downstream response are recognized in the literature. They correspond to the limiting flow conditions of the LCR, *i.e.*, to Froude number equal to zero (Hayami (1951), Dooge (1973)) and one (Strupczewski & Napiorkowski, 1988).

The Diffusion Analogy (DA) model is based on the solution of the linearized Saint Venant equation for semi-infinite, uniform channel with Froude number equalled to zero and consequently used for small Froude numbers. Note that both terms of the LCR contain a pure lag which vanishes for F=0 only. Consequently the impulse response given below is for the Lag Diffusion Analogy (LDA) model:

$$h(x,t) = \frac{x}{\sqrt{4\pi D(t-\Delta_{LDA})^3}} \exp\left\{-\frac{[x-a(t-\Delta_{LDA})]^2}{4\,D\,t}\right\} \qquad (2.17)$$

The Rapid Flow (RF) model is based on the solution of the linearized Saint Venant equation for uniform channel with Froude number equaled to one and consequently used for flow condition characterized by great Froude numbers (Strupczewski & Napiorkowski, 1990). The impulse response of the RF model is given by

$$h(x,t)=P_0(\Lambda)\cdot\delta(t-\Delta_{RF})+\sum_{i=1}^{\infty}P_i(\Lambda)\cdot h_i[(t-\Delta_{RF})/\alpha]\cdot 1(t-\Delta_{RF}), \qquad (2.18)$$

where

$$P_i(\Lambda)=\frac{\Lambda^i}{i!}\exp(-\Lambda) \qquad (2.19)$$

is a Poisson distribution,

$$h_i(t/\alpha)=\frac{1}{\alpha(i-1)!}(t/\alpha)^{i-1}\exp(-t/\alpha) \qquad (2.20)$$

is the impulse response of i-linear reservoirs with a time constant α and a time lag Δ_{RF}, and $1(t)$ is the unit step function.

Cumulants of these three impulse responses are given in Appendix A.

Having defined an impulse response of a model in time domain $h(x,t)$ or in Laplace transform domain $H(x,s)$ and the upstream boundary condition $U'(0,t)$ or $\bar{U}'(0,s)$, the $U'(x,t)$ hydrograph in any cross section downstream the river ($x>0$) can be obtained from the convolution integral

$$U'(x,t)= h(x,t)*U'(0,t) \qquad (2.21)$$

or in the transform domain

$$\bar{U}'(x,s)=H(x,s)\cdot\bar{U}'(0,s) \qquad (2.22)$$

The inverse linear flood routing problem is to find the perturbation potential $U'(0,t)$ in the upstream end of a reach for given impulse response and the potential somewhere downstream the river, *i.e.* to derive input hydrograph for given both IUH and output hydrograph. In particular it is the key problem of reservoir control during a flood period directed to protect valley downstream of reservoir against flooding. To achieve a prescribed flow condition at a downstream point it is necessary to predict the conditions required at the upstream end. Looking at equation (2.22) it is obvious that the solution of the inverse problem in the transform domain can be found from

$$\bar{U}'(0,s)=\bar{U}'(x,s)/H(x,s) \tag{2.23}$$

and then to invert it to the time domain numerical methods would be in use. Since the LCR shows that the system is highly dumped, any error of the upstream end hydrograph would be magnified at the upstream end. In particular, for non differentiable output, e.g. the pulse function , the corresponding function U'(0,t) does not exist. It is not possible to require the hydrograph to have a sharp form at the end of the long river reach of large roughness.

The second problem, *i.e.* the propagation of a flood generated downstream, was considered in oceanography as a problem of the propagation of an astronomical tide and the corresponding shallow water constituents (e.g. Dronkers, 1964) in a flat channel of uniform cross-section and storage width by the harmonic method with complex functions. The tidal equation of continuity is of the form (2.1) while the equation of motion:

$$\frac{\delta Q'}{\delta t} + \lambda Q' + gA_0 \frac{\delta h'}{\delta x} = 0 \tag{2.24}$$

where h' denotes vertical tide, A_0- a reference (mean) flow area and λ - the coefficient of resistance, *i.e.* a linear friction law is assumed. Eliminating from equations (2.1) and (2.24) one of the dependent variables and introducing the perturbation flow potential U'(x,t) we obtain the second order linear partial differential equation :

$$\frac{\delta^2 U'}{\delta t^2} + \lambda \frac{\delta U'}{\delta t} = c_0^2 \frac{\delta^2 U'}{\delta x^2} \tag{2.25}$$

where

$$c_0 = \sqrt{gA_0/b_0} \tag{2.26}$$

is the propagation velocity of sine waves if λ=0.

Dronkers (1964) presents the solutions of equation (2.25) for a propagation of a simple harmonic wave considering following cases:

(1) At the mouth of the river both the vertical tide

h(0,t) and the horizontal tide (*i.e.* flow discharge) Q(0,t) are given, then the solutions for stage and discharge at any upstream section are function of both lower boundary conditions;

(2) At the mouth of a channel the vertical tide h(0,t) is known but the channel has closed end at x=L so that Q(L,t)=0;

(3) At both ends of a channel reach the vertical tide is known, *i.e.* h(0,t) and h(L,t).

The problem of a propagation of a harmonic wave on a channel with run-off discharge is not considered, because the interference between this discharge and the tide is an essential factor in the propagation of the tide.

The third problem of the list is to solve equation (2.7) subject to the double initial condition (2.11) and to the boundary conditions:

$$U'(0,t)=U'_u(t) \qquad U'(L,t)=U'_d(t) \tag{2.27}$$

Since the solution in the time domain has a very complicated form and it was presented in a several publications (Dooge & Napiórkowski, 1987, Napiórkowski & Dooge, 1988, Napiórkowski, 1992) it will be omitted here. A slightly simpler form has the limiting case solution for the Froude number equaled one, *i.e.* for the diffusion routing (Dooge *et al.*, 1983, Dooge & Napiórkowski, 1984). It shall be emphasized that any pair of upstream-downstream hydrographs for which the linear channel response to both upstream and downstream input is applied shall be physical feasible, e.g. a steady state at the upper end of a reach of a finite length while a flood at the lower end may not be considered because they may result in a negative flow area along the reach. The problem of a finite reach with the steady state rating curve instead of the lower boundary condition is still waiting for solution.

A short review of the progress of solving the two last problems, *i.e.* the lumped and distributed inflow problems, we start from the statement that in the case of linear hydrological models of flood routing it is difficult to couple the simplicity of approach with the requirements of mathematics. As results from the nature of the linear Saint-Venant model, an element of lateral inflow would propagate both down streams and up streams. Due to a down stream channel disturbances caused by lateral inflow the convenient assumption of semi-infinite channel becomes irrelevant. Kirshen & Bras (1983) used the impulse response solution of complete linear Saint-Venant equations for semi-infinite channels (LCR) to route lateral inflow. Franchini and Todini (1988) did a similar thing using impulse response of the diffusion model. However, caution is needed when superimposing effects of different inputs. The convenient property of superimposing responses is valid here for one entry to a system only. This results from the boundary conditions requirements of linear partial differential equations representing the flow process. The classical solution of the linear-hydrodynamic model is valid for one input and semi-infinite channel only.

It seems that Van de Nes and Hendriks (1971) have been closer to the rigorous solution of the lateral inflow problem than other other investigators. They offered analytical solution for an impulse response for depth (y') and flow discharge (Q') of the diffusion model with lumped lateral inflow of a unit volume per unit width of channel added to the initially relaxed system at time t=0 at one point x=0. Since the paper is not easily available the final solution is presented here:
the impulse response of water depth

$$h_{y'}(x,t) = \frac{1}{2\sqrt{\pi D t}} \exp\left[-\frac{(x-at)^2}{4Dt}\right] \qquad (2.28)$$

and the impulse response of flow discharge

$$h_{Q'}(x,t) = \frac{(x+at)}{4\sqrt{\pi D t^3}} \exp\left[-\frac{(x-at)^2}{4Dt}\right] \qquad (2.29)$$

valid for $(x,t)\neq(0,0)$ and $t>0$, where a and D the translation coefficient and the diffusion coefficient respectively (compare equation (2.17) for $\Delta_{LDA}=0$).

However, they subsequently used above equations for flow routing in the multi-input framework. The multiple inputs considered were: lumped tributaries, distributed lateral inflows and inflow from upstream to the river reach (in other words: flow rate in the upstream terminating cross-section). Superposition of these effects is in contradiction with the diffusion equation describing the flow process. The situation would be different if a linear model is used, where the wave propagates downstream only, *i.e.* where the backwater effects do not exist. Then in the case of a lateral inflow directed downstream one could use the solution obtained in the single-input case also for the multi-input situation. There exist three linear hydrodynamic models with these properties, namely rapid flow model (Strupczewski & Napiórkowski, 1990a), kinematic diffusion model (Strupczewski & Napiórkowski, 1986, 1989) and kinematic wave, though possible an oversimplified one (pure translation). Naturally lumped conceptual models like Kalinin-Miljukov or Muskingum also fall under this category. Strupczewski & Kundzewicz (1992) have derived an analytical expression (in a series form) for impulse response of the rapid flow model with lateral distributed inflow. Therefore handling by RFM both lumped and distributed lateral inflows has been made possible.

3. Hydrodynamic Justification of the Lag Kalinin-Miliukov Model

To defend the leading position of the "conceptual models" one can argue that there are also some physical justifications of the "conceptual" flood routing models. In fact the Kalinin-Miliukov linear reservoirs cascade model and the length of characteristic reach for single linear reservoir were derived (Kalinin & Miliukov, 1958) by rough simplifications of the unsteady flow equation, *i.e.*, for the bottom slope close to zero ($S_o \simeq 0$) and long flood waves. However , if Froude number equals zero, *i.e.*, in particular if $S_o = 0$, the LCR is reduced to the diffusion analogy model (Dooge, 1973, Dooge & Napiórkowski , 1987) but not to the Kalinin-Miliukov model. The physical explanation of the Muskingum method (Mc Carthy, 1939) was given by Strupczewski & Kundzewicz (1980a) and Dooge *et al.* (1982). It was done under the assumption of a common looped rating curve for both the input and the output from the reach and linear changes of water table along the reach which limits the application to short reaches and long waves.

Both above mentioned "conceptual" models give response as soon as their input is activated. However, a time of flood waves transportation along river reach can be quite considerable. Therefore, in order to improve the ability of the Kalinin-Miliukov model to simulate flood wave propagation down stream a river channel, the linear channel providing pure translation is often supplemented in series. The task of this section is to show whether staying on the ground that the LCR is the best physical linear model there is any hydrodynamic reason of so great popularity of the Lag

Kalinin-Miliukov (LKM) model and to compare its accuracy for various flow conditions as well as an accuracy of two limiting cases solution of LCR, *i.e.* for the Froude number of zero and one. This has been done (Strupczewski, 1993) by an analysis of the impulse responses of the above mentioned models. Working on a criterion for choosing a linear three parametric conceptual flood routing models Strupczewski and Kundzewicz (1980b) went up to the fifth central moments' comparison finding a high compatibility of the LKM and LCR models for certain flow conditions. The derivation of LCR for the limiting case of the Froude number equal one, *i.e.* the Rapid Flow (RF) model made by Strupczewski and Napiórkowski (1990a) and of the general form for the R-th order cumulant of LCR by Romanowicz *et al.* (1988) (corrected in the present study) have created an opportunity of making a more precise and detailed approach to the problem. In fact it was a philosophical aspect of the problem concerning a relationship of practice and theory which stimulated this study. That is why the structural relations of the conceptual models with RF model are analyzed first.

3.1 IMPULSE RESPONSES OF CONCEPTUAL MODELS

In applications of the conceptual flood routing models, pure delay is considered as an important element considerably improving their ability to represent measured hydrologic responses. This is why the impulse responses presented here contain the pure time lag.

The simplest conceptual model is the Linear Kinematic Wave (KW) model, considered also as a hydrodynamic one, with the inflow-outflow impulse response in the time domain of the form

$$h(x,t) = \delta(t-\Delta_{KW}) \tag{3.1}$$

The impulse response of the Lag Muskingum (LM) model is given by

$$h(t)=-\frac{a}{1-a}\delta(t-\Delta_{LM})+\frac{1}{K(1-a)^2}\exp[-\frac{t-\Delta_{LM}}{K(1-a)}]\;1(t-\Delta_{LM}) \tag{3.2}$$

The impulse response of the Lag Kalinin-Miliukov (LKM) model is given by:

$$h(t)=\frac{1}{k(n-1)!}\left(\frac{t-\Delta_{LKM}}{k}\right)^{n-1}\exp\left(-\frac{t-\Delta_{LKM}}{k}\right)\;1(t-\Delta_{LKM}) \tag{3.3}$$

Cumulants of the above impulse responses together with physical estimates of parameters are given in Appendix B.

3.2 STRUCTURAL RELATIONS OF THE CONCEPTUAL MODELS AND THE RF MODEL

Since the RF model, which is a limiting case solution of the linearized Saint Venant equation, is build of the conceptual models' elements like a linear reservoir and linear channel which provides pure translation it seems to be a good starting point for a search of structural linkage between the conceptual models and the LCR.

Looking at equation (2.18) one can see that in terms of the Poisson distribution a flow along every path is directed downstream. From equation (A.23) it follows that the parameter Λ of the Poisson distribution is linearly related to the length of channel reach. Therefore, for a short reach a major portion of the unit volume of water flows along paths with small i-number, *i.e.* with small numbers of reservoirs in series. For $\Lambda<1$ the zero-path is most active which makes the KW model to be a reasonable approximation for a very short reach .

Let us express the LM model's response, given by equation (3.2) in the form comparable to equation (2.18):

$$h(t)=p_0\delta(t-\Delta)+p_1h_1[(t-\Delta)/\alpha] \tag{3.4}$$

where

$$p =-\frac{a}{1-a} \tag{3.5}$$

$$p_0+p_1=1 \tag{3.6}$$

$$\alpha=K(1-a). \tag{3.7}$$

The only difference between equation (2.18) and equation (3.4) is that the Poisson distribution is replaced by the two-point distribution. Therefore if for given Λ the sum of P_o and P_1 in the Poisson distribution is closed to one, the LM model will be agood approximation of the RF model.

To get a downstream transport along both paths defined by equation (3.4), the parameter a should be negative. From matching the first three cumulants of the LM model (equations(B.3-4)) with those of the RF model (equations (A.20-21)) we get

$$\Lambda = \frac{9}{8}\,\frac{(1-2a)^3}{[1-3a(1-a)]^2} \tag{3.8}$$

Solving equation (3.8) for a we find that it has real solutions for $\Lambda \le 9/8$ only. Substitution of $\Lambda=9/8$ into equation (A.23) gives the upper limit of length of a reach admissible for the LM model. For $\Lambda = 9/8$ which corresponds to LKM with a single reservoir (*i.e.* a=0) we get from equation (2.19) that

$$P_0+P_1=0.3245+0.3650=0.6895 \tag{3.9}$$

Since the LCR and the RF response have identical values of the first three cumulants, the above limit is in force for the LM response approximation of the LCR. These two solutions for the parameter a differ in sign.

The relative error of fourth cumulants can be taken as a measure of accuracy of approximation as proposed by Strupczewski & Kundzewicz (1980b). For the RF approximation by the LM the relative error of the fourth cumulants

$$\delta k_4(LM,RF)= \frac{k_4(LM)-k_4(RF)}{k_4(RF)} \tag{3.10}$$

is

$$\delta k_4(LM,RF)= \frac{9}{8}\,\frac{(1-4a+6a^2-4a^3)(1-2a)}{[1-3a(1-a)]^2} - 1 . \tag{3.11}$$

The choice of the solution of equation (3.8) which better fits the LCR model was discussed by Strupczewski & Napiorkowski (1990 b). They gave preference to the negative solution stating: "...for positive values of the parameter a, a good fit for the fourth cumulant excludes a good fit for pure delays...However for a<0 ...a good fit for fourth cumulants follows a good fit for pure lags." Since the RF is the LCR limiting case the conclusion remains valid here.

Comparing the LKM model with the RF model, *i.e.*, equation (3.3) with equation (2.18), it is clear that the LKM model corresponds to a transport of the whole unit volume along one path of the RF model. All cumulants in the Poisson are equal

$$k_R= \Lambda \qquad \text{for } R=1,2,... \tag{3.12}$$

Then in particular, the variance of the distribution grows

linearly with the length of the reach which results from equation (A.23). A share of most active path decreases with Λ according to

$$\max_i P_i = \frac{\Lambda^{[\Lambda]}}{[\Lambda]} \exp(-\Lambda) \tag{3.13}$$

Therefore, in spite of a certain structural similarity, the impulse response of LKM model differs considerably from that of the RF and does not converge to it for any particular condition. However , this does not exclude a possibility of a good fit with the LCR for any other than F=1 flow conditions. This is investigated in the next subsection of the paper.

3.3 RELATIONSHIP OF THE APPROXIMATED MODELS' CUMULANTS

Analyzing the expressions for the R-th cumulant of the LDA response (equation (A.16)), of the RFM response (equation (A.21)) and of the LKM response (equation (B.7)) we find that the R-th cumulant is related to the second and third cumulant by

$$k_R = \alpha_R \frac{k_3^{\,R-2}}{k_2^{\,R-3}} \qquad \text{for } R = 2,3,4,\ldots. \tag{3.14}$$

where the coefficient α_R depends on the model and the order of cumulant. Expressions for α_R together with the respective recurrence relationships for the three approximated models are given in TABLE 1.

Using the moment matching technique for physical estimation of the parameters of the above models we make their three first moments equal to those of the LCR. Substituting into equation (3.14) the expressions for k_2 and k_3 from equation (A.1) we get

$$k_R = \alpha_R \frac{\phi_3^{\,R-2}}{\phi_2^{\,R-3}} \frac{z^R}{D^{R-1}} \qquad \text{for } R = 2,3,4,\ldots \tag{3.15}$$

TABLE 1. The coefficients α_R of the LCR approximated models

Model	LDA	RF	LKM
$\alpha_R =$	$\frac{(1)(3)..(2R-3)}{3^{R-2}}$	$\frac{R!}{2\cdot 3^{R-2}}$	$\frac{(R-1)!}{2^{R-2}}$
Recurrence relation $\alpha_R =$	$\frac{2R-3}{3}\alpha_{R-1}$	$\frac{R}{3}\alpha_{R-1}$	$\frac{R-1}{2}\alpha_{R-1}$

3.4. APPLICABILITY REGIONS OF APPROXIMATED MODELS

Since the first three cumulants of the discussed models and the LCR are mutually equaled, all these models give the same response to a polynomial function of the third degree. Hence the differences between the fourth cumulants of the impulse responses of the approximated models and the LCR can serve as a criterion of a model choice as proposed by Strupczewski & Kundzewicz (1980b).

The border line between regions of applicability of two models, say M_1 and M_2, fulfills the condition of equal absolute relative errors of the fourth cumulants:

$$|\delta k_4(M_1,\text{LCR})| = |\delta k_4(M_2,\text{LCR})| \; . \tag{3.16}$$

Taking into account equation (3.15) we get for the LDA, RF, and LKM models

$$\delta k_4(\;\cdot\;,\text{LCR}) = \frac{\alpha_4(\;\cdot\;)\;\phi_3^{\;2}}{\phi_4\;\phi_2} - 1 \; . \tag{3.17}$$

Let us consider for the moment that the LDA and RF are the only models used as the LCR approximation. The border line between them fulfills the condition:

$$\delta k_4(\text{RF},\text{LCR}) = -\;\delta k_4(\text{LDA},\text{LCR}). \tag{3.18}$$

Substitution into it $\alpha_4(\cdot)$ from TABLE 1 gives

$$\phi_4 = \alpha_4 \frac{\phi_3^{\,2}}{\phi_2} \qquad \text{with } \alpha_4 = 3/2 \ . \tag{3.19}$$

Since ϕ_R are functions of F and m only, F and m are the onlyvariables of equation (3.19). Its solution for F takes the form:

$$F^2 = \frac{1}{(m-1)\,X}\left[0.1\ m\sqrt{m^2 - 20\ X} - (0.1\ m^2 - X)\right], \tag{3.20}$$

where

$$X = (m-1)(1-3\alpha_4/5) \tag{3.21}$$

and α_4=3/2 for the LDA/RF border line.

On the border line the absolute relative error (3.18) for alternative use of LDA and RF models reaches its maximum equal to 1/9.

Therefore a third approximated model shall be close to the border line defined by equation (3.19). Surprisingly, looking at TABLE 1 we find that just for the LKM the coefficient α_4 equals 3/2. That is, the LKM occupies on the (F,m) plane the most desirable position. One can get from equation (3.21) that for a wide rectangular channel with Manning friction (m=5/3) the Froude number equals 0.49301, while for Chezy friction (m=3/2) F=0.54018, which corresponds to the flow conditions that are frequently met in natural channel during floods.

Proceeding in a similar way, the LDA/LKM and the LKM/RF border lines were derived as shown in *Figure 1*. For the LDA/LKM line α_4 equals 19/12 with the error $|\delta k_4|$=1/19, while for the LKM/RF line α_4=17/12 with $|\delta k_4|$=1/17. That is, the inclusion of the LKM model has reduced the maximum error almost twice.

Relative errors for various sets of approximated models are shown in TABLE 2. A comparison of one model cases discloses the priority of the LKM model over the LDA and RF models. Furthermore, its maximum relative error is of a similar magnitude as the one for the LDA and RF used alternatively.

Although all three models, *i.e.* the LDA, RF and LKM models if used for any flow conditions are the LCR approximations there are differences between them in respect to their origins. While two first models are exact solutions of the LSV equation for the limiting cases, *i.e.*,

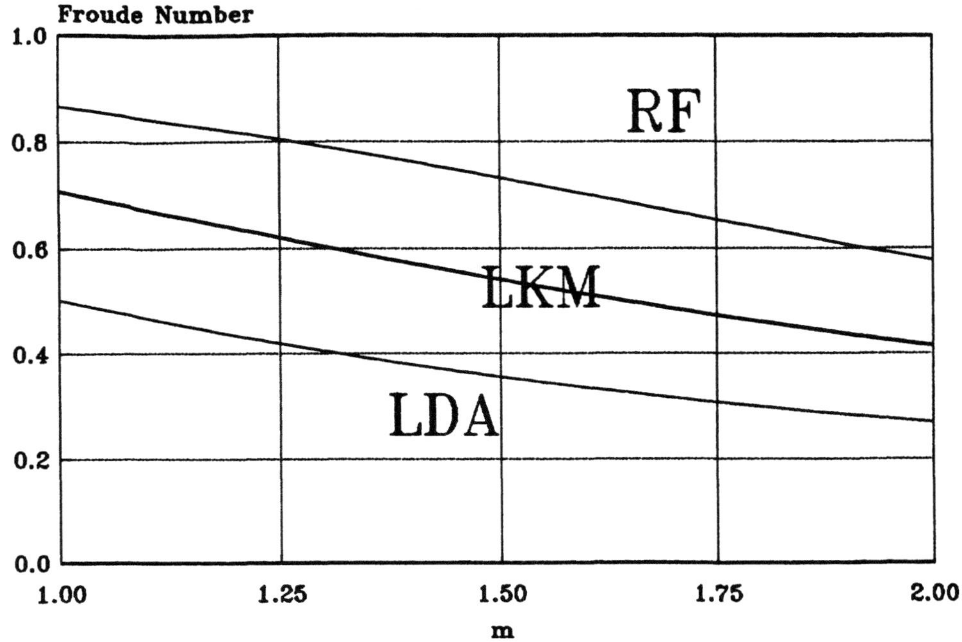

Figure 1. Regions of applicability of LDA, LKM, RF models as obtained from a comparison of fourth cumulants.

for F equal to zero and one respectively, the third one never provides the exact solution of the LSV equation being always an approximation only. It fits the LCR best for the flow condition given by equation (3.20). In the next part of the paper the quality of its approximation will be assessed.

TABLE 2. Relative error of fourth cumulant for various sets of approximated models, measured along the boundaries and the division lines.

Models		Boundary or division line					$\max\lvert\delta k_4\rvert$
		F=0	(LDA,LKM)	(LDA,RF)	(LKM,RF)	F=1	
Single	(LDA)	0	N.A.	N.A.	N.A.	-1/5	1/5
	(RF)	+1/4	N.A.	N.A.	N.A.	0	1/4
	(LKM)	-1/10	N.A.	N.A.	N.A.	+1/8	1/8
Two	(LDA/RF)	0	N.A.	±1/9	N.A.	0	1/9
	(LDA/LKM)	0	±1/19	N.A.	N.A.	+1/8	1/8
	(LKM/RF)	-1/10	N.A.	N.A.	±1/17	0	1/10
Three	(LDA/LKM/RF)	0	±1/19	N.A.	±1/17	0	1/17

"N.A." denotes "Not Applicable"

3.5. RELATIONSHIP OF THE LCR CUMULANTS

Since the LCR has four parameters, namely z, D, F and m, then the first four cumulants define all other ones. Substitution equations (A.9-14) for the parameters of equation (A.1) gives the LCR cumulants relationships:

$$k_R = f_R(k_1, k_2, k_3, k_4). \qquad (3.22)$$

There exists another, much simpler relationship for the LCR cumulants which is presented below.

For every pair of (F, m) values taken from the LCR validity region the values of the ϕ_R functions can be obtained from equation (A.2). Substituting of equation (A.1) into equation (3.14) and reordering we get

$$\alpha_R = \frac{\phi_R \, \phi_2^{R-3}}{\phi_3^{R-2}} \qquad (3.23)$$

Therefore for given F and m the coefficients α_R of the LCR are determined.

Let us put R=4 into equation (3.23)

$$\alpha_4 = \frac{\phi_4 \, \phi_2}{\phi_3^{2}} \qquad (3.24)$$

For the α_4 value taken from the interval (4/3 , 5/3), where the limits correspond to the limiting cases solution (see TABLE 1), the solution of equation (3.24) is given by equation (3.20). It describes the line on the (F,m) plane.

The newly discovered property of the cumulants' functions (3.23) is that for a fixed value of α_4 (or any α_R for R≥4) all other α_R values remain constant

$$\alpha_R = \phi_R(\alpha_4). \qquad (3.25)$$

The property was proved analytically for R=5 (Appendix C) and then confirmed numerically for higher orders R. The expression for the R-th LCR cumulant (Romanowicz *et al.*, 1988) has made the computation simple. For R=5 we get equation (C.6)

$$\alpha_5 = 5\,(\alpha_4 - 8/9), \qquad (3.26)$$

while for R>5 the relation is not linear.

Now one can look at the LCR as at an infinite number of responses corresponding to the different α_4 values. Two of them with relatively simple forms have been already recognized, *i.e.*, RF and LDA responses displayed as the horizontal lines in *Figure 1*. The LKM response will be analyzed in the next section as the approximation of the third one, *i.e.* for α_4=3/2.

3.6. LKM RESPONSE AS LCR APPROXIMATION

3.6.1. Comparison of Cumulants

The present task is to examine the accuracy of the approximation of the LCR by the LKM response based on a comparison of cumulants. First it will be done for the flow condition defined by equation (3.19), where the LKM fits best the LCR, and then extended to the vicinity of the line (3.20).

The relative error of R-th cumulant is defined as

$$\delta k_R(\text{LKM, LCR}) = \frac{k_R(\text{LKM})}{k_R(\text{LCR})} - 1 \qquad (3.27)$$

Substituting equation (3.15) and equation (A.1) with ϕ_R taken from equation (3.23) for $k_R(\text{LKM})$ and $k_R(\text{LCR})$ respectively, we get

$$\delta k_R(\text{LKM, LCR}) = \frac{\alpha_R(\text{LKM})}{\alpha_R(\text{LCR})} - 1 \qquad (3.28)$$

The values of coefficients α_R for both models and R=2,3,..,10 are given in TABLE 3. That is, as the effect of LCR property described in Subsection 3.5 the error for flow condition (3.19) depends on the order of cumulant only. The relative error (3.28) is shown in TABLE 2. The accuracy of estimation is surprisingly high. For R=5 the error is less than 2%. The error function tends to become linear with increasing R.

TABLE 3. Cumulants comparison of LKM impulse response with LCR for four first cumulants matched.

R =	1	2	3	4	5	6
α_R(LCR)	-	1	1	1.5	3.055	7.917
α_R(LKM)	-	1	1	1.5	3.0	7.5
δk_R (%)	0	0	0	0	-1.818	-5.263

Being aware of low accuracy of higher order cumulants estimated from hydrological time series one leans to the opinion that the LKM response can be considered as the third particular case solution of the linearized Sant Venant equation for semi-infinite uniform channel.

The sensitivity analysis of error of cumulants to change in F and m values around the point (F=0.49 , m=5/3) on the line (3.19) has shown that a high accuracy is kept in a wide interval of changes in F and m values.

3.6.2. Comparison of pure lags

The equality of the first few cumulants of two impulse responses secures a good fit of impulse responses for long times (Dooge (1973), Strupczewski & Kundzewicz (1980b)). Here a quality of LCR by the LKM approximation made by cumulants matching is assessed for short times.

Both the LKM response and the LCR contain a pure lag as the parameter defined by equation (2.15) and (B.10), respectively. Therefore, it is relatively simple to compare them. Let us take as a measure of fit the relative error of the pure delay

$$\delta\Delta(\mathrm{LKM},\ \mathrm{LCR}) = \frac{\Delta(\mathrm{LKM})}{\Delta(\mathrm{LCR})} - 1. \tag{3.29}$$

Substituting equations (2.15) and (B.10) we get

$$\delta\Delta(\mathrm{LKM},\ \mathrm{LCR}) = \frac{1+F}{m\ F}\left(1 - 2\,\frac{\phi_2^{\,2}}{\phi_3}\right) - 1 \tag{3.30}$$

The LCR contains the discontinous head of the wave as a delta function of the volume decreasing with length D (equation (2.13) while the LKM does not. This structural difference may caused that the forced equivalence of the first four cumulants is not followed by a perfect fit of pure lags while the error (3.29) is of negative sign. It equals -12.44 % for wide rectangular channel with Chezy friction and -11.28 % with Manning one.

4. Conclusions

The linear theory of an open channel flow has been tackled by hydrologists and oceanographers separetely. While hydrologists have worked on run-off modeling the main interest of oceanographers has been confined to tides propagation along a channel. Therefore for further progress of work on a general theory it seems necessary to combine results on both problems. The linear theory of an open channel flow is not closed chapter of hydrology yet. The lateral inflow problem has not been solved satisfactory except of the limiting case, or at least an accuracy of simplification by superimposing effects of different inputs has not been checked and found acceptable. It is worth to check how the linear theory can benefit from exchange initial and boundary conditions in a process of problems solving under assumption of a finite system memory. The problem of a stage-discharge rating curve as a lower boundary condition is also waiting for a solution. The problem of tides propagation in the presence of a steady or unsteady run-off discharge may be join task for hydrologists and oceanographers. A practical interest in linear open channel flow modeling is still considerable and a fascination for numerical methods of nonlinear modeling do not blind professionals working with real data. An educational purpose of a linear theory seems to be unquestionable. It is difficult to get understanding of unsteady flow process without getting first an insight into it which gives solutions of linear theory problems.

A comparison of the conceptual models of flood routing with the linearized Saint Venant model of unsteady flow due to an upstream impulse input in a semi infinite channel reveals that hydrological modelers - thanks to understanding of river flow process, scientific intuition and a work with real data - have managed to get the models which for certain flow conditions are close to the physical one but much simpler in structure. In fact, the impulse response of RF model, *i.e.*, the solution of the linearized Saint Venant equation for Froude number equal to one, consists of the elements of the "conceptual" models like a linear reservoir and linear channel. However, in spite of this none of the simple conceptual models with pure delay is satisfactory for a steep channel except of the Lag Muskingum with limit to a river reach not longer than the characteristic length of the LKM model.

Considering two limiting case solutions, *i.e.*, RF and LDA models, as the alternative approximation of the LCR in the whole range of (F,m) variability it was found that the approximation error measured as the relative difference of the fourth order cumulants reaches its maximum along the line (3.19) marking the position of LKM model. That is if one prefers to use three relatively simple models instead of LCR he will find the position of LKM on the (F,m) as the most desirable.

A comparison of the relative errors for various sets of approximated models (TABLE 2) reveals the priority of the LKM model over the LDA and RF models.

The property defined by Eq.3.25 of the α_R coefficients has made possible the classification of Lag two-parameter models. For RF, LDA and LKM models, α_4 equals 4/3, 5/3 and 3/2, respectively.

A comparison of the accuracy of LCR approximation by LKM model revealed amazingly good agreement (TABLE 3) for flow conditions given by equation (3.20) as well as in the vicinity while a match of pure lags is quite satisfactory. Taking into account the quality of the data used for calibration one can consider the LKM model as the third particular solution of the linearized Saint Venant equation in a semi-infinite channel fed by an upstream impulse.

In view of the presented results, the LKM model might have been derived by theoreticians working on fluid mechanics and then recommended for practice. However, it has happened the other way around and in fact it is not the only recognized case in water science of a direct contribution of practitioners in a theory development. Strupczewski (1973) reported a much more drastic case in water storage reservoir area. Long before the algorithms of approximate solution for square programming problem were adopted for reservoir regulation, the exact solution had been known and widely applied by practitioners. All these facts call for a removal of the barriers between practitioners and theoreticians and for conciliation and reconciliation in order to achieve a single-minded approach to handle all theoretical problems in water sector area.

Acknowledgement

The author is very grateful to Professor J.C.I. Dooge for his valuable comments on the first draft of the manuscript.

Appendix A. Cumulants and Physical Estimates of the LSV Flood Routing Models.

A.1. LINEAR CHANNEL RESPONSE

A.1.1. Cumulants

The cumulants of the Linear Channel Response (LCR) given by equations (2.13-14) have a general form (Dooge, 1972, Dooge *et al.*, 1987):

$$k_R = \phi_R(m,F)\,\frac{z^R}{D^{R-1}}\,. \tag{A.1}$$

The expression for the ϕ_R functions was derived by Romanowicz *et al.* (1988). After correction of its small algebraic error it would be of the form:

$$\phi_R = \left(\frac{1}{m}\right)^{R-1} \frac{[1+(m-1)F^2]}{1-F^2} \sum_{i=0}^{[R/2]} \gamma(R,i) \left[\frac{m\,F}{1+(m-1)F^2}\right]^{2i} \tag{A.2}$$

where

$$\gamma(R,i) = (-1)^i \frac{R!\,(2R-2i-3)!}{(R-i-2)!\,(R-2i)!\,i!} \left(\frac{1}{2}\right)^{R-2} . \tag{A.3}$$

Since the first five cumulants are frequently used in the paper it is convenient to define them separately:

$$\phi_1 = 1 \tag{A.4}$$

$$\phi_2 = \left[1-(m-1)^2F^2\right]/m \tag{A.5}$$

$$\phi_3 = 3\,\phi_2[1+(m-1)F^2]/m \tag{A.6}$$

$$\phi_4 = \frac{3(1-F^2)}{m}\,\phi_2^{\,2} + \frac{4}{3}\,\frac{\phi_3^{\,2}}{\phi_2} \tag{A.7}$$

$$\phi_5 = \frac{5\,\phi_3}{\phi_2}\left(\phi_4 - \frac{8}{9}\,\frac{\phi_3^{\,2}}{\phi_2}\right) . \tag{A.8}$$

A.1.2. The LCR Parameters in Terms of Cumulants

$$z = k_1 \tag{A.9}$$

$$D = 3\,\frac{k_1^{\,2}}{k_2}\,\frac{A-B}{B(A^2-3B)} \tag{A.10}$$

$$F^2 = \frac{(A-B)^2}{(A^2-3B)} \tag{A.11}$$

$$m = \frac{2A-B-3}{A-B} , \tag{A.12}$$

where

$$A = \frac{k_3 \, k_1}{k_2^{\,2}} \tag{A.13}$$

$$B = \frac{3 \, k_2 \, k_4 - 4 \, k_3^{\,2}}{3 \, k_2^{\,4}} \, k_1^{\,2} \, . \tag{A.14}$$

A.2. LINEAR DIFFUSION ANALOGY MODEL (LDA)

A.2.1. Cumulants

The cumulants of the Linear Diffusion Analogy Model impulse response given by equation (2.17) have a general form

$$k_1 = \frac{x}{a} + \Delta \tag{A.15}$$

$$k_R = \{(1)(3)(5)\ldots(2R-3)\} \left(\frac{2 \, D}{a \, x}\right)^{R-1} \left(\frac{x}{a}\right)^{R} \quad \text{for } R>1 \tag{A.16}$$

A.2.2. Matching of LDA to LCR

By matching the three first cumulants of the LCR and LDA responses the physical estimates of the LDA model's parameters have been derived

$$a = \frac{x}{3z} \; \frac{\phi_3}{\phi_2^{\,2}} = \frac{1+(m-1)F^2}{1-(m-1)^2F^2} \; m \, v_0 \tag{A.17}$$

$$D = \frac{1}{54} \; \frac{\phi_3^{\,3}}{\phi_2^{\,5}} \; \frac{x^2}{z \, D} = \frac{1}{2m} \; \frac{[1+(m-1)F^2]^3}{[1-(m-1)^2F^2]^2} \; \frac{x^2}{z \, D} \tag{A.18}$$

$$\Delta = \left[1 - 3 \, \frac{\phi_2^{\,2}}{\phi_3} \right] z = \frac{m(m-1)F^2}{1+(m-1)F^2} \, z \, . \tag{A.19}$$

A.3. RAPID FLOW MODEL (RF)

A.3.1. Cumulants

The cumulants of the impulse response of the Rapid Flow

Model (RF) defined by equation (2.18) have a form

$$k_1 = \Delta + \alpha\, \Lambda \tag{A.20}$$

$$k_R = R!\, \alpha^R\, \Lambda \qquad \text{for } R>1\ . \tag{A.21}$$

A.3.2. Matching of RF to LCR

$$\alpha = \frac{1}{3}\,\frac{\phi_3}{\phi_2}\,\frac{z}{D} = \frac{1}{m}\left[\,1 + (m-1)F^2\right]\frac{z}{D} \tag{A.22}$$

$$\Lambda = \frac{9}{2}\,\frac{\phi_2^{\,3}}{\phi_3^{\,2}}\,D = \frac{m}{2}\,\frac{1-(m-1)F^2}{[1+(m-1)F^2]^2}\,D \tag{A.23}$$

Appendix B. Cumulants and Physical Estimates of Parameters of the Conceptual Flood Routing Models

B.1. LINEAR KINEMATIC WAVE MODEL (KW)

B.1.1. Cumulants

Cumulants of impulse response of Kinematic Wave (KW) model defined by equation (3.1) have a form

$$k_1 = \Delta \tag{B.1}$$

$$k_R = 0 \qquad \text{for } R > 1 \tag{B.2}$$

B.1.2. Physical Estimate of the Parameter Δ

$$\Delta = \frac{x}{m\, v_0}\ . \tag{B.3}$$

B.2. LAG MUSKINGUM MODEL (LM) (equation (3.2))

B.2.1. Cumulants

$$k_1 = K + \Delta \tag{B.4}$$

$$k_R = (R-1)!\ [(1-a)^R - (-a)^R]\ K^R \qquad \text{for } R>1 \tag{B.5}$$

B.2.2. Matching of LM to LCR

Expressions for physical estimates of parameters were given

by Strupczewski & Napiorkowski (1990 b). Because of their long form they are not repeated here.

B.3. LAG KALININ-MILIUKOV MODEL (LKM) (equation (3.3))

B.3.1. Cumulants

$$k_1 = n\,k + \Delta \tag{B.6}$$

$$k_R = n(R-1)!\,k^R \qquad \text{for } R>1 \tag{B.7}$$

B.3.2. Matching of LKM to LCR

$$k = \frac{1}{2}\frac{\phi_3}{\phi_2}\frac{z}{D} = \frac{3}{2m}\left[1+(m-1)F^2\right]\frac{z}{D} \tag{B.8}$$

$$n = 4\,\frac{\phi_2^{\,3}}{\phi_3^{\,2}}\,D = \frac{4\,m}{9}\,\frac{[1-(m-1)^2F^2]}{[1+(m-1)F^2]^2} \tag{B.9}$$

$$\Delta = \left(1-2\,\frac{\phi_2^{\,2}}{\phi_3}\right)z = \left[1-\frac{2}{3}\,\frac{1-(m-1)^2F^2}{1+(m-1)F^2}\right]z. \tag{B.10}$$

Appendix C. Relationship between the Coefficients α_4 and α_5 of LCR.

Let us take equation (3.14) with R = 4 and 5

$$\alpha_4 = \frac{k_4\,k_2}{k_3^{\,2}} \tag{C.1}$$

and

$$\alpha_5 = \frac{k_5\,k_2^{\,2}}{k_3^{\,3}}\,. \tag{C.2}$$

Then due to equation (A.1) we get

$$\alpha_4 = \frac{\phi_4\,\phi_2}{\phi_3^{\,2}} \tag{C.3}$$

and

$$\alpha_5 = \frac{\phi_5\,\phi_2^{\,2}}{\phi_3^{\,3}}\,. \tag{C.4}$$

Therefore

$$\alpha_5 = \frac{\phi_5\,\phi_2}{\phi_4\,\phi_3}\,\alpha_4 \tag{C.5}$$

Substituting into equation (C.5) equations (A.5-8) for ϕ_2, ϕ_3, ϕ_4 and ϕ_5 we get

$$\alpha_5 = 5(\alpha_4 - \frac{8}{9}). \tag{C.6}$$

References

Becker, A. and Kundzewicz, Z.W. (1986)Nonlinear flood routing with multi linear models ,*Water Resour. Res.*, 22,1413-1421.

Brutsaert, W. (1973) Review of Green's functions for linear open channel, *J. Engin. Mech. Division Proc. ASCE*,99,EM6, 1247-1257.

Cowpertwait, P.S.P. and O'Connell (1992) A Neyman-Scott shot noise model for the generation of daily stream flow time series, *Advances in theoretical hydrology,* edited by J.P. O'Kane , European Geophys. Soc., Hydrol.Sc.Series,1, 75-94.

Deymié, Ph. (1935)Propagation d'une intumescence allongée (Propagation of a long wave). *Rev. Gen. l'Hydraulique* 3, 138-142.

Dooge, J.C.I. and Harley, B.M. (1967)Linear routing in uniform open channels,*Proc. Intern. Hydrol. Symp.,* Fort Collins, Colorado, September 6-8, 1, 57-63.

Dooge, J.C.I. (1973) *Linear theory of hydrologic systems,* Techn. Bull. No.1468, Agricult. Res. Serv., US Depart. of Agricult., Washington DC.

Dooge, J.C.I., Strupczewski W.G. and Napiórkowski, J.J. (1982) Hydrodynamic derivation of storage parameters of the Muskingum model,*J.Hydrol.*,54, 371-387.

Dooge, J.C.I., Kundzewicz, Z.W. and Napiórkowski, J,J. (1983) On backwater effects in linear diffusion routing, *Hydrol.Sc.J.*, 28, 391-402.

Dooge, J.C.I. and Napiórkowski, J.J. (1984) Effect of downstream control in diffusion routing, *Acta Geophys. Pol.* XXXII,363-373.

Dooge, J.C.I. and Napiórkowski, J.J. (1987) Applicability of diffusion analogy in flood routing, *Acta Geophys. Pol.*,35, 65-75.

Dooge, J.C.I., Napiórkowski, J.J. and Strupczewski, W.G. (1987a) The linear downstream response of a generalized uniform channel, *Acta Geophys. Pol.*, 35,3, 277-291.

Dooge, J.C.I., Napiórkowski, J.J. and Strupczewski, W.G. (1987b) Properties of the generalized downstream channel response, *Acta Geophys. Pol.*, 35,4, 405-418.

Dooge, J.C.I. and Napiórkowski, J.J. (1987) The effect of the downstream boundary condition in the linearized St. Venant equation. *Q.J. Mech. Appl. Math.*, 40,245-256.

Dronkers, J. (1964) *Tidal computations in rivers and coastal waters*, Chapter 8, North-Holland Publ. Co.

Franchini, M. and Todini, E. (1988) PABL: A parabolic and backwater scheme with lateral inflow and outflow, *Proceedings of Fifth IAHR Symposium on Stochastic Hydraulics*, Univ. of Birmingham, 2-4 August, paper 6.2, pp. 12.

Hayami, S.(1951) On the propagation of flood waves. *Kyoto Univ. Japan. Disaster Prevention Res. Inst. Bull.*1,1-16.

Kalinin, G.P. and Miliukov, P.I. (1958) Priblizennyj rascet neustanovivwsego dvizenija vodnyh mass, *Trudy C.I.P.*, 66, pp.71, one Tab.

Keefer,T.N. (1974) Desktop computer flow routing, *J.Hydraul.Div. ASCE*, 100, 1047-1058.

Keefer, T.N. and McQuivey, R.S. (1974) Multiple linearization flow routing model, *J.Hydraul.Div.ASCE*, 100,1031-1046.

Kirshen, D.M. and Bras, R.L. (1983) The linear channel and its effect on the geomorphologic IUH, *J.Hydrol.*,65,175-208

Lighthill, M.J.and Whitham, G.P. (1955) On kinematic waves. I - Flood movements in long rivers, *R. Soc. London Proc.* A 229, 281-316.

Massé, P.R. (1939) *Recherches sur la theorie des eaux courantes (Research on the theory of flowing waters)* John Wiley Sons Inc., New York, Intern. Cong. Appl. Mech., 5-th Mtg. New York, Proc., 545-549.

McCarthy, G.T.(1939) *The unit hydrograph and flood routing*, U.S. Corps Eng., Providence, R.I.

Napiórkowski, J.J. and Dooge, J.C.I. (1988) Analytical solution of channel flow model with downstream control, *Hydrol.Sc.J.*,33,269-287.

Napiórkowski, J.J. (1992) Linear theory of open channel flow, in J.P. O'Kane (ed.) *Advances in theoretical hydrology,* European Geophys. Soc., Hydrol. Sc. Series, 1, 3-15.

Philip, J.R. (1992) Hydrology and the Real World, in J.P. O'Kane (ed) *Advances in Theoretical Hydrology*, European Geophys. Soc., Hydrol. Sc. Series, 1.

Romanowicz, R.J., Dooge, J.C.I. and Kundzewicz, Z.W. (1988) Moments and cumulants of linearized St. Venant equation, *Adv. Water Res.*V.11, June, 92-99.

Saint-Venant, B. (1871) *Théorie du mouvement non-permanent des eaux crues des riviers et á l'introduction des marées dans leur lit (Theory of unsteady water flow, with application to river floods and to propagation of tides in river channels).* Acad. Sci., Paris, Comptes Rendus, 73, 148-154, 237-240.

Stewart, I. (1992) Warning - handle with care !, *Nature* V.355, 2 January, 16-17.

Strupczewski, W.G. (1973) Square programming model of reservoir regulation".(in Polish). *Przegl. Geofizyczny,* XVIII (XXVI),1/2, 89-110.

Strupczewski, W.G. and Kundzewicz, Z. (1979) Analysis of physical interpretation of parameters of linear conceptual models by means of moment matching method, *J.Hydrol. Sci.*,6, 143-159.

Strupczewski, W.G. and Kundzewicz, Z. (1980a) Muskingum method revisited, *J.Hydrol.*,48,3/4, 327-342.

Strupczewski, W.G.and Kundzewicz, Z. (1980b) Choice of a linear, three- parametric conceptual flood routing model and evaluation of its parameters, *Acta Geophys. Pol.* 28,2, 129-141.

Strupczewski, W.G., Napiórkowski, J.J. (1986) Asymptotic behavior of physically based multiple Muskingum model,*Proc.Fourth Intern.Hydrol.Symp.,* Forth Collins, Colorado,July 1985, 372-381. Edited by H.W.Shen, J.T.B. Obeysekera, V.Yevjevich & D.G.Decoursey, Publ. by H.W.Shen, Engin. Res. Center, Colorado State Univerity, Fort Collins, Colorado.

Strupczewski, W.G. and Napiórkowski, J.J. (1988) Solution of the complete linearized Saint Venant equation for limiting case of the Froude number equal to one, *Proc. Int. Conf. on Fluvial Hydraulics,* IAHR, Budapest,264-266.

Strupczewski, W.G. and Napiórkowski, J.J. (1989) Properties

of the distributed Muskingum model, *Acta Geophys. Polonica* 37,3-4,299-314.

Strupczewski, W.G. and Napiórkowski, J.J. (1990a) Linear flood routing model for rapid flow, *Hydrol. Sc.J.* No 35 (1,2),149-164.

Strupczewski, W.G. and Napiórkowski, J.J. (1990b) What is the distributed delayed Muskingum model, *Hydrol. Sc. J.* No 35 (1,2),165-178.

Strupczewski, W.G. and Kundzewicz, Z.W. (1992) Rapid flow model with lateral inflow, *Nordic Hydrol.*, 23,57-72.

Strupczewski, W.G. (1993) Approximation of the linearized Saint Venant model by the Lag Kalinin-Miliukov model,*Acta Geophys.Pol.* (in press).

Supino, G. (1950) Sur l'amortissement des intumescences dans les canaux (On the attenuation of steep waves in channels,*Rev. Gén. l'Hydraulique,* 57, 144-147.

Van de Nes, Th.J. & Hendriks, M.H. (1971) *Analysis of a linear distributed of surface runoff*, Landbouwogeschool, Wageningen, The Netherlands, Rapport 1, Jan., pp.129.

Yee, H.C., Sweby, P.K. & Griffiths, D.F.(1991) Dynamical Approach Study of Spurious Steady-State Numerical Solutions of Nonlinear Differential Equations. I. The Dynamics of Time Discretization and its Implications for Algorithm Development in Computational Fluid Dynamics, *Journal of Computational Physics* 97,249-310.

RELIABILITY OF THE MUSKINGUM METHOD FOR THE PREDICTION OF FLOOD WAVE PROPAGATION

by Aldo Giorgini† and Giuseppe Mendicino

Introduction

The Muskingum approach to the hydrologic routing of a hydrograph through a river reach has enjoyed considerable success among watershed modelers and it is one of the most frequently used options in the very well known program HEC-1 developed by the Hydrologic Engineering Center of the U.S. Corps of Engineers. In its most flexible form, the Muskingum approach to the flood routing through a river reach needs the determination of three parameters:

n	the number of subreaches into which the reach is broken;
K	the **travel time** of the wave through the whole reach;
x	the **wedge factor** of each subreach.

The program HEC-1 provides the option of calibrating these parameters by using an optimization scheme based on weighted square errors. In order to achieve this, the user must provide the reach's inflow and outflow hydrographs.

When the optimization option is applied to an actual river reach, difficulties arise due to the fact that lateral inflows to the reach cannot be accounted for in a distributed fashion, but only totally. For this reason HEC-1 requires the specification of a **pattern hydrograph** which serves as a templet for the actual lateral flow hydrograph.

In what follows we will assume that we do not have to reckon with such problem and we will be mainly interested in the routing of a floodwave through a river reach without lateral flow. The goal of this study will be therefore to ascertain how reliable the use of the Muskingum method is in the prediction of hydraulic routing and how reliable the program HEC-1 is in carrying out the Muskingum approximation.

In order to achieve this task we should specify the standards against which to compare the performance of the Muskingum approximation and its HEC-1 implementation. We will rule out laboratory or field experiments because we are dealing with a mathematical lumped approximation of the mathematical problem of solving the De Saint-Venant equations. We should confine our search for standards to reasonably realistic exact solutions of the De Saint-Venant equations. By **reasonably realistic** we mean solutions of problems of routing realistic looking, albeit simplified, inflow hydrographs through realistic looking channels. We have chosen as channel the one illustrated in Figure 1, where all the cross section's macro- and micro- geometric characteristics are given there. The bottom slopes is $S = .0002$ and the Manning's n values are $n = .030$ for the channel and $n = .080$ for the overbanks.

V. P. Singh and B. Kumar (eds.), Surface-Water Hydrology, 297–324.

The length of the channel reach is L = 10 miles. Our choice of inflow hydrograph has been

$$I(t) = Q_0 + \Delta Q_0 \left(\frac{t}{T_p}\right)^2 e^{2(1-t/T_p)} \tag{1}$$

where Q_0 is a constant flow rate present in the channel before t = 0 (baseflow) and continuing

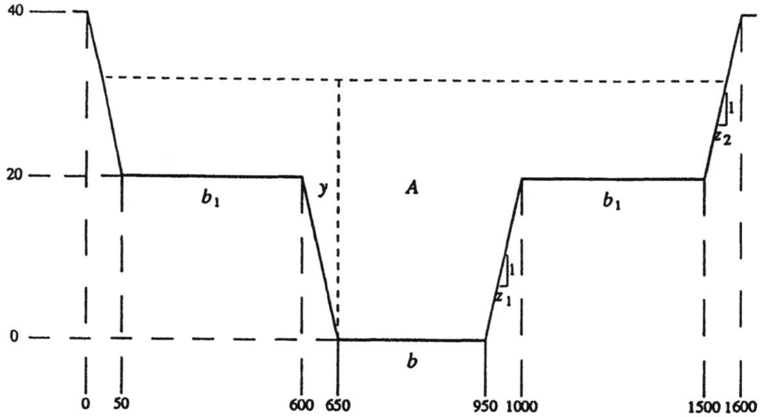

Figure 1. Channel Cross Section

$A_{ch} = 300\,y + 2.5y^2$	for 0<y<20 feet
$= 7000 + 400(y-20)$	for 20<y<40 feet
$P_{ch} = 300\,y + 5.385\,y$	for 0<y<20 feet
$= 353.85$	for 20<y<40 feet
$A_{ob} = 0$	for 0<y<20 feet
$= 1100\,(y-20) + 2.5\,(y-20)^2$	for 20<y<40 feet
$P_{ob} = 0$	for 0<y<20 feet
$= 1100 + 5.385\,(y-20)^2$	for 20<y<40 feet
$A_t = 300\,y + 2.5\,y^2$	for 0<y<20 feet
$= 7000 + 1500\,(y-20) + 2.5\,(y-20)^2$	for 20<y<40 feet
$P_t = 300 + 5.385\,y$	for 0<y<20 feet
$= 1453.85 + 5.385\,(y-20)^2$	for 20<y<40 feet

where :

A_{ch} : channel cross section

P_{ch} : channel wetted perimeter

A_{ob} : overbanks cross section

P_{ob} : overbanks wetted perimeter

after the effects of the flood wave have ceased. The flood wave component of equation (1) reaches a peak ΔQ_0 at time T_p. The initial conditions in the channel are the normal flow conditions for the discharge Q_0. The outflow hydrograph O(t) obtained by solving the De Saint-Venant equations with fixed values of Q_0 = 10,525 cfs and ΔQ_0 = 10,921 cfs and variable values of T_p = 1, 2, 4, 8, 16 hrs are our standard solutions. Unfortunately, regardless of the simplicity of the geometry and of the boundary and initial conditions, exact solutions of the De Saint Venant equations cannot be found and we have used therefore the program FEQ developed by D. Franz (1990), to obtain them. The FEQ program (which derives its name from the words Full Equations) has been demonstrated to be highly reliable by Bueno-Galdo and Giorgini (1991), and it has been deemed to be an excellent choice for the generation of the standard solutions of the problem. The five standard solutions considered in this study are presented in Figure 2.A,B,C,D,E. They correspond to the following values of the time to peak T_p = 1, 2, 4, 8, 16 hr. The figures present both inflow hydrograph and outflow hydrograph with a time step Dt = 15 min.

The Muskingum Approximation

The Muskingum lumped approximation consists in solving the continuity equation

$$\frac{dS}{dt} = I - Q \tag{2}$$

subject to the Muskingum condition

$$S = K\,(xI + (1-x)\,Q) \ , \tag{3}$$

where

S ≡ Water volume stored in the reach,
I ≡ Inflow hydrograph,
Q ≡ Outflow hydrograph,
K ≡ Wave travel time through the reach,
x ≡ Wedge factor of the reach.

Substituting (3) into (2) one obtains the equation

$$K(1 - x)\,\frac{dQ}{dt} + Q = 1 - Kx\,\frac{dI}{dt} \tag{4}$$

which is a first order ordinary differential equation in Q, where the forcing is a function of the known inflow hydrograph I.

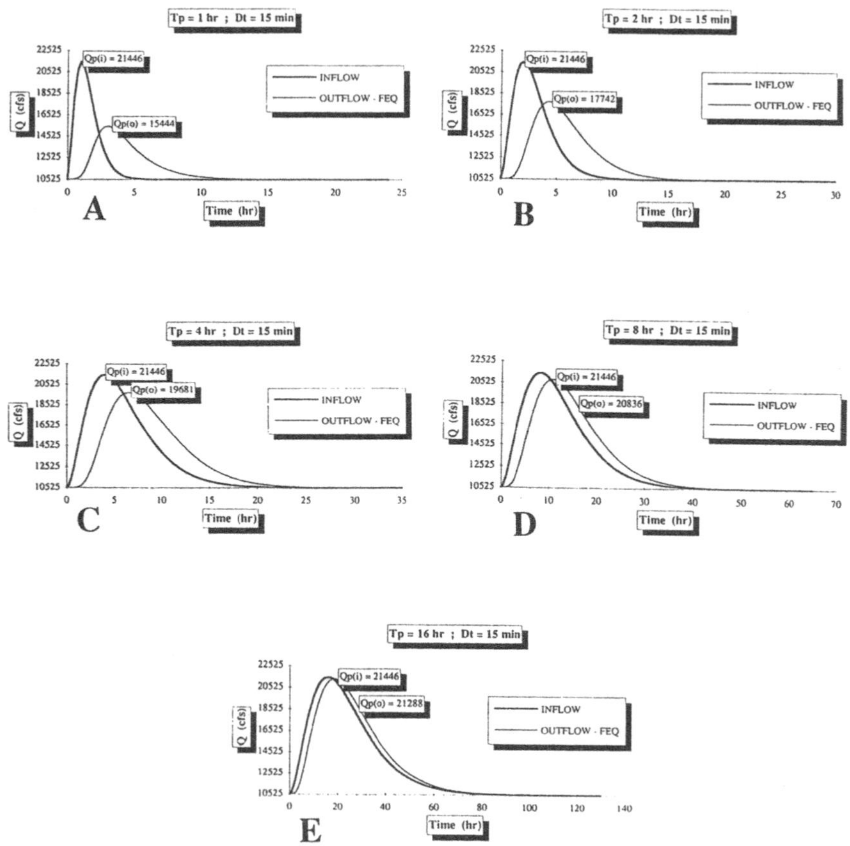

Figure 2. Standard solutions for T_p = 1, 2, 4, 8, 16 hrs.

The exact solution of (4), when the expression (1) is substituted for I, is

$$Q = Q_0 + \Delta Q_0 \, (b' + c' \, \tau + d' \, \tau^2) \, e^{2-2\tau} - \Delta Q_0 \, b' \, e^{2-\alpha\tau} \tag{5}$$

where

$$b' = 2\alpha \frac{1 + \xi}{(\alpha-2)^3} \; ; \tag{6.a}$$

$$c' = -2\alpha \frac{1 + \xi}{(\alpha-2)^2} \; ; \tag{6.b}$$

$$d' = \frac{\alpha + 2\xi}{\alpha-2} \; ; \tag{6.c}$$

$$\alpha = \frac{T_p}{(1-x)\, K} \; ; \tag{6.d}$$

$$\xi = \frac{x}{1-x} \; ; \tag{6.e}$$

$$\tau = \frac{t}{T_p} \; , \tag{6.f}$$

for $\alpha \neq 2$, and, for the particular case of $\alpha = 2$,

$$Q = Q_0 + \Delta Q_0 \left[-\xi\tau^2 + \frac{2}{3} (1 + \xi) \, \tau^3 \right] e^{2-2\tau} \; . \tag{7}$$

These solutions depend on the parameters K and x, which have not been determined yet. It is actually our task to find the values of K and x that make the solutions (5) and (7) as close as possible to the standard solutions. Since the standard solution O_i has been found at the instants $t = i\,\Delta T$, for $i = 0, N\,(T_p)$ with $\Delta t = 15$ min $= .25$ hr and

T_p	1 hr	2 hr	4 hr	8 hr	16 hr
$N\,(T_p)$	97	121	141	241	521

the task we have to accomplish now is to minimize a cumulative weighted square error function. We will perform a brute force calculation of the function for values of K in the range (0, 5) at increments $\Delta K = 0.1$ and for values of x in the range (0, 0.5) at increments $\Delta x = .02$.

$$\text{Error} = \left[\frac{1}{N}\sum_{1}^{N}(Q_i - O_i)^2\, w_i\right]^{1/2} \tag{8.a}$$

with

$$w_i = \frac{Q_i + Q_{ave}}{2\,Q_{ave}}\ , \tag{8.b}$$

$$Q_{ave} = \frac{1}{N}\sum_{1}^{N} Q_i\ . \tag{8.c}$$

The error function is presented graphically by means of contour lines in the plane x,K in Figure 3.A,C,E,G,I for the times to peak T_p= 1, 2, 4, 8, 16 hr respectively. The line crossing the contour lines in each one of the figures is the locus of the (x,K) points for which $\alpha = 2$, that is

$$K = \frac{T_p / 2}{1-x} \tag{9}$$

which represent hyperbolae with common vertical asymptote x = 1. The gray regions around the lines represent zones in which eq. (7) has been used instead of (5) in order to avoid large truncation errors due to the fact that (α-2) occurs at the denominators of equations (6.a,b,c). The optimal values of K and x can be found by a different procedure suggested by Giorgini (1992a) which yields the following formulas:

$$K = \frac{x\sum_{0}^{N} N_i A_i + \sum_{0}^{N} N_i B_i}{x^2\sum_{0}^{N} A_i^2 + 2x\sum_{0}^{N} A_i B_i + \sum_{0}^{N} B_i^2} \tag{10}$$

where

$$x = \frac{\sum_{0}^{N} N_i B_i \sum_{0}^{N} A_j B_j - \sum_{0}^{N} N_i A_i \sum_{0}^{N} B_j^2}{\sum_{0}^{N} N_i A_i \sum_{0}^{N} A_j B_j - \sum_{0}^{N} N_i B_i \sum_{0}^{N} A_j^2} \tag{11}$$

and where

$$A_i = I_{i+1} + O_i - (I_i + O_{i+1}) \tag{12.a}$$

$$B_i = O_{i+1} - O_i \tag{12.b}$$

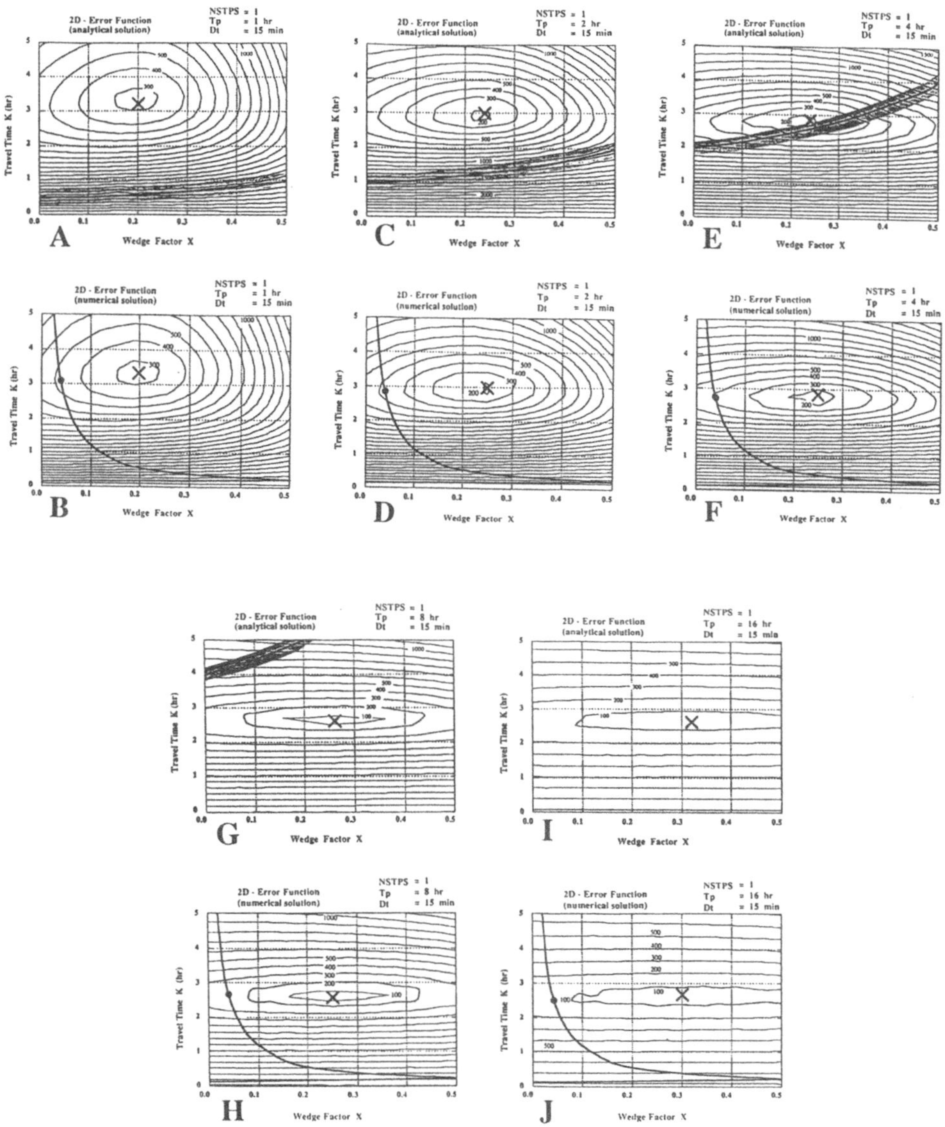

Figure 3. Error function contour lines in the parameter plane K, x for the analytical solution and for the numerical solution for the case of the one-subreach reach (NSTPS = 1).

$$N_i = .5\ \Delta t\ (I_{i+1} + I_i - O_{i+1} - O_i) \quad . \tag{12.c}$$

Table 1 shows the optimal values of K and x as obtained by using the minimization of the sum of the weighted square errors and those given by the formulas (10) and (11) for the solutions corresponding to the several values of T_p.

T_p (hr)	K		x	
	Min Err	Eq (10)	Min Err	Eq (11)
1	3.20	2.93	0.20	0.15
2	3.00	3.00	0.24	0.22
4	2.80	2.79	0.24	0.25
8	2.60	2.68	0.26	0.28
16	2.60	2.65	0.32	0.35

The values of K and x obtained with the two techniques are different because of the intrinsic differences of the functions to optimize and because of the coarse grid used in the generation of the error function. The values obtained with equations (10) and (11) are calculated with a higher degree of accuracy and they are used for the calculation of the optimal analytic solution.

The illustrations of Figure 4.A,B,C,D,E show the comparison between the standard solutions and the optimal exact solutions of the Muskingum equation. The figures clearly show that the outflow hydrographs obtained with the Muskingum equation tend to go below the base flow Q_0, which is obviously unphysical, as it has been shown by several authors, and which stems from the "rigidity" of the system represented by the Muskingum equation.

The fact is rather disappointing: the simplicity of the Muskingum equations make its appeal very high, especially in hydrologic simulations. On the other hand, the "undershooting" of the response, especially at low values of T_p is rather disturbing: in fact if the base flow were null, the undershooting would imply negative outflows ....!

A fortunate mathematical accident comes to our rescue. If we try to numerically integrate the Muskingum equation (4) by applying first order finite difference approximations we obtain:

$$K(1-x)\ \frac{Q_{i+1} - Q_i}{\Delta t} + \frac{Q_i + Q_{i+1}}{2} = \frac{I_i + I_{i+1}}{2} - Kx\ \frac{I_{i+1} - I_i}{\Delta t} \tag{13}$$

which yields

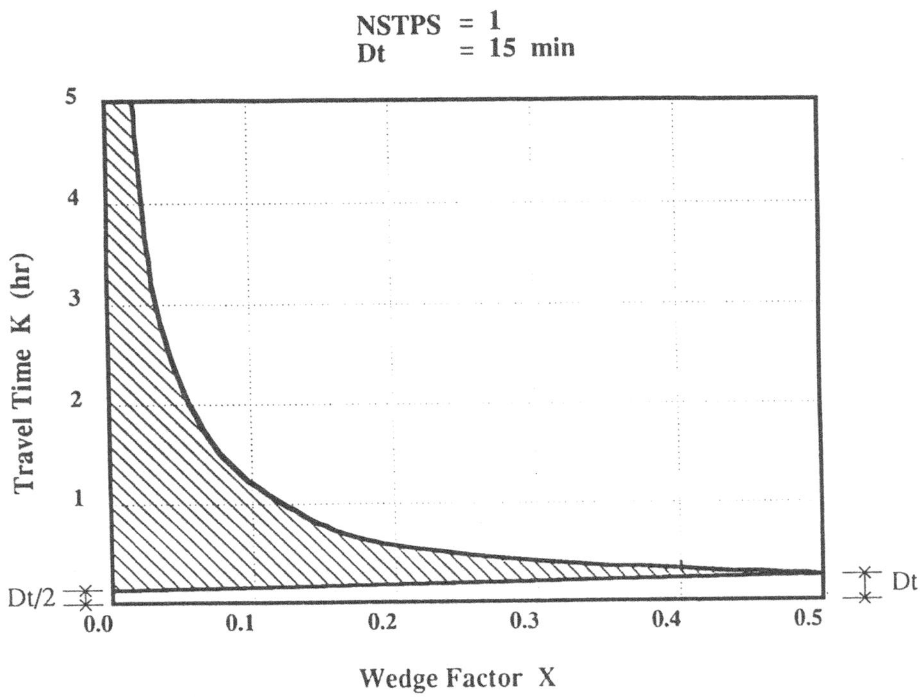

Figure 4. Acceptance region for non-undershooting solutions for NSTPS = 1.

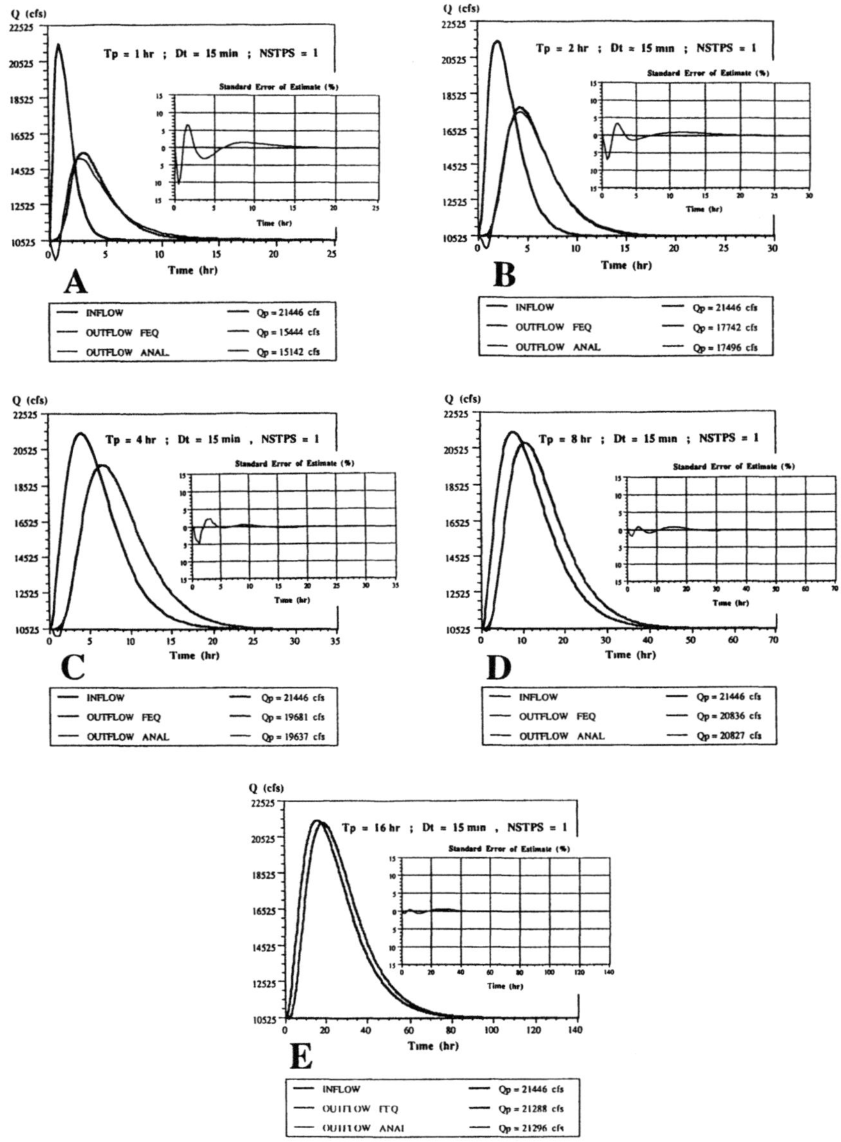

Figure 5. **Comparison of analytic solutions to standard solutions for NSTPS = 1.**

$$Q_{i+1} = \frac{K(1-x) - .5\Delta t}{K(1-x) + .5\Delta t} Q_i + \frac{Kx + .5\Delta t}{K(1-x) + .5\Delta t} I_i - \frac{Kx - .5\Delta t}{K(1-x) + .5\Delta t} I_{i+1} \ . \quad (14)$$

The coefficients of Q_i, I_i, and I_{i+1} are usually given the symbols C_0, C_1, and C_2 respectively. The numerical solution obtained by using (14) is capable of yielding non undershooting values if the following inequalities are satisfied

$$2xK < \Delta t < 2(1-x)\, K \ , \quad (15)$$

that is

$$\frac{\Delta t}{2(1-x)} < K < \frac{\Delta t}{2x} \ . \quad (16)$$

Assuming that we limit our interest to solutions for which x is positive, the condition of non-undershooting limits the values of K, x to the hatched region in Figure 4. The curious fact that the inequality (15) portrays is that, given K and x, Δt should be less than 2(1-x)K [so far so good] but should be *larger than* 2xK. Since numerical errors increase when Δt increases, that is tantamount to saying that we are ready to accept larger discretization errors for the sake of eliminating undershooting errors. While the instinctive repulsion toward negative values of the outflow is understandable (in the case where no base flow is present), it is not clear whether, by trying to avoid them, one does not choose larger errors for the sake of what may amount to no more than mathematical aesthetics.

In order to verify the above suspicion we have numerically integrated the Muskingum equation, by using equation (14) for the same values of K and x that were used for the exact solution and we have represented their error with respect to the standard solutions as contour lines in Figure 3.B,D,F,H,J, just below the corresponding error contours for the analytical solutions.

In all these figures the point of minimum square error is marked by a large x. In none of the aforementioned figures the x falls within the acceptance region. Since almost all the absolute minima of the error functions fall in the forbidden region, the "usable" values of K and x are those of the constrained minimum identified in the above figures with a bullet mark. The actual results and comparisons are shown in Figure 6.

Figures 6.A,a show the case T_p = 1 hr, Δt = 15 min. and juxtappose optimal numerical solution of the Muskingum equation (Figure 6.A) to the non-undershooting numerical solution (Fig. 6.a). The comparison shows clearly that the error in the "sanitized" numerical solution has considerably larger values.

Figures 6.B,b show the case T_p = 2 hr, Δt = 15 min and Figures 6.C,c show the case T_p = 4 hr, Δt = 15 min. For the particular case of T_p = 4 hr another experiment with Δt = 30 min

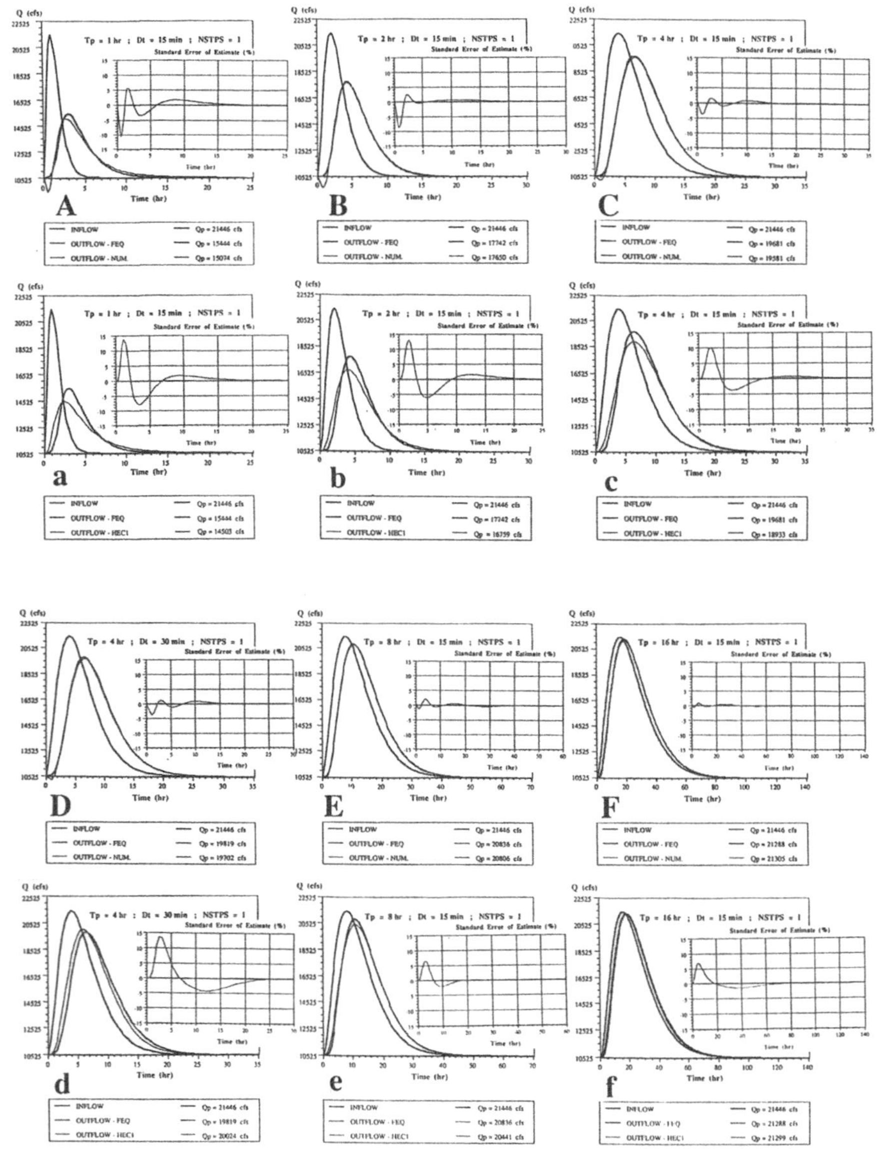

Figure 6. Comparison of numerical solutions and of HEC-1 solutions to standard solutions for NSTPS = 1.

is shown in Figures 6.D,d.

Figures 6.E,e show the case T_p = 8 hr, Δt = 15 min, and Figures 6.F,f show the case T_p = 16 hrs, Δt = 15 min. The error differential is seen to decrease as T_p increases, since in these cases the river reach behaves very much as a reservoir and the error function presents a very flat minimum along the x direction.

The Multiple Reach Extension

Since the constrained "optimal" solutions seen in the previous section are far from acceptable, it is obvious that hydrologists have strived to derive schemes that would ameliorate the performance of the Muskingum method. It is tempting to investigate what could be modified to bring the minimum of the error functions within the hatched region of Figure 4 (the acceptance region). This could be done either by raising the acceptance region or by lowering the position of the minimum. It so happens that both of these tasks are accomplished simultaneously by dividing the reach into an integer number n of identical subreaches.

It is very easy to show how this affects the position of the acceptance region. In fact, if we call x the wedge factor of any **subreach** and we call K the travel time for the **whole reach**, then the travel time for a subreach is K/n and the non-undershooting inequality (15) becomes (17)

$$2x \frac{K}{n} < \Delta t < 2(1-x) \frac{K}{n} \tag{17}$$

or

$$\frac{n\Delta t}{2(1-x)} < K < \frac{n\Delta t}{2x} \ . \tag{18}$$

Figure 7.a shows clearly that, as the number n of subreaches increases the acceptance region moves toward higher values of K, moving therefore toward the minimum values of the error function.

Simultaneously the minimum of the error function tends to move toward smaller values of x. This double process can be seen clearly in Figure 7.b,c,d that show the results for the case T_p = 1 hr, Δt = 15 min, for n = 1, 2, 3 (notice that on the figure the number of subreaches n is given the symbol NSTPS, because this is the symbol used in HEC-1).

While it is reasonable that the minimum of the error function should move along a horizontal line (because the travel time K in the reach does not change considerably by partitioning the reach in subreaches) the considerable variations of x, which could become negative in some instances (!) can be explained heuristically by recalling that, while K is a reach

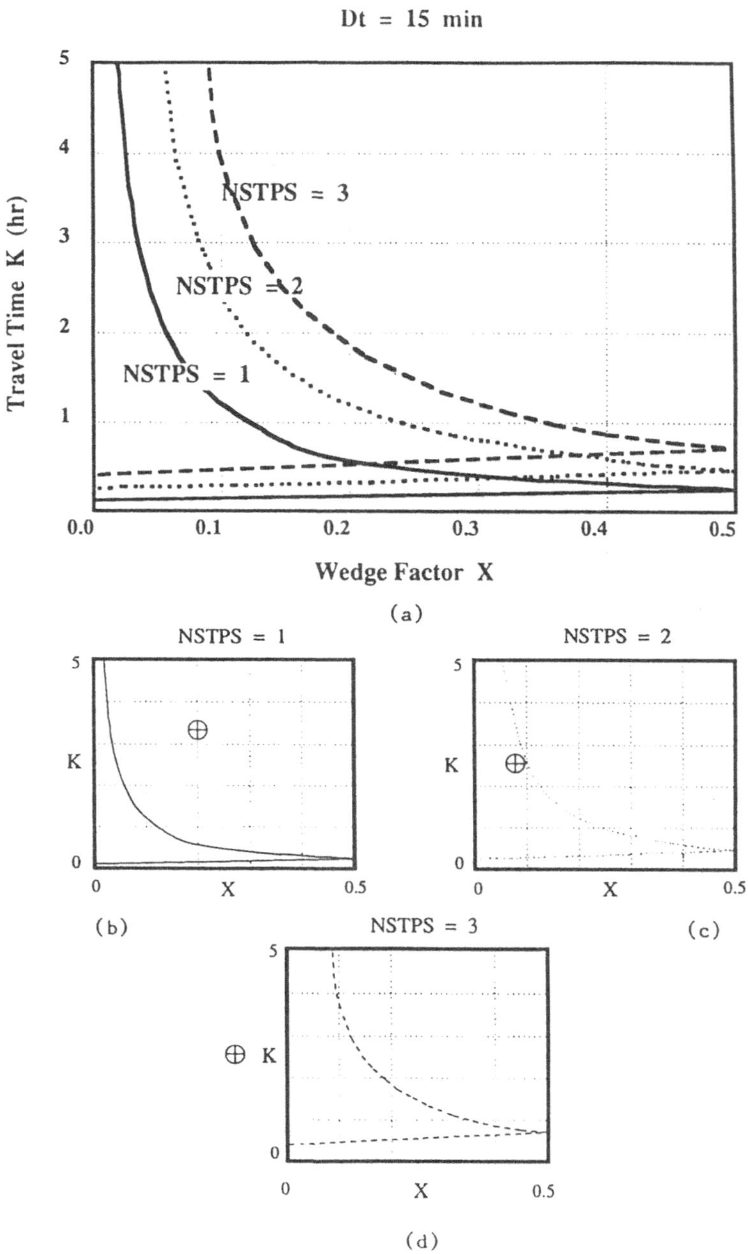

Figure 7. **Acceptance regions and error functions minimize locations.**

parameter, x is a subreach parameter and that therefore as NSTPS grows, and consequently as the subreaches become smaller and smaller, these tend to behave as reservoirs, that is their wedge factor tends to become zero.

The fact that sometimes x can take over negative values signifies that the Muskingum method has another unphysical characteristic. It is traditional to consider as "valid" only values of $x \geq 0$, but this tenet may be questioned, if, due to the imperfections of the method, the physical meaning of x has to be tempered. In this case we are simply dealing with a problem of best fit of a mathematical model to an analytic solution. Regardless of how physically sound the rationale of the method is, some unphysical consequences may arise, just because the model is not identical to the "reality" it wants to emulate. The problem is then to recognize whether these imperfections are tolerable or not.

The Case of the Bisected Reach

Where n = NSTPS = 2 the Muskingum equations can be written as

$$\frac{K}{2}(1-x)\frac{dQ}{dt} + Q = M - \frac{K}{2}x\frac{dM}{dt} \tag{19}$$

$$\frac{K}{2}(1-x)\frac{dM}{dt} + M = -\frac{K}{2}\times\frac{dI}{dt}, \tag{20}$$

where M is the hydrograph at mid reach. The solution of the above equations (only Q is of interest to us) is

$$Q = Q_0 + \Delta Q_0\,(b'' + c''\tau + d''\tau^2)e^{2-2\tau} + \Delta Q_0(-b'' + f_1''\tau)e^{2-\alpha\tau} \tag{21}$$

where

$$b'' = \frac{\alpha+2\xi}{\alpha-2}b' - \alpha\frac{1+\xi}{(\alpha-2)^2}c' + 2\alpha\frac{1+\xi}{(1\alpha-2)^3}d' \tag{22.a}$$

$$c'' = \frac{\alpha+2\xi}{\alpha-2}c' - 2\alpha\frac{1+\xi}{(\alpha-2)^2}d' \tag{22.b}$$

$$d'' = \frac{\alpha+2\xi}{\alpha-2}d' \tag{22.c}$$

$$f_1' = -\alpha(1+\xi)\,b' \tag{22.d}$$

$$b' = 2\alpha \frac{1+\xi}{(\alpha-2)^3} \tag{22.e}$$

$$c' = -2\alpha \frac{1+\xi}{(\alpha-2)^2} \tag{22.f}$$

$$a' = \frac{\alpha+2\xi}{\alpha-2} \tag{22.g}$$

$$\alpha = \frac{2T_p}{(1-x)\,K} \tag{22.h}$$

$$\tau = t \,/\, T_p \tag{22.i}$$

$$\xi = \frac{x}{1-x} \quad . \tag{22.j}$$

The solution (21) for the two-subreach reach has so far generic values of x and K. We will find the minimum of the error function as it was done for NSTPS = 1. The results are shown in Figures 8.A,B,C,D,E.

The optimal values of the parameters K and x are given in the following table.

T_p (hr)	K	x
1	2.60	0.08
2	2.60	0.06
4	2.60	0.04
8	2.60	0.08
16	2.60	0.16

Statistical formulas for K and x similar to formulas (10) and (11) have not been found yet at this time.

The numerical solutions corresponding to the aforementioned cases, with Δt = 15 min. have been calculated and are presented in Figure 8.a,b,c,d,e. The corresponding values of K and x which render minimal the error of the numerical solution with respect to the standard solutions are identical to those of the previous table.

Inspection of Figure 8.a,b,c,d,e shows that for NSTPS = 2 the numerical solution of the Muskingum equation has the minimum of the error function falling in the acceptance region for

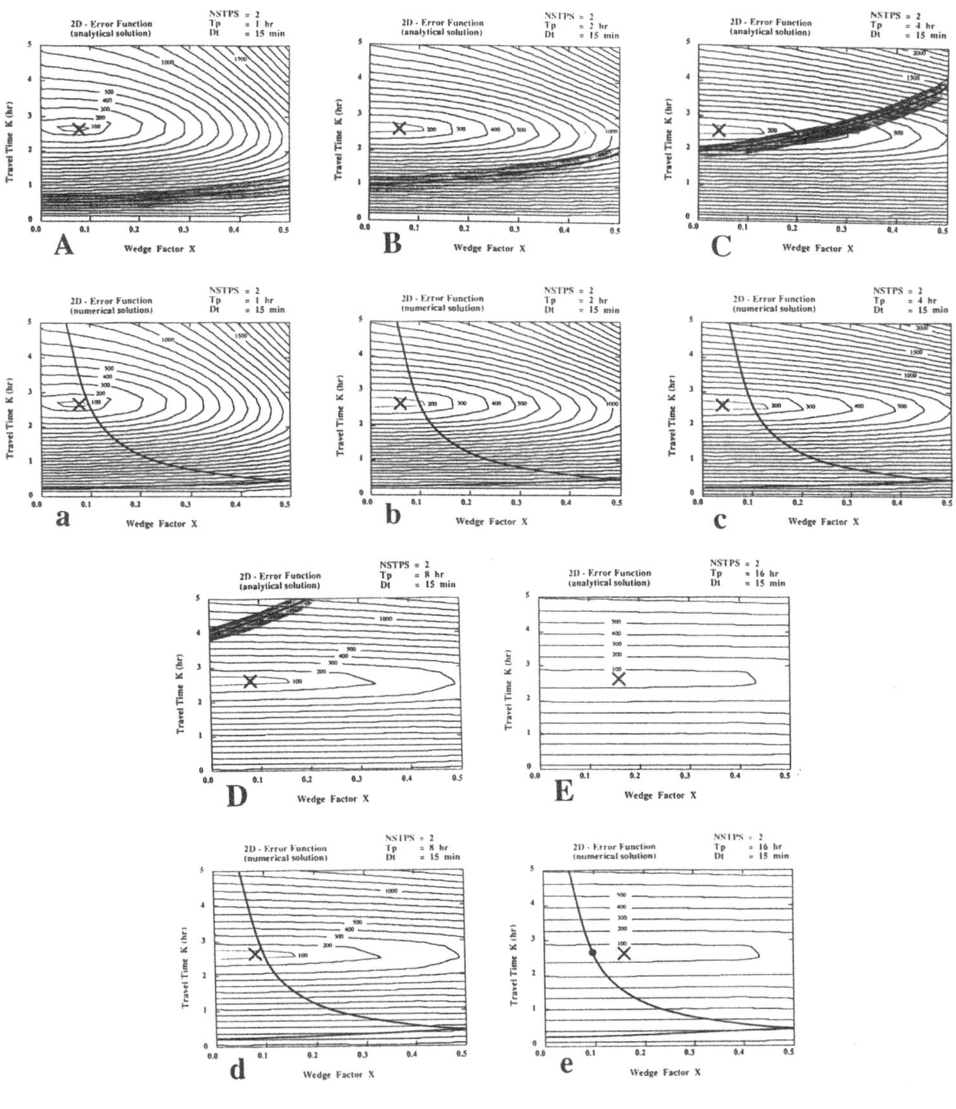

Figure 8. Error function contour lines in the parameter plane K,x for the analytical solution and for the numerical solution for the two-subreach reach (NSTPS = 2).

all values of T_p except for $T_p = 16$ hr. The value of K has not changed much from the unconstrained case NSTPS = 1, but the value of x has decreased considerably, according to our earlier observations.

HEC-1 runs have yielded the following values for the parameters K and x, for NSTPS = 2.

T_p (hr)	K	x
1	2.60	0.05
2	2.60	0.05
4	2.60	0.04
8	2.60	0.05
16	2.60	0.04

They are not identical to the values of the previous table because of the different procedure used in calculating the minimal value of the error function. Furthermore, the last value is very different because HEC-1 calculates the constrained minimum.

The Figures 9.a,b,c compare the exact solution of Muskingum equation, the numerical solution, and the HEC-1 solution to the standard FEQ solution for $T_p = 1$ hr and NSTPS = 2.

Analogous comparisons are accomplished by

Fig 9.d,e,f	for $T_p = 2$ hr
Fig 9.g,h,i	for $T_p = 4$ hr
Fig 9.j,k,l	for $T_p = 8$ hr
Fig 9.m,n,o	for $T_p = 16$ hr .

The errors for $T_p = 16$ hr are less than 1%.

The Case of the Trisected Reach

When n = NSTPS = 3 the Muskingum equations can be written as

$$\frac{K}{3}(1-x)\frac{dQ}{dt} + Q = M - \frac{K}{3}x\frac{dM}{dt} \tag{23}$$

$$\frac{K}{3}(1-x)\frac{dM}{dt} + M = N - \frac{K}{3}x\frac{dN}{dt} \tag{24}$$

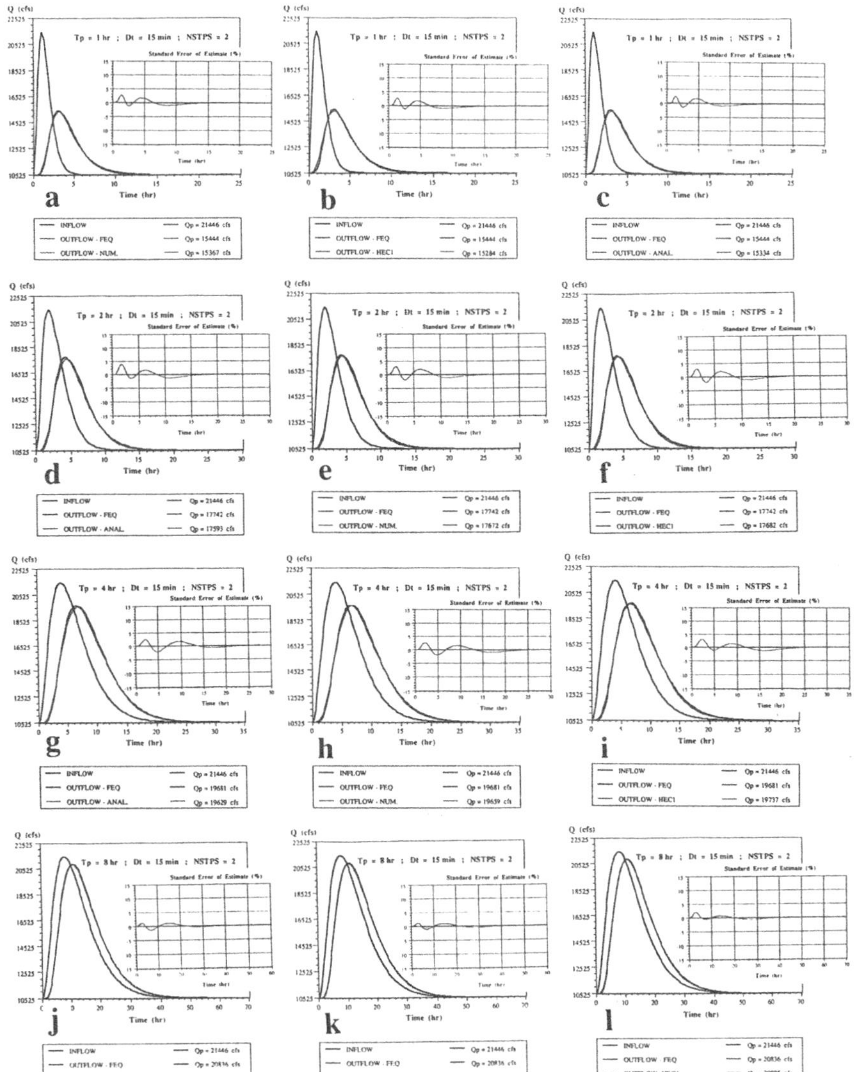

Figure 9. Comparison of analytical solutions, numerical solutions, and HEC-1 solutions to standard solutions for NSTPS = 2.

$$\frac{K}{3}(1-x)\frac{dN}{dt} + N = 1 - \frac{K}{3}x\frac{dI}{dt} \tag{25}$$

where M and N are the hydrographs at one thirds and two thirds of the reach. The solution of the above equations (only Q is of interest to us) is

$$Q = Q_0 + \Delta Q(b''' + c'''\tau + d'''\tau^2)\, e^{2-2\tau} + \Delta Q_0(-b''' + f_1'''\tau + f_2'''\tau^2)\, e^{2-\alpha\tau} \tag{26}$$

where

$$b''' = \frac{\alpha+2\xi}{\alpha-2} b'' - \alpha\frac{1-\xi}{(\alpha-2)^2} c'' + 2\alpha\frac{1+\xi}{(\alpha-2)^3} d'' \tag{27.a}$$

$$c''' = \frac{\alpha+2\xi}{\alpha-2} c'' - 2\alpha\frac{1+\xi}{(\alpha-2)^2} d'' \tag{27.b}$$

$$d''' = \frac{\alpha+2\xi}{\alpha-2} d'' \tag{27.c}$$

$$f_1''' = -\alpha(1+\xi)\, b'' - \xi f_1'' \tag{27.d}$$

$$f_2''' = \frac{\alpha}{2}(1+\xi)\, f_1'' \tag{27.e}$$

and where b'', c'', d'', f_1'', b', c', d', α, τ, ξ are given by equations (22.a to j).

The solution (26) must be optimized, in the sense of finding the values of K and x for which the error function between (26) and the standard solutions presents a minimum. This has been done and the results are presented in Figure 10.A,B,C,D,E and in the following table

T_p (hr)	K (hr)	x
1	2.40	-.10
2	2.40	-.10
4	2.40	-.10
8	2.40	-.14
16	2.40	0.00

Statistical formulas for K and x similar to formulas (10) and (11) have not been found yet at this time.

The numerical solutions corresponding to the aforementioned cases, with $\Delta t = 15$ min, have been calculated and are presented in Figure 10.a,b,c,d,e. The corresponding values of K and x which render minimal the error of the numerical solution with respect to the standard solution are given in the following table

T_p (hr)	K (hr)	x
1	2.50	-.10
2	2.50	-.16
4	2.50	-.18
8	2.50	-.14
16	2.50	-.04

Inspection of Figure 10.a,b,c,d,e shows that for NSTPS = 3 the numerical solution of the Muskingum equation has the minimum of the error function falling in the acceptance region. The value of K has not changed much from the unconstrained NSTPS = 1 case, but the value of x has become negative. HEC-1 runs have yielded the following values of the parameters K and x, for NSPTS = 3.

T_p (hr)	K (hr)	x
1	2.50	.04
2	2.50	.04
4	2.50	.04
8	2.60	.04
16	2.60	.04

They are not identical to the values of the previous table because of the different procedure used in calculating the minimal value of the error function. Furthermore the last value is very different because HEC-1 calculates the constrained minimum.

The Figures 11.a,b,c compare the exact solution of the Muskingum equation, the numerical solution, and the HEC-1 solution to the standard solution for $T_p = 1$ hr and NSTPS = 3.

Analogous comparisons are accomplished by

Fig. 11.d,e,f	for $T_p = 2$ hr
Fig. 11.g,h,i	for $T_p = 4$ hr
Fig. 11.j,k,l	for $T_p = 8$ hr .

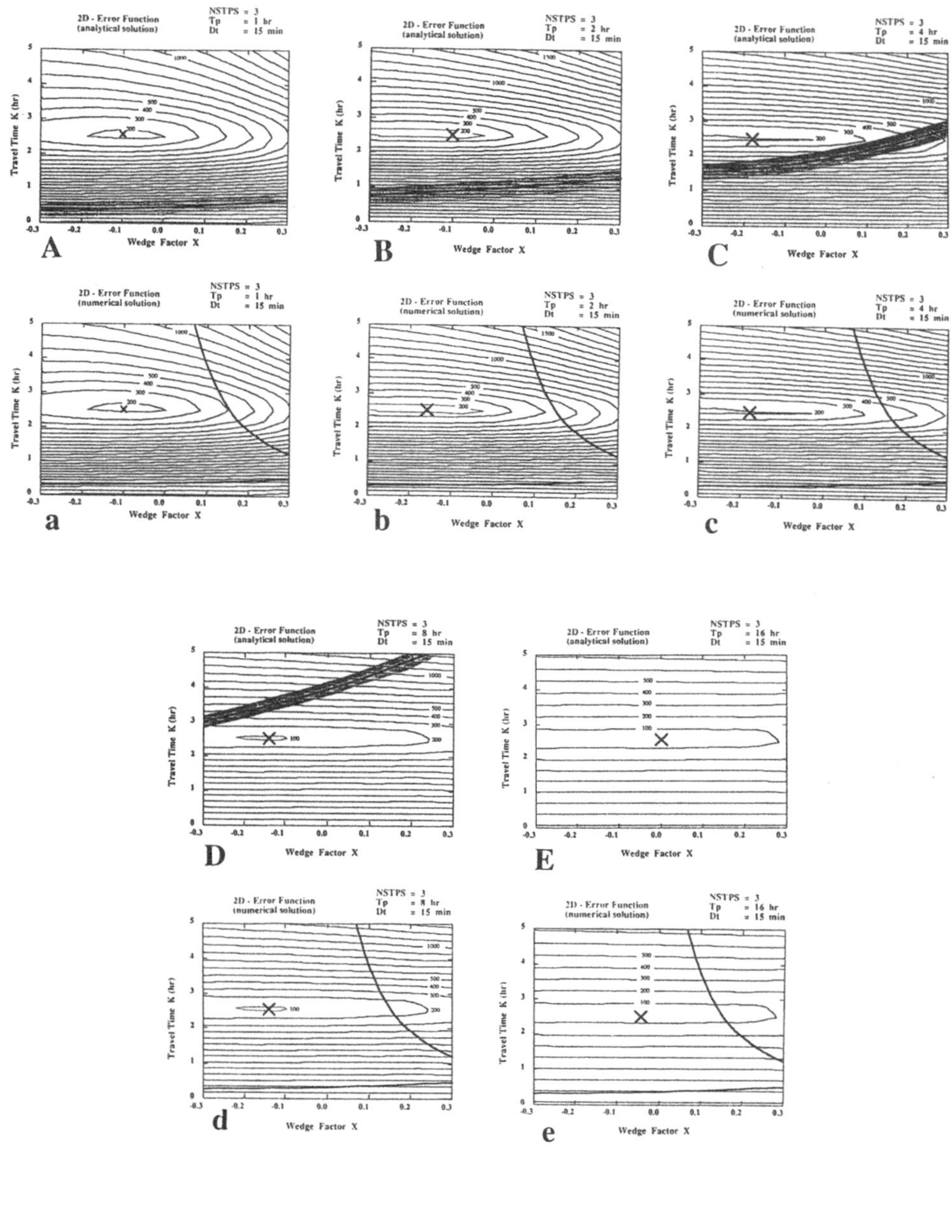

Figure 10. Error function contour lines in the parameter plane K,x for the analytical solution and for the numerical solution for the three-subreach reach NSTPS = 3.

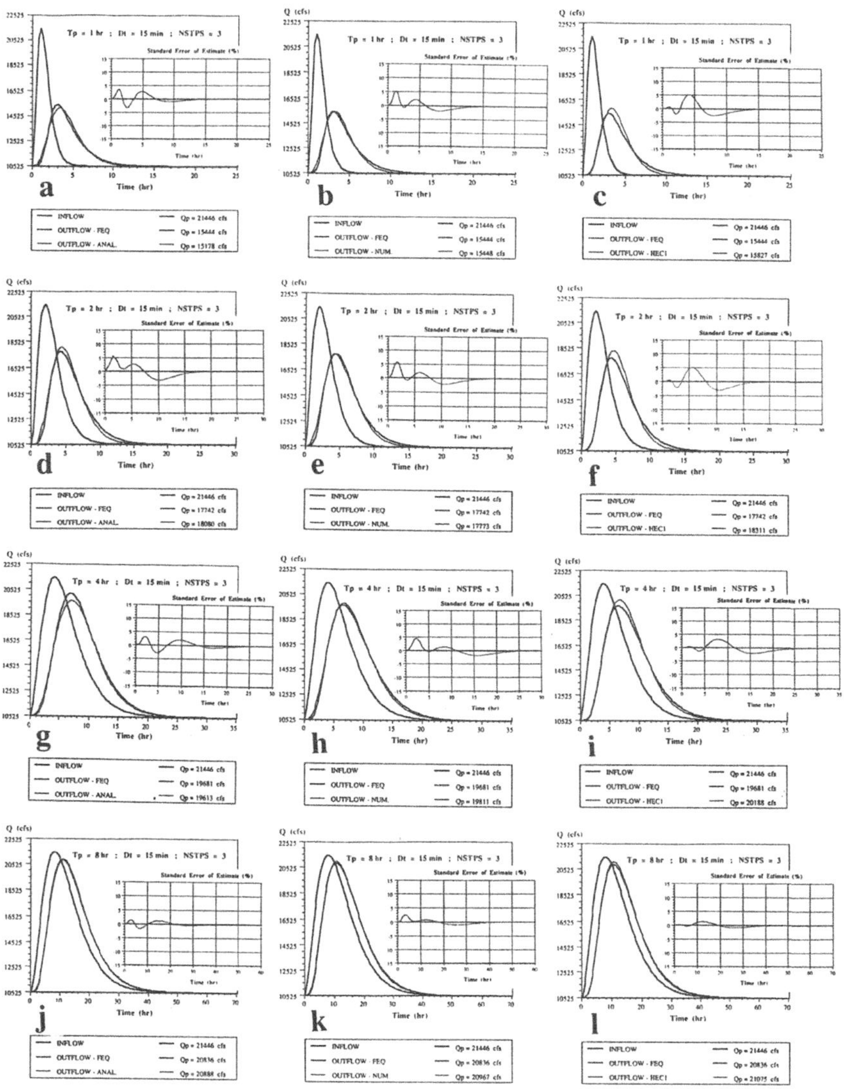

Figure 11. Comparison of analytical solutions, numerical solutions, and HEC-1 solutions to standard solutions for NSTPS = 3.

The errors for T = 16 hrs. are less than 1%

Conclusion

Our investigation has shown that the process of finding an optimal fit for a Muskingum solution to a "standard" solution of the De Saint-Venant equations using the parameters K, x and NSTPS presents some unphysical features.

When the values of the parameter NSTPS are fixed at 1, 2, and 3, and the absolute minimum of the error function is heeded to, undershooting of the base flow occurs as shown in Figure 12.A for T_p = 1 hr, in Figure 12.B for T_p = 2 hr, in Figure 12.C for T_p = 4 hr, in Figure 12.D, for T_p = 8 hr, and in Figure 12.E for T_p = 16. The close-ups of all the figures show clearly the undershootings of the optimal solutions.

The values of the error function in these cases are given in the following table

T_p	Minimum Value of Error Function		
(hr)	NSTPS = 1	NSTPS = 2	NSTPS = 3
1	260	96	155
2	155	133	174
4	99	138	157
8	63	81	90
16	36	37	47

The table shows clearly that, if undershooting is tolerated, the optimal solution occurs for NSTPS = 2 for T_p = 1, 2 hr and for NSTPS = 1 for T_p = 4, 8, 16 hr.

If the non-undershooting criterion is heeded to, the constrained minima, which coincide with the HEC-1 solutions, are obtained, with NO undershooting. The results are presented in Figure 13.

The values of the error function in these cases are given in the following table:

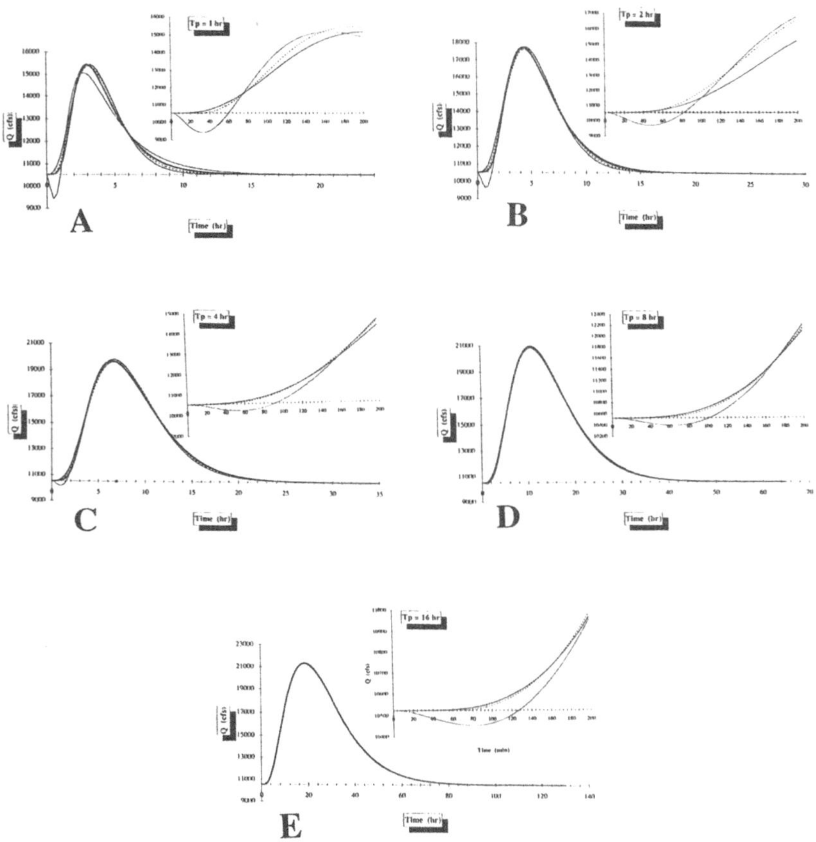

Figure 12. Comparison of analytical solutions, numerical solutions, and HEC-1 solutions to standard solutions.

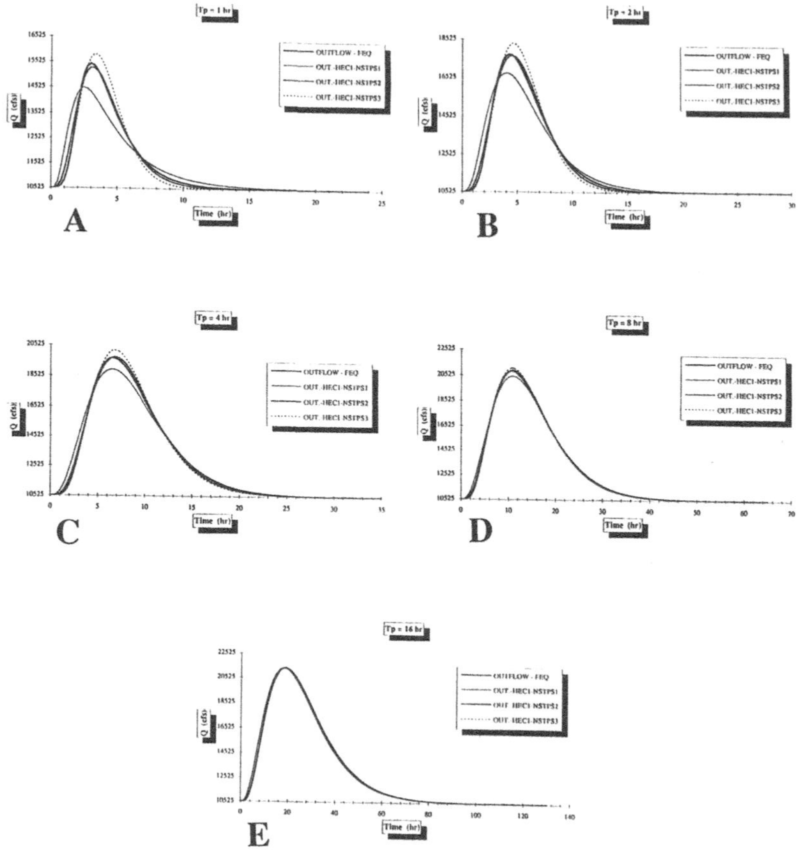

Figure 13. **Comparison of HEC-1 solutions (for NSTPS = 1, 2, and 3) to standard solutions.**

T_p	Minimum Value of Error Function		
(hr)	NSTPS = 1	NSPTS = 2	NSPTS = 3
1	469	114	231
2	466	129	276
4	374	127	225
8	200	72	97
16	300	48	43

In conclusion we can say that the algorithm in HEC-1 performs the operations suggested in this study in exemplary fashion. The Program, here in the optimization mode, performs searches for minimum of the error function, within the acceptance region (which includes the constraint $x \geq 0$) for increasing values of NSTPS and stops at the first value of NSTPS yielding an error larger than the preceding error. The preceding value of NSTPS is taken as the optimal value for simulation.

Since the optimal values of NSTPS and x vary with T_p, it seems unreasonable at this stage to assume that NSTPS, x, and K are parameters of the reach, independently of the flood event.

STRATIFIED MUSKINGUM ROUTING IN NATURAL CHANNELS

M.Karmegam
Professor in Civil Engineering
Centre for Water Resources
Anna University, Madras 600 025, INDIA.

ABSTRACT. Muskingum method is very popular in flood routing mainly for its simplicity and ease with which calculations can be done. The method receives considerable attention and is being constantly improved in its application. Falling in line, the author has successfully tested a procedure called "Stratified Muskingum Routing" employing a set of hypothetical data. This is now being extended to and tested in natural channels. Three reaches taken from two of the UK rivers, Wye and Eden, were considered in this study. Optimization techniques have been employed for the estimation of the model parameters. The results show that natural floods can be routed with improved parameters and better solutions than obtained before.

1. INTRODUCTION

Muskingum method, developed by McCarthy (1938), is a very simple flood routing procedure which is based on the continuity equation as

$$Q_i - Q_o + q_l\, L = dS/dt \tag{1}$$

where Q_i is inflow, Q_o is outflow, q_l is lateral inflow entering the reach per unit length, L is total reach length, S is storage in the reach and t is time. This equation is solved for Q_o in conjunction with the Muskingum storage equation as

$$S = K\,[\varepsilon Q_i + (1 - \varepsilon)\, Q_o] \tag{2}$$

where K is storage coefficient and ε is relative parameter.

According to McCarthy, the storage in a reach is directly proportional to the weighted flow. Hence, K and ε are determined by plotting S Vs $[\varepsilon Q_i + (1-\varepsilon)Q_o]$ for various trial values of ε taken between 0 and 0.5, picking up the narrowest loop among the plotted ones. This is severely criticised because the storage - discharge relationship of a natural channel is always a loop.

Identifying the numerical truncation errors in the finite difference scheme of the Muskingum equation and comparing them with that of the diffusion method, Cunge (1969) presented equations for K and ε as

$$K = \Delta x / V_w \tag{3}$$

$$\varepsilon = 1/2\, [1 - Q/(V_w\, \Delta x\, B\, S_o)] \tag{4}$$

V. P. Singh and B. Kumar (eds.), Surface-Water Hydrology, 325–331.

where Δx is sub-reach length ($L = (N - 1)\,\Delta x$, N - number of space sections), V_w is wave speed, B is top width of channel for the flow Q and S_o is bed slope of the channel. Considering V_w for the peak of a flood wave, Price (1974) found the above equations very useful to route floods in British rivers.

Ponce and Yevjevich (1978) proposed a slightly different procedure by replacing V_w by the celerity, c, given by

$$c = dQ/dA \tag{5}$$

and also replacing (Q/B) by what is called as unit width discharge, q, as

$$q = Q/B \tag{6}$$

where A is area of cross-section of flow. This procedure helps to calculate K and ε as functions of flow.

If the channel is rectangular, with Manning's equation for the mean velocity of flow and Kleitz - Seddon equation for the wave speed, it is possible to rewrite Equations (3) and (4), for K and ε, explicitly in terms of geometric and flow characteristics as

$$K = \frac{3\,\Delta x\, By\,(B + 2y)}{Q\,(5B + 6y)} \tag{7}$$

and

$$\varepsilon = \tfrac{1}{2}\left[1 - \frac{3y\,(B + 2y)}{\Delta x\,(5B + 6y)S_o}\right] \tag{8}$$

Use of the above equations may outwit and outwork the simplicity of the method itself. For instance, for every flow level, Q, the corresponding depth of flow, y, has to be assessed. This needs a depth - discharge relationship for every section used in the routing. Also, the equations may be explicit but the calculations have to follow a trial and error procedure because it concerns with four nodal points (Figure 1) involving two space sections and two time levels, at one of which the solution is sought. Realising this, Karmegam (1992) has gone in for a stratified approach of calculating the parameters and routing the flow, calling it as "Stratified Muskingum Routing". The method attempts to calculate K and ε as applicable to every stratum (Figure 2) bounded by different flow levels wherein the variations in Q, K and ε are all limited. Generating a hypothetical set of inflow and outflow hydrographs in a prismoidal rectangular channel he made an attempt to estimate K and ε (stratum-wise) through optimization and found that the estimated values to follow the theoretical values as given by Equations (7) and (8). Encouraged by this, an attempt is made here to extend the theory to flood routing problems in natural rivers.

2. RIVER REACHES USED IN THE STUDY

The study was carried out in the reaches as in Table I.

The River Wye flows from the Cambrian Mountains in Wales. It generally runs eastwards in Wales before entering West England and then flows southwards and finally joins the mouth of the River Severn. The River Eden in the north of England flows from the Pennines. It generally runs in a north-westerly direction through a valley, which in parts is fairly narrow, to Carlisle and finally into the Solway Firth. Flood Routing Studies (1975) can be referred to the catchment maps and for more details of the above reaches. Although the study was carried out in all the above three reaches, the results related to the first reach alone are presented here for want of space.

TABLE I Details of the reaches studied

Sl. No.	Reach	River	Length (km)	Bed Slope S_o ($x10^{-4}$)
1.	Erwood to Belmont	Wye	69.75	8.56
2.	Belmont to Redbrook	Wye	80.50	4.60
3.	Temple Sowerby to Warwick Bridge	Eden	37.00	16.50

The Erwood-Belmont reach has two large flood plains: one in the vicinity of Glasbury and Hay-on-Wye and the other between Witney and Bredwardine. These flood plains must play a crucial role in moderating the flow along the river. Ten floods which occurred between 1960 and 1969 have been used in this study (Table II).

TABLE II Flood events used in the study

Flood Number	Date of Occurrence	Peak flow (m^3/s) at		Computed latreal inflow ($m^3/s/m$) x 10^{-4}
		Erwood	Belmont	
1.	3-7, December 1960	1206	937	7.52
2.	25-28, November 1963	359	360	3.62
3.	8-12, December 1965	1090	642	1.62
4.	17-21, December 1965	685	623	5.00
5.	7-11, February 1966	275	254	2.95
6.	8-10, May 1966	349	381	7.60
7.	22-24, February 1967	331	311	2.39
8.	4-8, January 1968	296	269	1.59
9.	2-5, July 1968	245	330	6.37
10.	25-29, May 1969	304	332	4.92

Note: The lateral inflow was calculated by using the procedure employed by Price (1974), Flood Routing Studies (1975), Karmegam (1981) and Rangapathy et al (1988).

3. MODEL PARAMETER CALCULATION

The procedure used here is one of calculating the model parameters with the help of historical inflow and outflow data through optimization. The problem was thought of as calculating K and ε in a given river reach for a specified values of Δx and Q. A value of 6975.0m was used for Δx dividing the reach into ten sub-reaches. With regard to Q, the given flow hydrograph was divided into several strata as in Figure 2. A rise of 100.0m^3/s was used to mark the flow levels between the strata. The average flow in the reach was used to typify the stratification and hence to relate to the calculated coefficients. A value of 3600s was considered for the time increment, Δt, in solving the problem.

The objective function of the problem formulated here was to minimize the sum of the squares of errors between computed and observed values. A NAG Library (1977) subroutine, called "E04FBF" utilising a second order hybrid gradient method due to Powell (1968) which is based on an algorithm developed by Marquardt (1963), was used to solve the optimization problem.

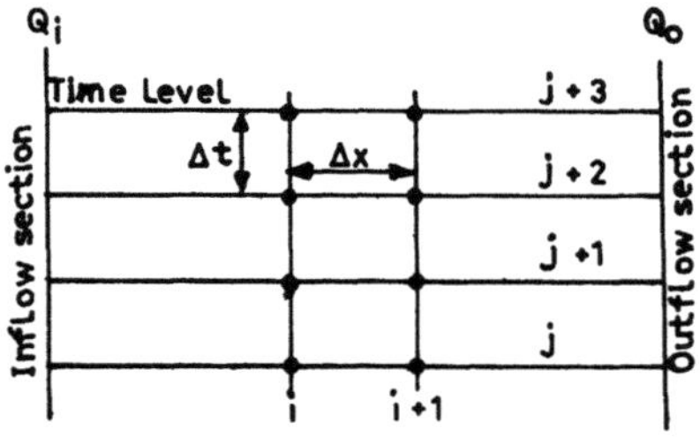

Figure 1 Rectangular grid

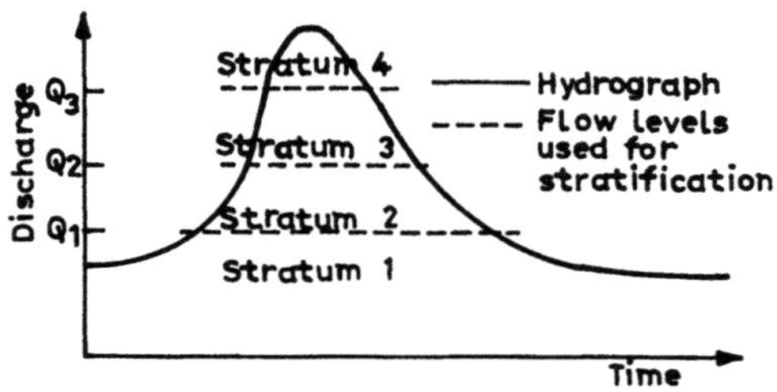

Figure 2 Stratification of a hydrograph

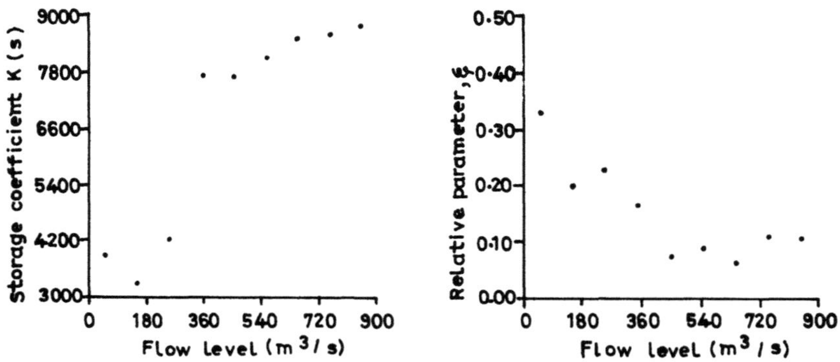

Figure 3 Calculated stratified Muskingum coefficients

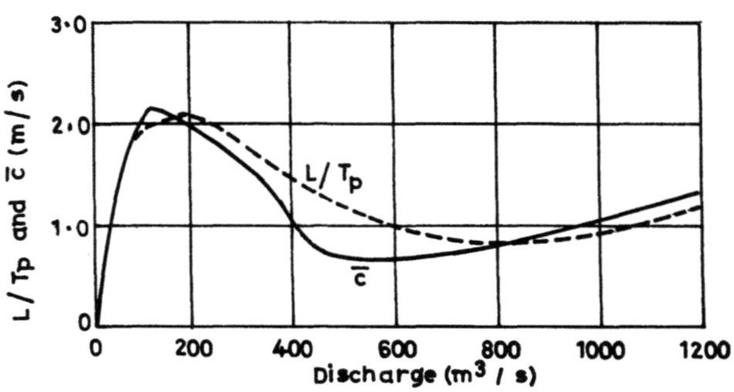

Figure 4 Speed-discharge relationship (Reproduced from Flood Routing Studies)

4. RESULTS AND DISCUSSION

The calculated K and ε values are presented in Table III as well as in Figure 3. Eight floods as in Table II except Floods 5 and 7 (which are reserved for flood prediction) were used in the calculation. As can be seen, there are a total of nine strata. But one has to realise that each flood will have only as many set of values of K and ε as the number of strata which will be decided on the basis of the peak flow of the flood. For instance, Flood 1 will have nine strata and hence nine set of values of K and ε, whereas Floods 6, 8, 9 and 10 will have only three strata each and three sets of K and ε values. Table III presents the mean of the calculated values (which are not presented here for want of space) for different strata.

TABLE III Calculated stratified Muskingum coefficients

Sl. No.	Flow level of stratum	Average flow (m^3/s)	Values of	
			K(s)	ε
1.	< 100	50	3861.4	0.3290
2.	100 - 200	150	3262.7	0.1973
3.	200 - 300	250	4195.3	0.2270
4.	300 - 400	350	7683.8	0.1628
5.	400 - 500	450	7687.8	0.0734
6.	500 - 600	550	8063.8	0.0894
7.	600 - 700	650	8454.8	0.0634
8.	700 - 800	750	8458.6	0.1095
9.	800 - 900	850	8757.5	0.1091

The study conducted by Karmegam (1992) clearly indicated the closeness with which the parameters can be calculated and the routing can be done. However, it may be worth in checking whether the calculated values as in Table III are in accordance with the basic principles as in Equations (3) and (4) or (7) and (8).

According to Equation (4) or (7), one can easily see that ε will decrease as Q increases. This trend is truely reflected in Figure 3 except a slight rise at higher flows which might be due to the influence of the flood plains present in the reach. As the flood rises and spreads in the flood plains, the area of cross-section of flow will suddenly rise making the speed of the wave to reduce drastically. The top width will also shoot up. Therefore, $Q/(V_w.B)$ will decrease and hence the increase in ε value, as Q overbanks and spreads into the flood plains, may be reasonable.

The variation of K basically depends on the variation of V_w with Q. If the channel section is prismatic, V_w will uniformly increase as Q increases and hence K will decrease. But if the channel is non-prismatic (especially when the cross-section changes erratically with large flood plains as in the study reach), V_w will decrease as the flow rises beyond the bankful level. Consequently K would be higher at higher flows. Unless it is possible to work out V_w accurately for different flow levels, it may not be possible to verify the variations in K as seen in Figure 3. However, it is possible to assess V_w reasonably with the help of the Flood Routing Studies (1975). The study has developed a relationship for the average convection speed as a function of discharge $\bar{c}(Q)$. This is found to be depended on the convection speed (ω) such as in a kinematic wave, speed of the flood peak (L/T_p, T_p is the time of travel of the peak) and the attenuation (Q_{at}). For the reach under study, $\bar{c}(Q)$ Vs Q and L/T_p Vs Q have been presented in the Flood Routing Studies which are reproduced here in Figure 4. Price (1973) used these relationships in the development of the variable parameter diffusion method. Following him and considering $\bar{c}$ in the place of V_w, it is now possible to check the calculated values of K as shown in Figure 5. As seen in the figure, it is possible to fit a line, with a reasonable correlation ($r^2 = 0.76$), of the form $K = C/\bar{c}$ where C is the constant of proportionality (i.e. slope of the line equal to 6189.0m). According to Equation 3, C should

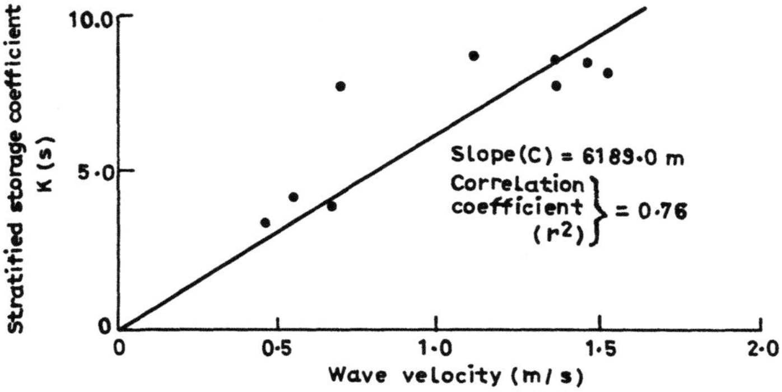

Figure 5 Stratified storage coefficient-wave velocity relationship

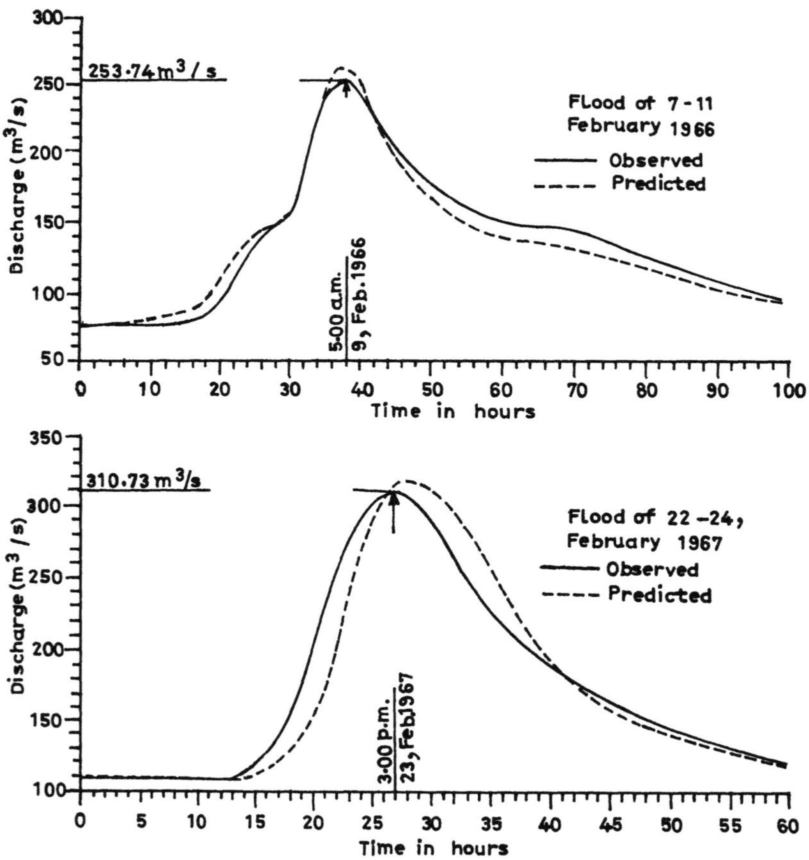

Figure 6 Flood prediction with stratified coefficients

be equal to the value of Δx, in this case it is 6975.0m. The value of C (i.e. 6189.0m) differs from that of Δx (i.e. 6975.0m) by 11.3%. Although this is on the higher side, one can accept it in view of the differences between the average convection speed ($\bar{c}$) and the wave speed (V_w).

Further to test the usefulness of the calculated stratified Muskingum coefficients in flood prediction, an attempt was made to use them in routing two of the floods presented in Table II (i.e. Floods 5 and 7). The respective routed hydrographs were compared with the observed ones as presented in Figure 6. Except for a slight time lag in the case of 22-24 February 1968 flood, the fit is generally good emphasising the fact that the Stratified Routing is useful for natural flood routing problems. It may be worth stating here that similar results were seen in all the three reaches studied.

5. CONCLUSION

The following are the important conclusions made out of this study:

i. Optimization techniques can be profitably employed in the model parameter calculation for the Muskingum-Cunge method eliminating the need for geometric and hydraulic details;
ii. Stratified approach helps to calculate the parameters as closely as possible and in accordance with the fundamental basis of the coefficients; and
iii. The calculated stratified coefficients are found to be very useful for flood prediction.

6. REFERENCES

Cunge, J.A., (1969), 'On the Subject of a Flood Propagation Method (Muskingum Method)', *Journal of Hydraulic Research, IAHR,* Vol.7, pp.205-230.

Flood Routing Studies, Vol.III, (1975), *Flood Studies Report,* Natural Environmental Research Council, 27 Charing Cross Road, London.

Karmegam, M., (1981), 'Parameter Identification in Flood Routing', *Ph.D. Thesis,* University of London, London.

Karmegam, M., (1992), 'Stratified Muskingum Routing', *VIII Congress of Asia and Pacific Regional Division of the IAHR,* Central Water and Power Research Station, Pune, India.

Marquardt, D.W., (1963), 'An Algorithm for Least-squares Estimation of Non-linear Parameters', *Journal of the Society for Industrial and Applied Mathematics,* Vol.II, pp.431-441.

McCarthy, C.T., (1938), 'The Unit Hydrograph and Flood Routing', *Unpublished manuscript presented at a Conference of the North Atlantic Division,* US Army Corps of Engineers.

NAG, (1977), 'Library Subroutine E04FBF', *NAG Fortran Library Manual,* Mark 6 (FLM 6).

Ponce, V.M. and Yevjevich, V., (1978), 'Muskingum-Cunge Method with Variable Parameters', Technical Notes, *Journal of the Hydraulics Division, ASCE,* Vol.104, Hy.12, pp.1663-1667.

Powell, M.J.D., (1968), 'A Fortran Subroutine for Solving Systems of Non-linear Algebraic Equations', *Harewell Report AERE, R5947,* HM Stationery Office, UK.

Price, R.K., (1973), 'Variable Parameter Diffusion Method for Flood Routing', *Report No. INT115,* Hydraulic Research Station, Wallingford, England.

Price, R.K. (1974), 'Comparison of Four Numerical Methods for Flood Routing', *Journal of the Hydraulics Division, ASCE,* Vol.100, Hy.7, pp.879-899.

Rangapathy, V., Karmegam, M. and Sakthivadivel, R., (1988), 'A monograph on Flood Routing Methods as Applied to Indian Rivers', *Anna University Publication,* Centre for Water Resources, Anna University, Madras, India.

TRANSFER FUNCTION MODELS FOR HYDROLOGIC FLOOD ROUTING

M.P.Cherian and Prabir Kumar Pal
Dept. of Civil Engineering
I.I.T., Bombay, Powai, Bombay 400076

ABSTRACT. In this paper, a new method for linear, hydrologic flood routing is introduced. The method uses a technique from time series analysis called transfer function modelling. Transfer function models are used to relate the flood discharge data at a downstream station with that at an upstream station. This new technique is demonstrated using two available sets of upstream and downstream discharge data. Application of the method is relatively straightforward since the required data analysis can be carried out using general-purpose mathematics and statistics libraries which are now widely available on mainframe and PC systems. The results obtained with the methodology developed herein are compared with those using the Muskingum method.

1. INTRODUCTION

Transfer function models, as used in the time series analysis, transfer an input data series through a system and generate an output data series from the same system [1]. The genesis of a new method for hydrologic flood routing introduced herein comes from the observation that hydrologic flood routing accomplishes the same task as a transfer function model; here the system is the river reach under considertion, and the input and output data series are the upstream and downstream discharge data, respectively. It can be demonstrated (as shown later) that the mathematical form of the general discrete, linear transfer function model is identical to that of the general, linear hydrologic flood routing model. The power of the transfer function modelling procedure adopted herein lies in its ability to identify the most parsimonious (compact) form of the general linear model applicable to the problem in hand. Thus the simplest and most accurate linear hydrologic flood routing model for the given pair of data series (inflows and outflows) can be established. There are other types of transfer function models based on purely deterministic approaches. These are, however, not as useful for hydrologic flood routing since records of discharge measurements invariably contain errors which, when compounded by small

V. P. Singh and B. Kumar (eds.), Surface-Water Hydrology, 333–342.

amounts of lateral flow that usually enter a natural river reach, may create sufficient noise to mask the deterministic response of the system.

2. METHODOLOGY

The relationship between two time series is essentially expressed by the nature of the cross correlation function between the two time series. However, since the two series may be non-stationary, and auto correlated, it is essential to reduce the two time series to a pair of stationary and uncorrelated time series so that the calculations of the the cross correlation coefficients between the two series is meaningful. The first step in the model building procedure is therefore to identify a prewhitening operator for the input series (upstream flows) by making the input series stationary, and then fitting an autoregressive moving average model (ARMA) model to the stationary series. The normal procedure to make a time series stationary is to difference it either once or twice; in the case of a time series of flood discharge values, a second difference is usually required. A first difference will yield a series of positive values (over the rising limb), followed by a series of negative values (over the recession limb) making the differenced time series look quasi-stationary; sometimes it is possible to work with a series differenced only once. Mathematically, the differencing operation on the input series can be expressed as

$$(1-B)^n X_t = x_t \tag{1}$$

where B is the backward differencing operator, n is the order of differencing (normally 1 or 2), X_t is the original input series, x_t is the differenced input series. The operator B is defined by

$$B\,X_t = X_{t-1} \tag{2}$$

where X_t and X_{t-1} are the values of the input series at the current and previous time steps, respectively. An appropriate ARMA model can be fitted to the x_t series by examining the auto correlation function (ACF) and partial auto correlation function (PACF) of the x_t series. The ACF and PACF are the fingerprints that can lead to identifying the correct ARMA model; the process, however is not always simple and straightforward. The ARMA model can be written as

$$\phi(B)\,x_t = \theta(B)\,\alpha_t \tag{3}$$

where $\phi(B)$ is the general AR operator, $\theta(B)$ is the general MA operator, and α_t is the required residual white noise series. The prewhitening operator is given by

$$\alpha_t = \frac{\phi(B)}{\theta(B)}(1-B)^n X_t \tag{4}$$

In the transfer function modelling procedure, which is valid for a pair of input and output series in which there is no feedback relation (i.e. the output does not affect the input), the same prewhitening operator is applied to the output series (downstream flows) to yield the following residual series

$$\beta_t = \frac{\phi(B)}{\theta(B)}(1-B)^n Y_t \tag{5}$$

In general, β_t is not a white noise series, since the prewhitening operation is strictly valid only for the input series; however in transfer function modelling parlance, β_t is called the prewhitened output series. Fortunately, hydrologic flood routing falls under the transfer function category, since the upstream flows "drive" the the downstream flows and as such there is no feedback relationship. In cases where feedback is present, a seperate prewhitening operator for the output series has to be identified; these cases are not considered herein. The prewhitened input and output series can now be cross correlated and a relationship of the following form can be developed:

$$\alpha_t = v(B)\beta_t + N'_t \tag{6}$$

where $v(B)$ are the impulse response weights (which are essentially the cross correlation coefficients normalised by the product of the standard deviations of the prewhitened series), and N'_t is a noise series. $v(B)$ is of the following form:

$$v(B) = (1 - v_1 B - v_2 B^2 - \cdots\cdots) \tag{7}$$

If the number of significant parameters in $v(B)$ (i.e. v_1, v_2, etc.) are too many, a parsimonious form of equation (6) as given below, can be identified:

$$(1 - \delta_1 B - \delta_2 B^2 - \cdots \delta_r B^r)\beta_t = (\omega_0 - \omega_1 B - \cdots \omega_s B^s)\alpha_{t-b} + N_t \tag{8}$$

where N_t is a noise series, and, δ_1 to δ_r and ω_0 to ω_s are coefficients, and the parameters r, s, and b can be determined using standard procedures involving certain features of the cross correlation function [3]. Equation (8) is the general transfer function model; the noise series N_t must be examined to see whether it is white noise or not. If it is not white noise, a suitable ARMA model must be fitted to N_t .After substituting the appropriate model for the noise series, and replacing the prewhitened series with the original input and output series, equation can be written as

$$(1 - f_1 B - f_2 B^2 - \cdots f_m B^m)Y_t = (g_0 - g_1 B - \cdots g_n B^n)X_{t-b} + a_t \tag{9}$$

where a_t is a white noise series (or, at the most contains, a MA dependence), and f_1 tof_m and g_0 to g_n , and m and n are constants. The hydrologic routing model is thus given by equation (9) minus the noise term. Notice the similarity with the general linear hydrologic routing model given by

$$O_i = \sum_{k=1}^{\infty} a_k\, O_{i-k} + \sum_{k=1}^{\infty} b_k\, I_{i+1-k} \tag{10}$$

where O_is, and I_is are the ouflows and inflows, respectively, and a_ks and b_ks are constants. This similarity is the key to the methodology developed herein. The remarkable feature is that the determination of the parameters r, s, and b in the general transfer function model leads to identifying the precise number of inflow and outflow terms (as well as values of the associated constants) that need to be considered in the general linear hydrologic model given by equation (10).

In the application of the above methodolgy, numerous tasks such as the differencing operation, and computations of the ACF, PACF, the cross correlation function and the impulse response weights were carried out using the International Mathematics and Statistics Library (IMSL), a comprehensive and easy-to-use computer package which is widely available on main frame computers and PCs. Appropriate subroutines from this package were used to perform the necessary functions.

3. APPLICATION

Transfer function models are fitted to two sets of flood routing data. The fitting procedure is described in the following subsections. Details can be found in [4].

3.1. Case 1

The inflow and outflow data consist of thirty daily discharge values as given in [5]. The raw inflow and outflow data are plotted and displayed in Fig.1. A first order difference of the inflow and outlow data is carried out to render the series quasi-stationary. The ACF & PACF of differenced inflows are calculated and plotted in Figs 2 and 3, respectively. The plot indicates that the differenced inflow series follows an ARMA(1,1) process. The inflow series can thus be modelled as

$$(1-\phi_1 B)\, x_t = (1-\theta_1 B)\, \alpha_t \tag{11}$$

where ϕ_1 is the AR parameter, and θ_1 is the MA parameter. ϕ_1 and θ_1 are estimated using the appropriate IMSL subroutines and these values are .987, with a standard error of 0.03, and -.514 with a standard error of 0.103., respectively. The inflow series is, therefore, modelled as

$$(1-.987B)\, x_t = (1+.515B)\, \alpha_t \tag{12}$$

The same prewhitening transformation is now applied to the outflow series. The outflow series is modelled as

$$(1-.987B)\, y_t = (1+.515B)\, \beta_t \tag{13}$$

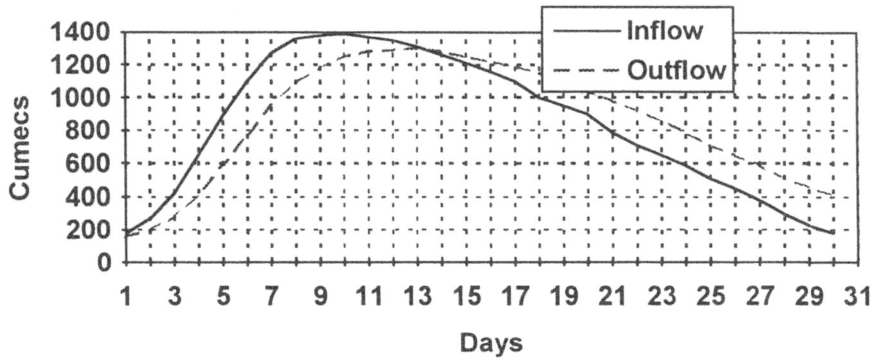

Fig 1. Plot of Inflow and Outflow Hydrographs - Case 1

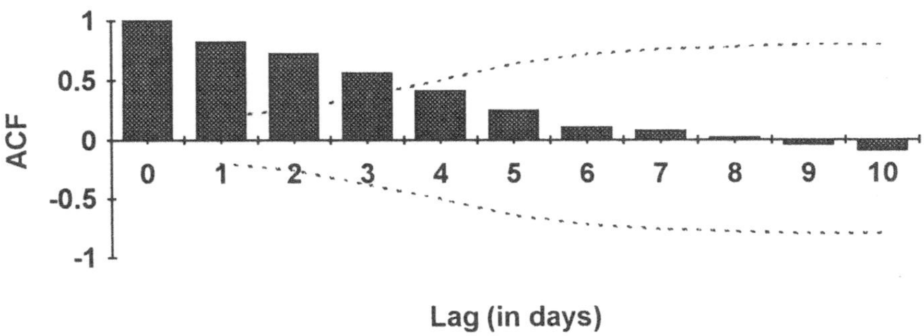

Fig 2. Plot of the ACF of the Stationary Inflow Series - Case 1

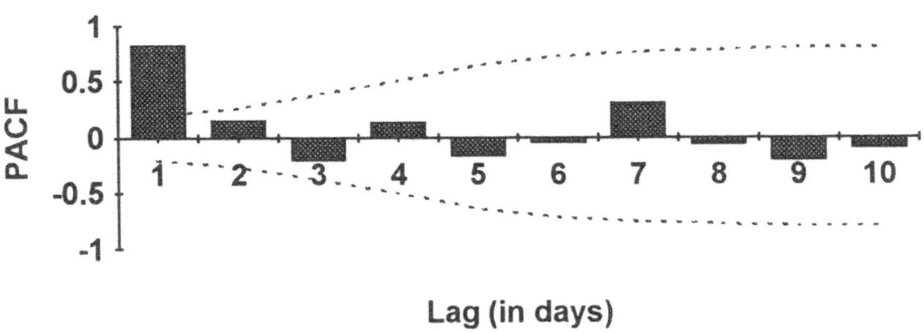

Fig 3. Plot of the PACF of the stationary input series - Case 1

The auto correlation values of α_t are calculated and plotted in Fig.4 which indicates that the series is indeed white noise. The prewhitening transformation of the inflow series is thus adequate. In the next step, the cross correlation values of the prewhitened inflow (α_t) and prewhitened outflow (β_t) are calculated for lags ranging from +15 to -15. It is observed that the cross correlation values are significant only for lags 0, 3, and 4, since they lie outside the range of plus or minus twice the corresponding standard error. Impulse response weights corresponding to these lags are calculated using the appropriate IMSL subroutine; thus, a model of the following form is arrived at:

$$\alpha_t = v(B)\beta_t + N'_t \tag{14}$$

where

$$v(B) = (0.006 + 0.092B^3 - 0.007B^4) \tag{15}$$

and N'_t is a noise series.This model is already in a nearly parsimonious form, so no attempt is made to estimate the parameters *r*, *s* and *b* of the general transfer function model. Substituting the prewhitening transformation and expressing in terms of the original data series, the appropriate transfer function model can be written as

$$(1-.987B)(1-B)Y_t = v(B)(1-.987B)(1-B)X_t + N_t \tag{16}$$

where N_t is another noise series. The ACF and PACF of N_t indicate that it is a white noise series. Thus the model given by equation (16) is the correct transfer function model for the given data set. The hydrologic routing model for this case is given by ignoring the noise term in equation (16) and is given by the following equation:

$$(1-.987B)(1-B)Y_t = v(B)(1-.987B)(1-B)X_t \tag{17}$$

Equation (17) can be expanded to generate the algebraic recursion relation that represents the routing equation; it may be noticed that this relation involves terms upto Y_{t-2} and X_{t-6}. To check the validity of the model, outflows are generated using the model and these are compared with the observed outflows (Fig. 7). The generated outflows are also compared with the outflows obtained using the Muskingum method for the same data (Fig. 7). The sum of the squares of errors between the observed and generated outflows is smaller using the transfer function model as compared to that using the Muskingum method.

3.2. Case 2

A transfer function model is fitted to a pair of upstream and downstream hourly discharge data given in [2]. Following essentially the same procedure as in the previous case, but with the addition of certain other features, a parsimonious transfer function model is identified for this data set. The plots of original inflow and outflow hydrographs are shown in Fig. 4 which indicates a double peaked hydrograph. Here, the inflow and outflow series are differenced twice in order to make the corresponding series stationary. The auto correlation and partial auto correlation values of the differenced inflow are calculated and

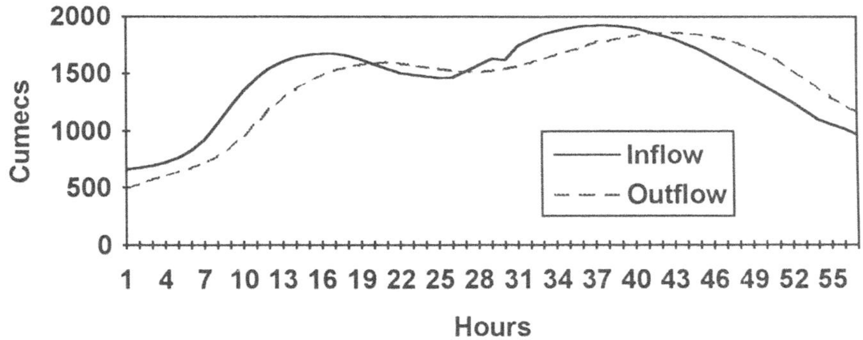

Fig 4. Plot of Inflow and Outflow Hydrographs - Case 2

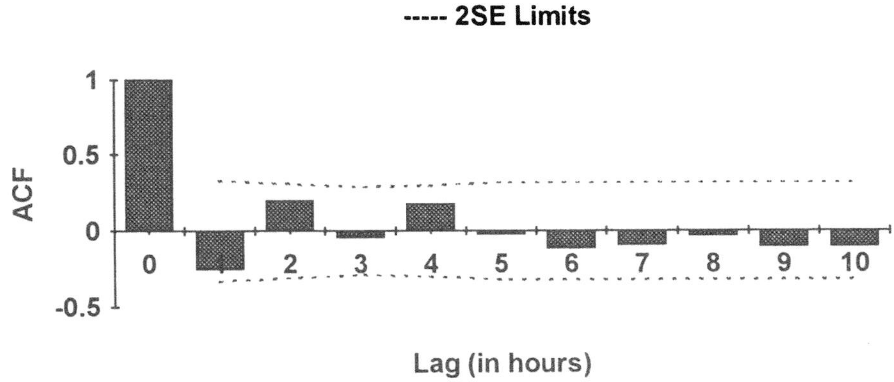

Fig 5. Plot of the ACF of the Stationary Inflow Series - Case 2

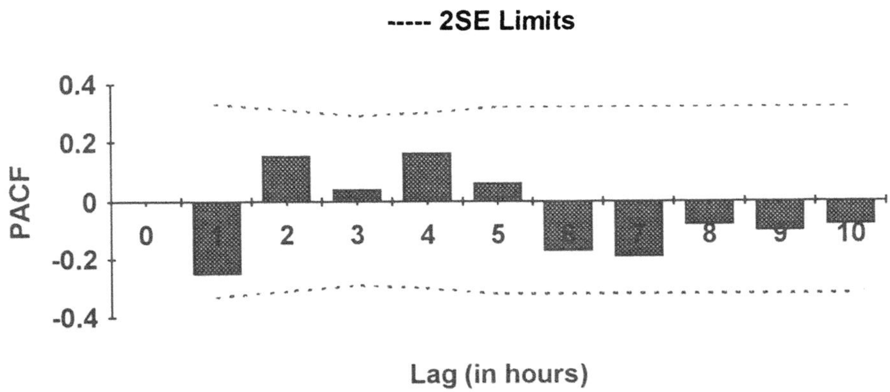

Fig 6. Plot of the PACF of the stationary input series - Case 2

these show no significant correlations on the basis of standard error estimates (Figs. 5 and 6). So the differenced inflow values are not required to be modeled as was done in the previous case. In other words, the prewhitened α_t and β_t series are themselves the x_t and y_t series, respectively. The differenced inflow and outflow series are directly cross correlated and the cross correlation values are calculated for a range of lags between -15 to +15. The cross correlation values are found to be significant at lags 0, 1, 2, 3, 4 and 5. Since the number of significant cross correlations here are many more than in the previous case, it was considered necessary to estimate the parameters r,s, and b.of the general transfer function model in order to arrive at a reasonably parsimonious form. The impulse response weights were computed for lags ranging from 0 to 5. The parameter b, which deals with the absolute delay before the inflow time series influences the outflow, is determined to be 1, since the cross correlation coefficient at lag 0 is just barely significant. Similarly, the parameters r and s were determined using the procedures illustrated in [3]. A parsimonious model of the following form is identified:

$$(1-0.816B)\beta_t = (0.108+0.502B+.021B^2)B\alpha_t + N_t \tag{18}$$

The ACF and PACF of the noise term in equation (18) indicated an ARMA(1,1) model with the AR parameter = 0.375 and the MA parameter = -0.448. Substituting for this model in equation (18), and expanding, the following is obtained:

$$(1-1.191B+.306B^2)\beta_t = (.108B+.010B^2+.002B^3-.008B^4)\,\alpha_t + a_t \tag{19}$$

where a_t is the final noise series. The ACF and PACF of the final noise series indicated that it is white noise. Equation (19), therefore, is the final form of the transfer function model. The hydrologic routing model is equation (19) with the noise term neglected and is as follows:

$$(1-1.191B+.306B^2)(1-B)^2Y_t = (.108B+.010B^2+.002B^3-.008B^4)(1-B)^2X_t \tag{20}$$

For the given inflows, outflows are generated using the above model and the Muskingum method; the comparison of these generated outflows with the observed outflow is shown in Fig. 8. Computations of the sum of the squared errors again indicate that the transfer function model is superior to the Muskingum method.

4. CONCLUSION

The linear discrete transfer function model is shown to be identical to the general linear hydrologic routing model of which the Muskingum method and several other hydrological methods are only special cases. The power of the methodology adopted in the study lies that its ability to identify the most succint or parsimonious structure of the general linear routing model which best fits the given data. Moreover, flood routing through transfer function models is independent of channel characteristics and channel parameters whereas

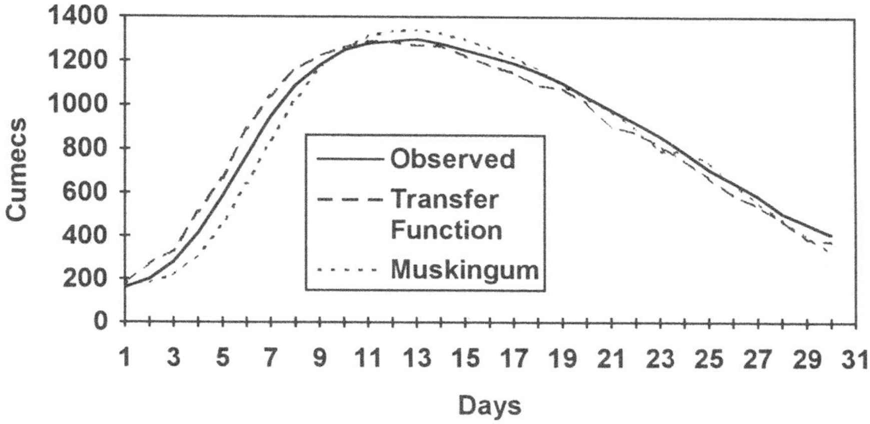

Fig 7. Observed and Generated Outflow Hydrographs - Case 1

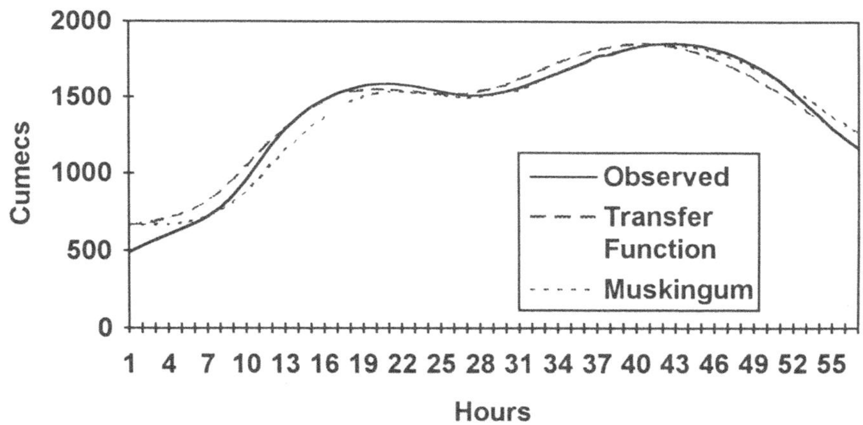

Fig 8. Observed and Generated Outflow Hydrographs - Case 2

all the conventional methods of hydrologic flood routing depend on channel parameters. Like all other methods of hydrologic flood routing, transfer function models have the following drawbacks: 1) inability to predict backflows, 2) loss of model validity in the event of physical changes in the river reach, and 3) inability to account for boundary conditions. Also the transfer function models utilized herein are linear models, thus it can only approximate the response of a river system which undoubtedly has nonlinear properties. Finally, this work can be extended for forecasting purposes. Once the inflow and outflow relationship is established, the outflow values for future time steps can be calculated knowing the past values of both series.

REFERENCES

1. Box and Jenkins, *Time Series Analysis - Forecasting and Control*, Holden Day, 1976.

2. Ciriani, T.A., Maione, U., and Wallis, J.R., Eds., *Mathematical Models for Surface Water Hydrology*, John Wiley and Sons Inc., 1977, pages 337-349.

3. Maridakis, Wright, and McGee, *Forecasting Methods and Applications*, John Wiley and Sons Inc, 1978.

4. Pal, Prabir Kumar, *Transfer Function Models for Hydrologic Flood Routing*, M.Tech Thesis, Dept. of Civil Engg., I.I.T., Bombay, 1993.

5. Singh, Vijay P., *Hydrologic Systems, Rainfall-Runoff Modeling, Vol. I*, Prentice Hall Inc., 1989, pages 405-407.

EFFECT OF CROSS-SECTIONAL SHAPE ON FREE-SURFACE INSTABILITY

VICTOR M. PONCE AND PEDRO J. PORRAS
Department of Civil Engineering
San Diego State University
San Diego, California 92182 USA

ABSTRACT. The effect of cross sectional shape on free-surface hydrodynamic instability has been analyzed. The characterizing parameter is the dimensionless relative kinematic wave celerity $c_{drk} = \beta - 1$, in which β is the exponent of the normal discharge-flow area rating. Two generic types of cross-sectional shapes are identified: (1) those of constant c_{drk}, and (2) those of variable c_{drk}. Three cross-sectional shapes of constant c_{drk} are: (1) hydraulically wide, with $c_{drkM} = 2/3$, (2) triangular, with $c_{drkM} = 1/3$, and (3) inherently stable, with $c_{drk} = 0$ (Manning or Chezy). (The subscript M refers to Manning friction). Cross-sectional shapes of variable c_{drk} include trapezoidal, rectangular, and circular shapes. Two asymptotic cross-sectional shapes are identified: (1) hydraulically wide, and (2) hydraulically narrow. These theoretical cross-sections set limits to the range of variation of c_{drk} for trapezoidal and rectangular shapes. The hydraulically wide channel sets the upper limit, with $c_{drkM} \rightarrow 2/3$; the hydraulically narrow channel sets the lower limit, with $c_{drk} \rightarrow 0$. Two types of stable channels are identified: (1) inherently stable, and (2) stable. An inherently stable cross section is such that the Vedernikov number **V** is identically zero for all Froude numbers. A stable cross section is such that $\mathbf{V} \leq 1$ for Froude numbers in the range $\mathbf{F} \leq \mathbf{F}_{ns}$, in which $\mathbf{F}_{ns}$ is a design neutral-stability Froude number. A stable cross-sectional shape is designed by setting c_{drk} and the related cross-sectional parameter δ to match a certain choice of $\mathbf{F}_{ns}$. The resulting stable cross-sectional shape is much narrower that the comparable inherently stable shape.

1. Introduction

The subject of free-surface hydrodynamic instability has interested many researchers and practitioners, beginning with the pioneering work of Vedernikov (1945), who identified the well-known criterion bearing his name. In effect, the Vedernikov number $\mathbf{V} = 1$ describes the condition of neutrally stable flow, which separates stable ($\mathbf{V} < 1$) from unstable ($\mathbf{V} > 1$) flow (Chow 1959). Craya (1952) enhanced the Vedernikov

V. P. Singh and B. Kumar (eds.), Surface-Water Hydrology, 343–354.

criterion by interpreting it as the ratio of relative kinematic and dynamic wave celerities. Liggett (1975) extended the Vedernikov criterion to channels of arbitrary cross-sectional shape. This led Ponce (1991) to define the "inherently stable channel" as that for which V is identically zero, regardless of flow depth or Froude number.

At the threshold of stability/instability, V = 1, i.e., the relative kinematic and dynamic wave celerities are equal. The relative kinematic wave celerity is (Chow 1959):

$$c_{rk} = (\beta - 1)\, v \qquad (1)$$

in which v = mean velocity, and β = exponent of the equilibrium or uniform flow rating

$$Q = \alpha A^{\beta} \qquad (2)$$

with Q = discharge, A = flow area.

The relative dynamic wave celerity, or Lagrange celerity, is (Chow 1959):

$$c_{rd} = (gD)^{1/2} \qquad (3)$$

in which g = gravitational acceleration, and

$$D = A/T \qquad (4)$$

is the hydraulic depth, and T = top width.

Since the Froude number is (Chow 1959):

$$\mathrm{F} = v\,/(gD)^{1/2} \qquad (5)$$

it follows that the Vedernikov number is:

$$\mathrm{V} = (\beta - 1)\,\mathrm{F} \qquad (6)$$

Equation 6 underscores the physical significance of β, or rather of

$$c_{drk} = \beta - 1 \qquad (7)$$

in which c_{drk} = dimensionless relative kinematic wave celerity. The latter is an intrinsic characteristic of the channel cross section and flow. Its definition as the ratio of Vedernikov and Froude numbers underscores its crucial role in free-surface hydrodynamic instability.

The objectives of this paper are thus twofold:

1. To evaluate dimensionless relative kinematic wave celerities c_{drk} (and related Froude numbers for neutrally stable flow) for a variety of cross-sectional shapes;
2. To develop criteria for the design of stable channels subject to physically realistic Froude number ranges.

2. Background

2.1 DIMENSIONLESS RELATIVE KINEMATIC WAVE CELERITIES

The Manning equation is (Chow 1959):

$$v = (1/n)\ R^{2/3}\ S_o^{1/2} \tag{8}$$

in which v = mean velocity ($v = Q/A$), n = Manning roughness coefficient, R = hydraulic radius ($R = A/P$), and S_o = bottom slope.

Assuming the existence of a functional relation $P = \kappa A^{\delta}$, Eqs. 7 and 8 lead to

$$c_{drkM} = (2/3)\ (1 - \delta) \tag{9}$$

in which c_{drkM} = dimensionless relative kinematic wave celerity under Manning friction.

A similar relation for Chezy friction is

$$c_{drkC} = (1/2)\ (1 - \delta) \tag{10}$$

Combining Eqs. 9 and 10, the relationship between dimensionless relative kinematic wave celerities under Chezy and Manning friction is

$$c_{drkC} = (3/4)\ c_{drkM} \tag{11}$$

2.2 RELATIONSHIP BETWEEN δ AND CROSS-SECTIONAL SHAPE

The cross sectional shape is characterized by the parameter δ, which varies in the range $0 \le \delta \le 1$. In general, δ varies with flow depth. There are, however, three special cases for which δ is a constant. These are recognized as:

1. The hydraulically wide channel: $\delta = 0$. This is a theoretical cross-sectional shape for which the wetted perimeter is constant ($P = \kappa$) and is independent of the flow area ($dP/dA = 0$). This cross-sectional shape is an asymptotic limit to all wide channels.
2. The family of triangular channels: $\delta = 1/2$. The actual shape of any triangular channel is a function of the side slope z (z horizontal to 1 vertical). All triangular channels feature $dP/dA = (1/2)\ (P/A) = (1/2)\ /R$.
3. The family of inherently stable channels: $\delta = 1$ (Ponce 1991). This is a cross-sectional shape for which the wetted perimeter is a linear function of the flow area: $P = \kappa\ A$. Since $R = A/P$, this cross-sectional shape has a constant hydraulic radius $R = R_o = 1/\kappa$.

From these three cases, it follows that

$$\delta = R\ (dP/dA) \tag{12}$$

For other cross-sectional shapes, including trapezoidal, rectangular, and circular, δ varies with the flow depth.

2.3 CROSS-SECTIONAL SHAPE AND FREE-SURFACE HYDRODYNAMIC INSTABILITY

The inherently stable cross-sectional shape implies that $\mathrm{V} \equiv 0$ for all flow depth and Froude number ranges. The original expression for the Vedernikov number is (Vedernikov 1945; Chow 1959):

$$\mathrm{V} = x\,\gamma\,\mathrm{F} \tag{13}$$

in which

$$x = (1 + b) / (2 - b) \tag{14}$$

with b = exponent of Reynolds number R in the frictional power law $f = a\ \mathrm{R}^{-b}$, and γ is a cross-sectional shape factor defined as

$$\gamma = 1 - R\ (\mathrm{d}P/\mathrm{d}A) \tag{15}$$

For Manning friction, $b = 1/5$, and with Eq. 14, $x = 2/3$. For Chezy friction, $b = 0$, and $x = 1/2$. Thus, given Eqs. 9, 10, and 12, the equivalence of Eqs. 6 and 13 is confirmed. Furthermore, given Eqs. 6 and 7:

$$c_{drk} = \mathrm{V} / \mathrm{F} \tag{16}$$

The neutral-stability Froude number F_{ns} is that corresponding to $\mathrm{V} = 1$. Thus

$$\mathrm{F}_{ns} = c_{drk}^{-1} \tag{17}$$

Thus, the neutral-stability Froude number and the dimensionless relative kinematic wave celerity are intrinsically related. Table 1 shows values of b, x, δ, γ, c_{drk}, β, and F_{ns} for the three cross-sectional shapes of constant δ and c_{drk}: hydraulically wide, triangular, and inherently stable.

3. Experimental Program

Given the role of the dimensionless relative kinematic wave celerity in characterizing free-surface hydrodynamic instability, a series of tests was devised to evaluate c_{drk} for a variety of cross-sectional shapes. A set of twelve cross-sectional shapes was selected for analysis. The set consists of eight trapezoidal/triangular and four inherently stable channels. The eight trapezoidal/triangular cross-sections are combinations of bottom widths B = 0, 1, 10, and 100 m, and side slope parameters z = 1 and 10. The four inherently stable cross-sections have bottom widths B = 0, 1, 10, and 100 m, and constant hydraulic radius. The design flow depth is H = 10 m.

Three additional cross-sections, consisting of circular, rectangular, and square shapes were considered. The diameter of the circular cross-section is D = 10 m. The bottom width of the rectangular cross-section is B = 10 m. The rectangular cross-section was used to study the asymptotic behavior of the hydraulically wide channel ($h/B \ll 1$) and its exact opposite, the hydraulically narrow channel ($h/B \gg 1$). Values of c_{drkM} were calculated for channels approaching these two limits. The square shape is a special case of the rectangular shape for which the flow depth is constant and equal to the bottom width (Flood Hydrograph Package 1990).

The calculated c_{drkM} values for hydraulically wide, triangular, and inherently stable cross-sectional shapes confirmed exactly the theoretical values shown in Table 1. Table 2 shows c_{drkM} values for the six trapezoidal cross-sections. Table 3 shows c_{drkM} values for the circular cross-section. Table 4 shows c_{drkM} values for the rectangular cross-section. In addition, the c_{drkM} value for a square cross-sectional shape was calculated to be $c_{drkM} = 2/9$. This value is slightly different from the widely accepted value of 1/3 (Flood Hydrograph Package 1990).

4. Stable Channel Design

In this paper, *stable channel design* refers to the design of a cross-sectional shape that remains stable for a specified Froude number range. Equations 9, 10, 16, and 17 are used for this purpose. To avoid confusion with prior usage of the phrase "stable channel" (Liggett 1975), an *inherently stable channel* is defined as that for which (1) $V \equiv 0$, (2) $c_{drk} = 0$, (3) $\delta = 1$, and (4) $F_{ns} = \infty$. In contrast, a *stable channel* is such that (1) $V \leq 1$, (2) $c_{drk} > 0$, (3) $\delta < 1$, and (4) $F_{ns} < \infty$.

4.1 INHERENTLY STABLE CHANNEL: THEORY

The differential wetted perimeter is (Liggett 1975):

$$dP = (dh^2 + dT^2)^{1/2} \tag{18}$$

in which h = flow depth. Dividing Eq. 18 by dh, and recognizing that $dA = T\, dh$,

$$T\,(dP/dA) = [1 + (dT/dh)^2]^{1/2} \tag{19}$$

The inherently stable channel has a constant hydraulic radius $R = R_o$, with R_o being the hydraulic radius corresponding to the initial flow depth h_o. With Eq. 12, Eq. 19 leads to:

$$T/R_o = [1 + (dT/dh)^2]^{1/2} \tag{20}$$

from which

$$dT/dh = [\,(T/R_o)^2 - 1\,]^{1/2} \tag{21}$$

TABLE 1. Characteristics of Cross-sectional Shapes of Constant δ *and* c_{drk}.

Type	b	x	δ	γ	c_{drk}	β	F_{ns}
Hydraulically wide							
Laminar	1	2	0	1	2	3	1/2
Manning	1/5	2/3	0	1	2/3	5/3	3/2
Chezy	0	1/2	0	1	1/2	3/2	2
Triangular							
Manning	1/5	2/3	1/2	1/2	1/3	4/3	3
Chezy	0	1/2	1/2	1/2	1/4	5/4	4
Inherently stable							
Manning	1/5	2/3	1	0	0	1	∞
Chezy	0	1/2	1	0	0	1	∞

TABLE 2. Average c_{drkM} Values for Trapezoidal Cross-sections.

B	z	c_{drkM} for selected flow depth range (in percent of maximum)			
(m)		1-25	1-50	1-75	1-100
1	1	0.363	0.347	0.341	0.325
1	10	0.355	0.346	0.343	0.328
10	1	0.520	0.460	0.427	0.393
10	10	0.391	0.367	0.357	0.338
100	1	0.644	0.624	0.607	0.576
100	10	0.552	0.498	0.465	0.430

TABLE 3. Average c_{drkM} Values for Circular Cross-sections.

Depth range (in percent of diameter)	c_{drkM}
0-25	0.446
0-50	0.422
0-75	0.377
0-98	0.028
25-50	0.377
25-75	0.312
50-75	0.247
50-98	-0.434
75-98	-0.876
94-98	-2.252

TABLE 4. Average c_{drkM} Values for Rectangular Cross-sections.

h/B range	c_{drkM}	Cross-sectional shape
$\rightarrow 0$	$\rightarrow 2/3$	Hydraulically wide
0-0.01	0.660	Wide
0-0.10	0.608	Wide
0-0.25	0.541	Transition to narrow
0-0.50	0.463	id.
0-0.75	0.408	id.
0-1	0.367	Narrow (Square)
0-2	0.269	Transition to hydraulically narrow
0-5	0.160	id.
0-10	0.102	id.
2-10	0.061	id.
8-10	0.038	id.
$\rightarrow \infty$	$\rightarrow 0$	Hydraulically narrow

subject to $T > R_o$.

Equation 21 represents a family of inherently stable cross sections, with R_o as the parameter. Assuming that $T = R_o$ for $h = h_o$, the solution of Eq. 21 is given by Liggett (1975):

$$h = h_o + R_o \ln \{(T/R_o) + [(T/R_o)^2 - 1]^{1/2}\} \tag{22}$$

In practice, an inherently stable channel comprises two subsections:

1. A lower subsection, of rectangular (or alternatively, trapezoidal) shape, of bottom width B, depth h_o, and hydraulic radius R_o corresponding to h_o; and
2. An upper subsection of inherently stable shape (Eq. 22). The greater h_o, the greater R_o, and the narrower the resulting inherently stable cross sectional shape.

4.2 DESIGN CRITERIA FOR STABLE CHANNELS

The inherently stable channel is stable for all Froude numbers. Since friction has a lower bound and cannot realistically approach zero, it follows that there is a practical upper bound to the Froude number. Therefore, a stable channel may be designed to remain stable within the Froude-number range $\mathbf{F} \le \mathbf{F}_{ns}$, where $\mathbf{F}_{ns}$ can be chosen as a Froude number not likely to be exceeded in a given design case. For example, a practical upper bound to the Froude number is $\mathbf{F} = 20$ (Chow 1959). Setting $\mathbf{F}_{ns} = 20$, the corresponding stable channel design parameters are: $c_{drk} = 0.05$, and $\delta = 0.925$. Thus, a cross-sectional shape designed to conform to a constant $\delta = 0.925$ will remain stable in the range $\mathbf{F} \le 20$.

The extension of Eqs. 19 to 21 to the stable cross-section, for which $\delta < 1$ while remaining constant for all flow depths leads to

$$dT/dh = [\, (\delta T/R)^2 - 1\,]^{1/2} \tag{23}$$

subject to $\delta T > R$.

Unlike the inherently stable channel (Eq. 21), for which the hydraulic radius is a constant, in the stable channel (Eq. 23) the hydraulic radius varies with flow depth. Therefore, Eq. 23 cannot be integrated analytically. Given a choice of $\mathbf{F}_{ns}$, and associated c_{drkM} and δ, and the hydraulic radius R_o corresponding to the depth h_o of the lower subsection, the shape of the upper subsection $T = f(\delta, R_o, R)$ can be obtained by numerical integration. As in the case of the inherently stable channel, the greater h_o, the greater R_o, and the narrower the resulting stable cross sectional shape.

Figure 1 shows examples of stable cross-sections, of rectangular bottom subsection, with $B = 5$ m: (a) $h_o = 0.5$ m, $R_o = 0.417$ m; (b) $h_o = 0.75$ m, $R_o = 0.577$ m; (a) $h_o = 1.0$ m, $R_o = 0.714$ m. (Due to symmetry, only half of the cross section is shown). Stable cross sections corresponding to $\mathbf{F}_{ns} = 5$, 10, and 100 are shown. Also included for comparison is the inherently stable channel, for which $\mathbf{F}_{ns} = \infty$.

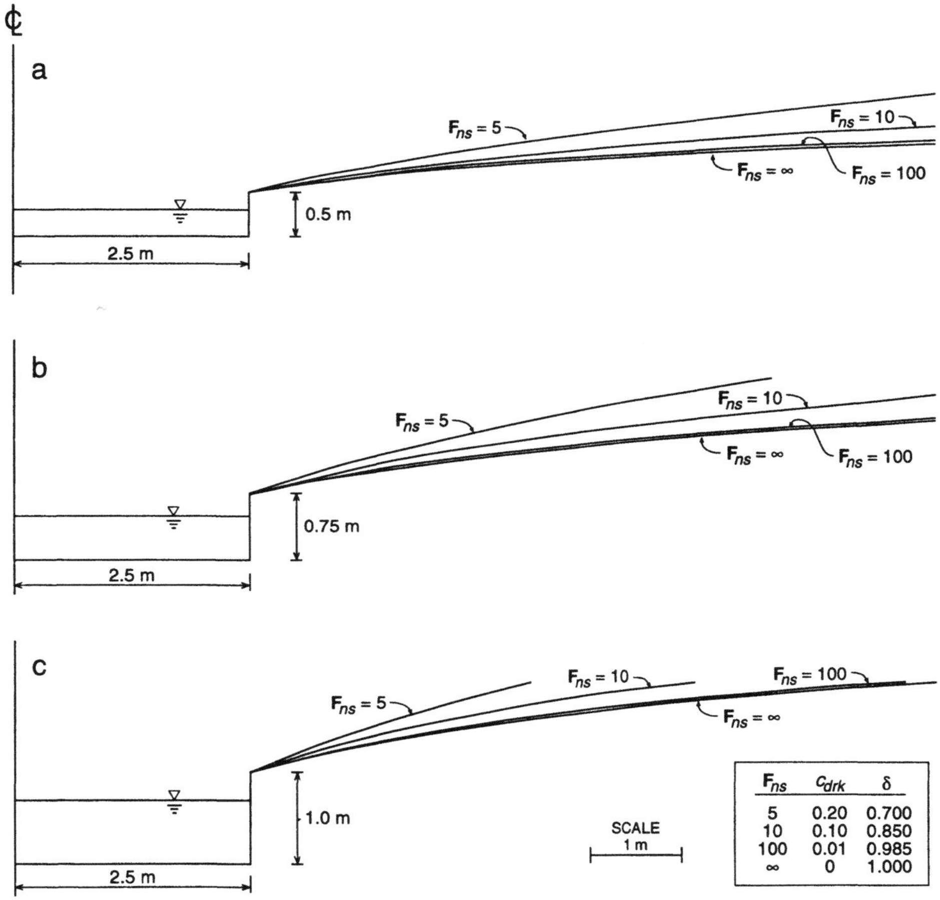

Fig. 1 Effect of Cross-sectional Shape on Free-surface Hydrodynamic Instability: (a) h_o= 0.5 m, R_o= 0.417 m; (b) h_o= 0.75 m, R_o= 0.577 m; (c) h_o= 1.0 m, R_o= 0.714 m.

5. Summary

The effect of cross sectional shape on free-surface hydrodynamic instability has been analyzed. The characterizing parameter is the dimensionless relative kinematic wave celerity $c_{drk} = \beta - 1$, in which β is the exponent of the discharge-flow area rating. Two generic types of cross-sectional shapes are identified: (1) those of constant c_{drk}, and (2) those of variable c_{drk}. Three cross-sectional shapes of constant c_{drk} are: (1) hydraulically wide, with $c_{drkM} = 2/3$, (2) triangular, with $c_{drkM} = 1/3$, and (3) inherently stable, with $c_{drk} = 0$ (Manning or Chezy). Cross-sections of variable c_{drk} include trapezoidal, rectangular, and circular shapes.

Two asymptotic cross-sectional shapes are identified: (1) hydraulically wide, with $c_{drkM} = 2/3$, and (2) hydraulically narrow, with $c_{drk} = 0$ (Manning or Chezy). These theoretical cross-sectional shapes set limits to the range of variation of c_{drk} for trapezoidal and rectangular shapes. The hydraulically wide channel is the upper limit, with $c_{drkM} = 2/3$. The hydraulically narrow channel is the lower limit, with $c_{drk} = 0$ (Manning or Chezy).

Two types of stable channels are identified: (1) inherently stable, and (2) stable. An inherently stable cross section is stable for all Froude numbers. A stable cross section is stable for Froude numbers in the range $F \leq F_{ns}$, in which F_{ns} is a design neutral-stability Froude number. A stable cross-sectional shape is designed by setting c_{drk} and the related cross-sectional parameter δ to match a design choice of F_{ns}. The resulting stable cross sectional shape is much narrower that the comparable inherently stable shape.

6. References

Chow, V. T. (1959). *Open-Channel Hydraulics.* McGraw-Hill, New York, N.Y.

Craya, A. (1952). "The Criterion of the Possibility of Roll Wave Formation." *Gravity Waves, National Bureau of Standards Circular No. 521,* 141-151.

Flood Hydrograph Package, User's Manual. (1990). U. S. Army Corps of Engineers, Hydrologic Engineering Center, Davis, Calif., Version 4.0, Sept. 1990.

Liggett, J. A. (1975). "Stability." Chapter 6, *Unsteady Flow in Open Channels,* Water Resources Publications, K. Mahmood and Vujica Yevjevich, eds., Vol. 1, 259-281.

Ponce, V. M. (1991). "New Perspective on the Vedernikov Number." *Water Resour. Res.,* 27(7), 1777-1779.

Vedernikov, V. V. (1945). "Conditions at the Front of a Translation Wave Disturbing a Steady Motion of a Real Fluid." *U.S.S.R. Acad. Sciences Comptes Rendus* (Doklady), 48(4), 239-242.

7. Notation

A	= flow area;
B	= bottom width;
b	= exponent of Reynolds number in the frictional power law $f = a\,\mathbf{R}^{-b}$;
c_{drk}	= dimensionless relative kinematic wave celerity;
c_{drkC}	= dimensionless relative kinematic wave celerity, Chezy friction;
c_{drkM}	= dimensionless relative kinematic wave celerity, Manning friction;
D	= hydraulic depth, circular channel diameter;
$\mathbf{F}$	= Froude number;
$\mathbf{F}_{ns}$	= neutral-stability Froude number;
f	= Darcy-Weisbach friction factor;
g	= gravitational acceleration;
H	= design flow depth;
h	= flow depth;
h_o	= depth of rectangular (or trapezoidal) lower subsection in inherently stable and stable channels;
n	= Manning roughness coefficient;
P	= wetted perimeter;
Q	= discharge;
R	= hydraulic radius;
R_o	= constant hydraulic radius for inherently stable channel;
$\mathbf{R}$	= Reynolds number;
S_o	= bottom slope;
T	= top width;
$\mathbf{V}$	= Vedernikov number;
v	= mean velocity;
x	= friction-type parameter; Eq. 14;
z	= side slope parameter (z horizontal to 1 vertical);
α	= coefficient of the normal flow rating;
β	= exponent of the normal flow rating;

γ = cross-sectional shape factor, Eq. 15;

δ = exponent of the wetted perimeter-flow area relation; and

κ = coefficient of the wetted perimeter-flow area relation

FLOW RESISTANCE ESTIMATION FOR LARGE-SCALE ROUGHNESS

V.A.Copertino, B.De Bernardinis, F.A.Sciancalepore, A.Sole
Dipartimento di Ingegneria e Fisica dell'Ambiente
Università della Basilicata
Via della Tecnica n.3
85100 Potenza
Italy

ABSTRACT. In channels when the size of roughness elements attains the same order of magnitude as the depth of the flow, uncertainties still exist to describe the flow resistance.

Because of the simultaneous action of different dissipative forms and the difficulty of a complete quantification of the morphological and geometrical characteristics of the bed, special care must be paid to analyse and model the physical processes.

The cobbles, which are just covered up by the free surface or which overhang it, give rise to distortions in the surface that strongly increase their drag forces, according to Froude number and relative submergence. They act as proper boundaries for the flows that are divided and funnelled on the nearly flat parts of the bed between the elements.

In order to take into account the effect of funnelling, a simple model has been developed to represent the geometry of roughness in a cross section in function of sediment size distribution; in this way it is possible to avoid the commonly diffused approximation to calculate the value of wetted perimeter simply as the channel width.

Finally, the estimation of coefficient of resistance has been carried out through the integration of a simple semilogarithmic velocity profile whose reference datum position is depending (on the grounds of hydrodynamic equivalence) on some parameters which can be easily measured.

The good performance of the proposed method has been subsequently tested by using a collection of flume data published by several authors.

V. P. Singh and B. Kumar (eds.), Surface-Water Hydrology, 355–370.

1. Introduction

Several paper on the behaviour of the flow resistance in natural channels have been written in the latest thirty years In spite of this, the knowledge of the physical processes involved has been shown to be very poor and the quantitative prediction completely failed when the size of the roughness elements attains the same order of magnitude as the depth of the flow

For the large-scale roughness some authors have suggested a power law [1] [6] [7] For the intermediate-scale all the authors agree on the semilogarithmic law [12] [16] [17] [19]. either in the classical form or in a modified one

In fact the estimation of the effective roughness size to be assumed in a classical semilogarithmic equation [3] [21] [22] is very difficult, because of the great number of factors affecting resistance However, even if this goal was somehow achieved, no physically based model would be constructed

In the present work an attempt to use the semilogarithmic equation has been performed to explain the flow resistance in the range of large-scale roughness shown by Bathurst et al [1] experiments

2. Physical features of the flow resistance for the large-scale roughness in gravel bed rivers

When the size of the roughness elements is comparable with the depth of the flow, the cobbles which are just covered by the free surface or overhang it, cause a big distortion in the surface and strongly increase the drag force due to themselves A study of Flammers et al [2] has shown that the dependence of the drag coefficient C_d on the relative submergence h/D and on the Froude number $Fr=U/\sqrt{gh}$ is different in three characteristic regions in the first one, for $h/D<1\ 2$ and $Fr<1\ 5$, the free surface effects are very high, in the second one, for $1\ 2<h/D<4$ and $Fr<1\ 3$, they gradually decrease, in the third one, for $h/D>4$. C_d shows no dependence on such variables

Then such a dependence may be neglected for $Fr>1\ 5$ or for $h/D>4$ These considerations suggest the following classification of roughness. proposed by Bathurst et al [1]

$h/D_{50} < 2$	or	$h/D_{84} < 1\ 2$	large scale
$2 < h/D_{50} < 7\ 5$	or	$1\ 2 < h/D_{84} < 4$	intermediate scale

To take into account the roughness geometry, several authors propose the concept of roughness concentration [1] [4] [5] [7], defined as

$$\lambda_e = \Sigma A_f/A_{bed}$$

where

ΣA_f = sum of the wetted frontal cross-sectional areas of the elements presented to the flow,

A_{bed} = area of bed on which the elements lie

Furthermore, some authors [1] point out that in channels where the roughness elements fill a significant part of the cross-section, the flow is funnelled between the elements They estimate this effect by the relative roughness area λ_w, which is supposed to be a function of roughness concentration and channel shape

$$\lambda_W = A_W/wH_o = (w/h)^{f(\lambda e)}$$

where
A_W = total wetted cross-sectional area of the elements,
w = flume width,
H_o = flow depth from the free surface to the bottom of the flume,
h = mean flow depth

3. Existing models for prediction of flow resistance

It has been gathered from literature [8] [9] [10] [11] [12] [13] [14] [15] [16] around four hundred data involving gravel-bed rivers A first investigation on these data [17] has shown the possibility of improving the accuracy by means of some parameters like Froude number or mobility number y

In this data collection, a hundred are concerned with the field of intermediate-scale roughness and more than forty with large-scale roughness Table I gives the computed values of standard error of estimate of resistance coefficient, using some of the existing models Fig 1 and Fig 2 show the variation of the cumulate standard error (i e standard error for all the data whose relative submergence is lower than a given value) with relative submergence The equations proposed by Bray [13] and Griffiths [10] were not obtained in the range of large-scale roughness and therefore do not show a very good behaviour in that region The equation suggested by Colosimo et al [9] exhibits a difference from the others for the presence of two linear terms involving Froude number and mobility number Nevertheless the influence of the latter term of the equation is negligible for the large-scale roughness

	R/D<1 2	1 2<R/D<4
Bathurst [12]	0 36	0 35
Bray [13]	0 74	0 38
Griffiths [10]	0 74	0 53
Thompson e Campbell [16]	0 28	0 29
Colosimo et alii [18]	0 27	0 30
Colosimo et alii [9]	0 50	0 42
Aguirre-Pe e Fuentes [19]	0 28	0 31

Tab I standard errors of estimate of resistance coefficient for some of the existing models

Flume data by Bathurst et al [1] have been considered to get a more complete and controlled test on the ability of the Froude number to describe the free surface effects

It is possible to calculate the distance Y_o between reference datum (i e the flume wall) and the surface on which velocity is equal to zero by the resistance equation

$$C= H_o/(H_o-Y_o)\ 1/k \log(H_o/Y_o) - 1/k \qquad (1)$$

where $C=U/(gRi)^{1/2}$, U is the mean velocity, R is the hydraulic radius and i is the slope

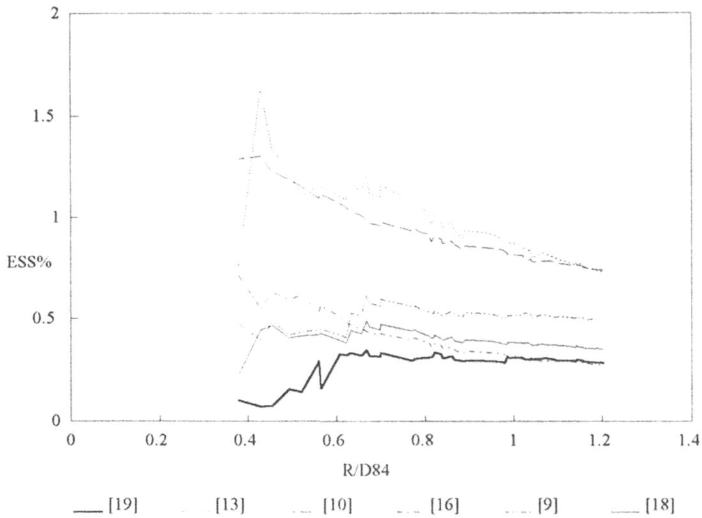

Fig. 1: Variation of standard error of estimate with relative submergence for the region of large-scale, referred to the different authors

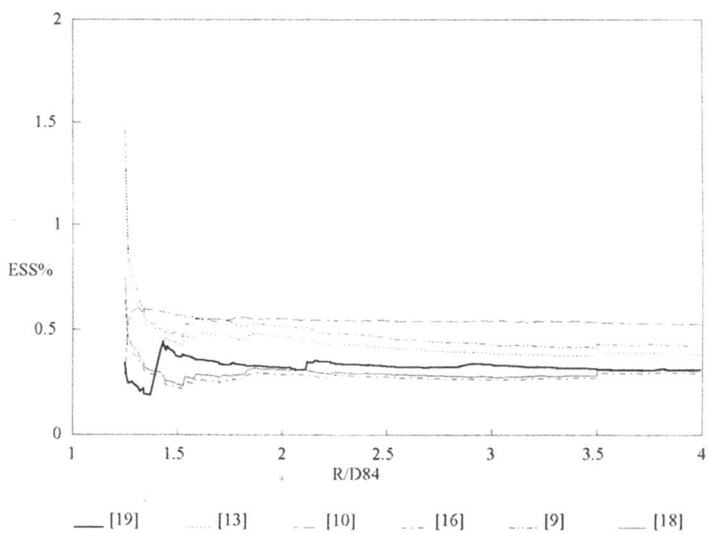

Fig.2: Variation of standard error of estimate with relative submergence for the region of intermediate- scale, referred to the different authors

Eq. (1) is obtained by integration of the Prandtl-von Karman logarithmic profile between the limits Y_o and H_o (fig. 3). Then, by the equation $Y_o=mY_r$, with Y_r roughness height, we get m.

It must be remarked that the integration often starts from Y_r so that the portion of flow below the top of the roughness is neglected and some error may occur in the case of large-scale roughness.

This procedure yields values of m that are strongly dependent on measured hydraulic radius. Several authors approximate the hydraulic radius R by mean depth h, since they guess the wetted perimeter equal to the channel width. This hypothesis gives a large error when the order of relative submergence h/D decreases to the unity. Therefore it would be useful to propose a simple model, which simulates the geometry of roughness in a cross-section.

4. Proposed method and results

A method has been constructed for generating an imaginary pattern of large-scale roughness in a cross-section if the size distribution of bed material is known, in order to put in evidence the implicit approximation hidden in the commonly diffused way to calculate the value of wetted perimeter simply as the channel width.

The roughness has been outlined by a succession of periodic classes of elements, as shown in Fig. 4, following a log-normal distribution, using the elements shape in Fig. 5. Moreover, the cobbles have been thought to be lying with their short axis normal to the bed surface, as already stated by Bathurst et al. [1].

The particular method adopted by the authors for gauging wetted and roughness areas (i.e. the trapezoidal rule) yields a systematic error in measured data. In order to reduce this error, a correction has been performed. Nevertheless a small error still persists. In Fig. 6 the measured, corrected and computed wetted areas are compared for each of the available data sets: the agreement seems to be acceptable, in consideration especially of the above mentioned error.

Fig. 7 shows the difference between the mean flow depth and the actual values of hydraulic radius yielded by the previous model.

It should be useful for applications to get the actual value of hydraulic radius without using the model for simulation of the roughness geometry. In fact, for h ranging from 0.01 m to 0.1 m, it has been found the linear regression

$$R = 0.547\,h + 0.0022 \tag{2}$$

whose correlation coefficient is 0.99. Fig. 8 shows this regression in dimensionless form.

Using the previously computed values of R and the procedure explained in section 3, C, Y_o, and consequently m, have been again determined from eq. (1). As concerns Y_r, it has been assumed to be equal to D_{84}, which is the most generally used in literature up to the present. In order to allow values of m to be compared with an available set of data [20], it has been turned to the following function of m: $B_r=-(1/k)\ln(m)$. Fig. 9 emphasizes that B_r decreases (and m increases) as Y_r/H_o increases; it supplies a confirmation on the trend showed by Graf [20] and extends its validity as far as the regions of large and intermediate-scale roughness.

Furthermore, fig. 10 shows that the lower is the Froude number, the stronger is the inverse dependence of B_r on Y_r/H_o. In particular, B_r reduces almost to a constant for Fr>1.6. This perfectly agrees with the results of Flammer et al. [2] since it is related to the influence of the free surface drag.

In table II, the variance of Br, calculated with reference to the mean value of each class, is shown for some different classes of Froude number.

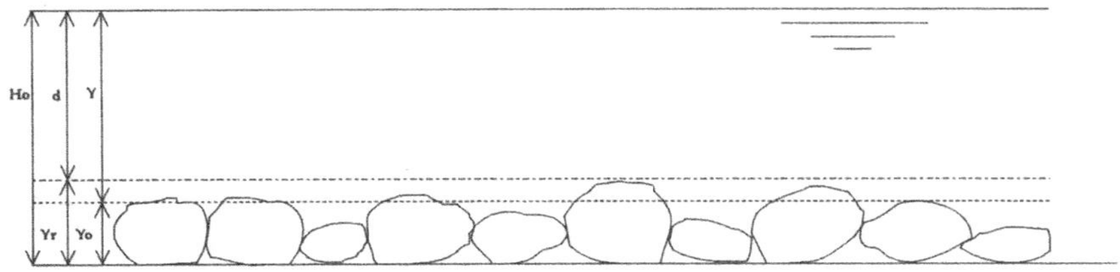

Fig 3 Meaning of some of the employed symbols

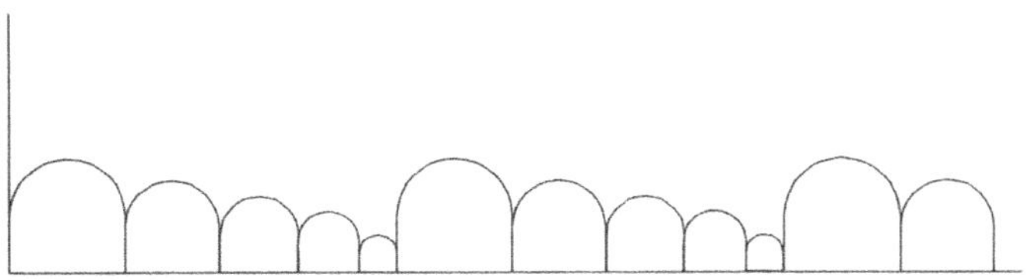

Fig 4 Imaginary pattern of roughness following a log-normal distribution

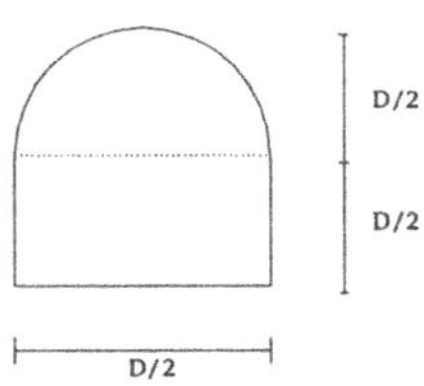

Fig 5 Imaginary shape of the elements

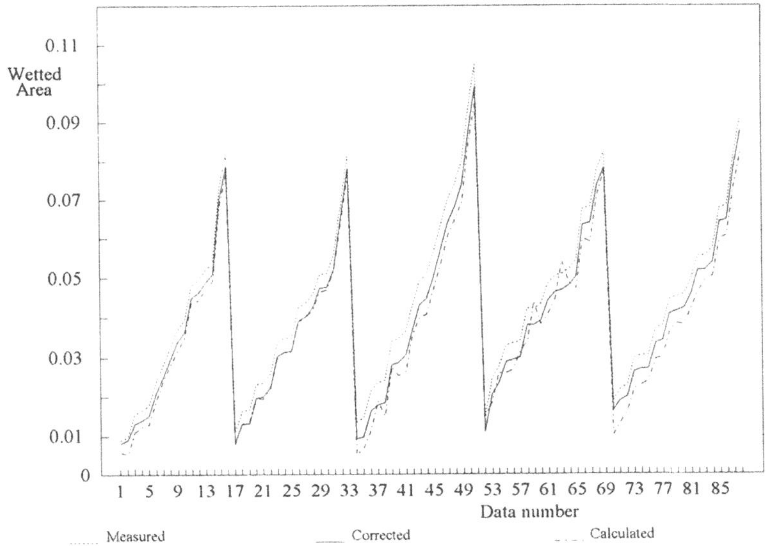

Fig 6.: Comparison of measured, corrected and calculated wetted areas

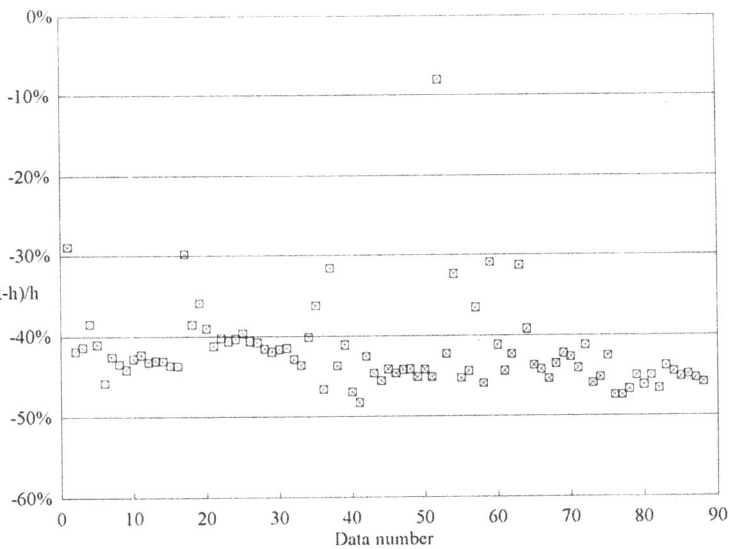

Fig.7· Deviation of the mean flow depth from estimated hydraulic radius

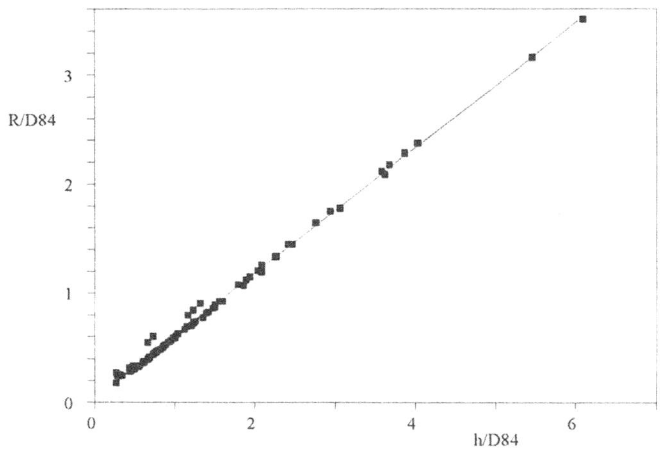

Fig.8: Regression of estimated hydraulic radius on measured mean flow depth in dimensionless form

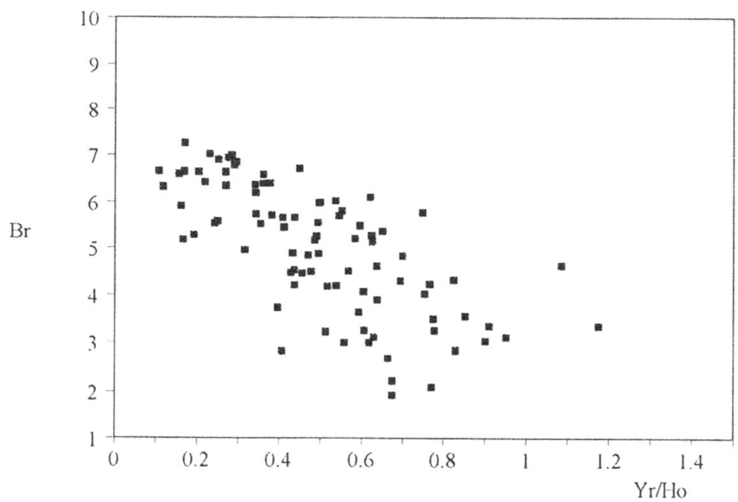

Fig.9: Variation of B_r with relative roughness Y_r/H_o

Fr	Variance of Br
< 0.8	0.853
0.8 ÷ 1.2	0.645
1.2 ÷ 1.6	0.403
> 1.6	0.068

Tab II Variance of Br for some different ranges of Froude number

It is plain that the equation (1), which is based on the classic theory of the wall turbulence, should not be able to describe the whole of the physical processes of energy dissipation that characterize the streams in the presence of large-scale roughness Nevertheless, the flow separation phenomena have been taken into account throughout this study as a simple perturbation of the logarithmic velocity profile That choice has allowed some degree of formal similarity with the traditional laws of resistance to be kept

Furthermore, it should be taken into account the effect induced by separation and funnelling of the flow In such a situation, the boulders are liable to act as proper boundaries for the branches of stream that flow on the flatter parts of the bed between them Thus, in those regions, the distribution of boundary shear stress and the extent of energy dissipation must also depend on size, shape and spacing of the elements

The foregoing considerations lead to take into account two different types of physical processes

- profile drag, mainly due to interaction between the free surface and the portion of cobbles which fall within the limits of large and intermediate scale,
- skin friction, generated by the fine fraction of sediment lying around the boulders and deeply affected by them

For investigating about the skin friction effect of the parts of bed between the boulders, it needs to reject those data for which free surface drag is not negligible In consideration of this, the only points with Fr>1.6 or $H_o/D_{84}>4$ have been plotted in fig 11 it can be noticed that B_r does not seem to vary much with roughness concentration λ_e This may be caused by the small variance of the size distribution of the bed material used in the Bathurst et al [1] experiments In this case, the effect of funnelling should be strongly reduced

The foregoing considerations suggest that an analytic expression of m should certainly include Y_r/H_o and Fr From fig 12 it appears as well that m is deeply influenced by the bottom slope As a matter of fact, a multiple regression yields

$$m = 0.9\, i^{0.75} Fr^{-1.6} (Y_r/H_o)^{-0.49} \tag{3}$$

with a correlation coefficient of 0.97 The standard error of estimate is 0.30 for large-scale and 0.18 for intermediate-scale

Nevertheless, it must be taken into account the presence of a spurious correlation between Froude number and m Infact values of m has been obtained by estimating Y_o from the equation (1), knowing the value of C For this reason any method in which the evaluation of C is carried out as a function of $Fr=C\sqrt{i}$ must show a very high spurious correlation coefficient, as already noted by Benson [26] and Copertino [25] Because of the spurious correlation effect, a lower value of error occurs when the Froude number is dropped in the above multiple regression This procedure yields

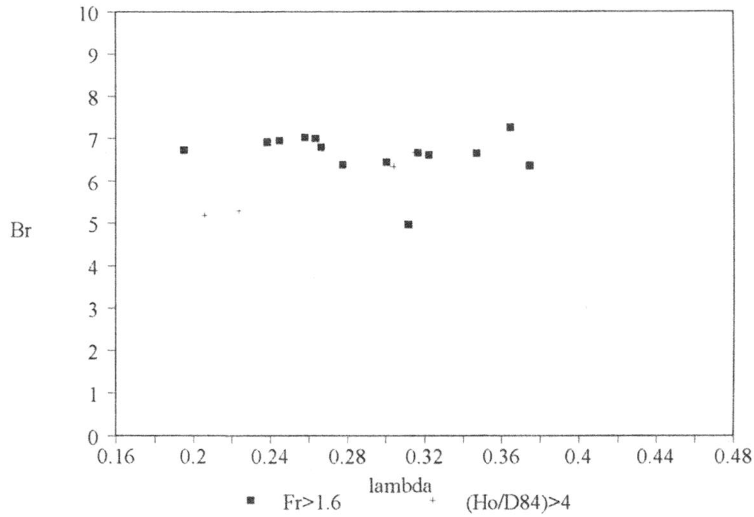

Fig.10: Variation of Br with relative roughness Yr/Ho, for different class of Froude number

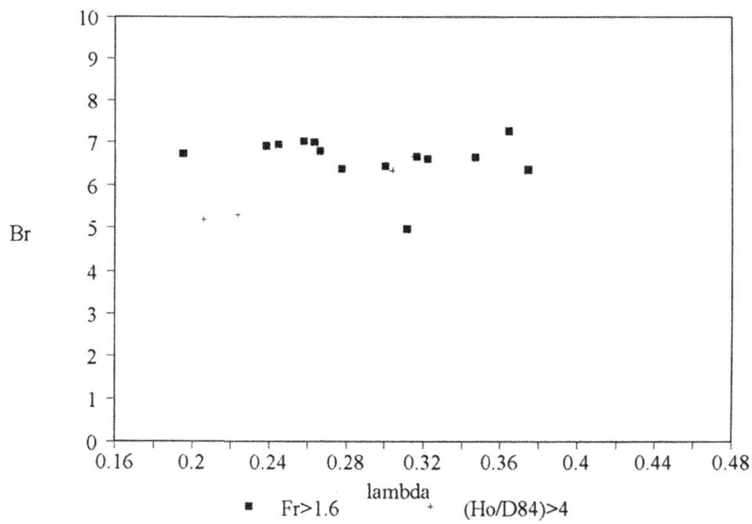

Fig.11:Effect of roughness concentration on Br for the flows having Fr>1.6 or R/D84>4

the following relationship

$$m = 0.144\, i^{-0.184} (Y_r/H_o)^{0.78} \tag{4}$$

The relation (4) allows, together with the equation (1) and the one defining resistance coefficient C, the mean velocity U to be estimated with good precision. In this case, values of standard error of estimate are 0.29 for large-scale and 0.15 for intermediate-scale.

A further improvement in precision can be obtained by using a parameter λ_e that could take into account the effect of roughness concentration, as already noted in section 2. This yields the relationship

$$m = 0.240 (Y_r/H_o)^{0.745} \lambda_e^{0.515}\, i^{-0.207} \tag{5}$$

Then the ability of the equation (5) to predict the flow resistance has been tested. The dimensionless Chezy coefficient C for all the flume data reported by Bathurst et al. [1] has been calculated and has been compared with the measured one in fig. 14. The standard error of estimate results to be 0.25 for the region of large-scale roughness and 0.13 for the one of intermediate-scale roughness, showing an almost perfect agreement with the performances of the Bathurst et al. equation [1], represented in fig. 13

$$C = (0.28 Fr/b)^{\log(0.755/b)} \left[13.434 (w/Y_{50})^{0.492}\, b^{1.025 (w/Y_{50})^{0.118}}\right] (A_w/wH_0) \tag{6}$$

where

$$b = \left[1.175 (Y_{50}/w)^{0.557} (h/S_{50})\right]^{0.648 \sigma^{0.134}} \tag{7}$$

For this equation the standard error of estimate is 0.26 for the region of large-scale roughness and 0.12 for intermediate-scale. This is a good result in consideration of the bigger simplicity of equation (5) and of its classical mathematical structure.

5. Conclusions

The present study, though is certainly far from an end, has anyhow brought some suggestions, both in knowledge of dissipative processes and in their quantitative interpretation, concerning the flows where the size of the roughness elements is comparable with the depth.

In particular, a model for evaluating flow resistance and mean velocity has been proposed. It is based on eq. 1, in which m and consequently Y_o=mD_{84} is given by eq. 4 and hydraulic radius R, to be introduced in U and Fr, is provided by eq. 5.

A test, which has been carried out using flume data, has shown the model to be characterized by a better performance in comparison with all the others taken here into consideration.

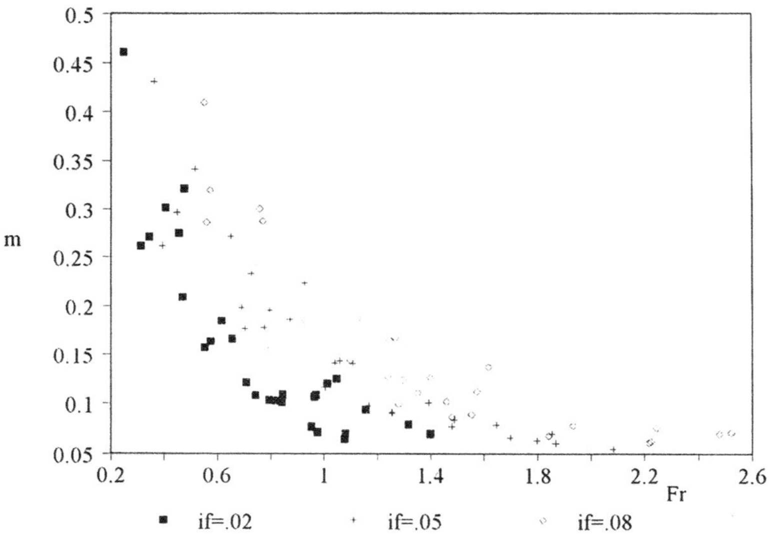

Fig. 12: Dependence of m on the bottom slope i

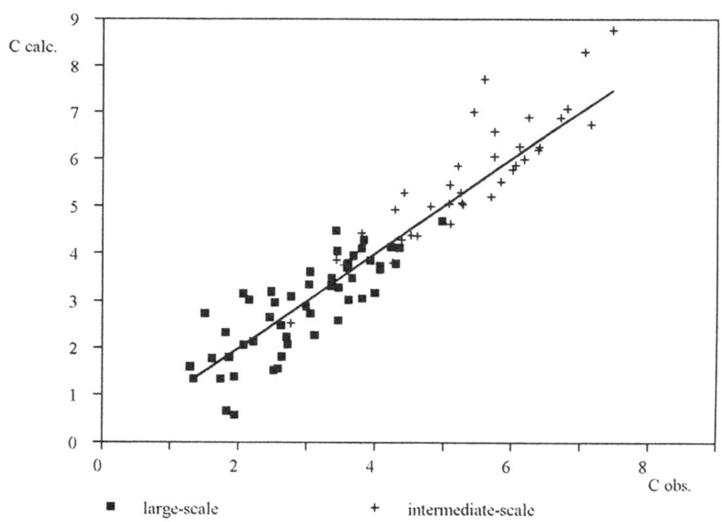

Fig. 13: Performance of the equation by Bathurst et al.[1]

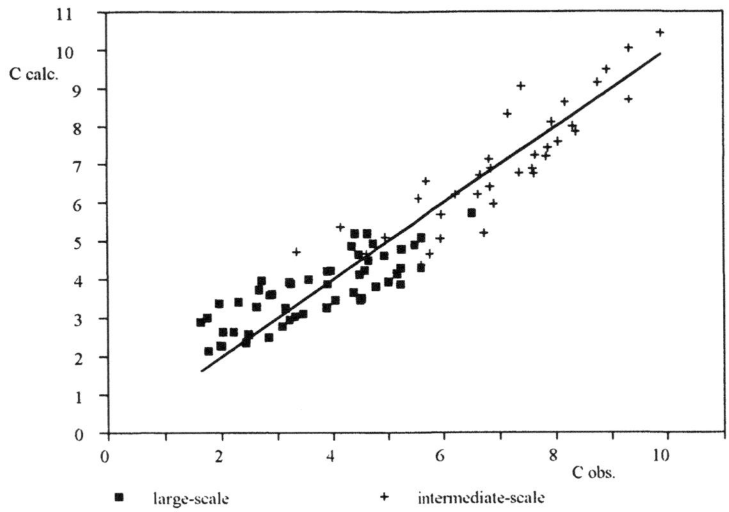

Fig. 14: Performance of the proposed equation [5]

Notation

h = mean depth of flow
R = hydraulic radius
w = surface width of the flow
g = acceleration due to gravity
U = mean velocity of flow
u_* = mean shear velocity = $\sqrt{(gRi)}$
$D_\varnothing$ = diameter of the median axis for which $\varnothing$% of sediments are finer
Fr = Froude number = $U/\sqrt{(gR)}$
$C_d = F_d/(1/2 r_o U^2 A)$ = drag coefficient
F_d = form drag
r_o = water density
A = frontal cross-sectional area of an object
Y_r = roughness height
Y_o = distance from the bed reference level of the surface on which velocity is equal to zero
$m = Y_o/Y_r$
H_o = depth of flow to the bed reference level
$d = H_o - Y_r$
$Y = H_o - Y_o$
f = friction factor
i = channel slope
C = dimensionless Chezy coefficient
k = universal von Karman constant = 0.4
K_s = Nikuradse standard grain size
h/D = relative submergence with reference to Flammer et al. [2]
R/D_{84} = relative submergence
y = mobility number
y_{cr} = critical value of mobility number
λ_e = effective roughness concentration
λ_w = relative roughness area
$Y_{50} = (L_{50}+D_{50})/2$
L_{50} = size of the long axis for which 50% of sediments are finer
S_{50} = size of the short axis for which 50% of sediments are finer
σ = standard deviation of the element size distribution.

6. References

[1] Bathurst, J C , Li R-M , Simons, D B , 'Resistance equation for large scale roughness', *Journal of the Hydraulics Division*, ASCE, vol **107**, No 12, 1981
[2] Flammer, G H , Tullis, J P , and Mason, E S , 'Free Surface, Velocity Gradient Flow Past Hemisphere', *Journal of the Hydraulics Division*, ASCE, vol **96**, No 7, 1970
[3] Bray, D I, 'A Review of flow resistance in Gravel Bed Rivers', Paper presented at the Italian National Group on Fluvial Dynamics, Cosenza, Italy, June 25-26, 1987
[4] Schlichting, H , 'Experimental investigation of roughness problem', Trans ASCE, nov 1937
[5] Pyle, R , Novak, P , 'Coefficient of Friction in Conduits with Large Roughness', *Journal of Hydraulic Research*, No 2, 1981
[6] Bathurst, J C , 'Flow Resistance of Large Scale Roughness', *Journal of the Hydraulics Division*, ASCE, vol **104**, No 12, 1978
[7] Judd, H E , and Peterson, D F , 'Hydraulics of Large Bed Element Channels', Report No PRWG 17-6, Utah Water Research Laboratory, Utah State University, Logan, Utah, 1969
[8] Colosimo, C , Copertino, V A , Veltri, M , 'La resistenza al moto nei corsi d'acqua naturali a letto fisso Primi risultati di una ricerca sperimentale', XVIII Convegno di Idraulica e Costruzioni Idrauliche, Bologna, 1982
[9] Colosimo, C , Copertino, V A , Veltri, M , 'Friction Factor Evaluation in Gravel Bed Rivers', *Journal of the Hydraulics Division*, ASCE, vol **114**, No 8, 1988
[10] Griffiths, G A , 'Flow Resistance in Coarse Gravel-bed Rivers', *Journal of the Hydraulics Division*, ASCE, vol **107**, No 7, 1981
[11] Jarret, R D , 'Hydraulics of High-gradient Stream', *Journal of the Hydraulics Division*, ASCE, vol **110**, No 11, 1984
[12] Bathurst, J C , 'Flow Resistance Estimation in Mountain Rivers', *Journal of the Hydraulics Division*, ASCE, vol **111**, No 4, 1985
[13] Bray, D I , 'Estimating Average Velocity in Gravel Bed Rivers', *Journal of the Hydraulics Division*, ASCE, vol **105**, No 9, 1979
[14] Hey, R D , 'Flow Resistance in Gravel Bed Rivers', *Journal of the Hydraulics Division*, ASCE, vol **105**, No 4, 1979
[15] Thorne, C R , Zevenbergen, L W , 'Estimating Mean Velocity in Mountain Rivers', *Journal of the Hydraulics Division*, ASCE, vol **111**, No 4, 1985
[16] Thompson, S M , Campbell, P L , 'Hydraulics of a large channel paved with boulders', *Journal of Hydraulic Research*, vol **17**, No 4, 1979
[17] Sciancalepore, F , 'Un nuovo modello per la valutazione delle resistenze al moto nelle correnti uniformi dei corsi d'acqua granulari', Tesi di Laurea in Ingegneria Idraulica, Università degli Studi della Basilicata, A A 1989-90
[18] Colosimo, C , Copertino, V A , Veltri, M , 'Un metodo per la valutazione delle resistenze al moto negli alvei granulari fissi', *L'Energia Elettrica*, No 11, 1983
[19] Aguirre-Pe, J , Fuentes, R , 'Resistance to Flow in Steep Rough Streams', *Journal of Hydraulic Engineering*, ASCE, vol **116**, No 11, 1990
[20] Graf W H ,'Flow resistance over a gravel bed, its consequence on initial sediment movement', International Workshop on Fluvial Hydraulics of Mountain Regions, Trento, Italy, 1989
[21] Marchi E e Rubatta A , *Meccanica dei fluidi*, UTET, 1981
[22] Keulegan G H , 'Laws of turbulent flow in open channels', *J Res National Bureau of Standards*, U S Dept of Commerce, vol **21**, 1938
[23] Simons D B , Al-Shaikh-Ali K S , Li R M , 'Flow resistance in cobble and boulder riverbeds',

Journal of Hydraulic Division, ASCE. vol.**105**, No.5, 1979.
[24] Becchi I., 'Considerazioni sulla scabrezza dei letti granulari piani', *L'Energia Elettrica*, No.11-12, 1978.
[25] Copertino V.A., 'Regimi morfologici e resistenze al moto negli alvei granulari', *Idrotecnica*, No.4, Luglio-Agosto, 1988.
[26] Benson M. A., 'Spurious correlatin in Hydraulics and Hydrology', *Journal of Hydraulics Division*, ASCE, vol.**91**, No. 4, 1965.

WARNING OF APPLICATION OF THE CHEZY-MANNING FORMULA REGARDLESS OF CHANNEL SHAPE

WITOLD G. STRUPCZEWSKI
Institute of Geophysics, Polish Academy of Sciences
Księcia Janusza 64, 01-452 Warsaw, Poland.

Abstract

The way of the Chezy formula derivation implies a straight uniform channel with an identical velocity distribution along cross-section, *i.e.* a wide uniform rectangular channel with uniform lining. However to apply it for natural river channels this restriction are relaxed. The reasons for that is the lack of alternative for depth-discharge rating curve derivation applicable for any channel shape in case of a small number of flow discharge measurements and an encouragement from authors of hydrologic books stating lack of definite evidence about the shape of a channel as an important factor affecting the value of n Manning coefficient. In fact the influence of shape may be overshadowed by other factors usually existing in natural channels like irregularities along their length, nonuniform lining,...

From human knowledge got during the ages of irrigation practice comes that for steady state in a channel with uniform lining and without flow obstruction:
(1) any increase of water level is related with increase of flow discharge;
(2) an increase of channel width above any level causes an increase in flow capacity for any other level above it.

The scope of the paper is a mathematical analysis of the Chezy formula for steady flow in an uniform symmetric channel with constant slope-friction factor focused on investigation of both axioms. A channel is assumed to be filled up to a certain level called an initial level and the analysis is made above this level with the corresponding initial shape defined by flow area, wetted perimeter and width of water surface. The problem of determination of the channel shape above initial level for given rating curve of

V. P. Singh and B. Kumar (eds.), Surface-Water Hydrology, 371–383.

stage-flow discharge or flow area- flow discharge, *i.e.* an inverse solution of the Chezy formula, is posed and then solved for selected forms of rating curves and of initial shapes. It is demonstrated that the Manning roughness shall be dependant on channel shape and on flow depth if both axioms are to be fulfilled by the formula.

The paper is addressed as a warning of unlimited in channel shape use of the Chezy formula and it may explain one of the reasons of difficulties faced in an interpretation of the roughness coefficient n - gauge height relationship in natural river channels.

1. Introduction

Approximate character of the Chezy's equation for uniform flow in open channel is widely acknowledged as can be seen from the numerous formulas for the Chezy coefficient quoted in the literature (e.g. Leliavsky, 1959, Chow, 1959, numerous papers of Conference Proceedings, 1989). Among them the Manning formula has become the most widely used for open-channel flow computations. However, as shown in numerous studies, the assumption of a single value of the Manning's roughness coefficient (n) rarely holds in practice. The blame for high variability of n with flow depth is put mainly on nonuniformity of both channel and surface roughness along its perimeter. The other reason for it may be that the distribution of shear along the wetted perimeter is not in general uniform as assumed in the Chezy's equation making the equation not adequately accurate for natural shapes. It is only in the case of a wide rectangular open channel or in the case of axi-symmetric semi-circular channel flowing full that the boundary shear will be the same at all points on the wetted perimeter and the length of normals from the boundary to the surface will all have the same value. However even between these two cases there is a difference 21% in value of mean velocity if the same value of the Manning roughness coefficient and the universal velocity-depth relationship

$$\frac{v}{v_*} = a \left(\frac{y}{k}\right)^p, \tag{1}$$

where v_* is the so-called shear velocity, k is a representative roughness height and p = 0 (*i.e.* constant velocity) for Chezy and p=1/6 for Manning, are assumed. For other shapes the distribution of shear along the wetted perimeter is not uniform and the isovels (lines of equal velocity) will not be similar as in two special cases above.

Since in reality the ideal uniform channel with constant surface friction does not exist, the potential influence of its shape on roughness coefficient is overshadowed by all kinds irregularities of natural channels. Chow (1959) named several factors affecting variability of n stating that "there is no evidence about the size and shape of a channel as an important factor affecting the value of n". In fact a significant dependence of n value on shape would act as strong argument against unlimited use of the Chezy equation with Manning roughness showing that either its two shape parameters unsatisfactorily describe a shape from hydraulic point or the form of equation is not adequately accurate for the practical purposes. Leliavsky (1959) has reviewed some experimental findings against the universal use of the hydraulic radius (R) in discharge formulae concluding that "their are still many reasons for believing that the estimate of the hydraulic efficiency of a canal profile may possibly be derived in future from a more sound basis than the hydraulic radius". At the International Conference for the Centennial of Manning's Formula and Kuichling's formula (Proceedings, 1989), Dr.Ben Chie Yen, meeting organizer, indicated that the longevity of the Manning formula is a mixed blessing. He questioned if the formula is fundamentally sound and practical or if it merely reflects our lack of progress.

From human knowledge got during the ages of irrigation practice comes that for steady state in a channel with uniform lining and without flow obstruction:

(1) any increase of water level is related with increase of flow discharge;

(2) an increase of channel width above any level causes an increase in flow capacity for any other level above it.

It is a pity that apart of a list of open problems in hydrology there is no list of the axioms with distinction of those got during the scientific hydrology era.

The idea of this work has come directly from practice. The author working with the Manning formula on deep, narrow channels of Liberian (West Africa) rivers with few measurements below a mean level only has found that neither of two axioms was fulfilled by the formula with constant slope-friction factor in case of encountered irregular cross-section shapes (Strupczewski & Sua, 1983). In fact it may have been anticipated since a violation of the first axiom is widely known in case of the direct use of the Manning formula for a compound channel and it is width/depth ratio which controls this paradox, i.e. it is more likely in a deep and narrow channel than in a wide compound one. Artificially division into subsections gives results conformable with both axioms but not necessary closer to the true values. In fact it is a weakness of hydrologists that they rarely refuse to answer practical question finding available information unsufficient. It brings them from structural designers jeering by-name"hydroliars". In the case of Liberia getting more discharge measurement data is quite expensive due to a long distance and bad roads hardly accessible during rainy season.

The standard applications of the Chezy equation with Manning friction are to determine depth of flow for specified flow rate, or to determine discharge for specified depths of flow and channel shape taking into account the resistance to flow in channel and friction slope. Therefore determination of a channel shape for the given flow depth (or area) - flow discharge relationship and both roughness coefficient and bottom slope can be considered as the inverse problems to those usually solved. For well posing the inverse problems, a shape shall be assumed to be symmetrical. Strupczewski and Szymkiewicz (1989) have shown that the constant Chezy or Manning friction can lead to quite unrealistic results like shape with the unvariable flow discharge with depth. As its continuation the inverse solution of the flow area-flow discharge problem and the depth - flow discharge are examined here.

2. Mathematical Formulation

For practical purpose, a turbulent uniform flow formula is usually expressed as

$$Q = K\, S^{1/2} = a\, A\, R^{r-1} S^{1/2} \qquad (2)$$

where a denotes resistance factor, Q and A flow discharge and flow area respectively, R - hyraulic radius and S - bottom slope. The term K is known as the conveyance of the channel section being a measure of the carrying capacity of the channel section. It was introduced by Bakmeteff (1932) and the related dimensionless factor K' was extensively tabulated by King (1954) and by the U.S. Corps of Engineers (1959).

Since both surface roughness and water surface slope are assumed here to be constant, it is convenient to express the

Chezy equation as the product of the slope-friction factor (SF) and the geometric factor (G)

$$Q = SF * G \tag{2a}$$

where SF as the constant would be further neglected while

$$G = A^r/P^{r-1} \tag{3}$$

where P denotes wetted perimeter. The exponent r equals 3/2 and 5/3 for Chezy and Manning friction law respectively.

Let channel is filled up to a certain level (h_o) called an initial level. Its geometry at the initial level is defined by A_o, P_o and T_o width of channel water surface or alternatively by A_o, hydraulic radius R_o and T_o.

For given above the initial level stage-geometric factor rating curve

$$G = f_h(h) \qquad \text{for } h \geq h_o \tag{4}$$

or flow area - geometric factor relationship

$$G = f_A(A) \qquad \text{for } h \geq h_o \tag{5}$$

and the initial conditions (A_o, P_o, T_o) the task is to find the symmetric cross-section profile above the initial level, i.e.

$$T = \phi(h) \qquad \text{for } h \geq h_o \tag{6}$$

The restriction of symmetry is put to get the unique shape for given by equation (6)width-depth curve.

Because equation (6) contains T and h variables only, A and P in equation (3) and its derivatives shall be expressed in terms of these two variables by means of following geometric relations from which two last are valid for symmetric channel only:

$$A = A_o + \int_{h_o}^{h} T\, dh \tag{7a}$$

$$dA = T\, dh \tag{7b}$$

$$P = P_o + 2 \int_{h_o}^{h} [\ 1 + (\ dT\ /\ 2\ dh\ )^2\ ]^{0.5}\, dh \tag{7c}$$

$$dP = 2\,[1 + (\,dT / 2\,dh)^2\,]^{0.5}\,dh \tag{7d}$$

3. Approximate Solutions for Given Depth - Discharge Curve

An alternate shape to the original channel above its initial level is being build up by equating the values of equation (3) for the original channel (1) and the alternate channel (2)

$$G_2(h) = G_1(h) \tag{8}$$

The solution is obtained numerically from the equivalence of terms

$$\frac{dG}{dh} = \frac{r}{A}\,\frac{dA}{dh} - \frac{(r-1)}{P}\,\frac{dP}{dh} \tag{9}$$

for the original channel (1) and the alternate channel (2)

$$\frac{dG_2}{dh} = \frac{dG_1}{dh} \tag{10}$$

The general forms of equations (7a-d) and equation (9) can be solved for dT/dh and T in the alternate channels.

For simplicity let assume the original channel with constant side slopes z over an initial level $h_0=0$. Then the relationships (7a-d) take form:

$$A_1 = A_0 + T_0\,h + z\,h^2 \tag{11a}$$

$$\frac{dA_1}{dh} = T_0 + 2\,z\,h \tag{11b}$$

$$P_1 = P_0 + 2\,h\,\sqrt{1 + z^2} \tag{11c}$$

$$\frac{dP_1}{dh} = 2\,\sqrt{1 + z^2} \tag{11d}$$

so that the value of the right hand side of equation (10) for the original channel can be determined for any value of h:

$$\frac{dG_1}{dh} = \frac{r(T\;+2zh)}{A_0+T_0h+zh^2} - \frac{2(r-1\;)\sqrt{1+z^2}}{P_0+ 2h\,\sqrt{1+ z^2}} \tag{12}$$

and then eliminating P, $\dot{P}$ and $\dot{A}$ from the left hand of equations (10) by substituting equations (7b-d) and reordering we get the differential-integral equation:

$$\frac{dT}{dh} = \frac{2}{\phi_1(h)\ \phi_2(A)} \sqrt{\left(\frac{r\ T}{A}\right)^2 - 2\ \frac{r\ T}{A}\ \phi_3(h) - \phi_1^2(h)\ \phi_2^2(A) + \phi_3^2(h)} \tag{13}$$

where

$$\phi_1(h) = 2\ (A_0+T_0h+zh^2)^{\ r/(r-1)}/(P_0+2h\sqrt{1+z^2}) \tag{14a}$$

$$\phi_2(A) = (\ r\ -\ 1\ )\ A^{-r/(r-1)} \tag{14b}$$

$$\phi_3(h) = r(T_o+2zh)/(A_0+T_0h+zh^2)-2(r-1)\sqrt{1+z^2}\ /(P_0+2h\sqrt{1+z^2}\ ) \tag{14c}$$

and A is given by equation (7a), with the initial condition $T = T_o$ for h=0.

In fact equation (13) has the only one solution describing the original channel, i.e. one of constant side slopes above the initial level. However allowing a small deviation from the depth-discharge curve, i.e. replacing equation (8) by

$$G_2(h) - \delta G(h)\cdot G_1(h) = G_1(h) \tag{8a}$$

where $\delta G(h)$ is the relative error assumed to be small, one gets an approximate alternate shape which can differ considerable from the original one. In order to show it equation (13) has been solved numerically by the Runge-Kutta fourth order method for various lengths of the depth step (Δh). The examples of the approximate alternate shapes for side slope z=0.25 and the initial semi-circular channel of 2 m. wide with the Manning friction (r=5/3) are shown at *Figure 1a* and the corresponding relative error $\Delta(h)$ of the geometric factor at *Figure 1b*. *Figures 1 a,b* show that according to the Manning equation a channel shape is highly sensitive to any deviation from the depth-flow discharge rating curve describing the original shape. The relative error $\delta G(h)$ of the rating curve increases with the distance from the initial level, i.e. with $(h-h_o)$. It also increases with the step of computation (Δh) in this case. Moreover it shall be noted from *Figure 1b* that the alternate shapes show lower flow capacity than the original one, which is in contradiction with the second axiom. In fact the contradiction can be shown in a much simpler way,

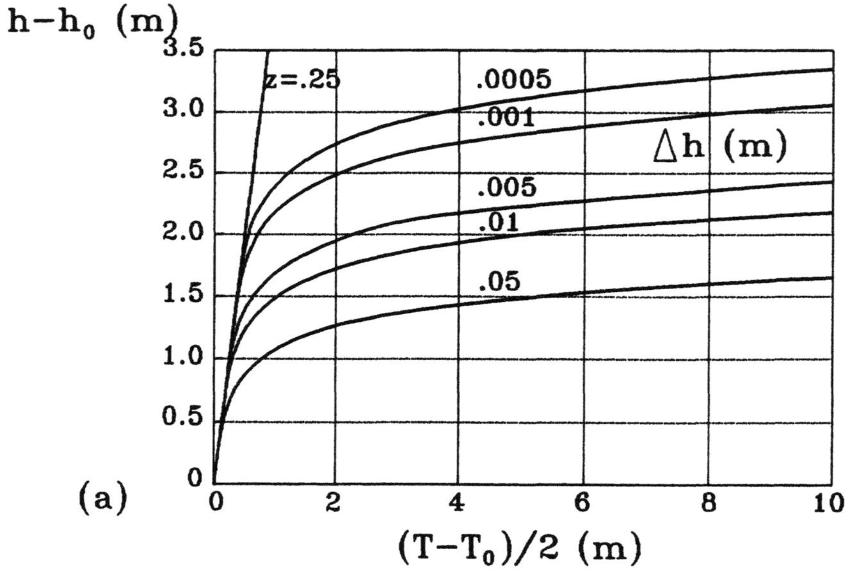

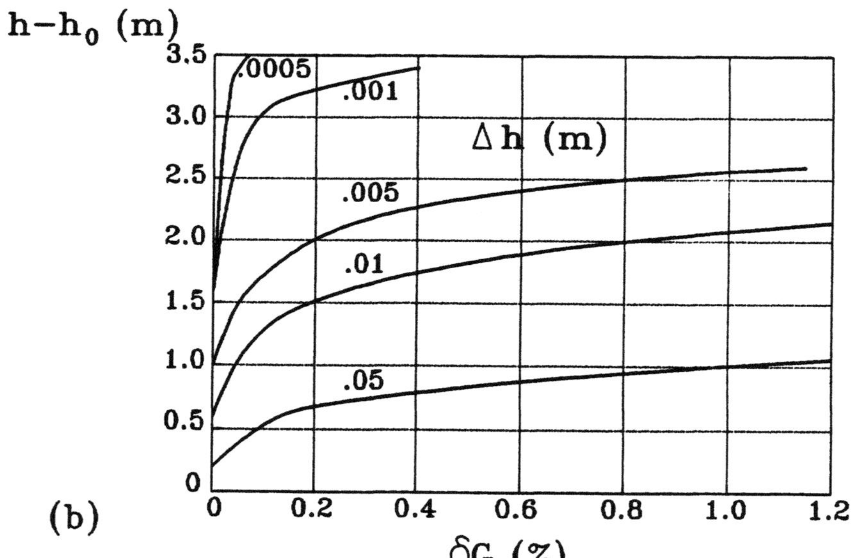

Figure 1. Solutions of equation (13) for depth-discharge relationship given by equation (12), various computation step (Δh) and Manning friction. Initial condition: $A_o = \Pi/2$, $P_o = \Pi$ and $T_o = 2$.

(a) Approximate alternate shapes

(b) The corresponding relative error of G

i.e. by a comparison of the depth-discharge rating curves above an initial level for various side slopes z.

4. Alternate Shapes for Given Area - Discharge Curve

Let us write the total derivative of the geometric factor G defined by equation (3)

$$\frac{dG}{G} = r\frac{dA}{A} - (r-1)\frac{dP}{P} \tag{15}$$

From the definition of the geometric factor of the steady flow formula in equation (2a) we can write

$$\frac{dG}{G} = \frac{dQ}{Q} = m\frac{dA}{A} \tag{16}$$

where m is is the ratio of kinematic wave speed to the average velocity of flow

$$m = \frac{dQ/dA}{Q/A} \tag{17}$$

Combination of equation (15) and (16) gives equation

$$(r - m)dA/A = (r - 1)\ dP/P \tag{18}$$

which shows the upper limit of m value. Because A, P and dP/dA are positive and furthermore $r>1$ for any friction law, we get $m<r$. For a rectangular channel the upper limit corresponds to the width/depth ratio (T/h) approaching to infinity, while for T/h going to zero m tends to one. One can deduce from equation (18) that the m value is not restricted in sign. The negative value of m means a decrease of flow discharge with an increasing flow area (or flow depth). It can happen in the case of a rapid increase of surface width with depth (e.g. Strupczewski & Szymkiewicz, 1989).

Solving equation (16) and then equation (18) with the initial conditions $A=A_0$, $P=P_0$, $G=G_0$ for $h=h_0$ and the constant m value we get

$$G = G_o\ (A\ /A_o)^m \tag{19}$$

and

$$(A/A_0)^{r-m} = (P/P_o)^{r-1} \tag{20}$$

respectively, both for $h \geq h_o$. Equation (19) for m equal one becomes linear, while for m=0 the geometric factor gets the constant value G_o. Eliminating P from equation (20) by equation (18) we obtain

$$A^{(1-m)/(r-1)}\, dA = B_o\, dP \tag{21}$$

where

$$B_o = [(r-1)/(r-m)]\ [A_o^{(r-m)/(r-1)} / P_o] \tag{22}$$

To express the variables in equation (21) by means of T and h only one shall use the geometric relations given by equations (7a-d). Then equation (21) takes form

$$dT/dh = \pm[(A_o + \int_{h_o}^{h} T\, dh)^{2[(1-m)/(r-1)]}\ T^2/B_0^2 - 4]^{0.5} \tag{23}$$

with $T = T_0$ for $h = h_o$.

There are two roots and therefore two solutions. The negative one corresponds to the narrowing surface width with increasing depth, while the positive one to the widening surface width with depth. Furthermore one shall notice that m is not restricted in sign.

In general there is no analytical solution of equation (23) except of the particular case when m =1 for $h \geq h_o$, i.e. when average flow velocity is constant and equaled to initial one and to the kinematic wave speed at the same time. Then equation (23) takes form

$$dT / dh = \pm\ [(\ T / R_0)^2 - 4\ ]^{0.5} \tag{24}$$

which does not depend on a friction law (r). Solving it with the initial condition T_0 for $h_0 = 0$ one obtains

$$h = \pm\ R_0\ \ln \left[\frac{T + [\ T^2 - (\ 2\ R_0)^2]^{0.5}}{T_0 + [\ T_0^2 - (\ 2\ R_0)^2]^{0.5}}\right] \tag{25}$$

Equation (23) has been solved by the Runge-Kutta fourth order method for various initial shapes, m values and both Manning and Chezy friction. The cross sections profiles for the rectangular initial shape with $T_o = 2$ m, $A_o = 2$ m^2 and Manning friction are displayed at *Figure 2*. Left hand branches correspond to negative sign of the square root in equation (23) and they have upper limits while right hand branches are unlimited. Although for conciseness the shapes for $m \geq 1$ are shown only, one can make the similar graph for $0 \leq m < 1$ or for negative values of m as done by Strupczewski & Szymkiewicz (1989) for the positive square root.

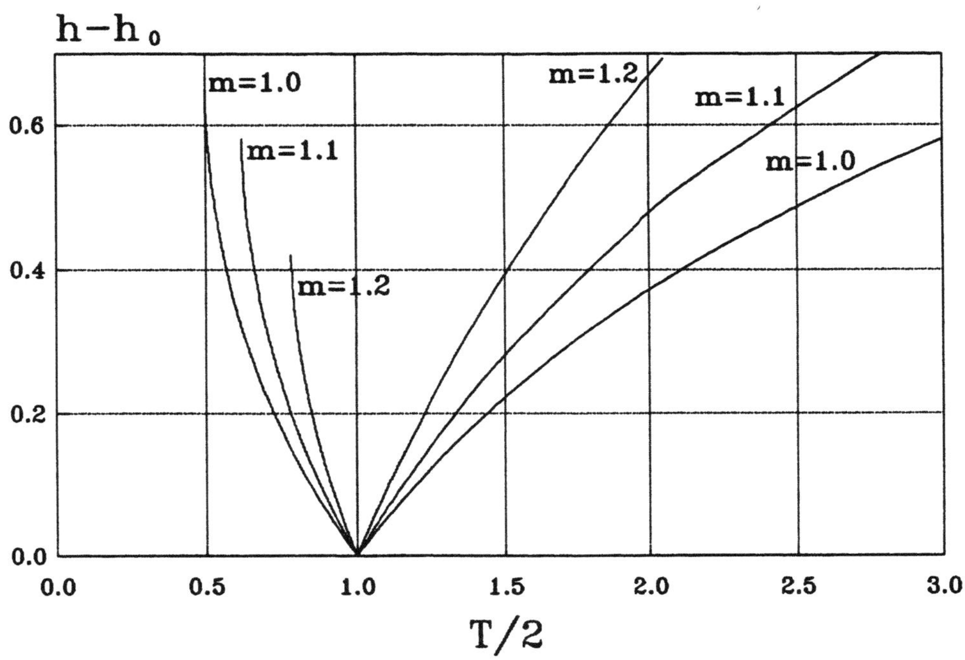

Figure 2. Double shapes for the area-discharge relationship given by equation (19), various m, Manning friction and the constant slope-friction factor. Initial condition : $A_o = 2$, $P_o = 4$, and $T_o = 2$.

5. Conclusions

The drawn comparison of the uniform flow formula for a wide rectangular channel and a semi-circular channel has pointed out a dependence of Manning's n on a channel shape amounts to 21% of mean flow velocity. One can expect for natural river channels differences to be greater than for the two model shapes. Presenting the inverse solutions it was proved theoretically that both channel shape and flow depth have significant influence on Manning's n value.

The foregoing analysis shows that for uniform symmetric channel with constant roughness there is double solution of the inverse Chezy problem for area-discharge curves while for depth-discharge curves there are numerous approximate solutions which may highly differ from each other. It is graphically demonstrated for the rating curve of a symmetric channel with semi-circular initial shape and a constant side slope above the initial level.

The existence of double solution, *i.e.* two shapes of channel, for area-discharge relationship can be intuitively accepted and moreover it is visible from the double root in equation (29). However the shapes derived for m close to zero or negative do not have anything common with reality.

In case of depth-discharge rating curve the approximate solutions present quite different shapes than the original one, *i.e.* one with a constant side slope above an initial level, and therefore they shall be considered solely as the property of the Chezy equation but not as the nature of flow in open channels. They will exist for any power r in the uniform flow formula (2).

Existence of the approximate solutions with channel shapes wider than the original one and having lower flow capacity than one of the original channel and the existence of solution for a negative value of m are in contradiction to the both uniform flow axioms pointing out limits to the application of the uniform flow formula for natural shapes of channel.

It is possible to extand the study to cover non-symmetric shapes of channel. To do it equations (7a-d) shall be modified by making distinction between left and right part of cross-section. However to get a unique solution one branch of cross-section shall be assumed. It would be interesting to compare presented theoretical findings with results of laboratory experiments. We believe that the paper adds the argument to support the idea that the problem of acceptable by practitioners a universal formula for uniform flow in open channel is still open.

Acknowledgement. The author would like to thank prof. J.C.I. Dooge for his valuable comments which significantly enhanced the quality of the paper. Thanks are given to Dr. R. Szymkiewicz from the Gdańsk Technical University who made the computer programs to this research.

References

Bakhmeteff, B.A.(1932) *Hydraulics of Open Channels*, McGraw Hill, New York.

Chow, V.T.(1959) *Open channel hydraulics*, McGraw Hill Book Co., New York.

King, H.W.(1954) *Handbook of Hydraulics.* McGraw Hill, New York.

Leliavsky, S.(1959) *Irrigation and hydraulic design.* Vol. I, Chapman & Hall Ltd., London.

Proceedings of the International Conference on Channel flow and Catchment Runoff: Centennial of Manning's Formula and Kuichling's Rational Formula. (1989) Editor Yen, B.C., Department of Civil Egineering, University of Virginia, 22-26 May.

Strupczewski, W.G. & Sua, D.Z. (1983) Rating curves of Liberian rivers. Part III. Hydraulic estimation. *Technical Report No.61, Liberian Hydrological Service, Department of Mineral Exploration and Research, Ministry of Lands & Mines,* Monrovia, pp.55.

Strupczewski, W.G. & Szymkiewicz, R. (1989) On direct applicability of the Chezy formula to natural channels. *Proc. Int. Conf. on Channel Flow and Catchment Runoff*, University of Virginia, 350-361.

U.S. Corps of Engineers (1959). Hydraulic Design Criteria. *Vicksburgs Experimental Station.*

SECTION 6:

STOCHASTIC MODELING OF STREAM FLOW

STREAMFLOW ANALYSIS AND SYNTHESIS BY PATTERN RECOGNITION TECHNIQUES

U.S. PANU
Department of Civil Engineering, Lakehead University,
Thunder Bay, Ontario, P7B-5E1, CANADA.

Analysis and synthesis of streamflows is required for a variety of purposes, such as streamflow forecasting, reservoir operations and management, and water resources planning and control. Among the various methods of streamflow analysis and synthesis, the pattern recognition based methods are the focus of this paper. A pattern, in general terms, is a shape representation of physical objects. In streamflows, objects can be conceived as groups of high or low flows. The pattern analysis includes collection of data on patterns and procedures used in inferring the deterministic and probabilistic structures that are present in data on patterns. The pattern synthesis generally involves procedures and techniques capable of utilizing the information obtained in pattern analysis phase to generate patterns which are likely to occur in the future.

In this paper, such procedures and techniques are utilized in analysis and synthesis of monthly streamflows of five rivers in Northwest Ontario. The correlogram and periodogram analyses of monthly streamflows of these rivers indicate the presence of two seasons, each comprising of six months. The segmentation of streamflows in a year into two types of objects is achieved by using the peak-valley method described by Panu and Unny (1980) and successfully applied by Kojiri, Unny, and Panu (1986), Wills (1987), Frenette (1989), Afza and Panu (1991), Goodyear and Panu (1992), Hong (1992), and Wu (1993). In other words, two types of streamflow objects, each corresponding to a season, are considered. Each streamflow object (or pattern) is numerically represented in a vector format by using the streamflow values corresponding to the object. The vectors thus obtained are analyzed using pattern recognition techniques for obtaining the deterministic and stochastic structures defining the relationships (i.e., intra-pattern-structure) within one type of pattern vectors and the relationships (inter-pattern-structure) among various types of pattern vectors. The deterministic and stochastic structures in this paper are obtained by using the pattern recognition system (PRS) developed by Panu (1978) and later modified by Kojiri, Unny, and Panu (1986).

Sixty realizations, each of 30 years long, have been synthesized for each river.

V. P. Singh and B. Kumar (eds.), Surface-Water Hydrology, 387–401.

Separate statistical tests have been conducted to examine whether the synthesized streamflow patterns are embedded with similar characteristics as observed in historical streamflow patterns. All such statistical tests indicate that synthesized streamflow patterns in each river (Table 1) are satisfactory for various hydrologic purposes.

Based on our results, it is concluded that pattern recognition based techniques provide an alternative to present philosophy of streamflow analysis and synthesis. It is noted that the pattern recognition based methodology inherently embeds the correlation structure, and the streamflow data analysis and synthesis is conducted in groups of data rather than as single data values. Pattern based analysis and synthesis can easily be extended to other hydrologic data, such as rainfall, sediment, water quality etc.

INTRODUCTION

A pattern in this paper refers to a specific sequence of monthly streamflows corresponding to an unique time period for example, a period of low flows or period of high flows. Recognition implies perception and identification. It is the process of recalling or retrieving knowledge of something (in our case streamflow patterns) that are known a priori such that a present pattern can be identified with one of them. Pattern analysis, in general, implies the process of gathering data on patterns and pattern classes in the environment (i.e., the monthly streamflow records) being examined and subsequently utilizing this data in a variety of procedures to infer the deterministic and probabilistic structures that are present in the data for future synthesis and analysis.

It is in this sense that a pattern is considered as a shape representation of physical objects. The monthly streamflows can be plotted to give a streamflow time waveform (STW) as described by Panu and Unny (1980). The rivers located in a geographically homogeneous region, where climate can be characterized by four seasons, summer, fall, winter, and spring; the variability of streamflows corresponding to these four distinct seasons can be represented by grouping streamflows into four distinct groups (i.e., the seasonal groups). On the other hand, one can also consider the representation of variability in streamflows by two seasons, i.e., a wet season where flows are relatively high, and a dry season where flows are relatively low. Obviously, such a representation will entail errors arising due to unspecified biased introduced through crude representation of seasonal variations. The part of the STW corresponding to each season forms an object. Thus an object is a shape representation of the STW in a particular season. Therefore, a hydrological year comprising of four seasons such that each season consists of three months will result in four objects. In thirty years, one would obtain (30x4=) 120 objects. A column vector $x_i^{(k)}$ containing the monthly streamflows corresponding to the ith object in the kth season is formed. This vector represents the pattern vector for the object. All of the pattern vectors $\{x_i^{(k)}, i=1, 2,\ldots,30\}$ in the kth season are collected together to form the kth pattern class. A pattern class is a group of patterns with certain distinct characteristics.

The determination (or extraction) of important characteristics is one of the most important aspects of efficient pattern recognition, commonly referred to as feature (i.e., the most important characteristic for a given problem) selection or extraction. Usually, the feature selection is preceded by preprocessing of the patterns i.e., normalization, filtering, and preliminary classification. A feature must meet certain requirements such as: (1) it properly describes the pattern, (2) it is invariant to translation and rotation of the pattern, and (3) it is easy to process. The number of features can be large, but each one should be the most effective one. In mathematical terms, each pattern (i.e., shape or object) is expressed by its measurements in a vector form, commonly referred to as a pattern vector. On the other hand, the vector containing only the important characteristics (i.e., features) is called the feature vector. In this process, the dimensionality of the measurement vector is reduced and consequently, the computational efforts are also reduced.

It is convenient to conceptualize the pattern recognition problem in three stages or spaces: the pattern space, the feature space, and the classification space [Figure 1]. The streamflows are measured at a specified location as inputs in the pattern space. Obviously, there are infinite ways of measuring streamflows and thus there exists a possibility of infinite dimensions. However, streamflows are measured in a specific way thereby effectively reducing the dimensionality of an object of streamflows that is into N scalar values. Further, it is desirable to reduce the dimensionality, N by considering the most important dimensions, R (such that the reduced dimensionality, R < N) for analysis and synthesis of a specific problem. Needless to add that the reduction in dimensionality must maintains the inherent discriminatory power of the data for classification purposes. The classification space then becomes the decision space in which one of the $1 \leq k \leq K$ classes has been selected, and thus this space is of K dimensionality. In summary, the pattern space is essentially the domain defined by the discretization of the data from the real world. The feature space is an intermediate domain between the data gathering space and the classification process. The feature space must be defined such that the inherent discriminatory power of data present in the pattern space is optimized for classification purposes. The classification space is K dimensional and contains the decisions implemented by the classification algorithms.

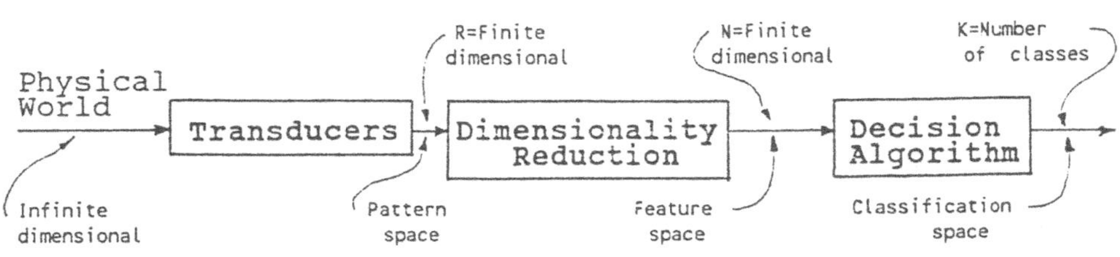

Figure 1. Conceptualization of a pattern recognition problem.

PATTERN ANALYSIS AND SYNTHESIS OF STREAMFLOWS IN NORTHWEST ONTARIO

The monthly streamflow records of five rivers, namely, the Neebing River at the gauging station near Thunder Bay; the Pigeon River at the gauging station at Middle Falls; the Nagagami River at the gauging station at Highway 11; the Turtle River at the gauging station near Mine Centre; and the English River at the gauging station at Umfreville in Northwest Ontario [Figure 2] are considered. The recording stations and their relevant characteristics are summarized in Table 1.

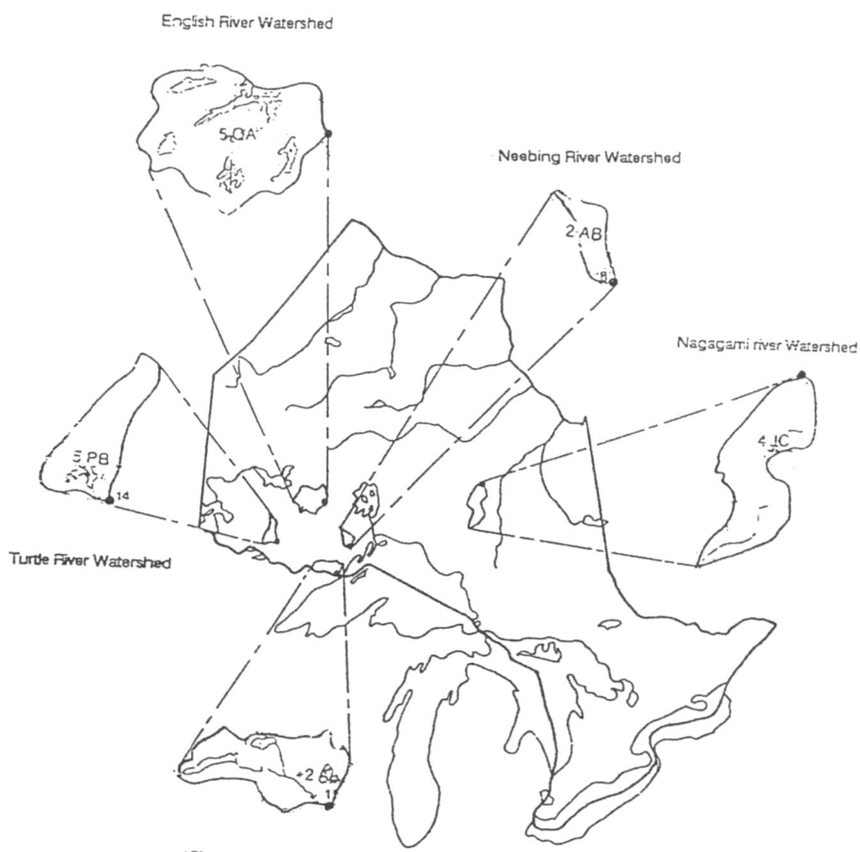

Figure 2. Location of Watersheds Used for Analysis in Northwest Ontario

Table 1 Streamflow gauging stations selected for analysis

Station No.	Name	Location	Drainage Area (Km^2)	Period (Years)	Period Used (Years)	Remarks
02AB008	Neebing River near Thunder Bay, Ont.	48-22-56 N 89-18-28 W	187	1953-1990	1954-1984	Natural Flow
02AA001	Pigeon River at Middle Falls, Ont.	49-46-44 N 84-31-48 W	1550	1921-1990	1952-1982	Natural Flow
04JC002	Nagagami River at Highway 11, Ont.	49-46-44 N 84-31-48 W	2410	1950-1990	1952-1982	Natural Flow
05PB014	Turtle River near Mine Centre, Ont.	48-51-00 N 92-43-30 W	4870	1914-1990	1949-1979	Natural Flow
05QA002	English River at Umfreville, Ont.	49-52-30 N 91-27-30 W	6230	1921-1990	1952-1982	Natural Flow

To assess the seasonality of streamflow records, the correlogram and the periodogram [Quimpo, 1968] provides a convenient procedure. The correlograms [Figure 3] of streamflow records have significant autocorrelations at lags (in months) 4, 9, 12, 15, 20, 24,... . In case of the Neebing river, for example, the autocorrelations at lags 3, 4, 5, 8, 9 are negative and at lags 11, 12, 13 are positive. One interpretation of this behaviour in the correlogram is that there are three types of streamflow groups within a geophysical year. In this paper, the geophysical year refer to the hydrological year. The flow in two groups, on the average, is lower than the mean flow and the flow in the third group, on the average, is higher than the mean flow. The two groups of low flows are followed by a group of high flows. Similar results are obtained for the Pigeon River, Nagagami River, Turtle River, and English River.

A similar conclusion with respect to the presence of a number of seasons in a year is reached through spectral analysis of the streamflow records of these rivers. The presence of 12-month and 6-month cycles is apparent from the periodograms (Hong, 1993) of these rivers. A 4-month cycle is not highly significant for these rivers, as was apparent in the correlograms, thus, a geophysical year for these rivers can be considered to comprise of two seasons, each season consisting of a 6-month duration. Therefore, for all these rivers, two seasons are considered to constitute a year.

AN EXAMPLE OF PATTERN ANALYSIS AND SYNTHESIS OF STREAMFLOWS

For illustrative purpose the Neebing River is chosen to exemplify the process of pattern analysis and pattern synthesis. The monthly historical record of the Neebing River is converted into the STW in the triangular format.

[a] Extraction of intra-pattern structure: A visual inspection of the peak-valley histogram for the Neebing River indicated that the most frequent peak and valley occur in the months of April and February, respectively. In order to include only one significant (or frequent) peak and valley in an object, it is appropriate to partition each hydrological year into two objects such that the first season consists of February, March, April, May, June, and July months; the second season consists of the August, September, October, November, and December months. Therefore, there are NK (=30x2) objects in the STW, comprising a time duration of N (=30) years with each year consisting of K (=2) types of objects.

A column vector, $x_i^{(k)}$ containing the monthly streamflows corresponding to the ith object in the kth season is formed. This column vector represents the pattern vector for the object. All the pattern vectors, $\{x_i^{(k)}, i=1, 2,\ldots,N\}$ in the kth season collectively form the kth pattern class for subsequent analysis. Therefore, for the given streamflow record there are two pattern classes numbered k=1, 2. Based on similar considerations, the STW of the Pigeon river, Nagagami river, Turtle river, and English river were partitioned, and appropriate objects and seasons were obtained in each case.

Table 2. Summary of Results of Multivariate Normality Analysis

Pattern Class (k)	$\chi^2_{4,\,0.95}$ (Calculated)				
	Neebing River Transformed Data	Pigeon River Transformed Data	Nagagami River Transformed Data	Turtle River Transformed Data	English River Transformed Data
1	3.667	1.667	3.667	4.000	2.333
2	1.333	4.667	5.000	1.667	1.000

Note: The theoretical $\chi^2_{4,\,0.95}$ value is 9.48773. All transformed pattern vectors within each pattern class are multivariate normal.

The data synthesis algorithm employs a multivariate normal distribution, therefore it is necessary to test in each river for multivariate normality of pattern vectors within each pattern class. The multivariate normality was tested based on a procedure given by Gnandesikan (1977). The calculated chi-squared values for each pattern class should be less than 9.48773, the theoretical chi-squared value at 95% confidence level and four degrees of freedom for 30 years of data. The results of this analysis are summarized in Table 2. Estimates of power transformations are obtained separately for each component of all pattern vectors exhibiting departure from normality within each pattern class. Such estimates of power transformations were obtained based on a procedure given by Hinkley (1977). It is apparent form Table 2 that the chi-square values for each pattern class in

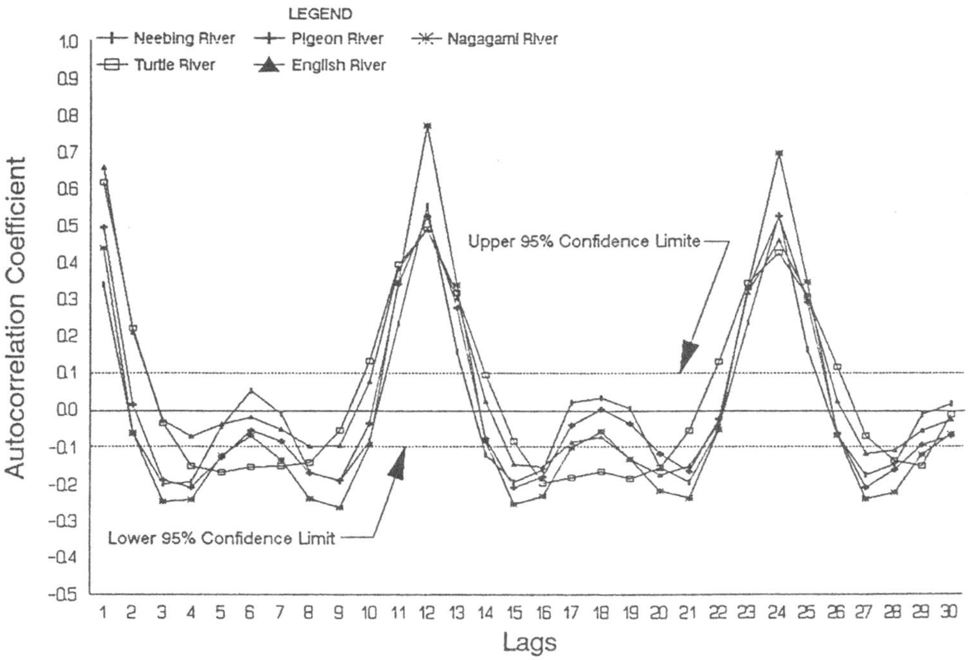

Figure 3. Correlograms of various streamflows.

Streamflow Pattern Analysis

The streamflow record for the time period February 1954 to January 1984 is analyzed in order to locate the points for segmentation of the STW and to identify the dimensionality of the pattern vector. The two seasons in a geophysical year are considered to be of equal time durations of six months each. To aid in the location of points for partitioning of the geophysical year into two seasons, a peak and valley histogram is plotted (Hong, 1993). The histogram provides a summary of the distribution of peaks and valleys in each month over the thirty year period. The starting and end-point of the time duration corresponding to a season within the geophysical year are established based on one or more of the following considerations.

(1) *The time of occurrence of the most significant peak or valley from a visual inspection of the STW.*
(2) *The time of occurrence of the most frequent peak or valley as obtained from the peak-valley histogram.*
(3) *Consideration of the hydrological year.*
(4) *The experience of the hydrologist with respect to the given STW.*

Based on above considerations, the partitioning of each STW into two seasons enables the formation of objects corresponding to each season and thus renders two objects in each hydrological year. Thus, an object is a shape representation of the STW in a season.

each of the five rivers are less than 9.48773, and thus the pattern classes are considered to have multivariate normal distributions for their pattern vectors.

In each pattern class (k=1, 2), the normalized feature vectors corresponding to their normalized pattern vectors are derived according to procedure given by Panu and Unny (1980). Each pattern class k consists of thirty feature vectors. Reference vectors are formulated such that one or more reference vectors define the details of all the feature vectors in a pattern class. Once a set of reference vectors is selected, the classification algorithm assigns feature vectors to groups such that each group is represented by a reference vector. The chosen set of reference vectors should be small in number but sufficient for recognition and classification. The reference vectors for all the rivers are summarized as follows. For example, it is evident from this table that in case of Neebing River, there are three reference vectors (1, 2, 3), one reference vector (1) for pattern class 1, and two reference vectors (2, 3) for pattern class 2.

Table 3. Reference Vectors for the Rivers

River Name	Pattern Class	Reference Vectors
Neebing River	1	-339.8310 -379.4125 2.0864 -11.1228 209.2637 490.9970
	2	-3.9068 0.1387 -2.3825 -3.2798 0.4058 0.9260
	2	-3.9068 -0.6366 -2.3825 -3.2798 0.4058 0.9260
Pigeon River	1	-4.6944 -2.1551 -3.0884 3.4849 -2.1998 -2.0479
	2	3.0354 -4.4339 -0.9083 2.3699 -2.9689 -0.6267
Nagagami River	1	-2.5644 1.5201 -0.8700 -1.4521 -0.8442 0.3164
	2	-0.7283 2.3079 -2.0966 -13.1845 -2.6909 -3.3241
Turtle River	1	-11.5279 10.1306 -1.8106 4.3546 0.2383 -3.6250
	1	-11.5279 10.1306 -1.8106 4.3546 -0.1760 -3.6250
	2	-1.2114 -3.6883 1.6230 1.0462 -1.0825 -0.7287
English River	1	-0.5543 -6.0196 -14.2794 9.2870 8.1285 -18.6843
	2	1.4980 -2.3136 0.4617 -2.7162 -0.1933 0.2050

Table 4. Summary of inter-pattern-structure

↗		Reference Vectors		
		1	2	3
Reference Vectors	1	0.0000	0.1667	0.8333
	2	1.0000	0.0000	0.0000
	3	1.0000	0.0000	0.0000

The feature vectors are then recognized and assigned to the group around any specific reference vector as follows. The distance $D_i^{(k)}$ between a specified reference vector $Y_i^{(k)}$ in the kth pattern class and any arbitrary feature vector F, is defined as follows:

$$D_i^{(k)} = [\,(F-Y_i^{(k)})^{T}(F-Y_i^{(k)})\,]^{1/2}$$

Considering all the feature vectors assigned to a given vector $Y_i^{(k)}$, each with its own distance D_i^k from the reference vector, a characteristic distance $D_i^{*(k)}$ can also be assigned to the particular reference vector. This characteristic distance can be the statistical mean value or any other appropriate statistic pertaining to all $D_i^{(k)}$.

Table 5a. Summary of Statistical Characteristics of Various Rivers at Series Level

Series Statistic	Neebing River			Pigeon River			Nagagami River			Turtle River			English River		
	Hist.	Synthetic Mean	S D	Hist.	Synthetic Mean	S D	Hist.	Synthetic Mean	S D	Hist.	Synthetic Mean	S D	Hist.	Synthetic Mean	S D
Mean	175 2	162 8	12 2	13 9	13 2	0 7	24 0	23 5	0 7	41 0	39 9	1 7	61 2	59 8	2 7
Std Deviation	264 7	281 4	37 1	16 7	15 8	1 7	25 0	24 4	1 1	32 1	29 7	2 3	45 7	41 5	3 2
Skewness	2 3	3 5	0 7	2 2	2 8	0 5	2 0	2 1	0 1	1 7	1 7	0 3	1 81	1 8	0 3
Kurtosis	8 5	19 0	7 9	7 8	13 6	5 0	7 1	7 2	0 8	6 2	6 6	1 9	6 7	6 8	2 2
Lag-1 Coeff	0 34	0 34	0 04	0 44	0 51	0 03	0 44	0 43	0 03	0 62	0 60	0 04	0 66	0 61	0 03
Maximum		181 8			15 0			24 8			43 3			66 0	
Mean	175 2			13 9			24 0			41 0			61 2		
Minimum		139 3			11 4			22 1			36 6			54 4	
Maximum		351 7			19 9			26 1			36 6			49 4	
S.D.	264 7			16 7			25 0			32 1			45 7		
Minimum		215 4			13 0			21 9			25 0			35 5	
Maximum		0 42			0 61			0 48			0 66			0 68	
Lag-1	0 34			0 44			0 44			0 62			0 66		
Minimum		0 25			0 43			0 38			0 42			0 56	

Note Hist , S D , Lag-1 Coeff , H Coeff respectively are the Historical, Standard Deviation, Lag-one Correlation Coefficient, and the Hurst Coefficient

[b] Extraction of inter-pattern structure: The task of extracting the inter-pattern-structure is accomplished by estimating the conditional probability of occurrence of each reference vector. The first-order Markov transition matrix for reference vectors of all pattern classes in each river is separately derived by employing equation (15) [Panu and Unny, 1980]. The Markov transition matrix for the Neebing River is given Table 4.

For example, in the above table, the probability of occurrence of the reference vector

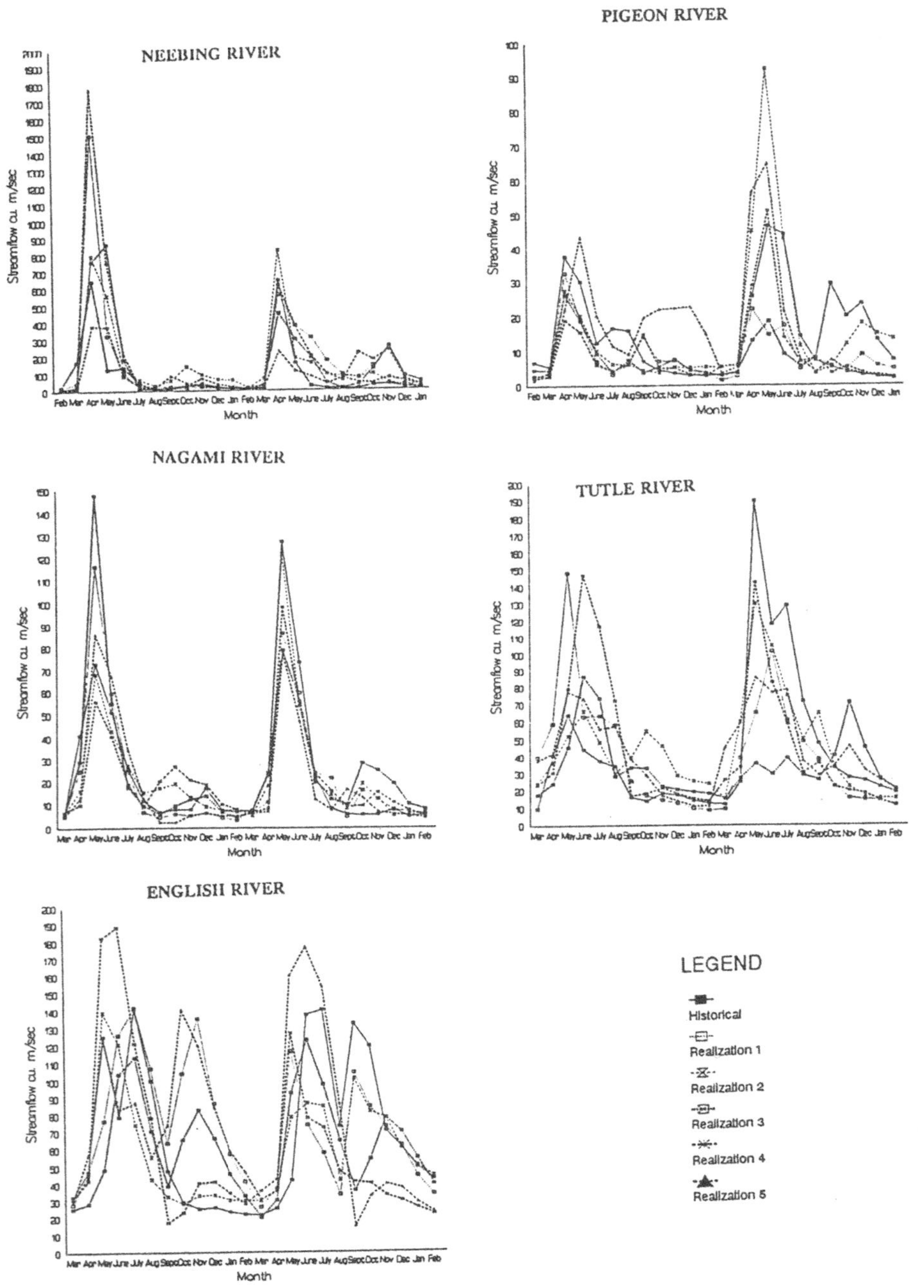

Figure 4. Comparison of historical and synthetic samples of various rivers.

3 following the reference vector 1 is 0.83333. Similar Markov transition matrices were obtained for other four rivers as given in Appendix.

Streamflow Pattern Synthesis: The synthetic realizations of length 30 years are obtained based on the data synthesis algorithm given by Panu and Unny(1980). Various inputs to the algorithm are: pattern vector dimensionality, power transformation parameters, inverse transformation matrices, and information on the intra- and inter-pattern structure for each pattern class. The algorithm also requires an initial reference vector for initiation. This reference vector is taken from the pattern class that had occurred in the most recent past.

Once the algorithm is initiated, the reference vectors are generated in succession according to the transition matrix of the river. Each reference vector is replaced by its appropriate synthesized feature vector obtained according to the multivariate normal distribution with proper bounds specified for the variate values. In turn, all synthesized feature vectors are transformed based on equation 21 [Panu and Unny (1980)] from the feature space to the pattern space. These synthesized pattern vectors are transformed into streamflow pattern vectors by taking the inverse of the particular power transformations. A sequence of synthesized streamflow pattern vectors is called a synthetic realization. In each river, sixty separate realizations of length 30 years have been synthesized for analysis and testing the applicability of the pattern based streamflow synthesis procedure in rivers of the Northwest Ontario.

RESULTS AND DISCUSSION

Synthetic streamflow realizations when compared to the historical streamflow sample (1954-1984) show a good agreement with respect to stochastic and seasonal characteristics of streamflows. Synthetic streamflow realizations appear to embed the general stochastic characteristics of the historical sample (Figure 4). In this figure five randomly selected synthetic streamflow realizations for a duration of two years are superimposed on the corresponding historical streamflow sample for visual comparison.

Comparison of Statistical Analysis of Short Term Characteristics: Various statistical characteristics such as the mean, standard deviation, and autocorrelation coefficient at lag-one are separately calculated for the historical streamflow sample as well as for the corresponding sixty synthetic streamflow realizations. There appears to be a good agreement between the historical and synthesized statistical characteristics for the five rivers considered in Northwest Ontario [Table 5a].

Similarly, such statistical characteristics at a monthly level are presented in Table 5b for the historical streamflow sample and the corresponding synthetic streamflow realizations. It is apparent from this table that at monthly level, the synthetic realizations embed the historical statistical values within the 95% confidence interval.

Table 5b. Summary of Statistical Characteristics of Various Rivers at Monthly Level

Month	Monthly Means														
	Neebing River			Pigeon River			Nagagami River			Turtle River			English River		
	Hist.	Synthetic Mean	Synthetic S D	Hist.	Synthetic Mean	Synthetic S D	Hist.	Synthetic Mean	Synthetic S D	Hist.	Synthetic Mean	Synthetic S D	Hist.	Synthetic Mean	Synthetic S D
January	24 5	22 0	2 4	4 4	4 4	0 6	6 0	5 9	0 2	17 9	17 4	1 2	30 4	29 9	1 6
February	15 8	14 4	1 9	3 8	3 7	0 3	6 1	6 0	0 3	16 6	18 3	1 8	27 6	27 6	1 2
March	57 9	41 8	7 1	4 3	4 2	0 3	22 5	21 4	2 0	43 3	43 6	3 9	38 2	38 3	1 7
Apnl	736 9	756 1	94 4	33 4	32 6	2 1	86 1	86 0	4 0	87 3	83 3	5 9	109 1	109 8	6 5
May	441 6	455 7	52 8	44 8	42 9	4 9	51 9	51 2	2 6	77 6	76 7	6 2	128 7	127 4	11 3
June	202 6	192 1	27 0	23 4	22 4	2 4	26 0	25 4	1 8	59 2	60 8	5 3	93 2	92 6	6 9
July	93 8	69 7	11 9	10 8	10 4	0 8	15 0	14 3	1 1	42 7	41 4	3 4	58 6	57 4	4 8
August	55 3	38 6	5 8	6 5	6 3	0 6	14 3	13 5	1 8	34 9	33 6	3 3	51 1	48 5	4 7
September	125 9	37 1	12 3	9 9	7 9	0 9	19 7	19 1	2 1	34 6	32 1	3 1	54 3	50 7	5 6
October	147 2	117 9	17 2	10 5	9 1	1 4	19 3	18 6	1 2	32 3	28 5	3 1	58 0	53 3	5 7
November	152 1	128 3	24 3	9 6	8 7	1 4	12 9	12 7	0 8	24 9	23 2	2 0	47 6	45 7	3 6
December	49 1	42 4	6 0	5 9	5 8	0 9	7 7	7 7	0 3	20 5	19 6	1 5	37 6	36 7	2 3

Month	Monthly Standard Deviatıons														
	Neebing River			Pigeon River			Nagagami River			Turtle River			English River		
	Hist.	Synthetic Mean	Synthetic S D	Hist.	Synthetic Mean	Synthetic S D	Hist.	Synthetic Mean	Synthetic S D	Hist.	Synthetic Mean	Synthetic S D	Hist.	Synthetic Mean	Synthetic S D
January	17 9	17 0	8 4	2 1	3 5	0 9	1 8	1 5	0 2	7 6	6 1	1 0	9 7	7 9	1 0
February	12 4	9 7	2 6	1 8	2 2	0 4	1 7	1 4	0 2	6 5	10 3	1 8	7 1	6 9	1 1
March	83 5	36 6	9 7	15 9	1 5	0 2	12 7	10 1	1 8	17 9	19 3	3 6	9 3	8 3	1 1
Apnl	340 5	449 1	113 6	29 9	14 3	2 6	24 0	21 2	2 6	45 5	28 9	4 2	39 3	31 7	3 7
May	274 1	352 5	88 5	11 3	25 8	5 3	15 4	14 1	2 3	35 4	33 9	6 0	62 6	56 3	9 3
June	160 3	154 0	40 1	5 7	12 7	2 4	11 9	10 6	2 8	26 1	30 2	9 9	43 9	36 3	4 5
July	138 0	65 7	17 2	3 7	5 0	0 9	7 9	6 2	0 8	20 6	19 3	4 0	29 0	26 6	4 9
August	254 4	35 6	10 5	15 3	2 9	0 5	10 4	9 1	2 2	20 0	16 4	4 8	32 9	25 7	4 7
September	144 9	80 7	21 4	12 1	5 5	0 9	13 0	10 4	1 3	22 3	18 2	5 4	36 2	28 0	4 3
October	151 0	112 4	30 8	9 7	7 3	1 9	11 0	9 5	1 2	26 8	16 6	4 4	48 2	28 1	4 8
November	42 4	126 3	40 0	5 2	6 7	1 7	5 7	5 1	0 7	14 7	10 3	2 0	26 4	18 0	2 6
December	17 9	33 4	7 6	3 1	4 7	1 3	2 4	2 0	0 2	10 0	7 5	1 3	15 4	11 6	1 5

Note Hıst and S D respectıvely are the Hıstorıcal and Standard Devıatıon

Comparison of Statistical Analysis of Long Term Characteristics: Synthetic realizations are found to embed the long term characteristics, for example, the historical value of the Hurst coefficient lies within the 95% confidence intervals obtained from the synthetic streamflow realizations [Table 6]. One interpretation is that the pattern based stochastic model of streamflow generation is capable of embedding in the synthetic realizations long term characteristics similar to those of the historical sample. In other words, synthetic realizations can be utilized for long term planning of water resources systems such as reservoirs and other long term water use allocations of water resources.

Table 6. Summary of Long Term Statistical Characteristics of Various Rivers

Long Term Statistic	Neebing River			Pigeon River			Nagagami River			Turtle River			English River		
	Historic	Synthetic Mean	S.D.	Historic	Synthetic Mean	S.D.	Historic	Synthetic Mean	S.D.	Historic	Synthetic Mean	S.D.	Historic	Synthetic Mean	S.D.
Hurst Coeff.	0.512	0.516	0.046	0.564	0.511	0.037	0.543	0.519	0.038	0.490	0.533	0.041	0.490	0.519	0.041
Maximum Hurst Coeff.		0.594			0.587			0.592			0.621			0.646	
Hurst Coeff.	0.512			0.564			0.543			0.490			0.490		
Minimum Hurst Coeff.		0.438			0.431			0.436			0.427			0.426	

Note: S.D. and Hurst Coeff. respectively are the Standard Deviation and Hurst Coefficient.

CONCLUSIONS

Pattern analysis has enjoyed enormous growth in the last decade. It is a discipline that has many real world applications in many different fields. Simply stated, pattern analysis, instead of transforming data to fit an existing model, allows similarity of patterns in data to suggest meaningful partitioning of data into homogeneous subgroups for further statistical analysis. The application of pattern analysis discussed in this paper deals with streamflow data, which can be very useful in the broad field of hydrology. The results of this paper demonstrates that in practical situations, a feature synthesis model is applicable in obtaining synthetic streamflow realizations for various rivers in Northwest Ontario.

ACKNOWLEDGEMENT

The partial support provided by Natural Sciences and Engineering Research Council in carrying out this investigation is gratefully acknowledged. The computational efforts by R.S. Hong, an undergraduate student, are thankfully appreciated.

REFERENCES

Afza, N. and U.S. Panu (1991) Infilling of Missing Data Values in Monthly Streamflows. A Technical Report, Department of Civil Engineering, Lakehead University, Thunder Bay, Ontario.

Frenette, R.J.(1989) Investigations into Pattern Recognition Techniques for Estimating Missing Streamflow Data. An Unpublished Undergraduate Technical Report, Civil Engineering Department, Lakehead University, Thunder Bay, Ontario.

Gnandesikan, G. (1977) Methods for Statistical Data Analysis of Multivariate Observations. John Wiley and Sons, New York.

Goodyear, C. and U.S. Panu (1992) Infilling Missing Monthly Streamflow Time Series Using Multivariate Approach. A Technical Report, Civil Engineering Department, Lakehead University, Thunder Bay, Ontario.

Hinkley, D.V. (1977) "On Quick Choice of Power Transformation", Applied Statistics, 26 (1), 67-69.

Hong, R.S. (1992) Analysis of Synthetic Streamflow Data by Pattern Recognition Program. An Unpublished Undergraduate Technical Report, Civil Engineering Department, Lakehead University, Thunder Bay, Ontario.

Kojiri, T., Unny, T.E., and U.S. Panu (1986) Streamflow Pattern Analysis Using Clustering Techniques. A Technical Report, System Design Department, University of Waterloo, Waterloo, Ontario.

Panu, U.S. (1978) Pattern Recognition Based Computer Programs for Hydrologic Data Analysis and Synthesis. A Technical Report, Civil Engineering Department, University of Waterloo, Waterloo, Ontario.

Panu, U.S. and T.E. Unny (1980) "Stochastic Synthesis of Hydrologic Data Based on Concepts of Pattern recognition, III Application to Natural watersheds", Journal of Hydrology, 46, 197-217.

Quimpo, R.G. (1968) " Autocorrelation and Spectral Analysis in Hydrology", Jour. Hydraul. Div. Amer. Soc. Civil Eng., 94 (HY2), 363-373.

Wills, R.(1987) Pattern Recognition With Applications to Generation of Synthetic Streamflow Data. An Unpublished Undergraduate Technical Report, Civil Engineering Department, Lakehead University, Thunder Bay, Ontario.

Wu, S. (1993) Application of Pattern Recognition System to Precipitation Time Series in Northwest Ontario. An Unpublished Undergraduate Technical Report, Civil Engineering Department, Lakehead University, Thunder Bay, Ontario.

Appendix

Summary of inter-pattern structure

Pigeon River

		Reference Vectors	
		1	2
Reference Vectors	1	0.0000	1.0000
	2	1.0000	0.0000

Nagagami River

		Reference Vectors	
		1	2
Reference Vectors	1	0.0000	1.0000
	2	1.0000	0.0000

Turtle River

		Reference Vectors		
		1	2	3
Reference Vectors	1	0.0000	0.0000	1.0000
	2	0.0000	0.0000	1.0000
	3	0.5333	0.4667	0.0000

English River

		Reference Vectors	
		1	2
Reference Vectors	1	0.0000	1.0000
	2	1.0000	0.0000

MOMENT EVOLUTION OF THE OUTFLOW-RATE FROM NONLINEAR CONCEPTUAL RESERVOIRS

Karmeshu
School of Computer and Systems Sciences
Jawaharlal Nehru University
New Delhi, 110067, India

and

H. Schurz
Institute of Applied Analysis and Stochastics
Mohrenstr. 39
Berlin, 10117, Germany

ABSTRACT. The temporal evolution of moments of outflow-rate is investigated in a stochastically perturbed nonlinear reservoir due to precipitation. The detailed stochastic behaviour of outflow is obtained from the numerical solution of a nonlinear stochastic differential equation with multiplicative noise. The time-development of first two moments is studied for various choices of parameters. Using Stratonovich interpretation, it turns out that the mean outflow-rate is above that given by the deterministic solution. Based on the set of 9000 simulation runs, 90 % confidence intervals for the mean evolution of outflow-rate are computed. The effect of stochastic perturbations with finite correlation time is investigated.

1. INTRODUCTION

The time distribution of outflows at the outlet of catchments fed by rainfall is usually studied on the basis of a single or a cascade of conceptual reservoirs, see Nash [13]. The outflows driven by deterministic inputs of rainfall have been extensively studied within the framework of cascade of conceptual reservoirs. Efforts have also been made to incorporate the nonlinear relationship between storage and discharge in Singh [16]. Unny and Karmeshu [19] have extended the Nash cascade of reservoirs to take into account the stochastic nature of input. The objective of such an extension is to provide a principal basis for the generation of stochastic stream flows. Incorporating stochastic input, the storage balance equations for the system of reservoirs turn out to be stochastic differential equations (SDE's). Recently Karmeshu and Lal [6] have further investigated the stochastic behaviour of storage in conceptual reservoirs based on storage balance equations. The explicit time–development of moments of outflow for a reservoir could be obtained only in the linear case. However, in a nonlinear reservoir the resulting nonlinear SDE leads to a rather intractable hierarchy of moment equations. In a recent paper Fujita, Shinohara, Nakao and Kudo [4] have also considered a stochastic nonlinear reservoir and have analyzed the linearized version with additive noise.

V. P. Singh and B. Kumar (eds.), Surface-Water Hydrology, 403–413.

The purpose of this paper is to investigate the evolution of moments of outflow in a nonlinear reservoir arising due to stochastic rainfall/precipitation. The resulting SDE with multiplicative noise can be interpreted in two different ways – Stratonovich and Ito prescriptions, compare Gardiner [3]. These are related to each other in the sense that one can transform results of one prescription to those of the other. We have adopted Stratonovich prescription as it is preferable for modelling a physical process, due to Wong and Zakai [22]. However, for the purpose of comparison we have presented results for Ito prescription as well.

The evolution of moments of outflow is obtained by numerical integration schemes, in general, based on the stochastic Taylor expansion described in Kloeden and Platen [7]. Unny [20] also discussed the numerical integration of SDE's in the context of catchment modelling. Resorting to numerical analysis has enabled us to take into account the time varying nature of the rainfall intensity.

The paper comprises seven sections. Section 2 deals with the formulation of the stochastic model involving a nonlinear stochastic differential equation. Section 3 and 4 are concerned with the stationary probability density function (pdf) and the evolution of moments of outflow-rate. After discussing briefly numerical methods for the solution of stochastic differential equations in section 5, we carry out the numerical simulation of moments in section 6. The last section is devoted to the study of effects of coloured noise fluctuations on the evolution of moments.

2. FORMULATION OF THE MODEL

The continuity equation for the storage $S(t)$ in a reservoir can be written as

$$\frac{dS(t)}{dt} = I(t) - Q(t), \tag{2.1}$$

where $Q(t)$ denotes the outflow-rate and $I(t)$ is the inflow-rate. The storage equation takes into account the nonlinearity of the conceptual reservoir considered, following the consideration of Singh [16]. Among several storage equations proposed in the literature, the simplest nonlinear reservoir is defined as

$$S = kQ^n, \tag{2.2}$$

k and n being positive parameters. The exponent n is generally found to lie between 0.4 and 3.3. Combining equations (2.1) and (2.2), we get the differential equation for the outflow-rate

$$\frac{dQ}{dt} = aQ^{1-n}[I(t) - Q], \tag{2.3}$$

where

$$a = 1/(kn). \tag{2.4}$$

The inflow-rate being usually expressible in terms of precipitation is stochastic in nature. Customarily, the stochastic fluctuations in the precipitation can be represented by a white noise process. Thus we set

$$I(t) = \bar{I}(t)[1 + \sigma\xi(t)] \tag{2.5}$$

where $\bar{I}(t)$ denotes the mean precipitation, $\xi(t)$ represents the stochastic fluctuations and σ measures their intensity.
Substituting equation (2.5) in equation (2.3), we get the SDE for the outflow-rate

$$dQ(t) = aQ^{1-n}(t)[\bar{I}(t) - Q(t)]dt + \sigma a\bar{I}(t)Q^{1-n}(t) \circ dW(t) \tag{2.6}$$

where we replace $\xi(t)dt$ by the differential $dW(t)$ of the Wiener process, and the SDE (2.6) is to be interpreted in the Stratonovich sense.
The Itô SDE corresponding to the Stratonovich SDE (2.6) is

$$\begin{aligned} dQ(t) \;=\; & \left[aQ^{1-n}(t)(\bar{I}(t) - Q(t)) + \frac{1}{2}\sigma^2 a^2(1-n)\bar{I}^2(t)Q^{1-2n}(t)\right] dt \\ & + \sigma a\bar{I}(t)Q^{1-n}(t)dW(t) \end{aligned} \tag{2.7}$$

The SDE (2.6) as well as (2.7) completely describes the stochastic evolution of the outflow-rate from a conceptual reservoir. The solution process $Q(t)$ is a diffusion process, and accordingly the SDE is subject to the theory of diffusion processes.

3. STATIONARY PROBABILITY DENSITY OF THE OUTFLOW-RATE

The complete probabilistic description of the model can be obtained in terms of the Fokker–Planck equation (FPE) which yields the probability density for the outflow-rate $Q(t)$. The FPE corresponding to the SDE (2.6) is

$$\frac{\partial p(Q,t|Q_0)}{\partial t} = -\frac{\partial}{\partial Q}\left[\{f(Q,t) + \frac{1}{2}g'(Q,t)g(Q,t)\}p\right] + \frac{1}{2}\frac{\partial^2}{\partial Q^2}\left[g^2(Q,t)p\right] \tag{3.1}$$

where

$$\begin{aligned} f(Q,t) &= aQ^{1-n}(\bar{I}(t) - Q), \\ g(Q,t) &= \sigma a\bar{I}(t)Q^{1-n}. \end{aligned} \tag{3.2}$$

The FPE is to be solved under appropriate boundary conditions, one of them is that at $Q = 0$ there is a reflecting barrier. This condition can be expressed by the requirement that the current probability vanishes at $Q = 0$, i.e.

$$\left\{f(Q,t)p - \frac{1}{2}\frac{\partial}{\partial Q}[g^2(Q,t)p]\right\}_{Q=0} = 0. \tag{3.3}$$

The other boundary condition is provided by imposing the natural boundary condition for $Q \to \infty$. The initial condition can be expressed as

$$\lim_{t\to 0} p(Q,t|Q_0) = \delta(Q - Q_0). \tag{3.4}$$

It may be mentioned that to derive the explicit solution of the FPE seems to be impossible. However, the stationary probability density function $p_s(Q)$ (pdf) can be obtained by setting $\frac{\partial p(Q,t)}{\partial t} = 0$. Following Horsthemke and Lefever [5], the stationary pdf $p_s(Q)$ is then found to be

$$p_s(Q) = \frac{N(n)}{g(Q)} \exp\left[2 \int^{Q} \frac{f(u)}{g^2(u)} du\right] \tag{3.5}$$

where $N(n)$ is the corresponding normalization constant.

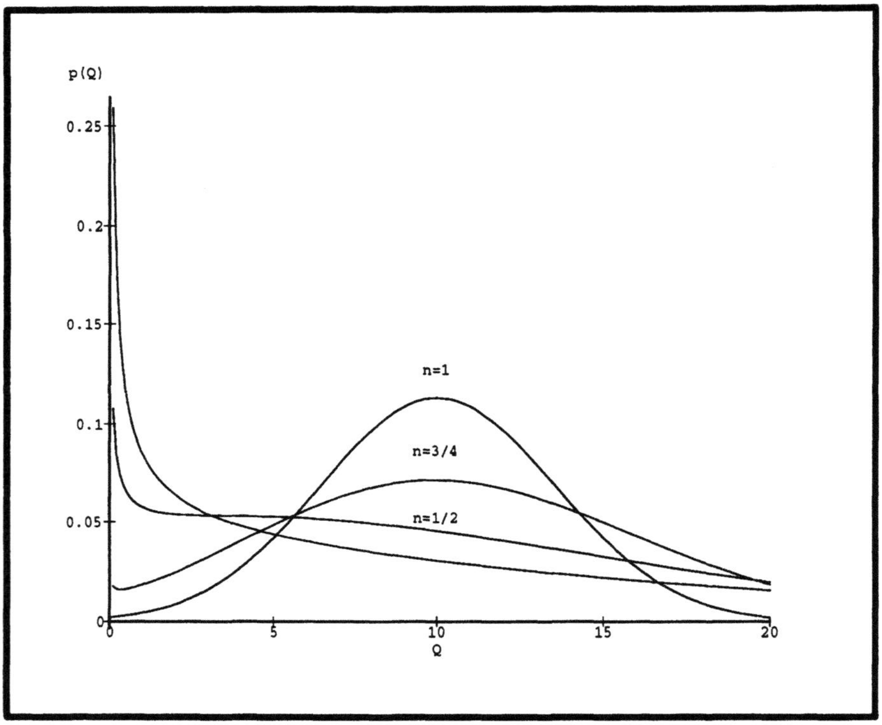

Figure 1. Stationary probability densities of the outflow-rate ($n = 1/3, 1/2, 3/4, 1$).

For the purpose of illustration we present in figure 1 graphs of $p_s(Q)$ for various values of the exponent n. We find that for $n < 1$, the shape of $p_s(Q)$ is highly skewed and tends to be symmetrical as n increases up to 1 and beyond, with more and more concentrating mass at $Q = \overline{I}$. Although it has not been possible to obtain the evolution of pdf $p(Q, t|Q_0)$, nevertheless significant insight could be gained from the evolution of moments pertaining to the pdf.

4. EVOLUTION OF MOMENTS FOR THE OUTFLOW-RATE

The differential equations governing the evolution of first two moments of the outflow-rate can be obtained from the SDE (2.6). Using Itô's formula (Arnold [1], Gardiner [3] or Kloeden and Platen [18]), the mentioned differential equations are

$$\frac{d\mathbb{E}[Q(t)]}{dt} = a\overline{I}(t)\mathbb{E}[Q^{1-n}(t)] - a\mathbb{E}[Q^{2-n}(t)] + \frac{1}{2}\sigma^2 a^2(1-n)\overline{I}^2(t)\mathbb{E}[Q^{1-2n}(t)];$$
$$\frac{d\mathbb{E}[Q^2(t)]}{dt} = 2a\overline{I}(t)\mathbb{E}[Q^{2-n}(t)] - 2a\mathbb{E}[Q^{3-n}(t)] + \sigma^2 a^2(2-n)\overline{I}^2(t)\mathbb{E}[Q^{2-2n}(t)]\,. \tag{4.1}$$

Proceeding in a similar manner, differential equations for higher order moments can be derived. It may be noted that except for the linear case ($n = 1$), the equations for the first (and second) moments involve moments of orders other than first (and second). It is easily seen that, in order to study moments, one is confronted with the hierarchy of moment equations (Soong [17]).
To make any progress one could truncate the hierarchy at some stage, but the truncation procedures based on the assumption that the pdf of the process $Q(t)$ is close to a Gaussian process are not suitable for our purpose. The reason is that the model considered here is described by a SDE driven by multiplicative noise which yields a probability distribution being quite different from the Gaussian one. The best one can do under these circumstances is to employ a numerical scheme and to simulate the evolution of the moments for the outflow-rate.

5. NUMERICAL METHODS FOR THE SOLUTION OF SDE'S

Numerous methods for the numerical treatment of stochastic differential equations can be found in the literature. For the sake of generality we only consider the following type of stochastic differential equations. Given a m-dimensional Wiener process $(W^j(t))_{j=1,\ldots,m}$ which drives the Itô differential equation

$$dQ(t) = a(Q(t))dt + \sum_{j=1}^{m} b^j(Q(t))dW^j(t) \tag{5.1}$$

starting at $Q(0) = Q_0 \in \mathbb{R}^d$ on the time interval $[0, T]$. As already mentioned the Itô and Stratonovich versions of (5.1) can be transformed each other in a natural way. Solutions $\{Q(t) : t \geq 0\}$ of (5.1) exists and are unique under the assumptions of Lipschitz continuity and of 'appropriate' polynomial boundedness of the functions $a(.)$ and $b(.)$. The simplest method to generate numerically such solutions is the Euler–Maruyama method constructed by the scheme

$$Y_{n+1} = Y_n + a(Y_n)\Delta_n + \sum_{j=1}^{m} b^j(Y_n)\Delta W_n^j. \quad (n = 0, 1, 2, \ldots). \tag{5.2}$$

Here Y_{n+1} means the value of the approximate solution using the step size $\Delta_n = t_{n+1} - t_n$ at the time point t_{n+1}. With $\Delta W_n^j = W^j(t_{n+1}) - W^j(t_n)$ we denote the current j–th increment of the Wiener process $W^j(t)$ which can be generated as a standard Gaussian random variable multiplied by $\sqrt{\Delta_n}$. At least for 'small enough' step sizes Δ_n, by corresponding convergence theorems the application of the method (5.2) to the equation (5.1) is justified to obtain an approximate solution depending on the practical purpose one is going to follow. In case of pathwise approximation (strong) one requires that it exists a constant $K > 0$ such that

$$\forall t_n : \mathbb{E}||Q(t_n) - Y(t_n)|| \leq K \cdot \Delta^{\gamma} \tag{5.3}$$

where $\Delta = \sup_n \Delta_n < +\infty$. In contrast to that, for momentwise approximation (weak) it is sufficient to demand only that

$$\forall t_n : ||\mathbb{E}(g(Q(t_n)) - g(Y(t_n)))|| \leq K \cdot \Delta^{\beta} \tag{5.4}$$

with respect to a class of 'sufficiently smooth' functions g (often $g \in C_p^{\infty}$). The weak convergence has more practical usage because one is mostly interested in the calculation of moments only. In those cases one may even simplify the generation of the random variables ΔW_n^j in (5.2). For equidistant approximations, it turns out to take any independent random variables $\xi_{j,n}$ instead of ΔW_n^j which satisfy the moment relation

$$|\mathbb{E}\xi| + |\mathbb{E}\xi^3| + |\mathbb{E}\xi^5| + |\mathbb{E}\xi^2 - \Delta| + |\mathbb{E}\xi^4 - 3\Delta^2| \leq C \cdot \Delta^3$$

for a constant $C > 0$. Thus we keep the weak convergence order $\beta = 1.0$ in (5.4). For example, this is true for the three–point distributed random variable ξ with

$$\mathbb{P}(\xi = \pm\sqrt{3\Delta}) = \frac{1}{6} \quad \text{and} \quad \mathbb{P}(\xi = 0) = \frac{2}{3}.$$

This simplification saves time and computational effort, but the same procedure cannot be applied to the scheme (5.2) approximating pathwisely the solution of (5.1) via the requirement (5.3). The method (5.2) possesses the strong convergence order $\gamma = 0.5$ and weak convergence order $\beta = 1.0$. Milstein has done one of the first trials of systematic construction of numerical methods and proved the convergence of the well–known Milstein schemes (with $\gamma = 1.0$ and $\beta = 1.0$). In general, corresponding higher order methods are derived from the stochastic Taylor expansion, which is due to the iterative application of Itô's lemma, by appropriate truncation. This approach has been firstly suggested in Wagner & Platen [21] and is described in Milstein [11] and Kloeden & Platen [7]. For further details, see Clark & Cameron [2], Pardoux & Talay [15], Newton [14], Talay [18] or Kloeden, Platen & Schurz [8]. In our experiments we used the scheme form (5.2) and obtained reasonable results for the outflow–rates. Note that higher order methods would not be always applicable to our models because of explosions in their numerical solutions close to zero. Furthermore, in Milstein, Platen & Schurz [12] one finds first attempts to achieve control in stochastically stiff situations, such systems where one observes slowly and rapidly varying stochastic components influencing decisively the dynamical behaviour.

6. NUMERICAL SIMULATION OF THE MOMENTS

As discussed in section 4, the numerical solution of SDE's is an effective means of obtaining detailed information about the stochastic behaviour of the outflow–rate. To this end we have simulated the system governed by SDE's for the following situations:

A. Constant mean precipitation,
B. Constant mean precipitation followed by exponential decay.

A. Constant mean precipitation

We have numerically solved the considered SDE for a few choices of the parameter σ when the mean precipitation is assumed to remain constant over the entire simulation run. In the following runs the time step size of the Euler approximation is $\Delta = 0.01$(hr). The initial time $t_0 = 0$ and the simulation is for 4 hrs so that $T = 4$. The other parameters of the model are fixed and their magnitudes are :
Mean precipitation $\overline{I} = 10, k = 1.0, n = 0.75$, Initial outflow-rate $Q(0) = 1.0$.
In figure 2 the deterministic evolution of $Q(t)$ is presented. This corresponds to the case $\sigma = 0$, when there are no fluctuations. While modelling consideration requires the Stratonovich interpretation, for comparison we have also computed the sample paths of $Q(t)$ when the SDE is interpreted in Itô sense as well. For small values of $\sigma = 0.1$, the two sample paths are very close to each other. Divergence between paths increases with σ caused by the growing up difference between the corresponding drift functions. We observe in figure 3 that the sample path under Itô interpretation gives a lower estimate of the outflow–rate as compared with the Stratonovich prescription. In figure 4 the temporal deterministic outflow–rate as well as the mean outflow–rate interpreted in Stratonovich and Itô calculus are given. One observes that Stratonovich prescription gives larger estimates of the mean outflow–rate than the deterministic values which are above the Itô estimates. The time–development of the second moment $\mathbb{E}[Q^2(t)]$ is viewed in figure 5, and we notice that, as time advances, it tends to settle down to a constant value. Figure 6 displays the 90% confidence intervals for the mean evolution in Stratonovich calculus compared with the deterministic values. The estimates are based on the mean of a set of 15 batches of trajectories repeatedly observed 600 times.

B. Constant mean precipitation followed by an exponential decay

In this situation we have assumed that the precipitation remains constant for $T = 2$ hrs and then it falls exponentially. This can be expressed as

$$I(t) = \begin{cases} \overline{I}_1 \,(1 + \sigma\xi(t)) \;, & 0 \leq t \leq 2 \\ \overline{I}_1 \exp(-\alpha(t-2))\,(1 + \sigma\xi(t)) \;, & t > 2 \end{cases} \tag{6.1}$$

where α (> 0) is a precipitation decay parameter. The model described by the equation 6.1 besides being more realistic is mathematically interesting. The process evolves till time $T = 2$ governed by (2.6) with given initial condition. Then, terminal points $Q(T)$ of the sample paths serve as random initial conditions for the outflow–rate which has to be incorporated in (2.6). Figure 7 shows the mean evolution ($\alpha = 2$) in both Stratonovich and Itô calculus compared with the deterministic values firstly increasing up to $T = 2$ and then starts declining.

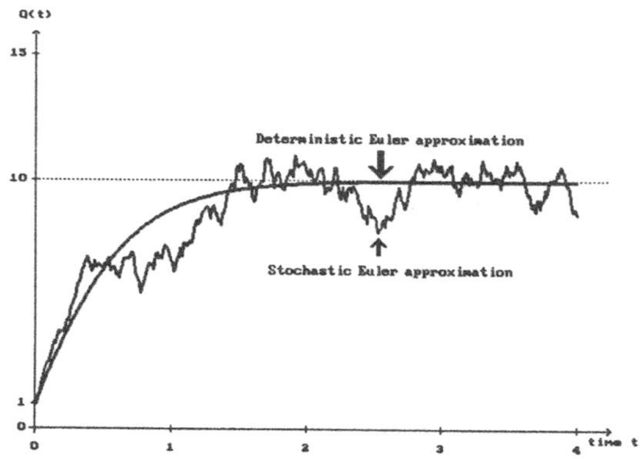

Figure 2. Sample path of the outflow-rate with $\sigma = 0.1$.

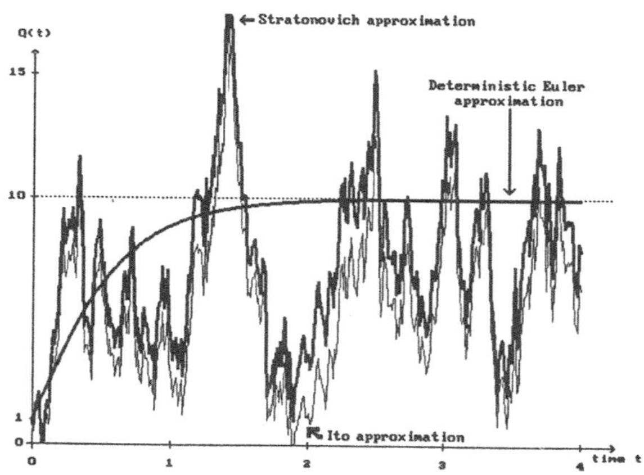

Figure 3. Sample path of the outflow-rate with $\sigma = 0.5$.

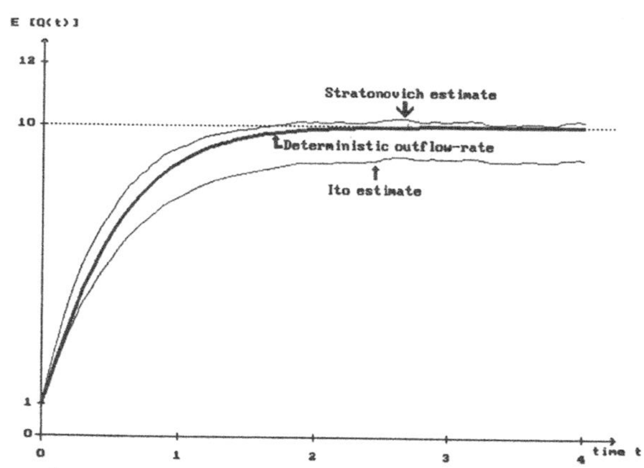

Figure 4. Estimate for the first mean evolution.

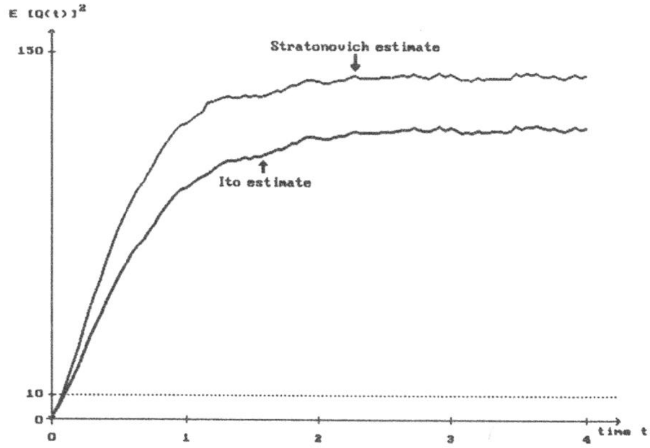

Figure 5. Estimate for the mean square evolution.

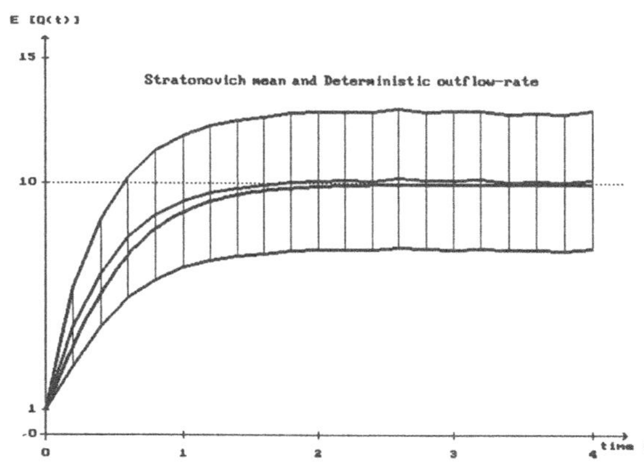

Figure 6. Confidence intervals (90%) for the mean evolution.

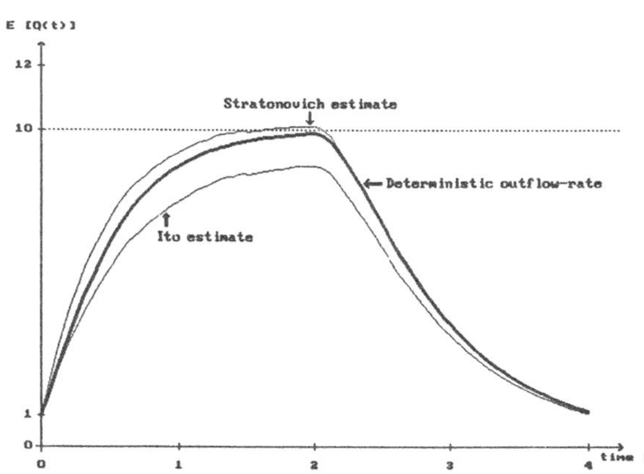

Figure 7. Estimated mean evolution with exponentially decaying precipitation.

7. EFFECT OF COLOURED NOISE FLUCTUATIONS

In the previous section we carried out the analysis to study the effect of white noise fluctuations in the precipitation. The assumption of white noise implies that the time scale of fluctuations is negligibly small compared with the macroscopic time scale of the system. This assumption is in general rather too restrictive, and in a more realistic situation the time scale of fluctuations may not be negligibly small. Accordingly, one has to assume that the stochastic perturbations in the precipitation are represented by coloured noise with a finite correlation time. A realistic version of noise with finite correlation time is the well-known Ornstein-Uhlenbeck process [3]. The auto-correlation of $\xi(t)$ is given by

$$\mathbb{E}[\xi(t)\xi(t')] = \frac{\nu}{2}\exp(-\nu|t-t'|)\,, \quad t > t' \tag{7.1}$$

which tends to the delta-correlated process as $\nu \to \infty$. Equations 7.1 can be used to define the correlation time as $1/\nu$ ($\nu > 0$). Now we rewrite the SDE (2.6) to

$$dQ(t) = aQ^{1-n}(t)(\overline{I}(t) - Q(t))\,dt + \sigma a\overline{I}(t)Q^{1-n}(t)\,\xi(t)\,dt \tag{7.2}$$

where $\xi(t)$ is a stationary Ornstein-Uhlenbeck process determined by

$$d\xi(t) = -\nu\xi(t)\,dt + \nu\,dW(t)\,. \tag{7.3}$$

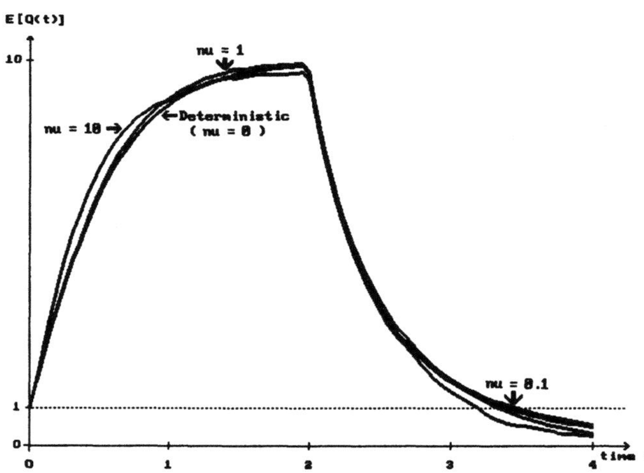

Figure 8. Estimated mean outflow–rate with exponentially decaying precipitation and various correlation parameters ν.

For various values of the parameter ν, the coupled SDE's 7.2 and 7.3 are solved numerically. Figure 8 gives the mean outflow–rate for the same values of parameters as considered in section 6.A. For large values of ν we find that $\mathbb{E}[Q(t)]$ tends to those viewed in the figure 4. However, for smaller values of ν the deviation is significant. In passing, we note that the framework proposed here can be easily extended to a cascade of stochastic nonlinear reservoirs.

REFERENCES

1. Arnold, L.: *Stochastic differential equations*, Wiley, New York, 1974.
2. Clark, J.M.C. & Cameron, R.J.: 'The maximum rate of convergence of discrete approximations for stochastic differential equations', Springer Lecture Notes in Control and Information Sc. Vol. **25**, p. 162–171, 1980.
3. Gardiner, C.W.: *Handbook of stochastic methods*, Springer, Berlin, 1983.
4. Fujita, M., Shinohara, N., Nakao and Kudo, M.: 'Stochastic response of storage function model for flood runoff', presented at Int. Conf. on Stochastic and Statistical Methods in Hydrology and Environmental Engineering, University of Waterloo, Waterloo, Ontario, Canada, 1993.
5. Horsthemke, W. and Lefever, R.: *Noise induced transitions*, Springer Series in Synergetics, Berlin, 1984.
6. Karmeshu and Lal, V.B.: 'Stochastic behaviour of storage in conceptual reservoirs', presented at Int. Conf. on Stochastic and Statistical Methods in Hydrology and Environmental Engineering, University of Waterloo, Waterloo, Ontario, Canada, 1993.
7. Kloeden, P.E. and Platen, E.: *Numerical solution of stochastic differential equations*, Springer Appl. of Math., Vol. **23**, pp. 632, 1992.
8. Kloeden, P.E., Platen, E. and Schurz, H.: *Numerical solution of stochastic differential equations through computer experiments*, Springer Unitext, pp. 292, 1993.
9. Kloeden, P.E., Platen, E. and Schurz, H.: 'The numerical solution of nonlinear stochastic dynamical systems : A brief introduction', Int. J. Bifurcation and Chaos, Vol. 1, p.277-286, 1991.
10. Milstein, G.: 'Approximate integration of stochastic differential equations', Theor. Prob. Appl. **19**, p.557–562, 1974.
11. Milstein, G.: *The numerical integration of stochastic differential equations*, Uralski Univ. Press, Sverdlovsk, p. 225, 1988.
12. Milstein, G., Platen, E. and Schurz, H.: 'Balanced implicit methods for stiff stochastic systems: An introduction and numerical experiments', Preprint No. **33**, IAAS – Berlin, 1992.
13. Nash, J.E.: 'The form of the instantaneous unit hydrograph', Int. Assoc. Sci. Hydrol. Monograph Pub. **45**, No. **3**, p.114-121, 1958.
14. Newton, N.J.: 'Asymptotically efficient Runge–Kutta methods for a class of Itô and Stratonovich equations', SIAM J. Appl. Math. **51**, p. 542–567, 1991.
15. Pardoux, E. and Talay, D.: 'Discretization and simulation of stochastic differential equations', Acta Appl. Math. **3**, p.23-47, 1985.
16. Singh, V.P.: *Hydrologic systems : Rainfall-runoff modelling*, Vol. **1**, Prentice Hall, 1988.
17. Soong, T.T.: *Random differential equations in science and engineering*, Academic Press, New York, 1973.
18. Talay, D.: *Analyse Numérique des Equations Differentielles Stochastiques, Thése 3ème cycle*, Univ. Provence, 1982.
19. Unny, T.E. and Karmeshu : 'Stochastic nature of outputs from conceptual reservoir model cascades', J. Hydrology, Vol. **68**, p.161-180, 1984.
20. Unny, T.E.: 'Numerical integration of stochastic differential equations in catchment modelling', Water Resorv. Res., Vol. **20**, p.360-368, 1984.
21. Wagner, W. and Platen, E.: 'Approximation of Itô integral equations', Preprint ZIMM, Acad. of Sc. GDR, Berlin, 1978.
22. Wong, E. and Zakai, M.: 'On the relation between ordinary and stochastic differential equations', Int. J. Eng. Sci., Vol. **3**, p.213-229, 1965.

COMPARATIVE STUDY OF STOCHASTIC MODELS FOR SEASONAL STREAMFLOW GENERATION

K. Srinivasan M. C. Philipose
Assistant Professor Research Scholar
Department of Civil Engineering
I. I. T., Madras - 600 036.

ABSTRACT

Lower order periodic stochastic models such as periodic autoregressive (PAR) and periodic autoregressive moving average (PARMA) models with and without harmonics are fitted to the monthly streamflows into Krishnarajasagara (KRS) reservoir located at the upper reaches of the river Cauvery in South India and a comparison is done between these models in terms of preservation of periodic statistics as well as over-year drought characteristics. Further, the comparison has been extended to the Thomas-Fiering lognormal 3-parameter (TFLN3) model which is supposed to be quite simple in structure and at the same time efficient in preservation of low flows. It is found that TFLN3 is able to reproduce the first two moments very well but, not the skewness or periodic correlations. On the other hand, PAR(2) and PARMA(1,1) models are able to preserve the periodic correlations. The periodic models which are not subjected to Fourier fitting, report lower relative root mean square error in overall standard deviation and correlations, compared to the ones which are subjected to Fourier fitting. From the over-year drought analysis, it is observed that all the stochastic models fitted underestimate severe drought durations corresponding to higher truncation levels. The over-year drought characteristics are computed using the method suggested by Wang and Salas (1989), based on one hundred periodic synthetic flow sequences generated from the different models. A comparative study of the preservation of drought characteristics corresponding to various truncation levels and return periods is presented.

INTRODUCTION

Application of stochastic modelling to seasonal and annual hydrologic time series started with the application of Markov models(Thomas and Fiering, 1962). Matalas (1967) discussed the detailed procedure of generation of synthetic sequences for two parameter and three parameter lognormal distributions using lag-one Markov generation scheme. Stedinger (1980) has given a detailed account of the various

V. P. Singh and B. Kumar (eds.), Surface-Water Hydrology, 415–427.

fitting procedures for two parameter and three parameter lognormal distributions to hydrologic data. He has used loss functions such as root mean square error in estimating quantiles to decide upon the best fitting procedure. In the recent decade, periodic models are being extensively used to model seasonal streamflows(McLeod and Hipel (1978), Hirsch (1979), Salas et al. (1980, 1982), Vecchia (1983), Vecchia et al. (1983), Noakes (1984), Vecchia (1985), Obeysekera and Salas (1986)). The periodicity in the parameters can also be represented by required number of significant harmonics in the respective parameters. This is known as the functional method of modelling and this has been dealt with in detail by Salas et al. (1980), Yevjevich (1984), Vecchia (1985) and Yevjevich and Harmancioglu (1989).

In this paper, lower periodic models with and without harmonic fitting and the popular, simple and efficient Thomas-Fiering model with 3-parameter lognormal distribution are fitted to the unregulated monsoon streamflows into KRS reservoir in the Cauvery river basin, South India. A comparative study of the models in terms of the preservation of basic periodic statistics and over-year drought characteristics is attempted.

PERIODIC STOCHASTIC MODELLING

Periodic models are designed for time series which contain periodic characteristics which may be caused by natural or man-made events. Often, the monthly or the weekly hydrologic time series display a periodic correlation structure. Hence, it becomes essential to go in for periodic autoregressive models (PAR) or periodic autoregressive moving average (PARMA) models. The main advantage of these models is that the parameters of each period can be estimated independently of the other periods.

Periodic Autoregressive (PAR) Model

The general structure of the PAR model is given by:

$$Z_{\nu,\tau} = \sum_{j=1}^{p} \phi_{j,\tau} Z_{\nu,\tau-j} + \varepsilon_{\nu,\tau} \quad (1)$$

$\nu = 1,2,\ldots,N$ years
$\tau = 1,2,\ldots,w$ seasons

where $\{Z_{\nu,\tau}\}$ is the time series suitably transformed and standardized and has an expected value equal to zero, ϕ is the autoregressive parameter and $\{\varepsilon_{\nu,\tau}\}$ is the error or noise term assumed to be uncorrelated and has zero mean. The order 'p' of the model is maintained to be constant through all the periods, though autoregressive parameter estimates are different for each period. The subscripts ν and τ denote the year and the period respectively.

Periodic Autoregressive Moving Average (PARMA) Model

The general structure of the PARMA model is given by:

$$Z_{\nu,\tau} = \sum_{j=1}^{p} \phi_{j,\tau} Z_{\nu,\tau-j} + \varepsilon_{\nu,\tau} - \sum_{j=1}^{q} \Theta_{j,\tau} \varepsilon_{\nu,\tau-j} \qquad (2)$$

where 'Θ' is the moving parameter of order 'q'.

Thomas-Fiering 3-parameter Model (TFLN3)

The general structure of the most widely used Thomas-Fiering model is given by:

$$Y_{i+1} = \mu_{j+1} + \rho_{j+1} (Y_i - \mu_j) * (\sigma_{j+1}/\sigma_j) + (1-\rho^2_{j+1})^{1/2} \; t_{i+1} \qquad (3)$$

(j=1,2,...,12: i=1,2,...,(N*12) for a monthly model)

where μ, σ and ρ are the mean, standard deviation and lag one correlation coefficient of the transformed flow series such that the normality assumption is satisfied, and t refers to a sampling variate from the normal distribution having mean zero and unit variance. If the flow is assumed to follow a LN3 distribution, then

$$y = \ln(X-A) \qquad (4)$$

where A is the location parameter, and X is the historical flow data.

The two variations of TFLN3 model considered in this paper are based on the method of parameter estimation namely: (1) Method of moments (MOM) (Matalas, 1967) and (2) Quantile lower bound (QLB) method (Stedinger, 1980). The moment estimates for each month are computed using the set of equations which relate the historical flow statistics to the transformed flow statistics (Charbeneau, 1978). The quantile lower bound for each month is estimated using the relationship given by Stedinger (1980):

$$\hat{A} = \frac{X_{min} X_{max} - (X_{med})^2}{X_{min} + X_{max} - 2^* X_{med}} \qquad (5)$$

in which X_{min}, X_{max} and X_{med} refers to the minimum, the maximum and the median of the respective monthly flows. The quantile lower bound estimated from eq.(5) is substituted for A in eq.(4) and the maximum likelihood estimates of mean variance, and lag-one correlation of the resulting transformed flows are determined. These estimates are used in the Thomas-Fiering generation algorithm given by eq.(3), and the monthly flows are generated.

HARMONIC ANALYSIS

The Fourier series representation of u_τ, a sample estimate of the population periodic parameter v_τ as given by (Yevjevich, 1972)

$$\hat{v}_\tau = \bar{u} + \sum_{j=1}^{h} [A_j \operatorname{Cos}(2\pi\tau/w) + B_j \operatorname{Sin}(2\pi\tau/w)] \qquad \tau = 1,\ldots,w \tag{6}$$

where $\bar{u}$ is the overall mean of u_τ, A_j and B_j are the Fourier series coefficients, j is the harmonic and h is the total number of harmonics which is equal to w/2 or (w-1)/2, depending on whether w is even or odd respectively. The mean $\bar{u}$ and the Fourier coefficients A_j and B_j are determined by

$$\bar{u} = \frac{1}{w}\sum_{\tau=1}^{w} u_\tau \tag{7}$$

$$A_j = \frac{2}{w}\sum_{\tau=1}^{w} u_\tau \operatorname{Cos}(2\pi j\tau/w), \qquad j=1,\ldots,h \tag{8}$$

$$B_j = \frac{2}{w}\sum_{\tau=1}^{w} u_\tau \operatorname{Sin}(2\pi j\tau/w), \qquad j=1,\ldots,h \tag{9}$$

When w is even the last coefficients A_h and B_h are given by:

$$A_h = \frac{1}{w}\sum_{\tau=1}^{w} u_\tau \operatorname{Cos}(2\pi h\tau/w) \tag{10}$$

$$B_h = 0 \tag{11}$$

$\hat{v}_\tau$ should be computed only using significant number of harmonics. The usual methods for the determination of the significant number of harmonics are namely, the cumulative periodogram test (Salas et al., 1980), the method of explained variance (Kottegoda, 1980) and the AIC criterion (Vecchia, 1985). In this paper, using the method of explained variance, significant harmonics in periodic mean, standard deviation and correlations have been fitted. The WHT-transformed flow data is subjected to the harmonic analysis, and Fourier fitting is done for mean, standard deviation and correlations at the periodic level. The Fourier coefficients only for the selected harmonics (criterion: 90% explained variance) are given in Table II. Using the Fourier fitted periodic mean, periodic standard deviation and periodic correlations, lower order periodic models have been fitted to the flow data. Under this category of periodic models, only PAR(2)(HAR)WHT and PARMA(1,1)(HAR)WHT have been found to pass the diagnostic checking, and enter the verification stage.

VERIFICATION

The verification of the models fitted concerns with the reproduction of the unbiased estimates of basic periodic statistics of the observed streamflows. One hundred synthetic monthly flow sequences of same length as that of the historical flows have been generated using the different candidate models selected based on diagnostic checks. The verification of the basic monthly flow statistics is given in Fig. I. From Fig. I, it is observed that the periodic means and periodic standard deviations are well reproduced by TFLN3(QLB) followed by TFLN3(MOM), while the periodic variation of skewness and correlations are not preserved. The PAR(2), PARMA(1,1) models with WHT transformations and also with harmonic fitting overestimate the standard deviation for July month. As far as the preservation of lag-one and lag-two correlations is concerned, PAR(2)WHT and PARMA(1,1)WHT models do a good job.

For a given model, a periodic statistic $\eta_\tau(i)$ is determined from a generated sample i, and the average η_τ and the mean square error $(mse)_\tau$ are obtained based on the 100 generated samples. Then, the overall average of η_τ and mse_τ (where $\tau = 1,\ldots,12$ months) are determined to get the overall generated $\eta(g)$ and mse(g). Now, the relative root mean square error (rmse) is computed by

Fig. I Preservation of Basic Statistics
Station: Krishnarajasagara River: Cauvery

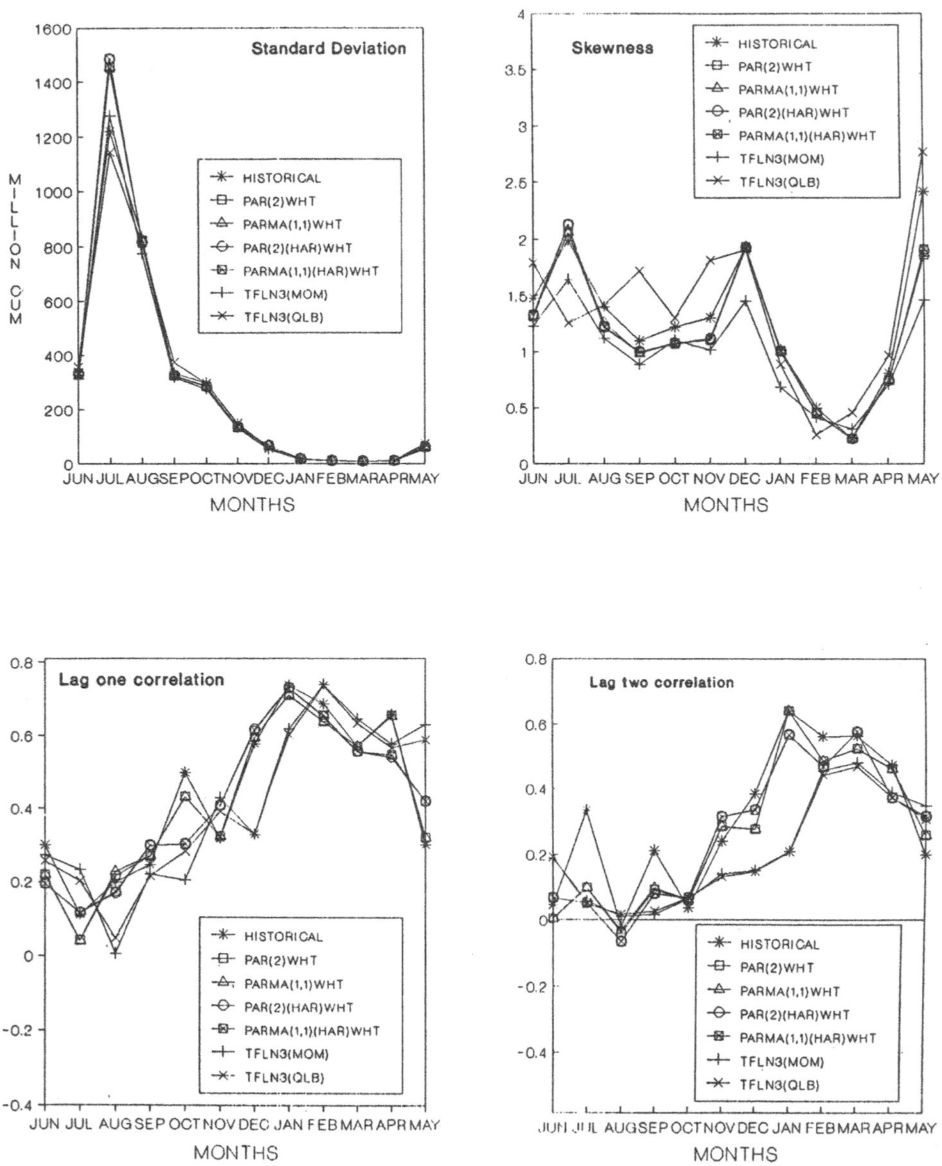

$$\text{rmse} = \frac{(\text{mse}(g))^{0.5}}{\eta(h)} \tag{12}$$

in which $\eta(h)$ is the overall historical statistic obtained by averaging $\eta_\tau(h)$, where $\tau = 1,\dots,12$ months. The rmse thus computed for basic periodic statistics for all candidate models is presented in Table III. It may be observed from Table IV that minimum flows are best preserved by TFLN3(QLB) model. While, TFLN3(MOM) consistently underestimates the minimum flows and PAR(2)WHT, PARMA(1,1)WHT models overestimate the minimum flow of the high flow month, namely, July.

DROUGHT ANALYSIS

The over-year drought characteristics namely, drought duration and deficit sum are computed for the historical sequence as well as for 100 synthetic sequences for three truncation levels(70%, 80% and 90% of mean monthly flow) and five return periods (10,20,30,40,50 years). This drought analysis is done using the following method suggested by Wang and Salas (1989), which is based on the averaged critical drought characteristics from possible number of overlapping subsamples, thus by eliminating the bias:

If the length of sequence is N years and the return period is T years, then, the possible number of overlapping sequences (N_s) can be given as:

$$N_s = (N-T+1) \tag{13}$$

From each of the above N_s subsamples, the longest drought is found out (for a specified truncation level expressed as a percent of mean monthly flow). Then the average of these values from the N_s subsamples gives an estimate of the T-year drought duration for the specified truncation level and return period. This procedure is repeated for all the five return periods and three truncation levels. The same method is adopted to estimate the deficit sum corresponding to the specified truncation levels and return periods.

RESULTS AND DISCUSSION

It is observed from Table I that the relative root mean square error for the overall mean and the overall standard deviation are the least in case of TFLN3(QLB), followed by TFLN3(MOM). However, these two models fail to preserve the periodic skewness, periodic correlations as well as overall skewness and overall correlations Fig. I, Table

Table I Periodic Statistics Station: Krishnarajasagara
River: Cauvery

	Mean (M cum)	Standard Deviation (M cum)	Skewness	Lag one correlation	Lag two correlation
JUN	433.18	338.74	1.477	0.300	0.046
JUL	2174.24	1219.41	1.986	0.110	0.334
AUG	1658.27	837.27	1.398	0.199	-0.021
SEP	640.13	334.86	1.094	0.247	0.212
OCT	498.71	295.92	1.217	0.496	0.036
NOV	193.29	146.68	1.302	0.318	0.240
DEC	82.87	55.45	1.931	0.577	0.384
JAN	45.00	17.31	1.017	0.736	0.641
FEB	29.82	11.32	0.501	0.686	0.562
MAR	19.27	8.62	0.228	0.570	0.563
APR	20.21	11.00	0.804	0.657	0.473
MAY	65.55	68.23	2.412	0.302	0.199

Table II Selected Fourier Coefficients

	Harmonic	A_j	B_j
Mean	2	0.012	-0.017
	3	0.015	0.013
	4	0.021	-0.003
	5	-0.009	0.014
Standard Deviation	2	-0.025	0.054
	3	-0.035	-0.033
	4	-0.048	0.001
	5	0.041	-0.043
Lag one Correlation	1	-0.058	-0.260
	3	0.071	-0.034
Lag two Correlation	1	0.005	-0.291
	6	0.064	0.000

Table III Relative Root Mean Square Error of Basic Periodic Statistics

Model	Mean (M cum)	Standard Deviation (M cum)	Skewness	Lag-one corre.	Lag-two corre.
PAR(2)WHT	0.158	0.444	0.437	0.282	0.493
PARMA(1,1)WHT	0.157	0.438	0.439	0.283	0.492
PAR(2)(HAR)WHT	0.158	0.480	0.438	0.333	0.535
PARMA(1,1)(HAR)WHT	0.158	0.479	0.437	0.333	0.535
TFLN3(MOM)	0.127	0.300	0.500	0.454	0.743
TFLN3(QLB)	0.122	0.276	0.606	0.440	0.752

Table IV Preservation of Minimum Flows (M cum)

	HISTO-RICAL	PAR(2) WHT	PARMA(1,1) WHT	PAR(2) (HAR) WHT	PARMA(1,1) (HAR) WHT	TFLN3 (MOM)	TFLN3 (QLB)
JUN	12.28	9.80	9.86	10.09	10.08	3.31	16.26
JUL	346.16	950.13	950.00	950.03	950.00	571.32	480.05
AUG	536.06	561.24	563.81	565.10	565.23	450.00	529.87
SEP	215.56	140.90	140.96	138.89	138.76	113.81	198.06
OCT	60.05	81.04	80.84	85.47	85.41	55.09	88.14
NOV	24.30	0.32	0.31	0.00	0.00	2.33	18.79
DEC	12.13	25.68	25.70	25.71	25.74	11.50	15.80
JAN	18.21	18.16	18.24	18.04	18.09	16.72	18.25
FEB	4.12	8.19	8.21	8.51	8.54	7.48	6.42
MAR	4.85	0.88	0.88	0.65	0.61	2.03	2.61
APR	2.07	1.52	1.50	1.55	1.53	1.80	2.57
MAY	6.95	9.15	9.15	9.15	9.15	0.38	7.50

III). The poor preservation of correlations in the case of these two models is only to be expected since they are basically of the PAR(1) type. On the other hand, PAR(2)WHT and PARMA(1,1)WHT models are able to preserve the periodic as well as overall skewness and correlations (Fig. I, Table III), while the standard deviations of July month is overestimated (Fig. I). Further, an interesting point to note from Table III is that the relative root mean square error in the overall standard deviation and the overall correlations are lower in case of PAR(2)WHT and PARMA(1,1)WHT models, compared to their counterparts with harmonic fitting.

It is observed from the Table V that at 70% truncation level (TL) the drought duration is preserved by TFLN3 models, while these models underestimate the same at moderate and higher truncation levels. On the other hand, the PAR(2)WHT and PARMA(1,1)WHT models are able to preserve the drought duration at moderate (80%) TL and the same is slightly overestimated at lower (70%) TL, especially for 20,30 and 40 year return periods. Another interesting point to be noted from the Table V is that severe historical drought durations corresponding to higher (90%) TL, are consistently underestimated by all the stochastic models fitted. Further, it may be observed from Table V that at 90% TL, the PARMA(1,1)WHT models followed by PAR(2)WHT models are able to preserve the deficit sum, while the TFLN3 models underestimate the same. At 70%, 80% TL, upto a return period of 40 years (upto moderate deficit conditions), TFLN3(MOM) followed by TFLN3(QLB) are able to preserve the deficit sum, while the PAR(2)WHT and PARMA(1,1)WHT models overestimate the same. It is to be noted that there is no significant change between models PAR(2)WHT, PARMA(1,1)WHT and PAR(2)(HAR)WHT, PARMA(1,1)(HAR)WHT in preserving drought duration and deficit sum.

CONCLUSIONS

Different stochastic models namely, PAR(2)WHT, PARMA(1,1)WHT, PAR(2)(HAR)WHT, PARMA(1,1)(HAR)WHT, TFLN3(MOM) and TFLN3(QLB) are fitted to the monthly streamflows into Krishnarajasagara (KRS) reservoir located at the upper reaches of the river Cauvery in South India, and a comparison is done between these models in terms of verification as well as over-year drought analysis. The periodic means and periodic standard deviations are well reproduced by TFLN3(QLB) followed by TFLN3(MOM), while the periodic skewness and correlations are not. The PAR(2)WHT, PARMA(1,1)WHT models with and without harmonic fitting, overestimate the standard deviation for July month. These models preserve the lag-one and lag-two periodic correlations. The relative root mean square error in the overall standard deviation and overall correlations are lower in the case of PAR(2)WHT and PARMA(1,1)WHT models compared to same models with harmonic fitting. The minimum flows are best preserved by TFLN3(QLB) model.

Table V Preservation of Drought Characteristics
Station: Krishnarajasagara River: Cauvery

Return Period (years)	Model	Drought Duration (years) Truncation Level* 70%	80%	90%	Deficit Sum (Mm^3) Truncation Level 70%	80%	90%
10	HISTORICAL	1.020	1.653	2.653	549	1349	2714
	PAR(2)WHT	1.096	1.731	2.505	692	1516	2812
	PARMA(1,1)WHT	1.082	1.732	2.506	688	1512	2805
	PAR(2)(HAR)WHT	1.106	1.729	2.514	681	1507	2805
	PARMA(1,1)(HAR)WHT	1.096	1.719	2.510	677	1499	2790
	TFLN3(MOM)	0.947	1.536	2.282	562	1279	2392
	TFLN3(QLB)	0.955	1.534	2.281	584	1306	2421
20	HISTORICAL	1.205	2.333	4.179	679	1628	3585
	PAR(2)WHT	1.465	2.312	3.364	1024	2097	3852
	PARMA(1,1)WHT	1.433	2.324	3.322	1011	2054	3767
	PAR(2)(HAR)WHT	1.480	2.323	3.376	1012	2079	3838
	PARMA(1,1)(HAR)WHT	1.483	2.310	3.333	1005	2062	3782
	TFLN3(MOM)	1.309	2.007	3.024	862	1763	3204
	TFLN3(QLB)	1.308	2.003	3.027	890	1795	3245
30	HISTORICAL	1.276	2.759	5.276	857	1891	4141
	PAR(2)WHT	1.646	2.645	3.885	1216	2439	4491
	PARMA(1,1)WHT	1.620	2.679	3.831	1195	2360	4346
	PAR(2)(HAR)WHT	1.670	2.675	3.890	1206	2410	4463
	PARMA(1,1)(HAR)WHT	1.689	2.658	3.836	1198	2384	4376
	TFLN3(MOM)	1.509	2.328	3.511	1036	2048	3734
	TFLN3(QLB)	1.512	2.330	3.512	1070	2088	3766
40	HISTORICAL	1.421	3.000	6.000	1186	2191	4390
	PAR(2)WHT	1.815	2.902	4.224	1340	2677	4943
	PARMA(1,1)WHT	1.788	2.954	4.172	1319	2596	4797
	PAR(2)(HAR)WHT	1.837	2.945	4.224	1325	2645	4917
	PARMA(1,1)(HAR)WHT	1.862	2.928	4.148	1314	2612	4820
	TFLN3(MOM)	1.662	2.557	3.869	1168	2261	4113
	TFLN3(QLB)	1.651	2.533	3.864	1207	2303	4135
50	HISTORICAL	1.862	3.000	6.000	1915	3022	4973
	PAR(2)WHT	1.950	3.049	4.466	1426	2828	5229
	PARMA(1,1)WHT	1.920	3.099	4.444	1408	2776	5112
	PAR(2)(HAR)WHT	1.963	3.078	4.433	1407	2794	5189
	PARMA(1,1)(HAR)WH	1.986	3.084	4.366	1395	2760	5101
	TFLN3(MOM)	1.813	2.757	4.167	1265	2453	4479
	TFLN3(QLB)	1.791	2.720	4.139	1309	2497	4515

* Truncation Level is expressed as % of Mean Monthly Flow

Note: Values of drought duration and deficit sum entered against the periodic stochastic models are averages from 100 synthetic sequences.

At 70% truncation level the drought duration is preserved by TFLN3 models, while these models underestimate the same at moderate and higher truncation levels. On the other hand, the PAR(2)WHT and PARMA(1,1)WHT models are able to preserve the drought duration at moderate (80%) TL and the same is slightly overestimated at lower (70%) TL, especially for 20,30 and 40 year return periods. At 90% TL, the PARMA(1,1)WHT models followed by PAR(2)WHT models are able to preserve the deficit sum, while the TFLN3 models underestimate the same. There is no significant change between models PAR(2)WHT, PARMA(1,1)WHT and PAR(2)(HAR)WHT, PARMA(1,1)(HAR)WHT in preserving drought duration and deficit sum.

ACKNOWLEDGEMENT

The authors are thankful to the Indian Institute of Technology, Madras for the facilities provided for this study. For parameter estimation of periodic models, the 1990 version of the program CSU001 developed by Salas et al. has been used and the authors are thankful to him for the same.

REFERENCES

1. Charbeneau, R. J., 1978. 'Comparison of the two and three parameter lognormal distributions used in Streamflow Synthesis'. Water Resources Research 14(1):149-150.

2. Kottegoda, N. T., 1980. Stochastic Water Resources Technology. Mac Millan Press, London, England.

3. Matalas, N. C., 1967. 'Mathematical Assessment of Synthetic Hydrology'. Water Resources Research 3(4):937-946.

4. McLeod, A. I., and Hipel, K. W., 1978. 'Developments in Monthly Autoregressive Modelling'. Technical Report. 45 - XM011178, Department of System Design Engineering, University of Waterloo, Ontario, Canada.

5. Noakes, D. J., 1984. 'Applied time series modelling, seasonality and forecasting'. Ph.D. dissertation, University of Waterloo, Waterloo, Ontario, Canada.

6. Obeysekera, J. T. B. and Salas, J. D., 1986. 'Modeling of aggregated hydrologic time series', Journal of Hydrology, 86, 197 - 219.

7. Salas, J. D., Delleur, J. W., Yevjevich, V., and Lane, W. L., 1980. Applied Modeling of Hydrologic Time Series. Water Resources Publications, Littleton, Colorado.

8. Salas, J. D.,Boes, D. C., and Smith, R. A., 1982. 'Estimation of ARMA models with seasonal parameters'. Water Resources Research 18(4): 1006 - 1010.

9. Stedinger, J. R., 1980. 'Fitting Lognormal Distributions to Hydrologic Data'. Water Resources Research 16(3):481-490.

10. Thomas, H. A., and Fiering, M. B., 1962. 'Mathematical Synthesis of Streamflow Sequences for the Analysis of River Basins by Simulation'. In Design of Water Resources Systems, A. Mass et al., 459-493, Harvard University Press, Cambridge, Mass.

11. Vecchia, A. V., 1983. 'Aggregation and Estimation for Periodic Autoregressive Moving Average Models'. Ph.D. dissertation, Department of Statistics, Colorado State University, Fort Collins.

12. Vecchia, A. V., 1985. 'Periodic Autoregressive Moving Average (PARMA) Modeling with Application to Water Resources'. Water Resources Bulletin 21(5):721-730.

13. Wang, D. C., and Salas, J. D., 1989. 'Stochastic Modeling and Generation of Droughts'. Hydraulic Engineering '89 Proceedings, National Conf. on Hydraulic Engg., ASCE, New Orleans, Los Angeles, USA.

14. Yevjevich, V., 1972. Stochastic Processes in Hydrology. Water Resources Publications, Fort Collins, Colorado.

15. Yevjevich, V., 1984. 'Structure of Daily Hydrologic Series'. Water Resources Publications, Littleton, Colorado.

ADAPTIVITY IN STOCHASTIC MODELLING AND FORECASTING

A. W. Jayawardena
Department of Civil & Structural Engineering
University of Hong Kong
Hong Kong

ABSTRACT : Time varying coefficient ARMA and ARMA-transfer function (ARMA-TF) models for improving forecasting of water quality and runoff are presented and applied. Kalman filter algorithms are used for recursively updating the coefficients in the models. In the Kalman filter algorithm, different schemes are used for estimating the noise covariance matrices R and Q. Their performances in tracking the variability of the coefficients are also examined. The methodology is then applied to obtain one-step-ahead water quality and runoff forecasts in Hong Kong. It was found that forecasting improvements can be achieved by using the time varying coefficient ARMA and ARMA-TF modelling approach with a Kalman filter algorithm.

1. Introduction

Auto Regressive Moving Average (ARMA) and Auto Regressive Moving Average Transfer Function (ARMA-TF) type models (Box and Jenkins, 1976) with constant coefficients have been widely used in hydrological and environmental time series analysis and forecasting. ARMA models have been used to model and forecast univariate time series (For example, Huck and Farquhar, 1974; Lohani and Wang, 1987; Jayawardena and Lai, 1989) whereas ARMA-TF models have used to model and forecast time series arising from cause effect type relationships (For example, Clarke, 1974; Anselmo and Ubertini, 1979; Novotny and Zheng, 1990; Baracos et al., 1981; Hipel et al., 1982; Snorrason et al., 1984). ARMA-TF models have the advantage over ARMA models in that they can incorporate other relevant information.

Most hydrological and environmental systems however exhibit uncertainties and/or non-linearities which cannot be described by constant coefficient models. Such behaviour can only be described by models with time varying coefficients which have the potential capability to reflect the inherent changes and non-linearities in the system. Recursive estimation of coefficients in such models can be carried out by the Kalman Filtering algorithm (Kalman, 1960). Examples of the application of the latter approach include forecasting of rainfall and runoff data using the time varying ARMA-TF model

V. P. Singh and B. Kumar (eds.), Surface-Water Hydrology, 429–442.

with least squares based algorithm (O'Connell and Clarke, 1981; Young, 1984) and forecasting sewer flow data (Capodaglio et al., 1990).

In this study, attempts have been made to model and forecast some univariate and input-output type data series using time varying coefficient ARMA and ARMA-TF models coupled with the Kalman filtering algorithm. The coefficients are obtained recursively and forecasting is carried out in an adaptive mode. Comparisons are made with forecasts made with corresponding constant coefficient models.

Implementation of the Kalman filter algorithm requires estimation of the noise covariances at each time step because the performance of the algorithm depends critically on these values. Although the problem of estimating noise covariance has been addressed in theoretical literature, the results have had limited use in practice. A discussion of the technique in the context of ARMAX modelling is given by Ngan (1985). In this study, different schemes for estimating the noise covariance are used with the Kalman filter algorithm and their performances investigated.

2. Constant Coefficient Model Formulation

Constant coefficient ARMA and ARMA-TF models of order p, q (Box and Jenkins, 1976) can be written as:

$$\phi(B)Y(t) = \theta(B)\eta(t) \tag{1}$$

and,

$$\phi(B)Y(t) = \theta(B)X(t-b) + \eta(t) \tag{2}$$

where

$$\begin{aligned} B^m Y(t) &= Y(t-m), \\ \phi(B) &= 1 - \phi_1 B - \ldots - \phi_p B^p, \\ \theta(B) &= 1 - \theta_1 B - \ldots - \theta_q B^q, \end{aligned}$$

and, Y(t) are the observations, $\eta(t)$ are independent random variables with zero mean and constant variance σ^2, and, ϕ_i and θ_j are the AR and MA coefficients. In the case of the ARMA-TF model, X(t) is the input series, Y(t) is the output series and b is the time delay.

3. Time Varying Coefficient Model Formulation : State-Space Representation

The ARMA and ARMA-TF models given in Eqs. 1 & 2 can be considered in their state-space forms by writing the observation and the state equation as follows:

$$\mathbf{Z}(t) = \mathbf{H}(t)\mathbf{C}(t) + \mathbf{V}(t) \qquad \text{(Observation equation)} \qquad (3)$$

$$\mathbf{C}(t+1) = \mathbf{T}(t)\mathbf{C}(t) + \mathbf{W}(t) \qquad \text{(State equation)} \qquad (4)$$

In general, **C**(t) and **Z**(t) denote the state and output vectors, and **V**(t) and **W**(t) are independent serially uncorrelated random vectors with zero mean and covariance matrices $\alpha^2\mathbf{R}$ and $\alpha^2\mathbf{Q}$, where α^2 is an unknown positive scalar. The state-space representation is obtained through the following definitions and equivalences:

For the ARMA model,

$$\mathbf{C}(t) = [\phi_1(t), \phi_2(t), \ldots \phi_p(t), \theta_1(t), \ldots, \theta_q(t)] \qquad (5a)$$

$$\mathbf{H}(t) = [Y(t-1), Y(t-2), \ldots Y(t-p), -\eta(t-1), -\eta(t-2), \ldots, -\eta(t-q)] \qquad (6a)$$

$$\mathbf{Q} = \text{diag}\,(\varepsilon,\varepsilon,\varepsilon, \ldots.)_{p+q} \qquad (7a)$$

For the ARMA-TF model,

$$\mathbf{C}(t) = [\phi_1(t), \phi_2(t), \ldots \phi_p(t), \theta_0(t), \theta_1(t), \ldots, \theta_q(t)] \qquad (5b)$$

$$\mathbf{H}(t) = [Y(t-1), Y(t-2), \ldots Y(t-p), X(t-b), -X(t-1-b), \ldots, -X(t-q-b)] \qquad (6b)$$

$$\mathbf{Q} = \text{diag}\,(\varepsilon,\varepsilon,\varepsilon, \ldots.)_{p+q+1} \qquad (7b)$$

and,

$$\mathbf{Z}(t) = Y(t) \qquad \text{(scalar)} \qquad (8)$$

$$\mathbf{V}(t) = \eta(t) \qquad \text{(scalar)} \qquad (9)$$

$$\alpha^2 = \sigma^2 \text{ and } R = 1 \qquad \text{(scalar)} \qquad (10)$$

where ε is a parameter that needs to be estimated.

Eqs. 3 & 4 represent time varying coefficient ARMA an ARMA-TF models in which the coefficients themselves follow stochastic processes that are modelled by Eq. 4. The noise term **V**(t) represents the measurement error which refers to the random error $\eta(t)$ from the model, and, **W**(t) accounts for the model inadequacy. When **T**(t) = **I**, the unit matrix, the coefficients are modelled by a simple random process. Eq. 10 shows that an approximate estimate of α^2 based on known observations is given by σ^2, which is the variance of the random error $\eta(t)$, and Eq. 7 implies an assumption that the variations in the coefficients of the models are of the same orders of magnitude.

In the past, studies have focused mainly on time varying coefficient AR models (Ledolter, 1982; Kahl and Ledolter, 1983; Young et al., 1991). Because of the moving average terms and the resulting non-linearity, the state-space representation of an ARMA model is considered difficult. However, if **H**(t) is allowed to contain both past observations and past random errors (Eq. 6a), an ARMA model can be cast into a linear form similar to Eq. 3 and the Kalman filter algorithm can then be used.

4. Decision Rule for Time Varying Coefficients

The question as to whether, and under what circumstances the time varying coefficient models lead to better forecasts has not been addressed in previous applications. Kahl and Ledolter (1983) proposed a rule tc decide whether it is worthwhile to consider models with time varying coefficients for univariate ARMA models. The decision rule is base on a concentrated log-likelihood function which is computed by,

$$L_c(\mathbf{Q} \mid y) = -N \log \hat{\sigma} - \frac{1}{2} \sum_{t=1}^{N} \log f(t) \tag{11}$$

where

$$\hat{\sigma}^2 = \frac{1}{N} \sum_{t=1}^{N} [\mathbf{Z}(t) - \mathbf{H}(t)\hat{\mathbf{C}}(t \mid t-1)]^2 / f(t) \tag{12}$$

in which

$$f(t) = 1 + \mathbf{H}(t)^T \mathbf{P}(t \mid t-1) \mathbf{H}(t),$$

$$\mathbf{Q} = \mathrm{diag}(\omega, \omega, \ldots, \omega),$$

and N is the number of observations considered. If it is found that

$$L_c(\varepsilon = 0.0) \geq L_c(\omega = 0.0001/\alpha^2) \tag{13}$$

then, the constant coefficient model is recommended. If on the other hand it is found that

$$L_c(\varepsilon = 0.0) < L_c(\varepsilon = 0.0001/\alpha^2) \tag{14}$$

then, forecast improvement from a time varying coefficient model can be expected.

5. Kalman Filter Adaptive Forecasting Algorithm

In the Kalman filter notation, the sequence of estimations are given in the following equations:

Initial estimate of coefficients

$$\hat{\mathbf{C}}(0 \mid 0) = \mathrm{E}[\mathbf{C}(0)] \tag{15}$$

Initial error covariance matrix

$$\mathbf{P}(0 \mid 0) = \mathrm{Var}[\mathbf{C}(0)] \tag{16}$$

<u>Prior to observing **Z**(t):</u>

State prediction

$$\hat{\mathbf{C}}(t+1 \mid t) = \mathbf{T}(t)\, \hat{\mathbf{C}}(t \mid t) \tag{17}$$

Output forecast

$$\hat{\mathbf{Z}}(t+1 \mid t) = \mathbf{H}(t+1)\, \hat{\mathbf{C}}(t+1 \mid t) \tag{18}$$

Predicted error covariance matrix

$$\mathbf{P}(t+1 \mid t) = \mathbf{P}(t \mid t) + \mathbf{Q} \tag{19}$$

Forecasting gain matrix (Kalman gain matrix)

$$\mathbf{K}(t+1) = \mathbf{P}(t+1 \mid t)\mathbf{H}^{T}(t+1)[\mathbf{H}(t+1)\mathbf{P}(t+1 \mid t)\mathbf{H}^{T}(t+1) + \mathbf{R}]^{-1} \tag{20}$$

<u>After observing **Z**(t):</u>

Coefficient estimation using the new observation

$$\hat{\mathbf{C}}(t+1 \mid t+1) = \hat{\mathbf{C}}(t+1 \mid t) + \mathbf{K}(t+1)[\mathbf{Z}(t+1) - \mathbf{H}(t+1)\, \hat{\mathbf{C}}(t+1 \mid t)] \tag{21}$$

Error covariance matrix after the new observation

$$\mathbf{P}(t+1 \mid t+1) = \mathbf{P}(t+1 \mid t)\, [\mathbf{I} - \mathbf{K}(t+1)\, \mathbf{H}(t+1)] \tag{22}$$

In the Kalman filter algorithm, the noise terms **W**(t) and **V**(t) are usually assumed to be normally distributed. However, it is worth noting that the algorithm given by Eqs.

15~22 has been originally derived within the framework of correlation theory with no assumptions made about the distributions of the noise terms (Kalman, 1960; Boguslavskij, 1988). Therefore the algorithm is applicable to non-normal data as well and the optimalities of Kalman filter can still be preserved in least squares and linear unbiased minimum variance senses.

6. Schemes for estimating the Noise Covariance

In the application of the Kalman filter algorithm it is necessary to estimate the noise covariance matrices **R** and **Q**, because both recursive coefficient estimation and adaptive forecasting are affected by these values. It is convenient if **R** and **Q** are treated as constants for the entire time period. However, this cannot be automatically assumed in many instances (Ngan an Russell, 1986). In this study, four schemes for estimating the noise covariance matrices are implemented. They include constant **R** and **Q**, recursive estimation of **R** and **Q** (Todini and Bouillot, 1976), multiplication of the updated error covariance matrix (Eq. 22) by a weighting factor $1/\lambda$ (Sastri, 1986; O'Connell, 1980) and smoothing of **R** and **Q** by a factor ξ which varies between zero and unity (Meade, 1985). The details of these are documented elsewhere (Jayawardena and Lai, 1991). The scheme which most efficiently tracks the variation of the model coefficients during the calibration period is taken as the optimum one.

7. Application

7.1 UNIVARIATE DATA - ARMA MODEL

Four data series consisting of 14-day surface DO and bottom NO_3 measured at 3 water quality monitoring stations (TM2, TM6, and TM8) in the Tolo Harbour area of Hong Kong during an eight year period were used (Table 1a). The data series were found to be trend free and non-seasonal. The coefficients of the appropriate ARMA models fitted to the four series are given in Table 2a.

The constant coefficient ARMA models were first used to obtain one-step-ahead forecasts. The forecasts seem to be not too different from the long term means implying that the constant coefficient model cannot successfully track the changes of water quality with time. Similar drawback can be seen in the water quality forecasts made by Lohani and Wang (1987).

In the adaptive mode, the ARMA models were cast into their state-space formulations using Eqs. 5~10. The Kalman filter was then used to obtain the recursive coefficient estimation and one-step-ahead adaptive forecasts.

7.2 INPUT-OUTPUT TYPE DATA : ARMA-TF MODEL

Two sets of daily rainfall and runoff data in Hong Kong were used for ARMA-TF modelling. Runoff data are from two gauging stations at Tai Lam Chung "A" (TLCA), and Nim Wan (NW) which drain catchments of areas 80 and 536 hectares respectively (Hong Kong Government, 1956~1986). The corresponding daily rainfall data are taken from the Royal Observatory records made at Tai Lam Country Park (Station No. 73) and Nim Wan Public Primary School (Station No. 118) rain gauge stations. Table 1b gives a summary of the data used.

Some insignificant linear trend components which cause non-stationarities in the mean were found in the two runoff series. The two rainfall series were found to be trend free. For the two sets of series, the annual periodicities were detected by using the smoothed spectral density function. The harmonic analysis method in which the daily means and daily standard deviations are estimated by Fourier series analysis failed to remove the periodicities in rainfall and runoff data. They were therefore removed by subtracting the daily means which were estimated from long term rainfall and runoff records. The de-seasonalised and de-trended runoff series and the de-seasonalised rainfall series were then found to be stationary. The Box-Jenkins procedure for identification of ARMA-TF model was then applied.

7.2.1 *Causality between rainfall and runoff*

Two methods were used to detect the causality between the rainfall and the runoff series and the results were found to be different. For example, for the Nim Wan catchment, the cross correlation function using the method proposed by Pierce-Haugh (1977) in which the rainfall and runoff series are pre-whitened by ARMA models showed a peak value at lag 8 with the cross correlation coefficient exceeding the 95% confidence limits at around lag 7. This implies that there is at least a 7-day (b = 7) time delay between the runoff and the rainfall that produced the runoff. For the same catchment, the cross correlation coefficient using the method proposed by Box and Jenkins (1976) in which the series are only de-trended and de-seasonalised, showed a peak value at lag zero, implying that there is no time delay (b = 0) between the rainfall and runoff. From field conditions for both these catchment, it is known that there is hardly any time delay between the two processes. Therefore, the zero time delay (b = 0) in the rainfall and runoff series is more realistic. The shift of peak value in the cross correlation function by the Pierce-Haugh method which is based on the properties of normal data is perhaps caused by the non-normality of the rainfall and runoff data. The skewness coefficients of the rainfall and runoff series are found to be quite large (9.227, 4.396, 5.299, and 3.973 respectively for TLCA runoff data, TLCA rainfall data, Nim Wan runoff data and Nim Wan rainfall data). Another possible reason for the shift is that the pre-whitening method can change the structure of the auto correlation function thereby affecting the structure of the cross correlation function (Diggle, 1990). Similar discrepancy was observed for the TLCA catchment. Therefore, for non-normal data, the Box-Jenkins method is suggested for detecting the causality.

The parameters of the appropriate ARMA-TF models for the two rainfall and runoff series are listed in Table 2b. It can be seen that the coefficients θ_0 are relatively larger than other θ_j and in the case of Nim Wan, larger than ϕ_i as well indicating that the present rainfall is a significant factor affecting the present runoff.

In this study, the decision rule given in Eqs. 11~14 suggests that forecasting improvements can be expected from time varying coefficient ARMA-TF models (Table 3).

7.2.2 *Sensitivity analysis of noise covariance estimating schemes*

The variability in the coefficients usually can be modelled by a random process (Wood and Szollosi-Nagy, 1978; Kahl and Ledolter, 1983; Ngan and Russell, 1986), and, therefore, the matrix **T**(t) in Eq. 4 is set to the unit matrix **I**.

A sensitivity analysis was carried out using the Kalman Filter with the four schemes to determine their relative performance. The one-step-ahead forecasting error as measured by the Mean Squared Error (MSE), Mean Absolute Error (MAE) and Mean Absolute Percentage Error (MAPE) were used as indicators of sensitivity (Eqs. 23~25).

$$\text{MSE} = \frac{1}{L}\sum_{l=1}^{L} [z_{t+l} - \hat{z}_t(l)]^2 \tag{23}$$

$$\text{MAE} = \frac{1}{L}\sum_{l=1}^{L} \mid [z_{t+l} - \hat{z}_t(l)] \mid \text{x } 100 \tag{24}$$

$$\text{MAPE} = \frac{1}{L}\sum_{l=1}^{L} \mid \frac{z_{l+1} - \hat{z}_t(l)}{z_{l+1}} \mid \text{x } 100 \tag{25}$$

In the case of ARMA-TF modelling, the forecasting error was measured by an additional indicator, namely, the variance of one-step-ahead forecasting error calculated in the stochastic component. The other indicators, namely, MSE, MAE and MAPE are calculated for complete forecasts after adding the seasonal and trend components.

The results of the sensitivity analysis show that when significant variations are introduced on the coefficients, the forecasting errors become large implying that the time variations of the coefficients are not very large. The schemes and initial settings for estimating **Q** and **R** corresponding to the three smallest error values in the sensitivity analysis period were chosen for further application.

8. Forecasting and Analysis of Results

Three combinations of the parameters **Q**, λ or ξ corresponding to the smallest error values referred to above were chosen for forecasting with the adaptive ARMA models. Final forecasts were obtained by combining these three forecasts using equal weights.

It can be seen that the adaptive one-step-ahead forecasts are closer to the observations than those by the constant coefficient ARMA models. The reductions in the forecasting error are listed in Table 4.

For one-step-ahead forecasts made using the adaptive ARMA-TF models, the corresponding results are given in Tables 5 and 6. It has been noted that the scheme and the initial settings for **Q** and **R** which gave the minimum forecasting error in the sensitivity analysis period does not always give the minimum forecasting error in the forecasting period. It is probably due to the fact that the variabilities in the coefficients are different from year to year. An attempt to obtain an optimal forecast was made by averaging out the individual forecasts corresponding to the three best "scheme-initial setting" combinations. Averaging was done by attaching equal or unequal weights to each individual forecast. Unequal weights γ_i are calculated as follows (Winkler and Makridakis, 1983):

$$\gamma_i = \frac{V_i^{-1}}{\sum_{i=1}^{k} V_i^{-1}} \tag{26}$$

where V_i is the variance of the one-step-ahead forecasting error in the sensitivity analysis period and k is the number of individual forecasts used to calculate γ_i, which in this case has been 3. In this study, the equal weight method is found to give slightly better combined forecasts than the unequal weight method. The results of combined forecasts using equal weights are shown in Table 6.

From the results of the sensitivity analysis and forecasting, it is seen that the "scheme-initial settings" which introduce small adjustments to the coefficients gave the best forecasting result. In the ARMA-TF models the coefficients θ_0 are relatively larger than the other coefficients ϕ_i and θ_j. Therefore the adjustments are mainly made in the coefficients ϕ_i and θ_j. That means the ratio of the present rainfall to present runoff does not change much, but some changes occur in other hydrological conditions. It was also seen that when an incorrect variation is introduced in the coefficients, the resulting forecasts are worse than those from a constant coefficient ARMA-TF model.

From Table 5, it can be seen that the performance of forecasting by ARMA-TF models are better than by ARMA models. This is due to the effect of combining the present rainfall in the ARMA-TF models. The error of one-step-ahead forecasting from constant coefficient ARMA-TF model can be reduced further by using time varying coefficient ARMA-TF models with Kalman filter based algorithm (Table 6).

9. Conclusion

Improvements in one-step-ahead forecasting of some water quality and hydrological data in Hong Kong were achieved by using time varying coefficient ARMA and ARMA-TF models with Kalman filter algorithm. Although the constant noise covariance estimate scheme gives the best performance for modelling and forecasting in this study, the Kalman filter algorithm with the four noise covariance estimate schemes provides a potentially valuable technique to track different variabilities in the coefficients of ARMA-TF models In detecting the causality between two time series, it is important to note that the pre-whitening method should be applied carefully It also has been noted that the Kalman filter algorithm can be applied to non-normal data

10. Acknowledgements

The contribution made by LAI Feizhou, while he was a graduate student under the direction of the author has greatly helped in the preparation of this paper The water quality data used in the study was provided by the Environmental Protection Department of Hong Kong while the rainfall data was provided by the Hydrometeorology Section of the Royal Observatory of Hong Kong Their co-operation is gratefully acknowledged

11. References

Anselmo, V and Ubertini, L (1979) Transfer function-noise model applied to flow forecasting, *Hydrological Science Bulletin IAHS* **24(3)**, 353-360

Baracos, P C, Hipel, K W and McLeod, A I (1981) Modelling hydrologic time series from the Arctic, *Water Resources Bulletin American Water Resources Association,* **17(3)** 414-422

Boguslavskij, I A (1988) *Filtering and control* Optimization Software, Inc New York, 133-134

Box, G E P and Jenkins, G M (1976) *Time series analysis Forecasting and control* Holden-Day, Oakland, Calif

Capodaglio, A G, Zheng, S, Novotny V and Xin Feng (1990) Stochastic system identification of sewer flow models *J Env Eng ASCE* **116(2)** 284-298

Clarke, R T (1974) The representation of a short period of experimental catchment data by a linear stochastic difference equation, *Proc of the Warsaw Symposium on Mathematical models in Hydrology July 1971 International Association of Hydrological Sciences,* **1** 3-16

Diggle, P J (1990) *Time series A Biostatistical Introduction,* Clarendon Press Oxford

Hipel, K W, McLeod, A I and Noakes, D J (1982) Fitting dynamic models to hydrological time series, In A H El-Shaarawi and S R Esterby(eds) *Time Series Methods in Hydrosciences* Elsevier Scientific Publishing Company, Amsterdam, Netherlands, pp 110-119

Hong Kong Government Water Supplies Department (1959~1986) Hong Kong Rainfall and Runoff

Huck, P M and Farquhar G J (1974) Water quality models using Box-Jenkins method *J Env Eng ASCE* **100(3)**, 733-753

Jayawardena, A W and Lai Feizhou (1989) Time series analysis of water quality data in Pearl River, China, *J Env Eng ASCE,* **115(3)**, 590-607

Jayawardena, A W and Lai Feizhou (1991) Water quality forecasting using an adaptive ARMA modelling approach, *Proc International Symposium on Environmental Hydraulics* Dec 1991, Hong Kong, 1121-1127

Kahl, D R and Ledolter J (1983) A recursive Kalman filter forecasting approach *Management Science* **29(11)**, 1325-1333

Kalman, R E (1960) A new approach to linear filtering and prediction problem *J Basic Eng*, **82**, 35-45

Ledolter, J (1982) Recursive estimation and adaptive forecasting in ARIMA models with time varying coefficients In D Findley (Ed) *Applied Time Series II* Academic Press, New York, pp 449-472

Lohani, B N an Wang, M M (1987) Water quality data analysis in Chung Kang River *J Env Eng ASCE,* **113(1)**, 186-195

Meade, N (1985) Forecasting using growth curves - An adaptive approach *J Operational Research Society*, **36(12)**, 1103-1115

Ngan, P (1985) Kalman Filter and its Application to Flow Forecasting, Thesis present to the University of British Columbia, Vancouver, B C , Canada, in partial fulfilment of the requirements for the degree of Doctor of Philosophy

Ngan, P and Russell, S O (1986) Example of flow forecasting with Kalman filter *J Hyd Eng ASCE*, **112(9)**, 818-832

Novotny, V and Zheng, S (1989) Rainfall-runoff function by ARMA modelling *J Hyd Eng ASCE*, **115(10)**, 1386-1400

O'Connell, P E (1980) (Ed) Real-Time Hydrological Forecasting and Control, *Proceedings of 1st International Workshop held at the Institute of Hydrology* Wallingford, Oxon, July, 1977

O'Connell, P E and Clarke, R T (1981) Adaptive hydrological forecasting - a review, *Hydrological Science Bulletin IAHS*, **26(2)**, 179-205

Pierce, D A and Haugh, L D (1977) Causality in temporal systems, *J Econometrics*, **5**, 265-293

Sastri, T (1986) A recursive algorithm for adaptive estimation and parameter change detection of time series models, *J Operational Research Society*, **37(10)**, 987-999

Snorrason, A , Newbold, P and Maxwell, W H (1984) Multiple input transfer function-noise modelling of river flow, In W H C Maxwell and L R Beard (Eds), *Frontiers in Hydrology* Water Resources Publishing, Littleton, Colorado

Todini, E and Bouillot, D (1976) A rainfall-runoff Kalman filter model, inter estimation algorithm for real-time hydrologic forecasting A case study, paper presented at the Workshop on Recent Developments in Real-Time Forecasting Control of Water Resource Systems *Int Inst for Appl Syst Anal* Luxembourg, Austria, Oct 1976, 18-20

Winkle, R L an Makridakis, S (1983) The combination of forecasts *J Roy Stat Soc Ser A*, **146**, 150-157

Wood, E and Szollosi-Nagy, A (1978) An adaptive algorithm for analyzing short-term structural and parameter changes in hydrologic prediction models, *Water Res Res* **14(4)**, 577-581

Young, P (1984) *Recursive Estimation and Time Series Analysis* Springer-Verlag, Berlin, Heidelberge

Young, P C , Ng, C N , Lane, K and Parker, D (1991) Recursive forecasting, smoothing and seasonal adjustment of non-stationary environmental data, *J Forecasting*, **10(1&2)**, 57-89

TABLE 1a: Summary of marine DO and NO_3 data

Data	Period of fitting ARMA model	Period of sensitivity analysis	Forecasting period
TM2, TM6 DO	May 1982 ~ May 1989 (N=173)	Jun 1989 ~ Nov 1989 (N=13)	Dec 1989 ~ May 1990 (N=13)
TM6, TM8 NO_3	May 1982 ~ Nov 1987 (N=149)	Dec 1987 ~ May 1988 (N=13)	Jun 1988 ~ Nov 1988 (N=13)

TABLE 1b: Summary of rainfall and runoff data

Data	Period of estimating daily mean	Period of fitting ARMA-TF model	Forecasting period
TLCA Flow	Apr 1959 ~ Mar 1985	Apr 1982 ~ Mar 1985	Apr 1985 ~ Mar 1986
NW Flow	Apr 1971 ~ Mar 1985	Apr 1982 ~ Mar 1985	Apr 1985 ~ Mar 1986
Rainfall	Apr 1980 ~ Mar 1985	Apr 1982 ~ Mar 1985	

TABLE 2a: Parameters of the constant coefficient ARMA models

	ARMA model coefficients				ARMA model test		
Data	ϕ_1	ϕ_2	θ_1	θ_2	Q	DOF	$\chi^2_{(0.95)}$
TM2 DO	-0.3593	0.1791	-0.3377	-	27.41	21	32.66
TM6 DO	0.0050	0.0971	-	-	23.08	19	30.14
TM6 NO_3	0.1045	0.0906	-	-	15.89	18	28.87
TM8 NO_3	-0.7049	0.1299	-0.7556	-	23.96	18	28.87

TABLE 2b: Parameters of the constant coefficient ARMA-TF models

	Coefficients ϕ_i			Coefficients θ_j				
Station	ϕ_1	ϕ_2	ϕ_3	θ_0	θ_1	θ_2	θ_3	θ_4
TLCA	0.1800	0.3254	-	0.2736	-0.0196	0.0478	0.0207	0.0160
NW	0.7815	-0.1480	0.1229	1.8453	0.9143	-0.2119	0.1864	-

TABLE 3 Concentrated log-likelihood function (Eq 11) for rainfall and runoff data during the period April 1984 ~ March 1985

Data Series	Concentrated log-likelihood function $\varepsilon = 0\ 0$	Concentrated log-likelihood function $\varepsilon = 0\ 0001/\alpha^2$	Variance σ^2
TLCA Runoff	337 05	-172 11	21 18
NW Runoff	-1289 32	-1268 61	2744 71
Note $Q = \mathrm{diag}(\varepsilon \quad \varepsilon)_{p+q}$ $\alpha^2 = \sigma^2$ = variance of residuals (Eq 10)			

TABLE 4 Comparison of DO and NO_3 forecasts using the long term mean, constant coefficient and adaptive ARMA models (The forecasting error reductions are between constant coefficient and adaptive ARMA models)

TM2 DO one-step-ahead forecasting					TM6 DO one-step-ahead forecasting				
Error	Long term mean	Constant ARMA	Adaptive ARMA	Error reduct	Error	Long term mean	Constant ARMA	Adaptive ARMA	Error reduct
MSE	264 98	259 00	160 68	38 0%	MSE	762 28	732 94	306 51	58 2%
MAE	1409 64	1401 99	1064 77	24 1%	MAE	2430 11	2401 43	1316 13	45 2%
MAPE	15 13	15 04	11 03	26 7%	MAPE	17 96	18 11	9 66	46 6%
TM6 NO_3 one-step-ahead forecasting					TM8 NO_3 one-step-ahead forecasting				
Error	Long term mean	Constant ARMA	Adaptive ARMA	Error reduct	Error	Long term mean	Constant ARMA	Adaptive ARMA	Error reduct
MSE	0 0005	0 0005	0 0003	40 0%	MSE	0 0001	0 0001	0 00005	50 0%
MAE	1 701	1 825	1 177	35 5%	MAE	0 817	0 799	0 562	29 7%
MAPE	95 43	101 56	73 62	27 5%	MAPE	88 35	84 52	53 59	36 7%

TABLE 5 Comparison of forecastıng errors using the daily mean, constant coefficient ARMA model, adaptive ARMA model and constant coefficient ARMA-TF model

Data	Indicator	Daıly mean	Constant ARMA	Adaptıve ARMA	Constant ARMA-TF
TLCA flow	MSE	19 79	13 69	13 18	4 37
	MAE	201 92	159 59	156 67	115 28
	MAPE	121 71	66 82	68 66	63 86
NW flow	MSE	3026 77	2082 98	2028 84	524 46
	MAE	2369 61	1674 90	1669 26	1049 17
	MAPE	265 54	151 01	162 41	41 66

TABLE 6 Comparison of one day ahead forecasting error using constant coefficient ARMA-TF model and time varying coetficient ARMA-TF model

Data	Indıcator	Constant ARMA-TF	Adaptıve ARMA-TF	Error reduction
TLCA flow	MSE	4 37	2 71	37 90%
	MAE	115 28	84 71	26 52%
	MAPE	63 86	38 74	39 34%
NW flow	MSE	524 46	451 19	13 97%
	MAE	1049 17	982 61	6 34%
	MAPE	41 66	40 19	1 61%

SHORT TERM DAILY FLOW PREDICTION FOR TROPICAL WATERSHEDS

Dr M V Somasundaram
Scientist F & Professor
Centre for Water Resources
Anna University, Madras 600 025
India

Dr S K Srinivasan
Professor of Applied Mathematics
Indian Institute of Technology
Madras 600 036
India

ABSTRACT. Short term daily flow prediction for tropical watersheds characterised by seasonal rain was carried out for Aliyabad watershed in Tamilnadu, India. Wiener Multiple time series approach and Venkataraman autoregressive approach were carried out for earlier data and analysis for a period of 14 years from 1958 to 1971 in 1978 and recent data and analysis of 17 years from 1972 to 1988 in 1991. C programs were developed for Fortran programs developed earlier. The major result of the study was the confirmation of the better behaviour of the raw time series in contrast to that of the decomposed time series signifying separately its trend,periodicity,persistence and the random residual components. Also it was observed in confirmity with the earlier work that no significant improvement was achieved by increasing the order of the model from 2 to 4 in the Venkataraman approach and that the error variances were less for southwest monsoon as compared to northeast monsoon prediction analysis. The complexity of rainfall prediction was also observed as compared to flow prediction. In the Wiener approach the error variances for southwest monsoon analysis were close for daily flows than for weekly flows. For prediction of daily rain and weekly flows Venkataraman approach yielded better results. In C programming the CPU time consumption was 96 seconds for both Wiener and Venkataraman approaches. Individually Venkataraman approach consumed 47 seconds and Wiener approach consumed 61 seconds, the total time being 108 seconds. For earlier data analysis in Fortran IV programming corresponding CPU time for Venkataraman approach was 62 seconds, for Wiener approach 75 seconds, the total time being 137 seconds.

1 INTRODUCTION

The process by which rainfall is transformed into flow is complex and uncertain in space and time. In the stochastic approach we have the two broad categories of analysis, namely, the Kendall approach(Kendall,1976) and the Wiener approach (Wiener,1949). In the Kendall approach the time series is decomposed into its components of trend, periodicity, persistence and random residual, studied and finally superposed for prediction purposes. On the other hand, the Wiener approach in the frequency domain is dynamic in nature and no such decomposition is done. Later Kalman developed the adaptive forecasting techniques in the time domain (Kalman and Busy,1961). Thomas Kailath has presented an excellent review of 3 decades of linear filtering theory (Thomas Kailath, 1974).

V. P. Singh and B. Kumar (eds.), Surface-Water Hydrology, 443–456.

The methods of prediction are dependent on the time span with which we are working on or of the purposes to which results are used. Various prediction methods used depend upon the circumstances under consideration. They may be broadly classified as short term, medium term or on long term basis depending on the time span under consideration. Long term prediction is difficult as compared to short term or medium term prediction.

It must be emphasized at this point that theoritical hydrologists all over the world are continuously developing new deterministic and stochastic models with increasing degree of sophistication and mathematical precision. This wave of mathematical modelling in pure hydrology is no doubt necessary for the ordely development of water resources on a long term basis, but as the practicing engineer cannot wait user oriented models must be given due attention. One such area where there is no attainment of significant progress is the prediction of watershed flow contributed by rain for semiarid regions characterised by seasonal rainfall. The absence of information on prediction is still more when the watershed is analysed as a stochastic system treating the input rainfall and output flow as uncertain phenomena. The reasons for lack of study in this much needed area in the past were due to inadequate data of sufficient length and quality, nonavailability of computer time and objective procedures for flow prediction and algorithms for saving computer time and space in data banks.

2 DESCRIPTION OF STUDY AREA

Aliyabad anicut was constructed in the year 1866 across Kamandalaru to facilitate diversion of water from the river to 16 tanks for irrigating a command area of 957 ha. A catchment area of around 190 Km^2 was observed to be under the influence of the Aliyabad rainfall station (Balaiya, 1992). The rainfall and flow gauging stations are located at a latitude of 12° 30' 50" north and longitude 79° 07' 24" east. The area receives rain from southwest about 45 %, northeast 40 %, winter 5 % and summer 10 % on the average. The average annual rainfall based on 35 years record from 1956 to 1990 was 1030 mm. A perusal of the surface flow gauge readings showed that surplus flows were realised mostly during the northeast monsoon period. Surplusses more than $10Mm^3$ during northeast monsoon season were noted in 13 years from a flow record of 45 years. Weathered and jointed Charnokite rocks dominate the study area with medium coarse grained peninsular gneiss including granitic gneiss at many places. Alluvium is restricted to the river courses only. Regional and amudi faults in the area are favourable for ground water extraction. It is inferred that the sustained flow in the river during lean months is due to base flow.

3 DIGITAL PROCESSING OF HYDROLOGIC DATA

A procedure for information storage and retrieval from data banks in the form of two subroutines DATA and ATAD complementary to each other was developed to process hydrologic information (Somasundaram and Srinivasan, 1991). DATA retrieves information from storage for processing. ATAD stores the input information in compact form. Indexed sequential file format was adopted. The procedure resulted in considerable saving of space and computer time in data analysis. Daily rainfall and flow data of Aliyabad anicut for a duration of 31 years from 1958 to 1988 was processed conveniently in Fortran and C using this procedure.

4 INITIAL STATISTICAL ANALYSIS

Basic concepts followed in this study were borrowed from the works of Kendall (Kendall, 1976) and Srinivasan (Srinivasan, 1974, 1976). Statistical analysis of rain and flow was conducted to know the number of rainy days, flow days, rainy weeks, flow weeks and maximum amount of rain and flow in a day and week for southwest, northeast, winter, summer and entire period of year every year for the duration of data of 31 years from 1958 to 1988 (Badarinath,1991). Some observed values for daily rain and flow are summarised in Table 1. The analysis revealed that 85 % of total rainy days and 68 % of total flow days were concentrated in the southwest and northeast monsoon period. Maximum rain occurred 11 times during southwest monsoon and 18 times during northeast monsoon. Maximum flow ocurred 4 times during southwest monsoon and 23 times during northeast monsoon. The minimum and maximum annual rainfall were 41 and 209 mm. The corresponding southwest values were 24 and 197 mm. The corresponding northeast values were 21 and 209 mm. The minimum and maximum annual flows were 2 and 162 m^3/s. The corresponding minimum and maximum southwest flows were 0.71 and 62 m^3/s. The corresponding minimum and maximum northeast flows were 0.43 and 162 m^3/s. This kind of analysis was useful in the sense that the prediction methods were insensitive to very low and high values of daily rain and flow and were able to accomadate values only in a bandpass range in the linear mean square sense.

TABLE I INITIAL ANALYSIS OF DAILY RAIN AND FLOW

	Annual				Southwest				Northeast			
	RD	FD	R	F	RD	FD	R	F	RD	FD	R	F
Minimum	27	63	41	2	8	3	24	0.71	8	2	21	0.43
Maximum	63	282	209	162	32	73	197	62	30	92	209	162
Average	46	147	103	33	19	28	72	11	17	70	87	33

RD : Rainy days
FD : Flow days
R : Maximum daily rainfall in mm
F : Maximum daily flow in m^3/s

In the earlier data analysis for the 14 year period from 1958 to 1971, compressed computer plots of daily rain and flow, parameters of central tendency, dispersion, symmetry, peakedness, flatness and other sensitive characteristics upto the sixth moment, time and ensemble averages, auto and cross covariance plots and corresponding power spectral density plots, absolute spectra and coherence functions were generated to assess the behaviour clearly before rushing into prediction.

The compressed computer time plots of daily rain and flow showed that the series were highly nonstationary endowed with periodicities, persistence, abnormal components and random residuals. There were no sudden jumps or visible trends in the plots visible to the naked eye but a subsequent cubic trend analysis in the least square sense exhibited a weak trend on a long term basis.

The covariance plots followed by the raw and smooth power spectral density plots revealed that the significant harmonics were only 3 in the data and it is not nececessary to go in for more than 5 or 6. A sudden decrease in covariances from zero time lag to other time lags was noticed in all the entire and seasonal plots. The sharp peaks at lower frequencies in all spectral plots also substantiated this information.

Studying the within-the year-movement, the daily rain and flow time series behaved like some typical biological processes which grow and die in a pattern. Four stages of growth namely the quiescent winter season (January 1 to March 31), summer season (April 1 to May 31), southwest monsoon season (June 1 to September 30) and northeast monsoon season were identified. This growth pattern was more prominent for rain than for flow which was artificially controlled depending on needs.

5 STATIONARITY ANALYSIS

Based on the assumption that every nonstationary stochastic process contains fairly good intervals of stationary subsequences attempts were made to extract the same from the original parent time series of rain and flow. At the present state of knowledge no clearcut procedure is available in this direction. The method usually followed is a least square procedure to identify and remove the trend, a harmonic analysis to identify and remove the periodicities and Markoff models to identify and remove the dependence thereby arriving at an artificial series which is subjected to further analysis. In most cases this artificial series is either random or of a smooth oscillatory kind as Kendall has reported. These series of transformation although useful in studying the behaviour of the underlying phenomenon in terms of its trend, periodicity and persistence separately, it has not been used sucessfully in flow prediction through autoprojection or multiple time series approach. The differencing scheme suggested by Box and Jenkins to take care of nonstationarities resulted in series with variance tending to infinity (Kashyap and Ramachandra Rao, 1976). Either the normalisation procedure advocated by deducting the ensemble mean and dividing by the ensemble standard deviation from the original series or the harmonic analysis for identifying and removing the periodicities in the ensemble mean and standard deviation (Yevjevich,1966) did not succeed in the analysis of Aliyabad watershed data as most of the days there were no rain or flow. The running of the computer programme was interrupted due to zero values of standard deviations and squre roots of arguements being negative in the evaluation of significant harmonics in periodicity removal. Subjective interpretations were necessary in circumventing these errors and making the programme work.

In this study using a trial and error procedure through the computer system and information from practicing engineers attempts were made to extract stationary subsequences from the original nonstationary time series of rain and flow. The southwest and northeast seasonal time series of rain and flow together and seperately drawn from the parent series ignoring the edge effects tended towards stationarity. While it was easy to identify the seasons it was difficult to fix the duration of the seasons. The onset and withdrawal dates of monsoons vary from year to year and monsoon vagarism needs detailed study (Somasundaram, Duraisaminathan and Pundarikanthan, 1993). To justify this approach to stationarity the time and ensemble averages were tested for significance using the distribution free nonparametric run test. It was seen that the parent series of rain and flow were nonstationary based on the runs for ensemble mean and ensemble root mean square values at 5 % significance level. But on the other hand the time values of mean and root mean square wre constant. The runs for the seasonal series indicated a strong tendency towards stationarity in the mean and root mean square values but still nonstationary at 5 % significance level.

6 VENKATARAMAN APPROACH

A class of linear stochastic models of autoregressive nature not necessarily stationary advocated by Venkataraman (Venkataraman, 1975) can be conveniently written as

$$X_{t+n} = \alpha_1 X_{t+n-1} + \dots + \alpha_n X_t + \Gamma_1 t^3 + \Gamma_2 t^2 + \Gamma_3 t + + \Gamma_4 + (t+b)\epsilon_{t+n} \quad \dots \; n = 2,3,4 \qquad (1)$$

where n is the order of the model, α's, the autoregressive parameters, Γ's the trend parameters, b, a constant and ϵ the the disturbance term. The disturbance term ϵ designate a class of independent identically distributed random variables with zero expectation and positive finite variance. The sinusoidal terms accounting for periodicities were not considered as already the seasonal series were extracted from the parent series reducing the cyclicities to a minimum. The trend terms were also not considered in the first instance as the compressed computer plots did not exhibit any long term trend and the subsequent cubic trend analysis revealed only a weak trend as already reported. Hence for simplicity the revised model can be written as,

$$X_{t+n} = \alpha_1 X_{t+n-1} + \ldots + \alpha^n X^t + \Gamma + \epsilon_{t+n} \qquad (2)$$

The autoregressive parameter, α's and the constant term Γ can be determined on the principle of least squares minimising the expression,

$$\sum_{t=1}^{N_p-n} [\ X_{t+n} - \alpha_1 X_{t+n-1} - \ldots - \alpha_n X_t - \Gamma\ ]^2 \qquad (3)$$

where N_p is the number of observations in the past. Differentiation by the α's and Γ yielded the required normal equations, the summations being carried out over (N_p-n) observations of time series X_t. Simultaneously solving these normal equations, the estimates of the parameters α's, and Γ were computed.

7 WIENER APPROACH

This approach is based on the assumption that both the rainfall and flow are strictly stationary stochastic processes pertaining to a physically realizable, stable, time invariant linear system. Mikio Hino fitted special functions for covariances and then derived the kernel function (Hino, 1970). The numerical application of Wiener-Hopf equation (Wiener, 1949) involves the determination of generalised unit hydrograph (response function) obtained by the product of inverted rainfall autocovariance matrix (input autocovariance matrix) and the cross covariance matrix of rainfall to flow (input to output) and convolving it with the rainfall function to arrive at the desired flow function.

Let the rainfall stochastic process X(t) and the flow stochastic process Y(t) obey the following equations.

$$X(t) = X_s(t) + Y_n(t) \qquad (4)$$

$$Y(t) = Y_s(t) + Y_n(t) \qquad (5)$$

where $X_s(t)$ and $Y_s(t)$ are the seasonal stocahstic processes of rainfall and flow and $X_n(t)$ and $Y_n(t)$ are nonseasonal stochastic processes of rainfall and flow. It is assumed that the seasonal and nonseasonal stochastic processes expressed as sums are independent and that they can be separated from the parent stochastic stochastic processes ignoring the edge effects which are a minimum. Only the seasonal stochastic processes are subjected to further analysis.

The numerical solution of the Wiener-Hopf equation in matrix form (Hino,1970) can be written as:

$$[\ h_{opt}\ ] = [\ R_{xx}\ ]^{-1}\ [R_{xy}] \qquad (5)$$

where h_{opt} is the required weighting fuction or the generalised unit hydrograph and R_{xx} is the autocovariance of rainfall evaluated from the equation:

$$R_{xx}(k) = 1/N\text{-}k \sum_{i=1}^{N-k} X_i X_{i+k} \quad \ldots\ k = 0,1,2,\ldots,m-1 \qquad (6)$$

and R_{xy} is the crosscovarince of rainfall to flow evaluated from the equation:

$$R_{xy}(k) = 1/N\text{-}k \sum_{i=1}^{N-k} X_i Y_{i+k} \quad \ldots\ k = 0,1,2,\ldots,\ m-1 \qquad (7)$$

m being the value of maximum lag considered and N is the total number of observations. After evaluating the weighting function h_{opt}, it is convolved with the rainfall to obtain the predicted flows through the following convolution criterion:

$$Y_j = h_1 X_j + h_2 X_{j-1} + \ldots + h_i X_{j-1+1} \qquad (8)$$

where Y_j is the value of the flow predicted at time j.

Initially the entire data of rainfall and flow was used in determining the covariance matrices. The size of the rain autocovariance matrix inverted was 366 by 366. Inversion of such large matrix presented problems like loss of significance in computations. This loss of significance was due to the nature of physical phenomena such as concentration of rain only on southwest and northeast monsoon and scanty or zero rainfall for most of the days for the rest of the year. Thus only seasonal rainfall were subjected to further analysis. Accordingly the size of the rainfall autocovariance inverted for southwest was 123 by 123 and for northeast 93 by 93 for daily analysis and 18 by 18 and 14 by 14 for weekly analysis. Sample results are presented for daily and weekly model representation and model forecasting.

8 MODEL REPRESENTATION AND FORECASTING

In the earlier analysis conducted in 1978 (Somasundaram, 1978) the 14 year data from 1958 to 1971 was conveniently split into two parts for model representation and analysis. The initial 10 year data from 1958 to 1967 was referred as the past data and was used for model representation and evaluation of model parameters.

Model representation was checked comparing the observed

series with the generated series for the past. Then using these parameters, series in the future for the next 4 years was generated and compared with the unused observed data from 1968 to 1971 to determine model validity in future. Statistical parameters and error variances for the generated past series and future series were compared with the observed series for model representation and model forecasting abilities.

This exercise was repeated for recent data of 17 years duration from 1972 to 1988.

8 RESULTS

Due to limitation of space only representative results are given. The error varinces to compare both the performance of model representation and model forecasting was conveniently defined as:

$$\sigma_p^2 = 1/N_p-n \sum_{t=1}^{N_p-n} (x_{t+n} - x_{t+n})^2 \quad (9)$$

$$\sigma_f^2 = 1/N-N_p \sum_{t=N_p+1}^{N} (x_{t+n} - x_{t+n})^2 \quad (10)$$

where N is the total number of records, N_p the number of records used in the past data for model representation, n the order of the model in autoregressive schemes if applicable otherwise zero, x_{t+n}, the values of observed data and x_{t+n} the generated data for model representation or model forecasting as the case may be. σ_p is the error variance for comparing model representation and σ_f, the error variance for comparing model forecasting performances.

Some typical values of error variances for seasonal analysis are presented in Tables 2,3,4 and 5 for comparison. A typical plot of comparison of observed and computed values one each for Venkataraman approach and Wiener approach are presented in Figs 1 and 2.

TABLE 3 ERROR VARIANCES OF VENKATARAMAN AUTOREGRESSIVE SCHEMES FOR SOUTHWEST MONSOON

	Daily Rain		Daily Flow		Weekly rain		Weekly rain	
Model	MR	MF	MR	MF	MR	MF	MR	MF
VAR2	109.3	132.2	2.5	1.1	128.4	152.1	8.9	8.9
VAR3	107.2	130.5	2.5	1.1	128.5	156.5	8.9	6.9
VAR4	106.6	132.8	2.5	1.1	128.3	151.9	3.9	3.0

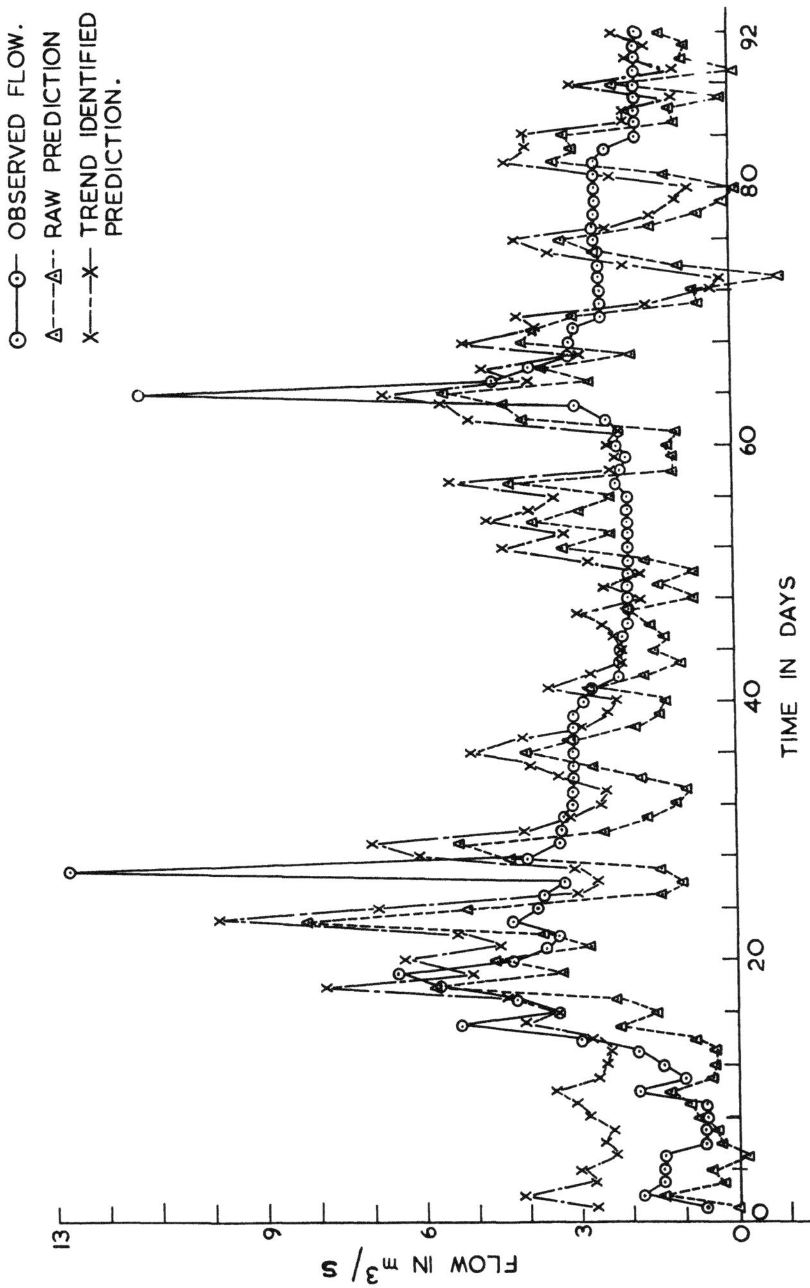

FIG.1 MODEL REPRESENTATION OF DAILY FLOW; YEAR 1962 WIENER APPROACH

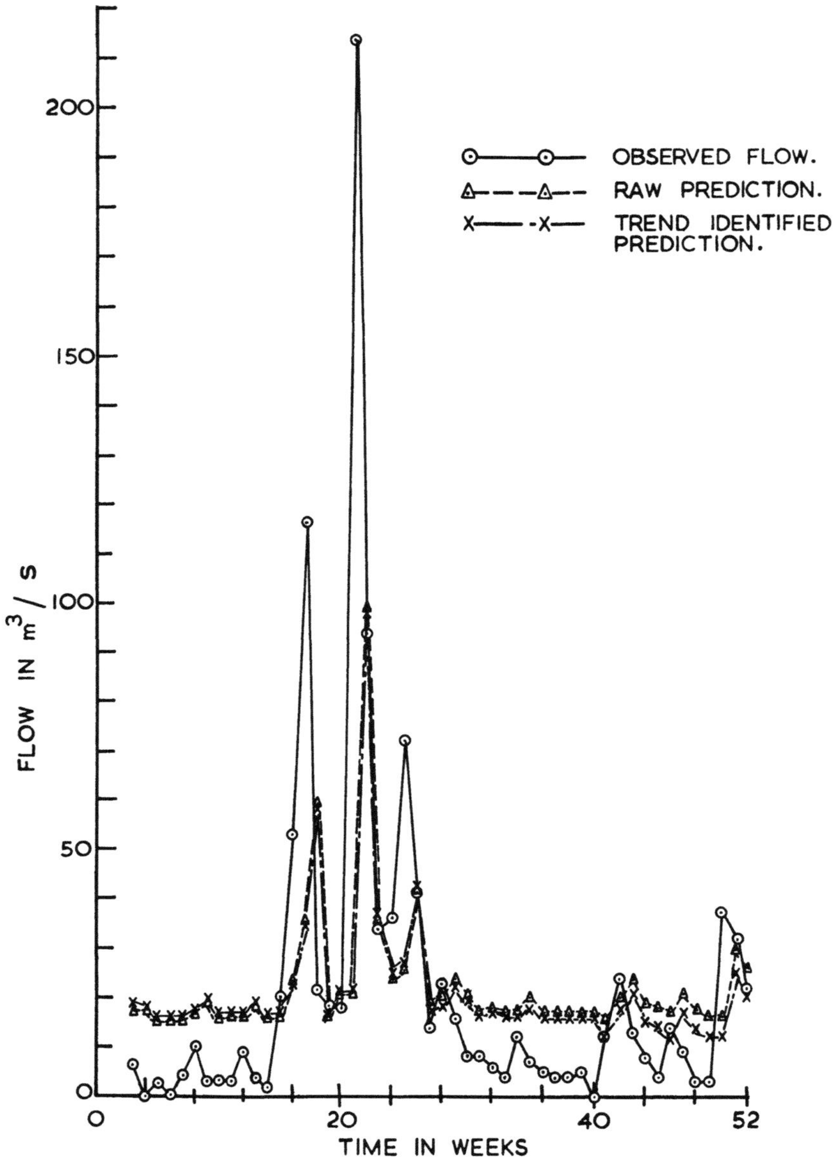

FIG. 2 EXTRAPOLATION OF WEEKLY FLOW: SECOND ORDER AUTOREGRESSIVE SCHEME.

TABLE 4 ERROR VARIANCES OF VENKATARAMAN AUTOREGRESSIVE SCHEMES FOR NORTHEAST MONSOON

Model	Daily Rain MR	Daily Rain MF	Daily Flow MR	Daily Flow MF	Weekly rain MR	Weekly rain MF	Weekly rain MR	Weekly rain MF
VAR2	215.8	153.7	42.4	11.7	3319.0	4363.8	1203.0	1178.3
VAR3	214.5	155.2	41.9	11.4	3330.3	4439.4	1187.7	1266.0
VAR4	211.5	161.7	41.8	11.3	3344.5	4520.4	1101.8	1082.1

TABLE 5 MODEL FORECASTING ERROR VARIANCES OF WIENER APPROACH FOR MONSOON DATA

YEAR	SOUTHWEST MONSOON			NORTHEAST MONSOON		
	Daily	Weekly	Weekly/Daily	Daily	Weekly	Weekly/Daily
1984	0.4	372.7	945.6	8.5	4.1	232.2
1985	1.4	742.2	730.4	4.5	1.0	97.2
1986	2.1	421.3	860.3	2.4	3.1	42.1
1987	0.8	192.8	2111.3	4.1	3.2	420.8
1988	2.9	812.7	1604.7	3.3	3.8	13.8

TABLE 6 COMPARISON OF CPU TIME FOR EARLIER DATA (1958-71)

Case	IBM 350/155 FORTRAN IV	MAGNUM II C (UNIX)
	CPU time in seconds	
1	62	47
2	43	33
3	75	61
4	32	40
5	37	25
6	34	28
7	16	11
8	15	9
9	NA	10
10	14	8

MR : Model Representation
MF : Model Forecasting
VAR2 : Venkataraman second order autoregressive scheme
VAR3 : Venkataraman third order autoregressive scheme
VAR4 : Venkataraman fourth order autoregressive scheme

Weekly/Daily : Weekly series extracted from predicted daily series.
Case 1 : Venkataraman Autoregressive scheme for trend identified data
Case 2 : Venkataraman Autoregressive scheme for raw data

Case 3 : Wiener approach for model representation of trend identified daily data
Case 4 : Wiener approach for model representation of raw daily data
Case 5 : Wiener approach for model forecasting of trend identified daily data
Case 6 : Wiener approach for model forecasting of raw daily data
Case 7 : Wiener approach for model representation of trend identified weekly data
Case 8 : Wiener approach for model representation of raw weekly data
Case 9 : Wiener approach for model forecasting of trend identified weekly data
Case 10 : Wiener appraoch for model forecasting of raw weekly data

NA : Not available

9 DISCUSSION

Compensating behaviour of southwest and northeast monsoon within the calender year was observed. The error varinces for southwest data analysis were generally less than for northeast data characterised by thunder storms.Also the error variances for daily analysis were generally less than for weekly analysis. No significant improvement was observed in increasing the order of the model from 2 to 4 in the autoregressive approach. The error variances for weekly rain analysis and prediction were considerably more than other cases indicating the complexity of rainfall prediction. For modelling weekly seasonal flows second order autoregressive schemes were better than multiple time series approach. For modelling daily seasonal flows Wiener approach is ideally suitable, compared to other approaches.

10 CONCLUSION

The modelling assumes gross rainfall as the input and ignores the abstraction which is seasonally time variant and highly nonlinear. The errors of prediction for a semiarid basin with predominant seasonal rainfall can be justified as a holistic approach is attempted in this regard. The daily rain and flow processes are closer to white noise processes than hourly or weekly processes (Hino, 1970) . The study further reveals that prediction with raw data is better compared to component identified predictions. The algorithms and programs developed for Wiener and Venkataraman approaches in this study can be easily fitted with any interactive general purpose river basin system analysis package which are user friendly and can be straight away applied to specific project planning studies and simulation (GAMS, 1992).

ACKNOWLEDGEMENT

The authors thank Professor Satish Chandra, Director, National Institute of Hydrology, Roorkee for the constant and sustained encouragement given for this user-oriented study from the early 1970's. Thanks are also due to Professor N V Pundarikanthan, Director, Centre for Water Resources, Anna University for the encouragement and facilities placed at their disposal for this study on a permanent basis.

REFERENCES

1 Anthony Brooke.,David Kendrick.,Alexander Meeraus (1992)., 'GAMS-A user's guide,Release 2.25, The scientific press,South San Francisco,USA.

2 Badarinath A (1991).,' Short term flow prediction for tropical watersheds using C language', ME thesis, Centre for Water Resources, Anna University, Madras, pp 1-59.

3 Balaiya K (1992).,'Study of Aliyabad anicut system',ME thesis,Centre for Water Resources, Anna University, Madras, pp 1-64.

4 Hino M (1970).,'Runoff forecasts by linear predictive filter' Journal of the Hydraulics Division, ASCE, Vol 96, Hy 3, pp 681-702.

5 Kalman R E and Busy R S (1961).,'New results in linear filtering and prediction theory', Transactions of ASME, series D,J, Basic Engineering, Vol 83, pp 95-107, Dec 1961.

6 Kashyap R L and Ramachandra Rao A (1976).,'Dynamic stochastic models from empirical data', Academic Press, New York

7 Kendall M G (1976).,'Time series', Charles Griffin and Company Ltd, London and High Wycombe.

8 Somasundaram M V (1978).,'Short term flow prediction for semiarid watersheds characterised by seasonal rainfall',Ph.D thesis, Hydraulics and Water Resources Department, College of Engineering, Guindy, Madras University, pp 1-183.

9 Somasundaram M V., Duraisaminathan V., Pundarikanthan N V (1993)., ' Temporal and spatial behaviour of monsoon rainfall in a semiarid tropical shadow region, Ppaper presented at the international conference on sustainable development strategies and global/regional/local impacts on atmospheric composition and climate', January 25-30, New Delhi.

10 Somasudaram M V., Sakthivadivel R., Srinivasan S K (1978)., 'Short term flow prediction using Wiener approach for semiarid watersheds characterised by seasonal rainfall', proceedings of

the Third world congress, Mexico.

11 Somasundaram M V., Srinivasan S K.,(1991).,'Digital processing of hydrologic data', Proceedings of the national seminar on use of computers in hydrology and water resources, Central Water Commision, New Delhi, pp 1-35 to 1-48.

12 Srinivasan S K (1974).,'General theory of stochastic processes', Stochastic problems in mechanics, University of Waterloo Press, study No 10, pp 1-36.

13 Srinivasan S K and Mehata K M (1976).,' Stochastic Processes' Tata McGraw-Hill, New Delhi.

14 Thomas Kailath (1974).,'A view of three decades of linear filtering theory', IEEE transactions on information theory, USA, Vol IT 20, No 2, March issue, pp 145-181.

15 Venkataraman K N (1975).,'Some convergence theorems of certain linear stochastic models of autoregressive nature', Journal of mathematival and physival sciences, Vol IX, No 4, pp 305-331.

16 Wiener N (1949).,'Extrapolation, interpolation and smoothing of stationary time series - with engineering applications' MIT Press.

17 Yevjevich V (1972).,'Stochastic Processes in hydrology', Water resources Publications, Fort Collins, Colorado.

Role of the Rainfall Input in the Real Time Flood Forecasting

Krashnendra Singh & Mandakinee Majumdar

Department of Civil Engineering,
Indian Institute of Technology, Kharagpur - 721302. INDIA

ABSTRACT

The problem of real time flood forecasting for a system with ungauged tributaries and scanty rainfall information is taken up for study. An ARMA type channel routing model is implemented for 3 hourly (single step) ahead flood forecasting on river Barakar, a tributary of river Damodar, India. The channel routing model yields better results than a simple rainfall runoff model mainly because of inadequacy of the available rainfall data, its associated uncertainties, and complexity of the hydrologic processes involved. However, the routing type model for the system reported herein shows a typical unit lag in the estimation of flood peaks and also underestimates the discharge on the rising limbs of flood hydrographs. A model based on filter separation AR method, a rainfall runoff model of ARX type, which is reported to effectively take care of the system nonlinearity, is implemented. The rainfall prediction based on the inverse estimated rainfall components is found to substantially improve the results. The simulation of daily inflows is done using a rainfall runoff model, the nonlinearity of the process being handled through rainfall transformation. The low pass and threshold type rainfall filters are implemented and the results compared.

INTRODUCTION

The real time flood forecasting in a river basin is a difficult task because of the nonlinearity and nonstationarity of the complex stochastic processes involved. The unavailability or lack of reliable representative information of some of the system inputs, e.g. the tributary inflow, rainfall, etc., make the task of accurate forecasting even more complicated.

The rainfall runoff models [1,2,3,13] depend heavily on the quality of the rainfall data. Further, it is difficult to faithfully model the temporal and spatial variabilities of within the basin

V. P. Singh and B. Kumar (eds.), Surface-Water Hydrology, 457–471.

processes. It is observed that the real time forecasting based on channel routing of flood hydrographs is generally more accurate compared to the one based on rainfall runoff models [2]. In recent times, river flow routing type models with single and multiple inputs [8,9,11] are proposed in the literature for the purpose. The accuracy of a routing type model depends upon availability of a properly placed upstream gauging station and the information regarding the lateral inflow contributions to the reach.

In a situation where there is no suitable upstream station available or there is substantial lateral inflow into the reach, the rainfall runoff type models are indispensable in spite of low accuracy. The filter separation AR method [5,6] is a rainfall runoff model in which the streamflow is separated into its components, the quick flow and base flow, and rainfall corresponding to each component can be estimated once the model is built up. Simple rainfall prediction technique is reported to yield good results for a real time forecasting problem.

The rainfall runoff models offer the only possible flood simulation approach for small and medium size catchments for relatively long term forecasting. Such models are widely used for daily (24 hourly) discharge simulation [1,13,14]. In view of the nonlinearity of the system, mainly attributable to the infiltration process - separation of rainfall contributing to various run off components, various rainfall transformations have been tried out. Among these, threshold (ϕ index) type [14], low pass filter [7,14] type, separation into multiple series [3] are the important ones.

In the present study, the performance of a channel routing type model and a filter separation AR method are studied for the purpose of real time forecasting (time step of 3 hours). The worth of the observed and the predicted rainfall is also studied. In addition, a daily (24 hour time step) simulation study is also reported wherein various rainfall transformations are examined.

METHODOLOGY

THREE HOURLY STREAMFLOW FORECASTING

The channel routing type model for real time flood forecasting implemented herein is an ARMA (auto regressive moving average) model given below :

$$y_t = a_1 y_{t-1} + a_2 y_{t-2} + \ldots + a_p y_{t-p} + \\ + b_1 u_{t-1} + b_2 u_{t-2} + \ldots + b_q u_{t-q} + \varepsilon_t \qquad \ldots (1)$$

where y : discharge at the base station,
u : discharge at an upstream station,
a and b : the model coefficients,
p and q : the model order, and

ε : the noise term.

Such a model when implemented for the system identified earlier, underestimates the discharge on the rising limbs of flood hydrohgraphs and shows a unit time lag in the estimation of flood peaks. This is mainly because of the substantial inflow contribution into the reach and the apparent travel time for the reach is less than the prediction time step.

An effort is made to incorporate the rainfall information in model (1) thereby resulting into an ARMAX model of the form:

$$y_t = a_1 y_{t-1} + a_2 y_{t-2} + \dots + a_p y_{t-p} + b_1 u_{t-1} + b_2 u_{t-2} + \dots + b_q u_{t-q} + c_1 x_{t-1} + c_2 x_{t-2} + \dots + c_r x_{t-r} + \varepsilon_t \quad \dots(2)$$

where the exogenous component in model (2), that is x, is the average rainfall corresponding to the lateral contribution, r the order and c's, the associated coefficients. In absence of reliable real time rainfall data for this, the average rainfall for the entire Barakar river basin may be used in model (2) presuming that the rainfall pattern in the basin is more or less uniform. This model results in marginally improved flood forecasting. Inadequacy of the raingauge network, which is quite sparse, real time availability of the data, and highly nonlinear nature of the hydrologic processes involved may be held responsible for such a behaviour.

The filter separation AR method [5,6] presumes that the two subsystems involved in the rainfall runoff process, giving rise to surface and interflow or quick flow component and the ground water or base flow component, can be adequately represented by an ARX model of the form :

$$y_t^{(l)} = a_1^{(l)} y_{t-1}^{(l)} + a_2^{(l)} y_{t-2}^{(l)} + \dots + a_p^{(l)} y_{t-p}^{(l)} + \lambda b^{(l)} x_{t-1}^{(l)} + \varepsilon_t \quad \dots(3)$$

$$(l=1,2 \; ; \; t=1,2,\dots)$$

where $x^{(l)}$ and $y^{(l)}$ are the lth rainfall and streamflow components, λ the conversion factor given by

$$\lambda = A/3.6 \quad \dots(4)$$

if x and y are expressed in mm/hr and m^3/s and the catchment area A, in square km. The base flow component corresponds to l=1 and the quick flow to l=2. The above ARX model (3) can also be expressed in the Impulse Response form, i.e in a unit hydrograph type of expression.

This model (3) reduces to a simple AR model after the cessation of rainfall. Hence, the AR coefficients a's of the model can be determined from the recession data when there is no rainfall. The coefficient b is given by the continuity equation :

$$b^{(1)} = 1 - a_1^{(1)} - a_2^{(1)} - \dots - a_p^{(1)} \qquad \dots(5)$$

An improved filter for base flow separation [4,10] is used in this study. The quick flow component is given by:

$$y_t^{(2)} = \frac{3\alpha - 1}{3 - \alpha} y_{t-1}^{(2)} + \frac{2}{3 - \alpha} (y_t - \alpha y_{t-1}) \qquad \dots(6)$$

subject to

$$y_t^{(2)} \geq 0 \qquad \dots(7)$$

where y is the original streamflow, and α is the recession constant. The base flow component is the residual, given by:

$$y_t^{(1)} = y_t - y_t^{(2)} \qquad \dots(8)$$

The above base flow separation filter (6)-(8) is much simpler than the low frequency-pass filter [5,6] which simulates the response of a mass-dashpot-spring filtering system. This filter involves only one parameter that can be determined from the streamflow data whereas the latter involves two, one of them being an empirical constant.

Even if reliable and accurate rainfall information is available, the problem of its separation into components corresponding to the subsystems remains since the infiltration is a very complex phenomenon. In this study, the rainfall for the lth component is inverse estimated sequentially from the runoff by

$$\hat{x}_{t-1}^{(1)} = [y_t^{(1)} - a_1^{(1)} y_{t-1}^{(1)} - \dots - a_p^{(1)} y_{t-p}^{(1)}]/(\lambda b^{(1)}) \qquad \dots(9)$$

where $\hat{x}$ stands for the inverse estimated value of the rainfall. The variation in the base flow component is slow due to damping nature of the infiltration process. Thus a simple extrapolation model is used to predict the corresponding single step ahead rainfall component:

$$x_{t+1}^{,(1)} = \rho^{(1)} * \hat{x}_t^{(1)} \qquad \dots(10)$$

The rainfall component for the quick flow, which shows relatively violent variations, can be estimated using a similar approach based on average rainfall component:

$$x_{t+1}^{,(2)} = \rho^{(2)} * [\sum_{i=1}^{m} \hat{x}_{t-i+1}^{(2)}]/m \qquad \dots(11)$$

This component can also be predicted from the observed rainfall as below, subject to nonnegativity.

$$x'^{(2)}_{t+1} = \mathbb{F}\, x_t - x'^{(1)}_{t+1} \qquad \ldots(12)$$

where $\mathbb{F}$ is the runoff coefficient. When the observed rainfall is not highly reliable, the inverse estimated rainfall can be used to predict the future effective rainfall component using equation (12) and dropping the last term from it.

DAILY STREAMFLOW SIMULATION

The daily modelling of streamflow process is particularly difficult because of the nonlinear and time variant hydrologic processes involved. The soil moisture, evapotranspiration, storage, etc., affect the system behaviour in a very complex fashion. Two types of rainfall transformations, a low pass filter and a threshold type, are considered with a view that it will allow the system to be satisfactorily handled through a linear model. The daily streamflow may be represented by an ARMA model:

$$y_t = a_1 y_{t-1} + a_2 y_{t-2} + \ldots + a_p y_{t-p} + $$
$$+ b_0 x_t + b_1 x_{t-1} + \ldots + b_q x_{t-1} + \varepsilon_t \qquad \ldots(13)$$

where y is the streamflow, x the filtered (effective) rainfall and ε is the noise term.

The low pass filter empirically makes allowance for varying soil moisture conditions. It is given by [7,14]:

$$s_t = s_{t-1} + (r_t - s_{t-1})/T \qquad \ldots(14)$$

$$x_t = s_t r_t / u_f \qquad \ldots(15)$$

where x : transformed rainfall,
r : observed rainfall,
s : soil moisture parameter,
T : time constant, and
u_f : the scaling factor.

In the threshold transformation, a constant amount is deducted from the rainfall for each time step. Mathematically,

$$\left.\begin{aligned} x_t &= r_t - \phi && \text{if } r_t > \phi \\ x_t &= 0 && \text{if } r_t \le \phi \end{aligned}\right\} \qquad \ldots(16)$$

where ϕ is the threshold rainfall.

The statistics used to evaluate the forecasts should be separated according to periods of hydrograph and rise and recession [2]. Two separate indices are evaluated in this study :

EF - the normalized RMS (root mean square) *E*rror for the *F*ull series, and

ER - the normalized RMS *E*rror for *R*ising limbs only

The normalization is done by dividing the series by its average value.

APPLICATION

The proposed models are set up for the base station Nandadih on river Barakar, a tributary of river Damodar, India. The discharge of the immediate upstream station Palganj is used in the channel routing model. The catchment making lateral contribution to the reach is around 15% of the total catchment area (5400 square km). A major tributary river Usri drains a part of this area into the reach. Three hourly discharge records of Nandadih are available for seven years, numbered 1 to 7 for convenience, and corresponding discharge records of Palganj are available for six years, namely 1, 2, 3, 4, 5 and 7. The rainfall records, also 3 hourly, are readily available for years numbered 2, 3, 4, 5, 6 and 7 for five stations in this catchment. Since the records are available for the monsoon months only, the analysis has been carried out on a season to season basis.

RESULTS AND DISCUSSION

THREE HOURLY MODEL

An ARMA(2,3) model is found to be suitable for 3 hourly Nandadih discharge forecasting [12]. This model underestimates the flood discharge during rising parts of the hydrographs and almost invariably shows a unit (3 hours) lag in the estimation of flood peak magnitude. The correlogram in figure 1 shows the autocorrelation function of Nandadih discharge and its cross-correlation with Palganj (upstream station) discharge and the average observed rainfall for 3 hourly data of year No. **7**. This typical correlogram clearly indicates that inclusion of rainfall into an ARMA model given by equation (1) can bring about only marginal improvement in model performance.

Extensive studies carried out using an ARMAX(2,3,3) model confirm the above observation, however in the rising limb region, the improvement in performance is substantial if the average basin rainfall is used in model (2) in place of vary limited information for the reach alone. Indices EF and ER are compared for ARMA(2,3) and ARMAX(2,3,3) models in table I. It is obvious that an ARMAX model results in much improved flood forecasting, particularly in the rising limb region. It may be noted that the rainfall data for year **4** is available only for four sites and two of these have six hourly records. For this year, the ARMAX model is just as good as ARMA model. Thus it can be concluded that the ARMAX model performance is heavily dependent on the quality of rainfall information.

TABLE I

Comparison of ARMA(2,3) and ARMAX(2,3,3) models

Year ↓	Serial No. Index	ARMA model Normalised RMS Error	ARMAX model Normalised RMS Error	Percent Change in RMS
2	EF	0.259	0.239	-7.6%
	ER	1.049	0.972	-7.3%
3	EF	0.306	0.291	-4.9%
	ER	0.992	0.913	-8.0%
4	EF	0.305	0.305	nil
	ER	0.603	0.604	nil
5	EF	0.367	0.351	-4.3%
	ER	1.630	1.221	-25.1%
7	EF	0.237	0.223	-5.7%
	ER	0.944	0.824	-12.7%

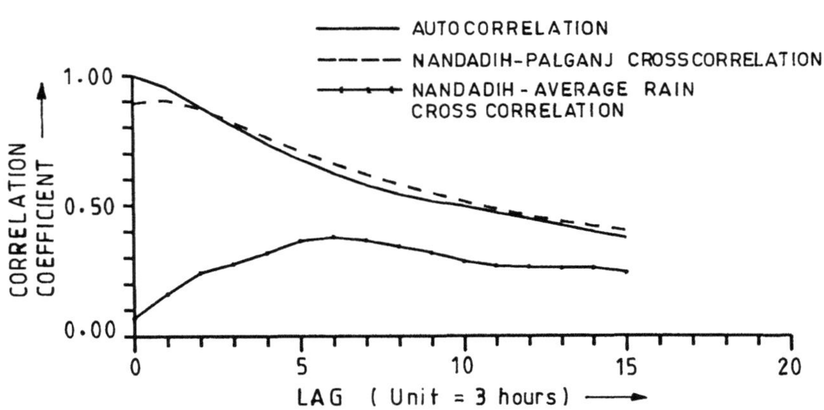

FIGURE 1: A TYPICAL CORRELOGRAM FOR 3 HOURLY NANDADIH AND PALGANJ DISCHARGE AND AVERAGE RAINFALL DATA.

The recession constant for the basin is obtained from analysis of recession curves of flood hydrographs, and a value of 0.925 is selected. The observed streamflow is decomposed into two components, namely, the base flow component and the quick flow component, using the filter earlier outlined. To the recession data of these subsystem outputs, i.e. flow components, are fitted AR models since the input X is zero. An AR(5) adequately represents the ground water subsystem and an AR(3), the surface/subsurface subsystem. The coefficient corresponding to input X in the ARX model is found using the continuity equation as suggested. The ARX models thus identified are converted to the impulse response or unit hydrographs and the same are shown in figure 2. In concurrence with Hino [6], the ground water unit hydrograph is found to be more or less invariant. However, the impluse response function for quick flow is not so. Typical 3 hourly unit hydrographs for the subsystems shown in figure 2 are used in further studies. This filter separation AR method based model is implemented with rainfall prediction being done using equations (10) and (12), as suggested earlier. The extrapolation coefficient ρ is taken to be 0.80 for both the quick flow and the base flow components. The run off coefficient $\mathbb{F}$ is taken to be 0.60.

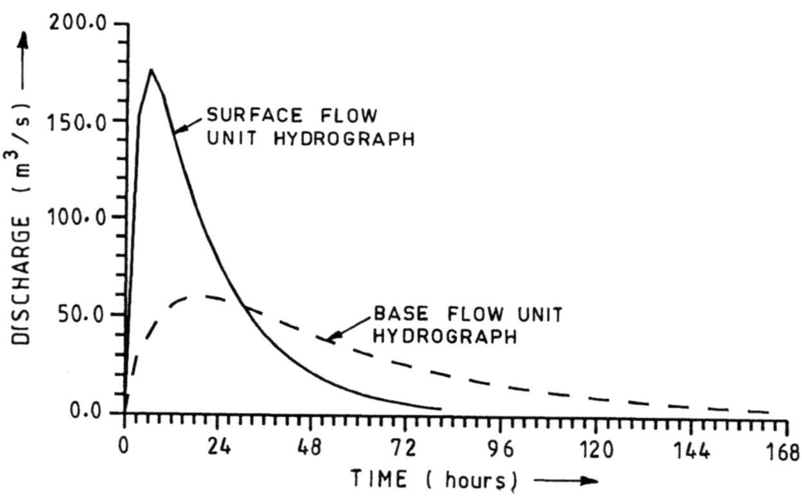

FIGURE 2 : TYPICAL 3-HOUR UNIT HYDROGRAPHS FOR SURFACE FLOW AND BASE FLOW AT NANDADIH

The filter separation method results in conspicuously improved rising limb tracking as compared to the ARMA model. As mentioned earlier, the observed rainfall information is hardly usable for 3 hourly modelling, the inverse estimated rainfall is used in the filter

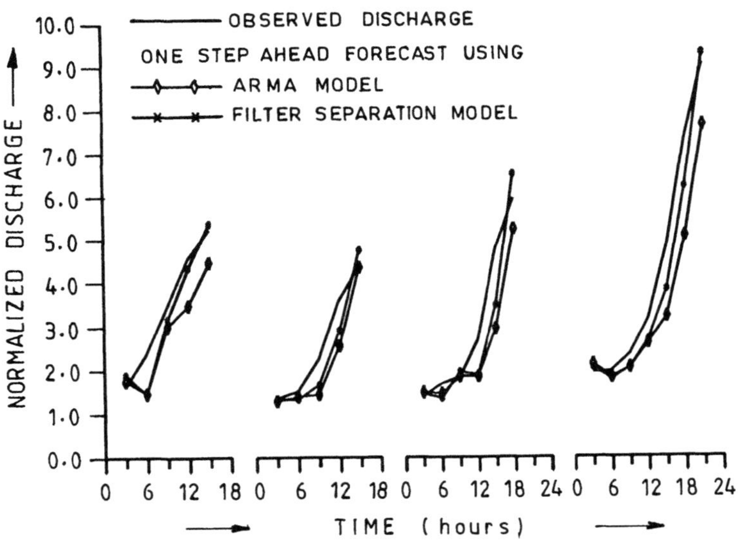

FIGURE 3 : A COMPARISON OF PERFORMANCE OF THE ARMA(2,3) MODEL AND THE FILTER SEPARATION ARX METHOD IN THE RISING LIMB REGION OF FLOOD HYDROGRAPH.

TABLE II

Comparison of ARMA(2,3) and filter separation ARX method

Year ↓	Serial No. Index	ARMA model Normalised RMS Error	ARX model Normalised RMS Error	Percent Change in RMS
1	EF	0.307	0.366	+19.2%
	ER	0.737	0.531	-28.0%
2	EF	0.259	0.328	+26.6%
	ER	1.049	0.865	-17.5%
3	EF	0.306	0.451	+47.4%
	ER	0.992	0.646	-34.9%
4	EF	0.305	0.459	+50.5%
	ER	0.603	0.668	-10.8%
5	EF	0.367	0.493	+34.3%
	ER	1.630	0.816	-49.9%
7	EF	0.237	0.302	+27.4%
	ER	0.944	0.822	-12.9%

separation method and still the results are much better as evident from figure 3. The peak is slightly over estimated in most of the cases unlike the ARMA model which almost invariably underestimates the same. The performance indices EF and ER for ARMA(2,3) and the filter separation ARX model are compared in table II.

A higher EF value for ARX model than ARMA model results from the over estimation immediately after the flood peak (refer figure 5) and also from higher ARX model sensitivity (refer figure 4 and 5). However, in spite of the increase in index EF it is considerably smaller than corresponding ER value, and hence acceptable. The discharge over estimation on the falling crest may primarily be attributed to this basin's behaviour as a hydrologically small catchment. Further, more accurate and representative rainfall information may improve the scene.

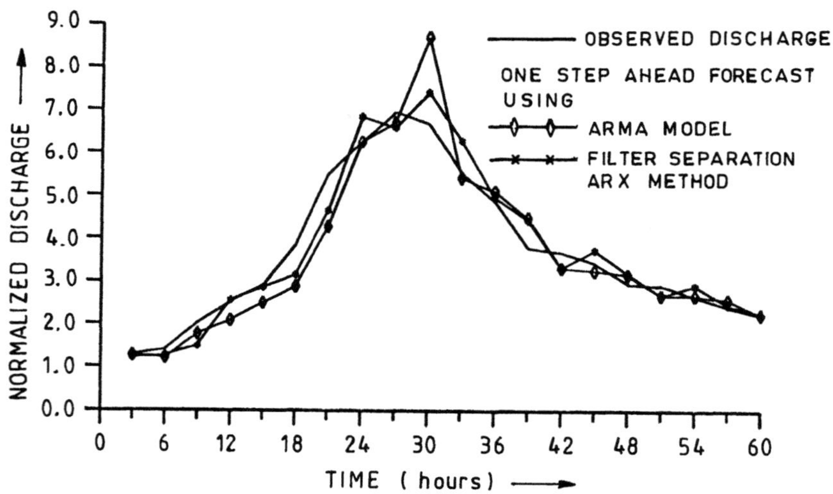

FIGURE 4 : COMPARISON OF (3 HOURLY) ARMA MODEL AND FILTER SEPARATION ARX METHOD

DAILY MODEL

An ARMA(1,2) model, MA having zero lag, as in equation (13), is found to be acceptable for daily discharge modelling at Nandadih. The average rainfall, raw or filtered, forms the MA component in this model. The correlation structure for a typical record is shown in table III.

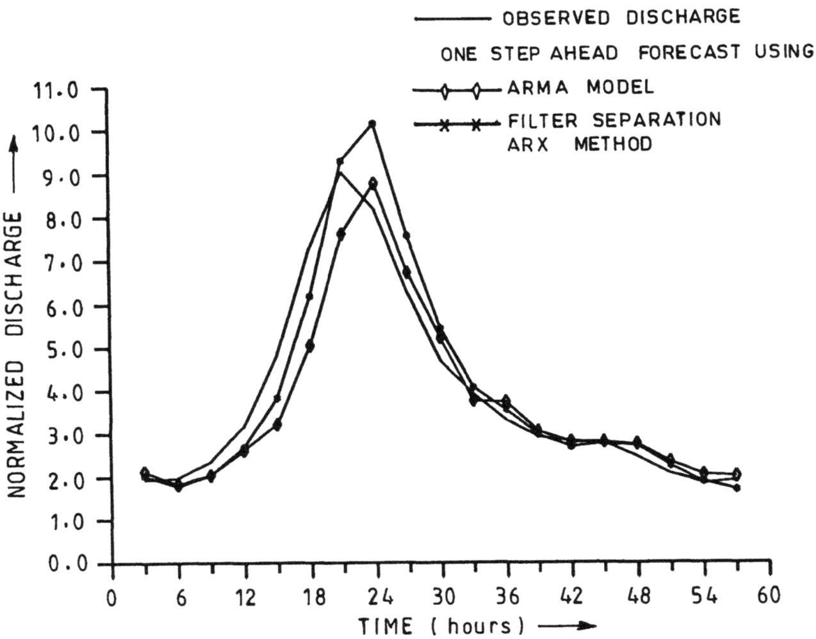

FIGURE 5 : COMPARISON OF (3 HOURLY) ARMA AND FILTER SEPARATION ARX METHOD.

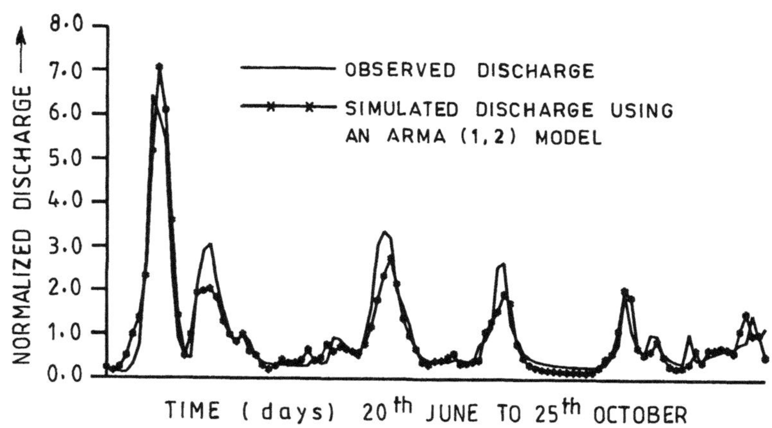

FIGURE 6 : DAILY SIMULATION OF NANDADIH DISCHARGE USING A RAINFALL RUN OFF MODEL

TABLE III

Correlation structure for Nandadih (daily data) and the basin average rainfall.

Lag	Auto-correlation coefficient	Cross-correlation coefficient
0	1.000	0.457
1	0.650	0.794
2	0.368	0.548
3	0.185	0.309
4	0.110	0.082
5	0.060	-0.051
6	0.052	-0.169
7	-0.043	-0.026

The mean and RMS error of residuals for the relavent years are shown in table IV. The model fitting is done using Recursive Least Square technique.

TABLE IV

Statistics of the residuals

Year Number	Mean	Root mean square error (normalised)
2	0.146	0.539
3	0.067	0.399
4	0.110	0.431
5	0.017	0.381
6	0.096	0.409
7	0.008	0.434

The low pass filter and the threshold type transformations are tried out with a view to handle the system nonlinearity. These transformations are reported to result in improved simulation for many other systems. However, for the system studied herein, there is at best only a marginal improvement, as is evident from table V. For all years except 2, 4 and 7, the RMS error is the least for no rainfall transformation, i.e. ϕ = 0.0. Even for the exceptional years, the minimum RMS is within -3% of the RMS for ϕ = 0.0 model. Thus, either of the rainfall transformations does not make any significant difference and hence can be dispensed with. Further, the comparison of weighted averages of the normalized RMS errors, in the last column of table V, clearly indicate that the model without any rainfall

TABLE V

Normalized RMS error of residuals for an ARMA(1,2) model with various rainfall filter options

Year No.-> Filter type	2	3	4	5	6	7	Weighted average
Threshold Filter (ϕ in mm)							
ϕ = 0.0	0.539	0.399	0.431	0.381	0.409	0.434	0.432
ϕ = 1.5	0.544	0.407	0.436	0.384	0.417	0.432	0.436
ϕ = 3.0	0.550	0.419	0.441	0.384	0.426	0.435	0.443
ϕ = 4.5	0.557	0.432	0.445	0.387	0.435	0.439	0.449
Low Pass Filter							
T = 10	0.530	0.530	0.441	0.382	0.422	0.421	0.456
T = 20	0.530	0.484	0.425	0.384	0.411	0.420	0.443
T = 30	0.534	0.454	0.419	0.389	0.409	0.422	0.438
T = 40	0.537	0.443	0.419	0.394	0.412	0.425	0.439
T = 50	0.538	0.444	0.420	0.397	0.416	0.427	0.441
T = 60	0.540	0.451	0.422	0.400	0.419	0.429	0.444

transformation should be preferred. The daily discharge simulation for the monsoon months of the year numbered 3, using an ARMA(1,2) model without any rainfall filter, is shown in figure 6. It may be noted that the matching of flood peaks is excellent.

CONCLUSIONS

A channel routing type ARMA(2,3) model, its extensionARMAX(2,3,3) and a filter separation ARX model are implemented for real time (3 hourly) flood forecasting in Damodar river basin. The ARMA model invariably under estimates the flood discharge in the rising limb region of flood hydrographs.

Since no representative rainfall information is available for the interbasin, to account for the lateral inflow into the reach, the average rainfall over the Barakar river basin is used in the ARMAX model presuming that the pattern over the entire basin is uniform. The forecasting performance of the ARMAX model is much better than that of the ARMA model, particularly in tracking the rising limbs of flood hydrogrphs. The ARMAX model is heavily dependent on the reliability of rainfall information to justify the additional computational effort.

The ground water or base flow subsytem for the catchment studied herein can be taken to be invariant. The behaviour of the surface/subsurface subsytem does not allow such an assumption. The choice of right impulse response function is very important if the streamflow components and the inverse estimated rainfall are to have any physical meaning. The filter separation ARX method results in a considerably improved tracking of the rising limb. However tracking of the falling limbs of flood hydrographs is much degraded. This is mainly due to the fact that the available rainfall information is not directly usable in this model and rainfall prediction based on the inverse estimates of the same is relied upon. It is surmised that better reliable rainfall information would definitely go a long way in alleviating this problem. Further, the catchment under consideration behaves as a hydrologically small basin and thus the multiple step ahead forecasting with a shorter prediction time step may be preferrable.

The study conducted on the applicability of various rainfall transformations reveals that the low pass filter type and the threshold (ϕ index) type rainfall transformations do not afford any visible advantage for this particular system for daily discharge modelling.

ACKNOWLEDGEMENT

The authors are extremely grateful to the Central Water Commission and the Damodar Valley Corporation , at Maithon for supplying the data used in this study.

REFERENCES

[1] Bergman, M. J. and Delleur, J. W., 1985. 'Kalman filter estimation and prediction of daily streamflows: II. Application to the Potomac river'. *W R Bull.*, **21**,827-832.

[2] Bertoni, J. C., Tucci, C. E. and Clarke, R. T., 1992. 'Rainfall based real time flood forecasting'. *J Hydrol.*, **131**, 319-339.

[3] Bolzern, P., Ferrario, M. and Fronza, G.,1980. 'Adaptive real time forecasting of river flow rates from rainfall data'. *J Hydrol.*, **47**, 251-267.

[4] Chapman, T. G., 1991. comment on 'Evaluation of automated technique for base flow and recession analysis' by R J Nathan and T A McMahon. *W R R*, **27(7)**, 1783-1784.

[5] Hino, M. and Hasebe, M., 1984. 'Identification and prediction of nonlinear hydrologic systems by the filter separation Autoregression (AR) method: Extension to hourly hydrologic data'. *J Hydrol.*, **68**, 181-210.

[6] Hino, M. and Kim, C. H., 1986. 'Nonlinear flood forecasting by the filter separation AR method'.*J Hydrol.*, **88**, 165-184.

[7] Jakeman, A. J., Littlewood, I. J. and Whitehead, P. G., 1990. 'Computation of the instantaneous unit hydrograph and identifiable concept flows with application to two small upland catchments'. *J Hydrol.*, **117**, 275-300.

[8] Kachroo, R. K. and Liang, G. C., 1992. 'River flow forecasting Part 2. Algebraic development of linear modelling techniques'. *J Hydrol.*, **133**, 17-40.

[9] Liang, G. C. and Nash, J. E., 1988. 'Linear model for river flow routing on large catchments'. *J Hydrol.*, **103**, 157-188.

[10] Nathan, R. J. and McMahon, T. A., 1990. 'Evaluation of automated technique for base flow and recession analysis'. *W R R*, **26(4)**, 1465-1473.

[11] Papamichail, D. M. and Papazafiriou, Z. G., 1992. 'Multiple input single output linear functional models for river flow routing'. *J Hydrol.*, **133**, 365-377.

[12] Singh, K. and Majumdar, M., 1993. 'Model Identification studies in Damodar basin'. Proceedings of *58th R&D Session* (Bangalore), Vol I, Publication No. **233**, Central Board of Irrigation of Power, New Delhi.

[13] Todini, E. and Bouillot, D., 1976. 'A rainfall runoff Kalman filter model'. In: G C Vansteenkiste (Ed.), *System Simulation in Water Resources*. North-Holland Publishing Company.

[14] Weeks, W. D. and Boughton, W. C., 1987. 'Tests of ARMA model forms for rainfall runoff modelling'. *J Hydrol.*, **91**, 29-47.

CHANGES IN FLOW CHARACTERISTICS OF THE RIVER MURRAY IN THE LAST 100 YEARS

B.L. Maheshwari[1], K.F. Walker[2] and T.A. McMahon[3]

[1]School of Agriculture & Rural Development, University of Western Sydney, Richmond, NSW 2753, Australia.

[2]River Murray Laboratory, Department of Zoology, University of Adelaide, SA 5005, Australia.

[3]Centre for Environmental Applied Hydrology, Department of Civil & Environmental Engineering, University of Melbourne, Parkville, VIC 3052, Australia.

ABSTRACT. Flow regulation in the River Murray and its tributaries has increased during the last 100 years by construction of storages, locked weirs and changes in operation strategies. This has resulted in considerable changes in the flow characteristics of the river system with implications for the river ecosystem. The main aim of this study was to examine the effects of the various regulatory structures on the flow characteristics of the River Murray and to investigate the effects of the weirs and locks on the water-level fluctuations along its lower reaches. The study indicates that the average monthly and annual flows are lower under the present conditions than those that existed under the natural conditions. The water-level regime of the lower Murray has also been considerably affected by weir operations, particularly through increases in average water-level elevations, decreased fluctuations at upstream pool gauges and increased fluctuations at downstream pool gauges.

1. Introduction

The Murray-Darling Basin is a vital resource for Australia, supporting 25% of the nation's cattle and dairy farms, 50% of sheep and cropland and 75% of irrigated land, and returning more than A$10 billion in annual production (e.g. MDBMC, 1989; Walker, 1992). The catchment is 1.073 million km^2. Over 90% of this area is arid or semi-arid land, and rainfall and streamflow are generally low and erratic. Indeed, annual flows in the system are among the most variable in the world (Finlayson and McMahon, 1988), although the total discharge is comparatively small. For example, the River Murray at Blanchetown during 1950 to 1980 had mean annual runoff 151 mm and mean discharge 27500 ML/d, with a range 1730 to 135000 ML/d.

Flow regulation inevitably changes the physical and biological environments of rivers (e.g. Petts, 1984). These changes may be exaggerated in semi-arid rivers like the Murray, because of the disparity between natural and regulated patterns of flow. Whereas the natural regime is prone to droughts and floods, the regulated regime is engineered to provide reliable supplies for irrigation and other forms of water use. Irrigation accounts for about 90% of annual water use in the Murray-Darling Basin and so largely determines the pattern of regulation.

V. P. Singh and B. Kumar (eds.), Surface-Water Hydrology, 473–481.

Flow regulation has meant profound environmental changes to the Murray. In the lower Murray, for example, changes are evident in problems of salinisation and channel erosion and deposition associated with the weirs, and in continuing declines of native plants and animals (Walker and Thoms, 1993). These changes will affect the long-term utility of the river as a resource and, if we are to develop a more secure relationship with the river, we will need to manage flows in a manner less damaging to the environment. This will require a basic understanding of the natural flow regime and the ways that it has been changed by regulation. Thus, the aim of this study was to examine the effects of regulation on the flow characteristics of the River Murray and to investigate the effects of the weirs on water-level fluctuations in the lower Murray.

2. Murray-Darling River System

The Murray-Darling Basin is drained by over twenty major rivers. The two principal rivers, the Murray and Darling, have a combined river length of 5500 km, making the fourth largest river system in the world. The Murray may be divided into upper, middle and lower sections, corresponding to the river between its source and Yarrawonga, between Yarrawonga and the South Australian (SA) border, and between the border and the Murray Mouth (see Figures 1 and 2). About half of the total discharge from the system is from the Murray catchment above Yarrawonga, due to rainfall and snow in winter and spring. The middle Murray itself provides a much smaller contribution, but its flows are augmented by the Goulburn, Broken, Campaspe, Murrumbidgee and Darling rivers. The Darling drains the northern part of the Murray Basin and is fed by erratic summer rainfall. The lower catchment yields negligible runoff.

3. Flow Regulation

Flow regulation of the Murray-Darling system first became significant in 1922-40. Key developments included a regulator at Lake Victoria (1928), Lake Hume (1936), Yarrawonga Weir (1939), the Murray Mouth barrages (1940), and 13 low-level locked weirs (called Lock 1, Lock 2 etc. in Figure 2) between Blanchetown and Torrumbarry (1922-37). Lake Hume was later enlarged (1961), and flows to the upper reaches were supplemented by diversions from the Snowy Mountains Hydroelectric Scheme (1955-74). Note also that the volume of annual diversions increased sharply after 1950 (see Close, 1990). Storages at Menindee Lakes on the Darling River were completed in 1968, and Lake Dartmouth was completed on the Mitta Mitta River in 1979. Eight historical stages may be recognised: (i) natural (unregulated, 1860), (ii) Pre-Lake Victoria (1926), (iii) Pre-Hume (1929), (iv) Post-Hume (1939), (v) Post-Yarrawonga (1948), (vi) Post-Hume Enlargement (1962), (vii) Post-Menindee (1977), and (viii) Post-Dartmouth (1990).

Flows in the lower Murray in summer and autumn are maintained by mixing water from the Murray and Darling rivers, with a disproportionate contribution from the Darling. The lower river is impounded by ten 3-m weirs, constructed in 1922-35 and forming a continuous series of pools from Blanchetown (Lock 1) to Wentworth (Lock 10) (see Figure 2). Except in times of flood, goal of regulation is to maintain flows at bankfull through out the year. This involves maintaining upper weir-pool levels within 50 mm of a design level, requiring the lockmasters to adjust the panels and stoplogs as necessary. This causes changes in the river level downstream of each weir (the "lower pool"); these are typically 200 mm daily and are progressively diminished in the approach to the next weir.

4. Simulation Model

With the rapid development of regulation, the short periods between new developments and the inherently high variability of flows in the system, it is virtually impossible to evaluate the impact of regulation on flow characteristics by direct examination of historical data. Consequently, the Murray Darling Basin Commission has developed the Murray-Darling Monthly Simulation Model (MSM) (Brewsher, 1980; Close, 1990). This employs a water-

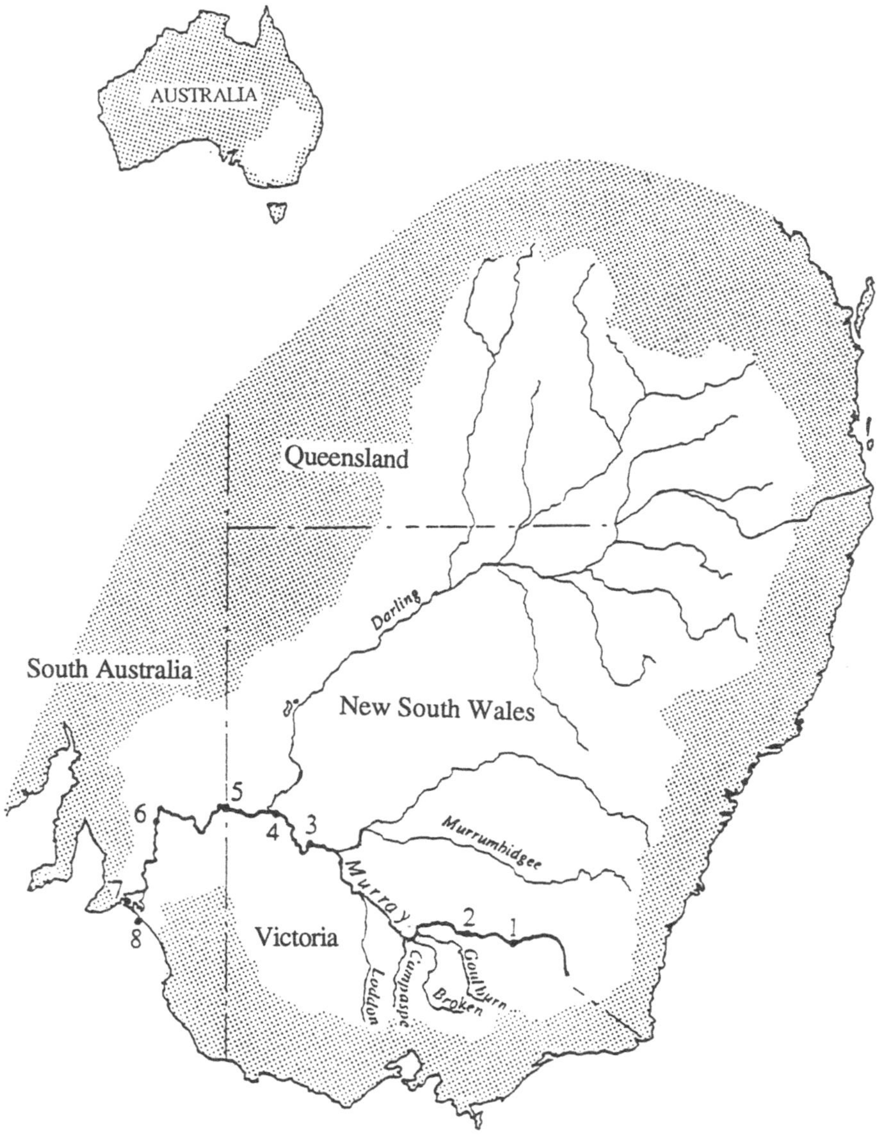

Stations on the Murray

1	Albury	5	South Australian Border (Lock 6)
2	Yarrawonga	6	Lock 1
3	Euston	7	Barrages
4	Upstream of Murray-Darling Junction	8	Murray Mouth

Figure 1. Location map of the Murray-Darling basin and river system.

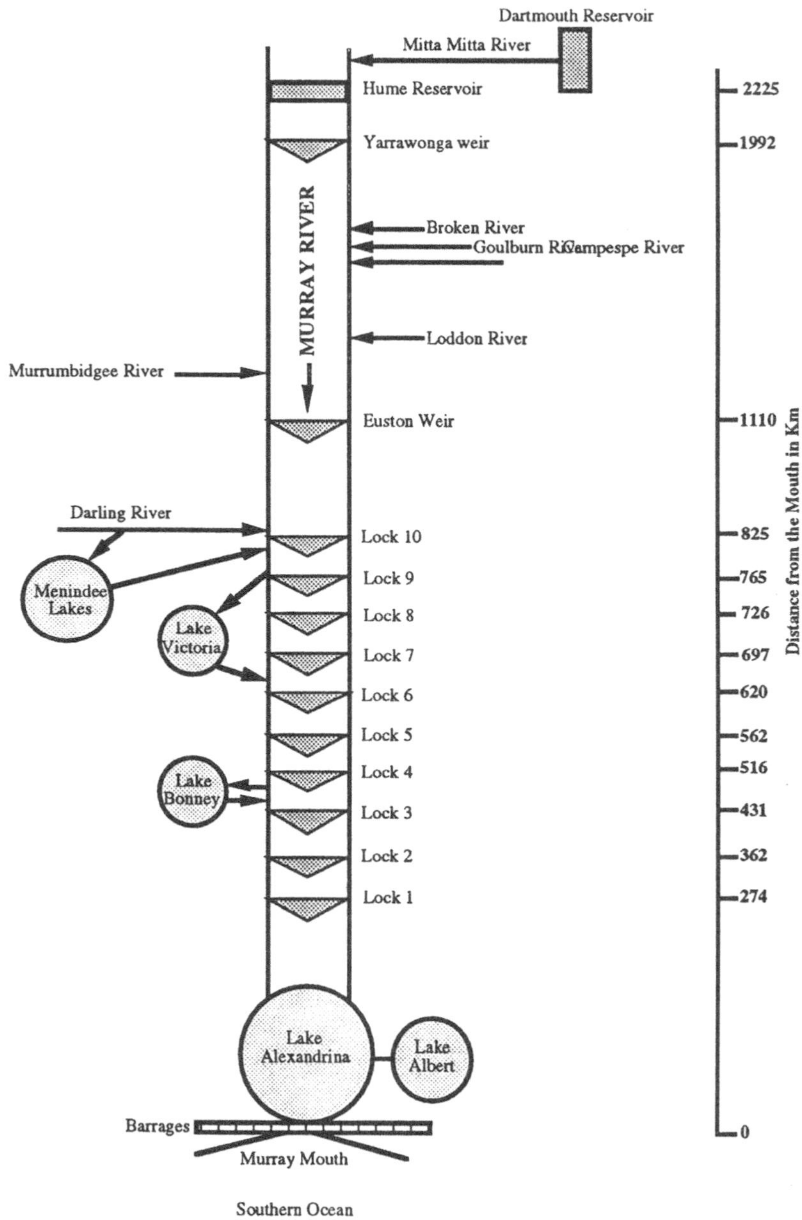

Figure 2. Schematic representation of the River Murray showing major tributaries and regulating structures (not drawn to scale).

balance approach, with a time step of one month, to account for runoff, reservoir storage, inputs from the Snowy Scheme and losses to groundwater, evaporation and diversions. The simulation period is from 1891 to 1986. The model simulates flows for any given stage of development by appropriately adjusting the storages and diversions. Natural flows are simulated by assuming no storages or diversions. No verification of the model is possible; assessments of reliability rest entirely on the apparent reasonability of the assumptions and approaches used in the model.

5. Simulation Experiments

Eight stages of development (as mentioned in Section 3) are used for simulations to determine the effects of regulation on the natural flow characteristics of the system. For each stage of development, the Monthly Simulation Model was used to estimate flows at 8 stations selected along the Murray. The stations were Albury, Yarrawonga, Euston, upstream of Murray-Darling Junction, South Australian Border, Lock 1, Barrages and Murray Mouth. The model simulated monthly flows that would have occurred, assuming that a given stage of regulation prevailed throughout the period for which the simulations are carried out (i.e. 1891 to 1986).

6. Analysis

The simulated flow time series at each station and stage of development was analysed by (i) calculating the monthly and annual values of average flow, coefficient of variation and coefficient of skewness, and (ii) constructing curves of the peak-flow frequency, low-flow frequency and flow duration. The peak and low- flow frequency curves were constructed from ranked minimum and maximum monthly flows for each year, plotted on log-normal axes. The construction of flow duration curves involve sorting of flow data into class intervals and expressing the number in each class as a percentage of the total number of records. Beginning with the largest, the percentages were then summed to show the cumulative percentage of time that flows exceed the lower limit of each class.

The water-level fluctuations along the lower Murray for the upper and lower pools of Locks 1 to 6 were studied. Observed daily water-level elevations for both the pools at these lock sites over >60 years were used. The data were analysed, for both pre- and post weir periods, by determining the minimum, maximum and average water-level elevations, and rates of water-level fluctuations (rises and falls) of different durations. Periods of continuous rise and fall were extracted from the data, and the change in level was divided by the duration of continuous rise or fall to yield average rates of rise or fall in mm/day over 1, 3 and >5 days. The rates at Locks 3 and 6 were examined using Box plots, showing the distributions of records across the 10, 25, 50, 75 and 90th percentiles. The analysis was for two time periods: before and after weir construction. The post-weir period was further divided into six decades: 1930-39, 1940-49, 1950-59, 1960-69, 1970-79 and 1980-89.

7. Results and Discussion

The following is a brief outline of the outcomes of this study; details of them are provided in Maheshwari et al. (1993).

7.1. GENERAL FLOW CHARACTERISTICS

The average monthly and annual flows of the Murray are considerably lower under present conditions than under natural conditions, mainly due to storages and diversions for irrigation and other uses (e.g. see Figure 3). The major impact of flow regulation became evident at Stage 4 (post-Hume), and brought significant changes to monthly and annual minimum and average flows, variability and skewness. Seasonal patterns of flow have changed markedly in the upper section of the Murray where the predominant influence has

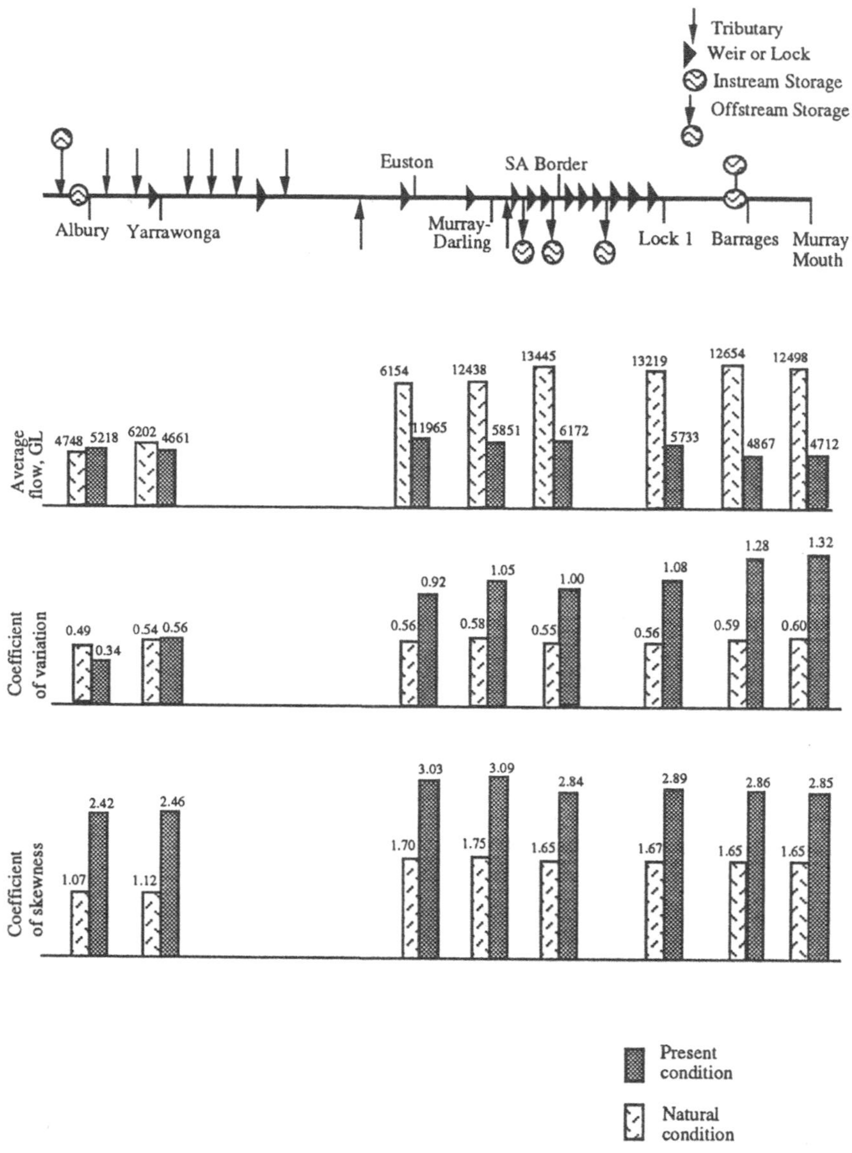

Figure 3. A comparison of annual values of average flow, coefficient of variation and coefficient of skewness under natural and present conditions along the Murray.

been due to dams, but comparatively little in the middle and lower sections where weirs are most influential.

7.2. EXTREME FLOW AND FLOW-DURATION CHARACTERISTICS

Regulation has considerably changed the peak-flow and low-flow characteristics of the system (see Figure 4a). The effects are minimal at Albury, but increase with distance downstream and the stage of regulation. The magnitude of small and medium floods has been reduced markedly at all stations, whereas large floods (those occurring every 20 years or more) are little affected.

In general, natural flow-duration characteristics change little with location along the Murray, whereas there is considerable variation under regulated conditions (see Figure 4b). The effect of regulation is least at Albury and becomes more pronounced downstream. It is most apparent in regard to medium flows (those exceeded 20-80% of the time).

7.3 WATER-LEVEL REGIMES

The water-level regime of the lower Murray has been greatly affected by weir operations, particularly through increased average water-level elevations, decreased fluctuations at upstream gauges and increased fluctuations at downstream gauges. Water-level changes during the post-weir period are characterised by higher rates of falls than rises at both upstream and downstream gauges.

7.4 ENVIRONMENTAL IMPLICATIONS

Under both natural and regulated conditions most of the flow in the Murray is from the river and its tributaries above Euston. Downstream, the Murray flows as an exotic stream in a semi-arid region where both rainfall and runoff are sparse. The Darling's contribution is minor, as it often has little or no flow for half the year.

Storages and irrigation diversions substantially reduce average annual flow in the middle and lower Murray, and alter the trend of downstream variation (see Figure 3). The flow in the lower river is now being maintained at or near bankfull capacity (about 1100 to 1300 GL/month in South Australia), although this is partly due to ponding effects associated with the weirs. These changes have had profound effects on the lower river environment. For example, the littoral zone no longer experiences the erratic wetting/drying cycle that prevailed before regulation, and the change in river stage associated with major floods has been significantly reduced.

A seasonal shift in flows is most apparent in the upper Murray in the region downstream of Lake Hume, but of minor consequence in the middle and lower sections of the Murray. Summer monthly flows at Albury under present conditions are more than three times those under natural conditions. Again, this change has had major environmental effects. The seasonal shift of flow patterns in the Murray between Lake Hume and Yarrawonga Weir is believed to be one of the factors responsible for the local extinction of golden perch and catfish (e.g. Walker, 1980). With flow regulation now in place, the present study shows that the flow characteristics along the Murray have changed considerably in the last 100 years, and due to this some stress on the river eco-system can not be ruled out.

8. Conclusions

(i) The average monthly and annual flows of the Murray are considerably lower under the present conditions than under natural conditions, mainly due to storages and diversions of river flows for irrigation and other uses.

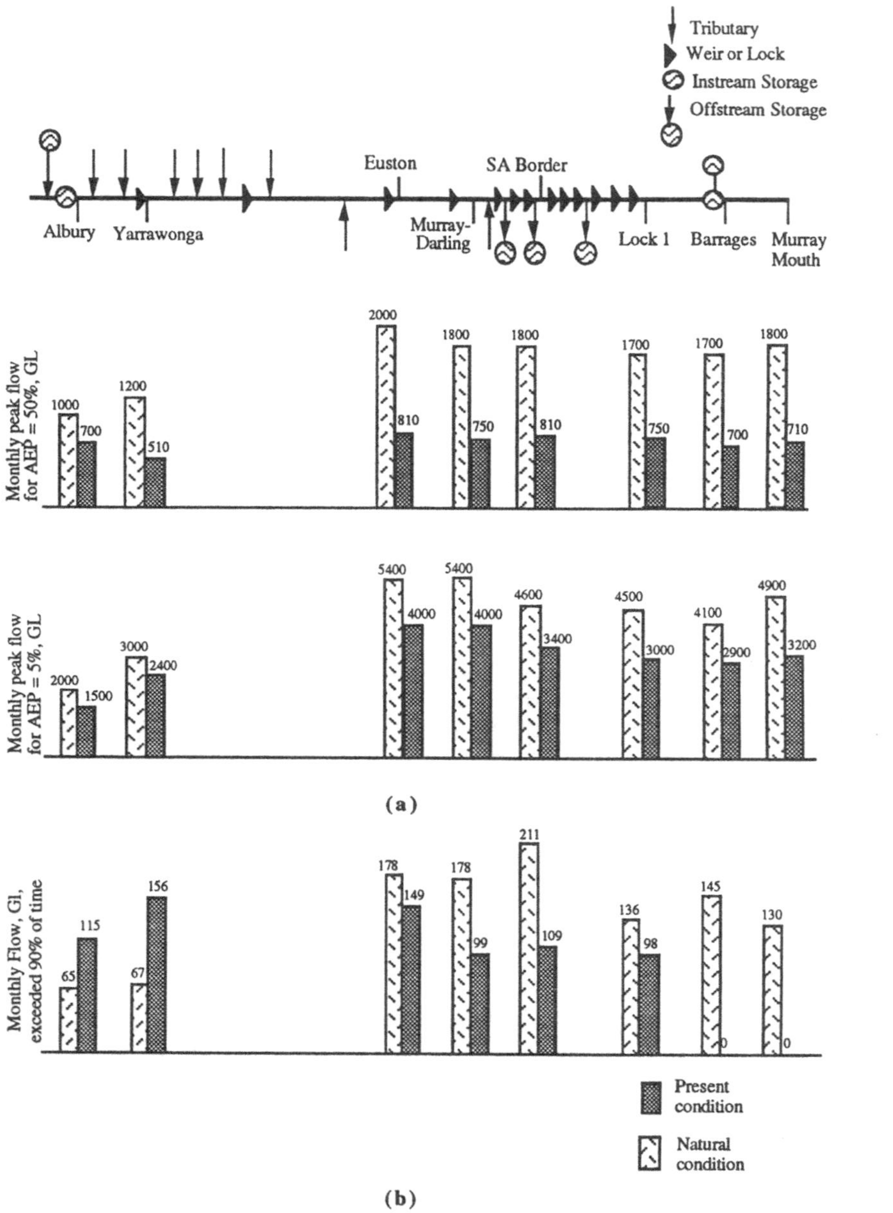

Figure 4. (a) Comparisons of monthly peak flows at annual exceedance probability (AEP) of 50% and 5% under natural and present conditions along the Murray, (b) A comparison of monthly flows that are exceeded 90% of time under natural and present conditions along the Murray.

(ii) Natural flow-duration characteristics change little with location along the Murray, whereas there is considerable variation under regulated conditions. The effect of regulation is minimal at Albury (near the headwaters) and becomes more pronounced downstream.

(iii) Regulation has considerably changed the peak-flow and low-flow characteristics of the system. The effects are minimal at Albury, but increase both with distance downstream and the stage of regulation.

(iv) The water-level regime of the lower Murray has been affected by weir operations, particularly through increases in average water-level elevations, decreased fluctuations at upstream gauges and increased fluctuations at downstream gauges.

9. Acknowledgments

Our thanks to the staff of Engineering & Water Supply Department, Adelaide and the Murray-Darling Basin Commission, Canberra for their assistance. The financial assistance of the Land and Water Resources Research and Development Corporation also is gratefully acknowledged.

10. References

Brewsher, A., 1980. Description of the River Murray System computer simulation program. Murray Darling Basin Commission, Canberra, 51p.

Close, A, 1990. The impact of man on the natural flow regime. In N. Mackay and D. Eastburn (Editors), The Murray, Murray Darling Basin Commission, Canberra, 61-76.

Finlayson, B.L. and McMahon, T.A., 1988. Australia *vs.* the world: a comparative analysis of streamflow characteristics. In R.F. Warner (Editor), Fluvial Geomorphology of Australia, Academic Press, Sydney, 17-39.

Maheshwari, B.L., Walker, K.F. and McMahon, T.A., 1993. Impact of Flow Regulation on the Hydrology of the River Murray and Its Ecological Implications. Centre for Environmental Applied Hydrology, University of Melbourne, Parkville, Australia, 154p.

MDBMC, 1989. Background Papers: Murray-Darling Basin Natural Resources Management Strategy. Murray-Darling Basin Ministerial Council, Canberra, 172p.

Petts, G.E. 1984. Impounded Rivers: Perspectives for Ecological Management. Wiley, Chichester, 326p.

Thoms, M.C. and Walker, K.F., 1992. Channel changes related to low-level weirs on the River Murray, South Australia. In P.A. Carling and G.E. Petts (Editors), Lowland Floodplain Rivers: Geomorphological Perspectives. Wiley, Chichester, 235-249.

Walker, K.F., 1992. A semi-arid lowland river: The River Murray, Australia. In P. Calow and G.E. Petts (Editors), The Rivers Handbook, Vol.1, Blackwell Scientific, Oxford, 472-492.

Walker, K.F. and Thoms, M.C., 1993. A case history of the environmental effects of flow regulation on a semi-arid lowland river: The River Murray, South Australia. Regulated Rivers: Research and Management, 8:103-119.

*****

SECTION 7:

FLOOD FREQUENCY ANALYSIS

ON THE STATISTICAL FREQUENCY ANALYSIS OF HYDROLOGICAL EXTREMES

Fahim Ashkar
Professor, Department of Mathematics, University of Moncton, Moncton, New-Brunswick, Canada, E1A 3E9

ABSTRACT. Statistical frequency analysis has been extensively used in hydrology for estimating probabilities of occurrence of extreme events such as high or low streamflows, extreme rainfall intensities, high snow depths, and so on. One approach widely employed for estimating these probabilities uses only one event of the basic hydrological variable within a conveniently chosen time interval, such as the year. This is the "annual series" (AS) approach to extreme-value hydrological analysis. Another approach uses an average of m values of the basic hydrological variable per year of record; e.g. those "peak" values that exceed a certain threshold level, X_b. This is the "peaks over threshold (POT)" approach. A large number of D/E procedures (Distribution D, with method of parameter estimation E) for analyzing annual maximum series (AMS), and POT series, have been proposed. However, their insufficient physical justification has often led to some confusion among practitioners. Attempts at surveying and comparing these procedures has failed to effectively eliminate this confusion.

A reflection is made on certain aspects of AMS and POT extreme-value analysis which we consider important from a practical and scientific viewpoint. We review and briefly discuss some of the classical statistical hypotheses that have been used for analyzing hydrological AMS and POT series such as randomness, independence, homogeneity and stationarity. We also comment on the problem of handling "outliers" (values in the record that are far from the bulk of the data), which can frequently be found in AMS or POT data sets. We also discuss the choice of D/E procedures for AMS- / POT- series analysis, giving some general ideas about the choice of Distribution (D), and parameter-estimation methods (E). Some of the difficulties in choosing a D/E procedure for estimating hydrological extremes are identified and briefly discussed.

The significant improvements brought about in the past number of years in the computer field, have resulted in interest in a group of statistical methods known as "resampling techniques". Probably the most popular of these methods is the "bootstrap" method (Efron, 1979, 1987). In our judgement these "computationally intensive" techniques have important potential practical applications in hydrology, which merit more extensive research and investigation. We also list some important points that we feel need to be pursued further with regard to POT modeling in hydrology.

Key words: Statistical frequency analysis, hydrological extremes, annual maximum series, peaks over threshold, probability distribution, parameter estimation, outlier, resampling techniques.

V. P. Singh and B. Kumar (eds.), Surface-Water Hydrology, 485–503.

1. Introduction

Statistical methods have been widely used in the areas of hydrology and water resources for analyzing hydrological data sets. Statistical frequency analysis using probability distributions has been a particularly employed procedure for estimating the relation between the magnitude and frequency of occurrence of various hydrological events. It has been used for studying such hydrological phenomena as rainfall, water demand for residential use, evaporation, sediment transport, tides, floods, droughts, ground-water levels, and so on. According to Yevjevich (1972, p. 3) "hydrology cannot be fully understood, described, and best applied to water resources conservation, development and control without the extensive use of methods of probability theory."

In hydrologic design, one often needs to estimate the probability of occurrence of some *extreme* or *rare* event, such as a severe flood or drought (or, vice versa, the event corresponding to a given probability of occurrence). The adequate estimation of such "hydrological extremes" may be needed for managing a hydro-electrical facility, establishing the flow capacity of an urban sewer network, planning a hydraulic work on a water stream, or for similar purposes.

The most frequently used statistical frequency analysis procedure for studying hydrological extremes, involves: (1) selecting a sample of values of the basic hydrological variable of interest (e.g., streamflow level, snow or ice accumulation or melt, lake levels, sediment discharge) that satisfies certain statistical criteria (to be outlined later); (2) fitting a probability distribution to this sample using the best fitting technique available; and (3) using this fitted distribution to make statistical inferences about the underlying "population", such as estimating a well defined extreme event related to the basic hydrological variable of interest, and assessing the error of estimation.

Nonparametric methods have also been proposed for the frequency analysis of hydrological data. These methods do not fit a parametric probability model (distribution) to the sample values in the same way that has just been outlined. Due to some difficulties that still exist in applying these methods to the frequency analysis of hydrological extremes, we shall not consider them further in the present study.

2. Data Series Used in the Estimation of Hydrological Extremes

The typical sample considered in the statistical frequency analysis of hydrological extremes contains n elements corresponding to the maximum or minimum values ("extrema") of the basic hydrological variable under consideration (precipitation depth, river discharge, or the like). These extrema are usually taken over time intervals of equal length (e.g., $\{X_i; i=1,...,n\}$ = monthly, seasonal or yearly maximum or minimum values of a basic hydrological variable under consideration). These extrema define the random variable, X, to be analyzed statistically.

For the sake of discussion, let us consider the time interval over which the random variable X is defined, to be the year. The extreme-value frequency analysis concerns, in this case, a set of n values of X corresponding to an n-year period of record of the basic hydrological variable (sample of annual maxima or minima, of size n). This "annual-series (AS)" type of analysis, corresponds to the case most employed in hydrological practice, where only one event of the basic hydrological variable is retained, per year. In the remainder of the discussion, we shall assume that interest is in annual *maximum* series (AMS) rather than in annual *minima*, although much of the discussion could also be extended to annual-minimum-series-extreme-value analysis.

Another approach for estimating hydrological extremes uses an average of m values of the basic hydrological variable per year of record, which are those peak values that exceed a certain "truncation", "threshold", or "base" level, X_b. These peak values, or some other variable of interest related to them, are called "exceedances". This is the "peaks over threshold (POT)", or "partial duration series (PDS)" approach to extreme-value hydrological frequency analysis. Again, we shall restrict discussion to POT-series analysis, although many of the ideas could also extend to the study of "peaks-*below*-threshold".

Rasmussen et al. (1993) note that POT modeling is a compromise between AMS analysis and a more detailed analysis of, for example, the complete daily-data series using classical time series modeling. Note that POT analysis requires that peaks be modeled in both time and magnitude domains. Whereas the more detailed daily-data analysis focuses on modeling the nonstationarity and dependence structure of *all* the daily values, POT modeling concentrates only on the higher daily values, which contain most of the information about the extreme-value process, and which are in certain respects easier to model than the complete daily series. In comparison to AMS extreme-value modeling, on the other hand, POT modeling considers a wider and more selective spectrum of events than only the annual maxima. In many respects, it is also a more difficult type of modeling, mathematically, than AMS modeling, which is probably one of the principal reasons for its restricted popularity among practitioners.

3. Quantiles as Design Events

In the planning, design, and operation of water resources systems, it is often useful to base decisions on conveniently defined quantities, such as events X_T of "return period" T. In AMS analysis, X_T is defined according to the probabilistic relationship:

$$p = \text{Prob}\ (X \geq X_T) = 1/T.$$

where p is the probability ("risk") that the "T-year event" X_T (also often denoted by X_p), is exceeded in any particular year. In other words, X_T is the (1-p)-th quantile in the distribution of annual maxima. It can be shown that $T = 1/p$ is the average number of years separating two annual events equalling or exceeding X_T ($X \geq X_T$).

In the POT case, on the other hand, it can be shown that if m is the annual average number of exceedances of the threshold, then the T-year event is the (1-1/mT)-th quantile in the distribution of exceedances.

4. Statistical Hypotheses Used for Analyzing AMS and POT Series

In estimating extremes from AMS or POT series by classical procedures, four statistical hypotheses are commonly made. These are briefly discussed below.

(1) Randomness. This condition means that the fluctuations in the hydrological process arise from natural causes. For example, flood flows appreciably altered by dam operation are not random. No general statistical tests exist to check for randomness in hydrological data sets.

(2) Independence. Prior to fitting a statistical distribution to an AMS or POT sample, it should be verified that the sample values are independent from each other. This means that the occurrence and magnitude of one observation should not influence in any way the occurrence and magnitude of any other observation. A possible non-parametric statistical test to check for independence is the Wald-Wolfowitz test (1943) (Bobée and Ashkar, 1991). In the POT approach, a sufficiently high threshold value, X_b, has usually to be chosen if the assumption of independent exceedances is to be justified. The reason for this is that the serial correlation between peaks increases with the average number of exceedances per year. However, there is still a need in POT research to study the basic meaning of an independent exceedance (Rasmussen et al., 1993).

(3) Homogeneity. In fitting a statistical distribution to an AMS or POT sample, the sample's observations must all be drawn from a common statistical population. A change in the environment around a precipitation gauge, for example, or a movement of the gauge from one site to another, may significantly influence rain measurements, so that data collected before and after the change might be coming from two significantly different populations. In this case, it is better to analyze the sub-samples separately, and then combine them to provide the desired extreme-value estimate corresponding to the entire population. It has been argued by some authors that POT modeling provides a better tool for describing seasonality in hydrological phenomena (e.g., for combining rainfall and snowmelt floods into one single model), than AMS modeling (Todorovic and Rousselle, 1971; North, 1980; Waylen and Woo, 1982; Cruise and Arora, 1990; Rasmussen and Rosbjerg, 1991b).

The non-parametric Mann-Whitney (1947) test can be used to check for non-homogeneity in two sub-samples (Bobée and Ashkar, 1991). In the case where more than two seasons are believed to have significant influence on the hydrological process giving rise to the POT or AMS data (case of several nonhomogeneous sub-populations), the Kruskal-Wallis test (Noether, 1976) is one possible test to use. It cannot be over-emphasized, however, that to determine the existence, type and amount of non-homogeneity in AMS/POT series, reliance should not be solely on statistical tests, but all the physical factors that are judged to be important, should also be investigated.

(4) Stationarity. The condition of stationarity means that, excluding random fluctuations, the data series is invariant with respect to time. Two important forms of nonstationarity often present in hydrological data series, are "jumps" and "trends". In the case of river discharge for instance, these may be caused by the construction of a dam or a levee, or by the occurrence of a forest fire in the basin. Statistical tests for detecting nonstationarities include the Mann-Whitney (1947) test for jumps, and the Wald-Wolfowitz (1943) test for trend (Bobée and Ashkar, 1991). On the other hand, nonstationarity caused by long-term changes in hydrologically-related variables, such as global changes in weather, or very slow evolution in nature, cannot be detected by statistical tests, because of the general shortness of available records of these variables.

5. The Problem of "Outliers"

Values that are far from the bulk of the data are quite frequently found in AMS or POT data sets. These "outlying" observations, as they have often been called, can cause significant difficulties in fitting a statistical distribution to the sample. It has been shown that high- or low-valued "outliers" can lead to X_T event estimates that are significantly underestimated or overestimated.

Statisticians have sometimes recommended the elimination of outliers before applying certain types of parametric statistical procedures to data sets. However, two fundamental questions are not yet clearly answered with regard to this problem of "handling outliers" (Ashkar et al., 1993b): (1) how do we define an outlier? (in fact, there is no generally accepted definition for this term); and (2) what action should be taken when an outlier exists?

It is important to note that statisticians who have recommended the elimination of outliers from data sets, have not set generally accepted procedures for rejecting them. Moreover, the aim of these recommendations of minimizing the influence of outliers, was to improve the estimation of such distribution characteristics as the mean, which are closely associated with the central part of the probability density curve. It is obvious that these recommendations cannot be extended to the estimation of extreme events that are situated well out in the tails of a distribution.

Some hydrologists have suggested that AMS or POT distributions should be "insensitive" to outliers, especially in the tail in which the extremes under investigation are situated. It has been suggested, for instance, that certain methods of parameter estimation are "good" because they are "robust to outliers in the data". In the context of estimating hydrological extremes, however, this view of favouring statistical methods that are insensitive to outliers, is wrong, and dangerous, from both the statistical and hydrological standpoints. For estimating rare or very rare hydrological events, it is essential to handle outliers with much more care. It is necessary to gather and evaluate all the physical and statistical evidence, before a decision about what to do with "outliers" is taken (Ashkar et al., 1993b). Moreover, it is advisable in our judgement to avoid altogether the labelling of uncommon or outstanding values in hydrological data sets, as "outliers". We believe that the use of this strong term tends to elicit some negative feeling towards these data values (which are probably in most instances observations that have actually occurred, and not necessarily the result of any errors in data reporting). The use of this term may unjustifiably lead a practitioner to reject "outlying observations", or to give them lesser weight than needed for the proper estimation of extreme events. We think that a more appropriate (and somewhat "milder") term should be used, such as "singular", "uncommon", or "exceptional" data values.

To be able to handle samples with singular or exceptional values in data series more adequately, it might be helpful to be able to "detect" or identify them by quantitative test. The Grubbs-Beck (1972) test (Bobée and Ashkar, 1991) is an example of one possible statistical test to use.

6. Choice of D/E Procedures for AMS- / POT- Series Analysis

After having: (1) checked the AMS/POT sample for the fore-mentioned conditions of randomness, independence, homogeneity, and stationarity and (2) decided about how to handle any singular or exceptional values that might be present in the data, the usual next step is to fit the sample by an "appropriate" distribution (D), using a method of parameter estimation (E). This "D/E-procedure" is then used to make inferences about extreme events such as high or low quantiles, and their associated confidence intervals.

6.1. Generalities about the Choice of Distribution (D): Associated with the choice of distributions (D) for fitting an AMS or POT sample are some complex considerations such as: (1) choice of the number of distribution parameters; (2) specification of "desirable" properties that the distribution should possess; (3) deciding about the criteria to be used for selecting

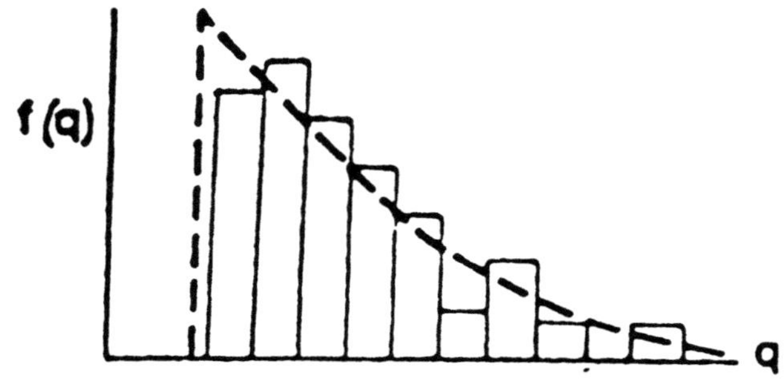

Figure 1. Distribution pdf for POT exceedances (WMO, 1989)

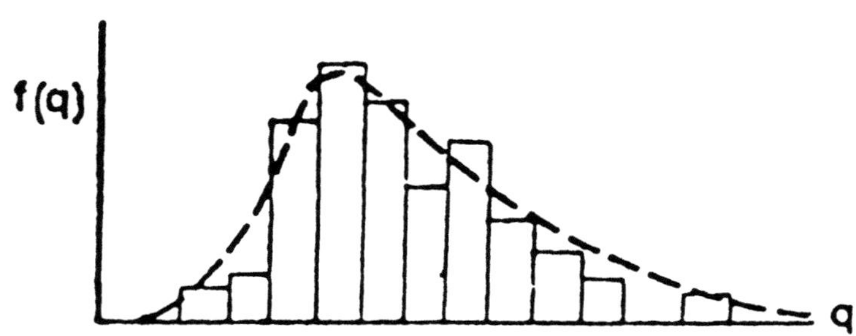

Figure 2. Distribution pdf for AMS series (WMO, 1989)

between two or more distribution pdf's (especially, if these differ in their number of parameters); and (4) identification of the factors that determine the "descriptive" and "predictive" abilities of distributions (Cunnane, 1987).

A general feature considered desirable when picking a POT exceedance distribution (but which is generally not so when choosing an AMS distribution), is the monotonic decrease of the distribution pdf (i.e., pdf with no mode). This concept is illustrated in figures 1 and 2. Some common distributions that have this characteristic are the exponential, generalized Pareto (GP), Weibull and gamma, for certain choices of their shape parameter. However, it should be mentioned that even in modeling POT exceedance distributions, it may be necessary to consider modal distributions, especially when the distribution is to be used for descriptive purposes (Rasmussen et al., 1993). For descriptive purposes of AMS series, it is often convenient to have at least three distribution parameters, in order to provide sufficient flexibility to the large variety of AMS data found in practice. However, with the typically small AMS samples available, two-parameter distributions may in many cases be preferred for prediction, because a lower number of parameters tends to reduce sampling variance in the extrapolation range (Ashkar et al., 1993b).

It has been suggested by some authors that distributions to be used for estimating hydrological extremes from POT or AMS series should have a cumulative distribution function (cdf) and an inverse cdf that can be expressed in closed form, and that these distributions should not be computationally complex. In our opinion, these distribution characteristics are desirable and attractive, but they should not form a strict basis for choosing any one type of distribution for estimating hydrological extremes over another.

A recent book by Bobée and Ashkar (1991) highlights the functional relationships between a group of distributions recommended in a number of countries for hydrological frequency analysis, and discusses the shape flexibility of these distributions. In particular, the book discusses: (1) how one single diagram can be used to show the functional relationships between some of the most commonly used distributions in estimating hydrological extremes (see also Ashkar et al., 1993a); and (2) how moment-ratio diagrams (e.g., diagrams that show the relationship between the coefficient of skewness and coefficient of kurtosis for different distributions) can be used to show the flexibility in shape of these various distributions (see also Bobée et al., 1993b). Ashkar et al. (1993a) also classify, in a tabular form, some of the important distributions used in hydrology, according to their right-tail characteristics (asymptotic properties).

6.2. Notes about Parameter-estimation methods (E): Although the method of maximum likelihood (ML) is favoured by many statisticians for fitting a wide variety of statistical distributions, it has often failed to give the desired accuracy in estimating extremes from hydrological data. It is known that for many statistical distributions, the ML method yields asymptotically optimal parameter estimators (unbiased, minimum variance), but it may produce bad estimators in small samples (especially when the random variable is restricted to an interval that depends on the parameters). This has frequently led hydrologists to resort to other fitting methods, among which the method of moments (MM) was often favoured for its simplicity. In the case of three-parameter distributions, the classical use of this method is based on the sample mean, variance, and coefficient of skewness. Much hydrological research has been directed to the study of higher order moments of distributions (moments of order 3 or more) because of their usefulness as shape indices for AMS distributions, and because of their large influence on the distribution tails. The sample coefficient of skewness (which is function of moments of orders 3 and 2) has been given special attention, but recent tendencies have been to avoid the use of moments of order 3 or more, in favour of moments that are less variable. Among the lower-order moments that have been suggested are the first- and second- order moments in log space ("method of mixed moments"; Rao, 1980; Phien and Hira, 1983), and the geometric and harmonic means in real space ("generalized method of moments"; Ashkar et al., 1988). Greenwood et al. (1979) have also proposed the use of "probability weighted moments" (PWM), which are linear combinations of order statistics (Landwehr et al., 1979; Hosking et al., 1985b, Hosking and Wallis, 1987). The use of "L-moments", which are derived from PWM's, but which have a clearer statistical interpretation, have been proposed by Hosking (1990).

These are only examples of methods that have been suggested for fitting distributions to hydrological data. Several other methods have been proposed, such as least squares, principle of maximum entropy, incomplete means, best linear combination of order statistics, and others (see, e.g., Jain and Singh, 1987; Arora and Singh, 1989; Phien, 1987; Raynal and Salas, 1986). We shall not go into detail about all these parameter estimation methods, but an important point to make about them is that they can differ greatly in the weights they give to the different elements of the sample. Methods that use moments in log space, for example, differ significantly from those that use moments in real space. This difference between methods can have pronounced effect on the estimation of quantiles situated out in the tails of a distribution (extremes), especially when exceptional values are present in the data. Needless to emphasize also, that the optimal choice of the distribution fitting method to be used in extreme quantile estimation, is closely associated with the type of distribution being considered (log-normal, gamma, Weibull, ...).

It is also very important to keep in mind that the ultimate interest of extreme-value hydrological analysis is the estimation of rare events (such as extreme quantiles), and not in the variability of any intermediate statistics that enter into the estimation of these extremes, such as sample moments. In light of this fact, recommendations that have sometimes been made that higher order sample moments be avoided in estimating hydrological extremes (because of their large variability), need to be examined more critically. One important point that has not been sufficiently emphasized is that the variability of quantile estimators of a distribution is a function not only of the variability of the sample statistics that are used in the estimation, but also of the *correlation* that exists between these statistics (Ashkar and Bobée, 1987).

6.3. Difficulties in choosing a D/E procedure for estimating hydrological extremes: The shortness and diversity of the commonly available AMS and POT hydrological records can pose serious difficulties in selecting an "appropriate" D/E procedure for analyzing them (typical record lengths of river discharge measurements around the world, for example, rarely exceed fifty years). Uncertainty in the estimation of hydrological extremes comes from many sources, such as errors due to incorrect model specification, sampling errors, and other factors such as measurement, recording, and processing errors and inconsistencies. The combined uncertainty from all of these sources becomes especially pronounced when quantiles beyond the observed sample need to be estimated ("extrapolation" into the range of rare or very rare events; Ashkar and Bobée, 1987). Experience has clearly shown that when interest lies in this tail area of a distribution, frequency estimates can differ substantially under alternate models that may fit the data well in the central area ("interpolation" range). In extrapolations, modeling error occupies an important portion of the global error, but in interpolation, the global uncertainty is mainly due to sampling (Bobée et al., 1993a). Due to these compounded errors that enter into extrapolation from short hydrological data sets, Canadian practice in the area of flood frequency analysis limits specification of return periods that can be estimated from currently available streamflow data sets, to periods of 200 years or less, "except in special circumstances" (ACH, 1989).

Selecting between two D/E procedures for estimating hydrological extremes poses special difficulties when the two distributions, D, differ in their number of parameters (differences in flexibility). Three complementary tools have been commonly used for choice of D/E procedure [see, for example, Bobée and Ashkar (1991) for more detail]: (1) visual evaluation of the data (e.g., plotting the sample values and the fitted probability Distribution on appropriate probability paper), (2) use of moment-ratio diagrams (MRD), and (3) use of statistical tests of goodness of fit. These are all weak tools, however, when small samples are involved. Goodness-of-fit tests tend to put more emphasis on goodness of fit than on uncertainty due to any additional parameters that have been entered into the model. In this way, they tend to be biased towards the most flexible distribution, or the one that has the largest number of parameters (Ashkar et al., 1993b; Chow and Watt, 1992). Therefore, a choice of D/E procedure based on an "after-the-fact" examination of a distribution's goodness of fit to the data, should be expected to be less and less reliable as the need for extrapolation beyond this range becomes more pronounced (estimation of rare or very rare events).

Two important things of interest to the hydrologist when estimating an extreme event from a set of observations are: (1) the most appropriate D/E procedure to use for estimating this extreme event, and (2) some measure to assess the statistical accuracy of the obtained estimate. The two main approaches that have been used to compare D/E procedures are (1) analytical, and (2) simulation methods. One major disadvantage of these two methods, however, is that they use the unverifiable assumption that the data (usually one single sample) comes from some well known Distribution. For data sets that do not satisfy this assumption, the degree of confidence in these analytical and simulation studies becomes very much diminished.

Improvements brought about in the past number of years in the computer field have resulted in interest in a group of statistical methods known as "resampling techniques". Probably the most popular of these methods is the "bootstrap" technique (Efron, 1979, 1987; Ashkar and Ouarda, 1993). We believe that these "computationally intensive" techniques have certain important practical advantages over classical simulation and analytical statistical methods which deserve to be examined more intensively in hydrology. One main advantage of these methods is their freedom from limiting assumptions about the data, and from the need to rely on exact simple analytical manipulations. They allow the numerical exploration of complex properties of the samples that are beyond the reach of exact mathematical analysis, thereby opening the way for a wider array of statistical tools that can be explored and used. Two important possible applications of these procedures in hydrological frequency analysis are: (1) to estimate the strongly needed statistical accuracy of quantile estimates obtained by applying a D/E procedure to a single observed hydrological data set, and (2) to choose between two candidate D/E procedures for extreme-value analysis, on the basis of one observed sample.

With regard to the POT model, various of its elements need to be pursued further, as stressed recently by Rasmussen et al. (1993). The main strength of this model is that by appropriate selection of the threshold, it allows for a better inclusion/exclusion of extreme events in the analysis, as compared to AMS modeling. Some important points that need to be pursued further with regard to POT modeling are:

- Search for the best criteria to determine which peaks of the hydrological variable under study should be included in the extreme value analysis. The establishment of such criteria is particularly important because it concerns the very definition of POT series (Rasmussen et. al., 1993).

- Analysis of different approaches to defining events or "peaks", and evaluating the advantages and disadvantages of these approaches. In particular, analysis and comparison should be done between: (1) methods which define peaks by using a predetermined threshold; (2) methods which do not make any reference to a threshold.

- Search for the best criteria to be used in selecting appropriate thresholds. Some procedures that have been suggested include: (1) choice based on fixing a priori the number of exceedances to be included in the analysis (an average of 2-3 exceedances per year, for example), (2) using a given quantile calculated from the AMS series as a basis for threshold selection, (3) choice based on putting special attention to the fit of exceedances to a specific distribution type (exponential, Weibull or generalized Pareto, for example). The advantages and disadvantages of these different criteria should be investigated and assessed in an aim to reach the largest possible flexibility in the POT model.

- Study of methods of visual inspection of the hydrological time series prior to the selection of a threshold. Also documentation and explanation of these methods should be done, for ease of use by practitioners (e.g., how to carefully inspect a discharge hydrograph for the purpose of flood frequency analysis, with the least subjectivity possible).

- Intensive study of hydrological POT series from many different hydrological regions (humid, arid, semi-arid, etc.) in order to better understand their stochastic characteristics and to help achieve POT modeling that best describes the realities of hydrological data.

- Search for best measures to ensure independence between exceedances, by imposing for example restrictions on the time gap between successive events ("peaks"), or by similar measures.

- Study of different types of candidate distributions to model exceedance magnitudes, with an analysis of their advantages and disadvantages, especially with respect to the prediction of rare events.

- Additional studies on how to include seasonality into POT modeling.

- More comparisons between POT and AMS modeling, using short and long data series, with emphasis on the ability of these types of models to predict rare events.

7. Some Controversies

The attention that has been given over the past 40 to 50 years to the problem of frequency analysis of hydrological extremes, and the comparisons that have been made of the various AMS and POT methodologies used in this area, have resulted in some general recommendations concerning "appropriate" D/E procedures to be used. However, these recommendations have often failed to satisfy the needs of practitioners because they still contain points that need further statistical and/or physical justification, or that require further discussion. In the area of flood frequency analysis, for example, Bobée et al., (1993a) suggest that there is a state of confusion among practitioners as to which procedures are best to use. According to Ashkar et al., (1992b, 1993b), many of the procedures that have been developed, still lack precision and often use misleading terminology. The information generally available to the practitioner on the advantages and disadvantages of the different proposed methods is often not sufficient. According to Potter (1987), much of the research on flood frequency analysis has "lacked balance", and has not put sufficient emphasis on the physical characteristics of the problem. According to Lettenmaier (1988), researchers have spent more time on developing new methods than on testing existing ones.

It always serves an instructive purpose to review from a historical perspective the commonly used procedures in a certain area of research, or engineering practice, in order to better understand the strengths and weaknesses of these procedures. A detailed historical review of D/E procedures used in estimating hydrological extremes is out of the scope of the present work; however, Tables 1 & 2 are intended to provide a quick look at the diversity of some of the procedures that have been used in the special area of flood frequency analysis. With an emphasis on AMS- and POT-based methodologies, these two tables highlight some of the important issues that have been addressed in this area, but they are not intended to make an extensive review of AMS and POT procedures.

There is an indication from Tables 1 & 2 that the methods used in different countries for flood frequency analysis have been influenced by historical traditions. A gap still exists between theory and practice in this area, and in similar areas pertaining to the estimation of hydrological extremes, which has sometimes led to hesitation among practitioners in applying new methods. So, if it is true that a state of confusion exists in the area of flood frequency analysis, as suggested by Bobée et al. (1993a), then it might be safe to conclude also that many elements of this state extend to similar areas of hydrological extreme-value analysis. The attempts that have been made in the past to survey and compare the various frequency estimation methodologies of hydrological extremes, have been useful in focusing on the problem as a whole, but have also pointed up the complexity of the issues under consideration. They have confirmed that some controversies still exist, and that the discussion needs to be pursued on many of today's employed hydrological extreme-value estimation techniques.

8. Conclusions

Yevjevich (1972, p. 3) states that: "treatment of data for the objectives of finding regularities in hydrology may be either by a deterministic approach, by a statistical approach, or by both simultaneously. The use of one approach independently of the other may lead to an analysis of data without a sound theoretical background or to a theoretical treatment without the proper verification of hypotheses and of discovered regularities. Future progress of hydrology may depend to a large extent on how these two approaches are combined for discovering, understanding, and generalizing hydrologic regularities of nature".

The problem of frequency analysis of hydrological extremes is a perfect example of a challenging area in hydrology in which all physical and statistical evidence should be gathered and analyzed before any wise decisions can be made about how to use these extremes. This area of estimating rare or very rare hydrological events still has some controversies that need to be sorted out and dealt with. Some misleading concepts have been used in this area, such as "robustness" or sensitivity to "outliers", or the notion of "separation of skewness", as outlined in Table 1. There is a need for continual review and discussion of analysis procedures of

Table 1. Annual-Flood-Series (AFS) Approach for Flood Frequency Analysis

1888	US Geological Survey started its stream flow gaging program (Kirby and Moss, 1987)
Horton (1913)	Was among the first to suggest the use of probability methods in runoff studies He discussed the application of the Gaussian (normal) distribution to study annual floods
Hazen (1914)	Found good agreement when fitting the log normal (LN) distribution to floods (unlike the normal distribution, which is symmetrical, the LN distribution is positively skewed)
Fuller (1914)	Used semi-log paper to plot flood flows
Foster (1924)	Presented procedures for fitting the Pearson Types 1 and 3 (P1 and P3) distributions to hydrological data
Gumbel (1941)	Applied the theory of extreme values to flood data, and proposed the method of moments to fit the Type 1 extremal distribution (EV1)
Chow (1954)	Studied the mathematical and statistical properties of the LN distribution and proposed a third parameter to be added to it for modeling hydrological events He justified the use of this LN3 distribution in that these events are often the result of the joint multiplicative action of a large number of independent causative factors By comparison, the normal distribution is justified for modeling phenomena that result from the joint *additive* action of a large number of independent factors
Jenkinson (1955)	Presented the three types of extreme-value distributions (Types 1, 2, and 3), that are used in hydrology, under one single formula By doing so, he established the 3 parameter generalized extreme value distribution (GEV)
Gumbel (1958)	Gumbel published his well-known textbook on the statistics of extremes, which contains many applications from hydrology
1960s & 1970s	A number of countries embarked on large scale studies of flood frequencies within their national boundaries, and proposed base methods to be used by their government and private agencies to achieve uniformity in flood frequency estimation In the USA (USWRC, 1967, Benson, 1968) the log Pearson Type 3 (LP3) distribution was recommended, and a study in the UK (NERC, 1975) led to the recommendation of the GEV distribution The P3 and LP3 distributions were later recommended in West Germany (Mosonyi et al , 1975), and the LP3 distribution was recommended by the Australian Institution of Engineers (1977)
Kritsky & Menkel (1969)	Proposed the use of the generalized gamma (GG) distribution for flood frequency analysis in the Soviet Union This distribution is also known among hydrologists as the Kritsky-Menkel distribution
Matalas & al (1975)	Designed a study in the U S A for identifying appropriate distributions for flood frequency analysis in which they considered a number of hydrologically "homogeneous" regions and examined the regional variability of the coefficient of skewness of annual-maximum-flood (AMF) series in these regions The study concluded that regional estimates of skewness coefficients tend to be more variable than those generated randomly from a group of distributions commonly used in hydrology This phenomenon was termed the "separation of skewness" or "separation effect" Some authors have set a criterion that for a distribution to be adequate for flood frequency analysis, it must be able to reproduce as much variability in skewness values, as is observed in empirical data sets
Houghton (1978), Rossi & al (1984)	Introduced two distributions to explain the "separation phenomenon" the 5 parameter Wakeby distribution, and the 4-parameter TCEV
Klemes (1976), Ashkar & Bobée (1989), Ashkar & al (1992a)	Criticized the use of the separation phenomenon as a basis for choosing the type of probability distribution to be used in a flood frequency analysis Ashkar et al (1992a) provided evidence that a primary cause of this phenomenon was the spatial mixing of different values of coefficient of skewness within *homogeneous regions* (in which flood series are assumed to have constant skewness), which was not accounted for in the study by Matalas et al (1975) Their main conclusion was that the separation of skewness cannot be used as a criterion for choosing the type of probability distribution a flood frequency analysis

Table 2. POT Approach for Flood Frequency Analysis

Langbein (1949)	Was among the first to use probabilistic concepts to determine a relationship between AFS and POT series
Borgman (1963), Shane & Lynn (1964), Bernier (1967), Todorovic & Zelenhasic (1970)	Were instrumental in developing the structure of the POT approach
Miquel (1984)	Published a book in France that discusses the POT approach in some detail
British Flood Studies Report (NERC, 1975), Canadian Flood Guide (ACH, 1989)	Gave guidelines for the application of the POT method The NERC report contains a whole chapter on this method
Todorovic & Woolhiser (1972), Gupta & al (1976), Karr (1976), Todorovic (1978)	Studied the joint distribution of floods in time and magnitude
Kirby (1969), Cunnane (1979), Kavvas (1982a b), Ashkar & Rousselle (1983b, 1987), Ben Zvi (1991),	Studied the arrival structure of flood events (e g , number of exceedances in a given time interval, Poisson assumption, application of the negative Binomial distribution, modeling flood events that occur in bunches, etc)
Zelenhasic (1970), Pickands (1975), Taesombut & Yevjevich (1978), Ashkar & Rousselle (1983a), Miquel (1984), Smith (1984) Van Monfort & Witter (1985, 1986), Hosking & Wallis (1987), Fitzgerald (1989), Jin & Stedinger (1989), Rosbjerg et al (1992), Rosbjerg et al (1991),	Discussed the modeling of exceedance magnitudes (distributions that are sensitive/insensitive to the choice of truncation, the exponential and generalized Pareto (GP) distributions, etc)
Smith (1987), Wang (1991)	Selected a high threshold level and focused on a few of the largest flood peaks
Yevjevich & Taesombut (1979), Rosbjerg (1985), NERC (1975), Cunnane (1979)	Studied the dependence among exceedances, and its implications on the estimation of extremes when peak values are separated by small time gaps
Rosbjerg & Madsen (1992)	Reviewed various methods to choose a threshold level in the POT model
Todorovic & Rousselle (1971), North (1980), Ouarda et al (1993)	Studied and proposed models to describe seasonal variations in flood arrivals and magnitudes
Cunnane (1973), Rousselle & Hindie (1976), Ashkar & Rousselle (1981), Konecny & Nachtnebel (1985), Nachtnebel & Konecny (1987), Rasmussen & Rosbjerg (1989), Buishand (1990), Rasmussen & Rosbjerg (1991 a,c)	Studied the precision of flood quantile estimators
Langbein (1949), Cunnane (1973), Yevjevich & Taesombut (1979), Tavares & Da Silva (1983), Rosbjerg (1985), Ashkar et al (1987), Rasmussen & Rosbjerg (1989), Jin & Stedinger (1989), Wang (1991)	Compared POT and annual maximum flood (AMF) methodologies for precision in estimating flood quantiles

hydrological extremes, in order to arrive at schemes and criteria for comparing them, that are as broadly agreed upon among hydrologists as possible. Bobée et al., (1993a) argue that a comprehensive and systematic approach to the comparison of flood-frequency-analysis procedures is needed. Criteria for comparing these D/E procedures should not be only statistical in nature, such as mean square error or bias of quantile estimates, but it would be more beneficial to consider a multiple-criteria-decision approach that allows for other criteria to be incorporated such as computational ones (e.g., computer program availability, ease of use), or a user-related criterion such as acceptability (Duckstein et al., 1991; Ashkar et al., 1993b).

We have reflected on certain aspects of AMS- and POT-data-analysis procedures which we consider important from a practical and scientific viewpoint. However, it was not our attempt to make an extensive review of this subject. Two particularly important areas that were not touched upon are "regional estimation methods of hydrological extremes" and the use of "historical information". These methods have been proposed and found very useful in reducing model and sampling uncertainties inherent in hydrological extreme-value estimations when short "local" or "at-site" records are involved, or to obtain estimates from regional information when there is no local record at all. One notes that up till now the POT method has rarely been used in a regional context, so we feel there is a need for regional procedures based on this method to be developed and applied.

9. Acknowledgements

The author would like to thank Dr. Taha B.M.J. Ouarda for his kind help and suggestions concerning this paper. Financial support for this research was provided by the Natural Science and Engineering Research Council of Canada (NSERC).

10. References

ARORA, K. and V.P. SINGH (1989). A comparative evaluation of the estimators of the Log Pearson type (LP) 3 distribution. Journal of Hydrology, 105: 19-37.

ASHKAR, F. and BOBÉE, B. (1986). Comment on: variance of the T-Year event in the log-Pearson Type 3 distribution. J. Hydrol., 84(1), 181-187.

ASHKAR, F. and BOBÉE, B. (1987). The generalized method of moments as applied to problems of flood frequency analysis: some practical results for the log-Pearson Type 3 distribution. J. Hydrol., 90(3-4), 199-217.

ASHKAR, F. and BOBÉE, B. (1989). Reflection on some problems of flood frequency analysis. EOS: Trans., American Geophysical Union 70(15), 325 (abstract).

ASHKAR, F. and OUARDA, T. B.M.J. (1993). Computer intensive resampling techniques in hydrologic frequency analysis: Application of the bootstrap method. Proceedings of the 11th Canadian Hydrotechnical Conference, CSCE/ASCE, Fredericton, N.B., Canada, Vol. I: 149-158.

ASHKAR, F. and ROUSSELLE, J. (1981). Design discharge as a random variable: A risk study. Water Resour. Res., 17: 577-591.

ASHKAR, F. and ROUSSELLE, J. (1983a). Some remarks on the truncation used in partial flood series models. Water Resour. Res., 19: 477-480.

ASHKAR, F. and ROUSSELLE, J. (1983b). The effect of certain restrictions imposed on the interarrival times of flood events on the Poisson distribution used for modeling flood counts. Water Resour.

Res., 19: 481-485.

ASHKAR, F. and ROUSSELLE, J. (1987). Partial duration series modeling under the assumption of a Poissonian flood count. J. Hydrol., 90: 135-144.

ASKHAR, F., EL-JABI, N., and BOBÉE, B. (1987). On the choice between annual flood series and peaks over threshold series in flood frequency analysis, in A.D. Feldman (ed.), Proc. Eng. Hydrol., Virginia, August 3-7, 1987, ASCE, New York, pp. 276-280.

ASHKAR, F., BOBÉE, B., LEROUX, D. and MORISETTE, D. (1988). The generalized method of moments applied to the generalized gamma distribution. Stochastic Hydrol. Hydraul., 2, 161-174.

ASHKAR, F., BOBÉE, B. and BERNIER, J. (1992a). Separation of skewness: reality or regional artifact? J. Hydraul. Eng., ASCE, 118(3), 460-475 (with discussion).

ASHKAR, F., BOBÉE, B. and F. FODDA (1992b). Review of statistical criteria and techniques used in flood frequency analysis with computer applications. International Journal of Modelling and Simulation 12(4): 139-142.

ASHKAR, F., BOBÉE, B. and T. B.M.J. OUARDA (1993a). Functional relationships and asymptotic properties of distributions of interest in hydrologic frequency analysis. International Conference on Hydrology and Water Resources, New Delhi, India, Dec. 20-22.

ASHKAR, F., BOBÉE, B., RASMUSSEN, P. and D. ROSBJERG (1993b). A perspective on the annual maximum flood approach to flood frequency analysis. International Conference on Stochastic and Statistical Methods in Hydrology and Environmental Engineering, Waterloo, Ont., Canada, June 21-23.

ASSOCIATE COMMITTEE ON HYDROLOGY (ACH) (1989). *Hydrology of Floods in Canada: A Guide to Planning and Design, Chapter 5: Statistical Frequency Analysis of Hydrologic Data* (Ottawa: National Research Council Canada).

AUSTRALIAN INSTITUTION OF ENGINEERS (1977). *Australian Rainfall and Runoff: Flood Analysis and Design.*

BEN-ZVI, A. (1991). Observed advantage for negative binomial over Poisson distribution in partial duration series. Stochastic Hydrol. Hydraul., 5: 135-146.

BENSON, M.A. (1968). Uniform Flood-Frequency Estimating Methods for Federal Agencies. Water Resour. Res., 4(5), 891-908.

BERNIER, J., (1967). Sur la th orie de renouvellement et son application en hydrologie. HYD. 67(10), Elec. France.

BOBÉE, B. and ASHKAR, F. (1991) The Gamma Family and Derived Distributions Applied in Hydrology. Water Resources Publications, Littleton, Colo., 217 p.

BOBÉE, B., CAVADIAS, G., ASHKAR, F., BERNIER, J. and RASMUSSEN, P. (1993a). Towards a systematic approach to the comparison of distributions used in flood frequency analysis. J. Hydrol., 142, 121-136.

BOBÉE, B., ASHKAR, F. and L. PERREAULT (1993b). Two kinds of moment ratio diagrams and their applications in hydrology. Stochastic Hydrology and Hydraulics, 7: 41-65.

BORGMAN, L.E. (1963). Risk criteria. J. Waterw. Harbors Div., 80: 1-35.

BUISHAND, T.A. (1990). Bias and variance of quantile estimates from a partial duration series. J. Hydrol., 109: 35-49.

CHOW, K.C.A. and WATT, W.E. (1992). Use of Akaike information criterion for selection of flood frequency distribution. Canadian Journal of Civil Engineering, 19, 616-626.

CHOW, V.T. (1954). The log-probability law and its engineering applications. Proc. ASCE, 80(5): 1-25.

CRUISE, J.F. and ARORA, K. (1990). A hydroclimatic aplication strategy for the Poisson partial duration model. Water Resour. Bull., 26(3): 431-442.

CUNNANE, C. (1973). A particular comparison of annual maxima and partial duration series methods

of flood frequency prediction. J. Hydrol., 18: 257-271.
CUNNANE, C. (1979). A note on the Poisson assumption in partial duration series models. Water Resour. Res., 15: 489-494.
CUNNANE, C. (1987). Review of Statistical Models for Flood Frequency Estimation. In V.P. Singh (ed.), *Hydrologic Frequency Modeling* (Dordrecht, Holland: Reidel), 49-95.
DUCKSTEIN, L., BOBÉE, B. and ASHKAR, F. (1991). A multiple criteria decision modelling approach to selection of estimation techniques for fitting extreme floods. Stochastic Hydrology and Hydraulics, 5(3), 227-238.
EFRON, B. (1979). Computers and the theory of statistics: thinking the unthinkable. Society for Industrial and Applied Mathematics, 21(4): 460-480.
EFRON, B. (1987). Better bootstrap confidence intervals. Journal of the American Statistical Association, 82(397): 171-200.
FITZGERALD, D.L. (1989). Single station and regional analysis of daily rainfall extremes. Stochastic Hydrol. Hydraul., 3: 281-292.
FOSTER, H.A., (1924). Theoretical frequency curves and their application to engineering problems. ASCE 87, 142-203.
FULLER, W. E., (1914). Flood Flows. Trans. Am. Soc. Civ. Engrs., 77, 564-617.
GREENWOOD, J.A., LANDWEHR, J.M., MATALAS, N.C. and WALLIS, J.R. (1979). Probability weighted moments: Definition and relation to parameters of several distributions expressable in inverse form. Water Resour. Res., 15(5), 1049-1054.
GRUBBS, F. E. and BECK, G. (1972). Extension of sample sizes and percentage points for significance tests of outlying observations. Technometrics, 14, 847-854.
GUMBEL, E. J., (1941). The Return Period of Flood Flows. Annals of Mathematical Statistics, 12, 163-190.
GUMBEL, E. J., (1958). *Statistics of Extremes* (New York: Columbia University Press).
GUPTA, V.K., DUCKSTEIN, L., and PEEBLES, R.W. (1976). On the joint distribution of the largest flood and its time of occurrence. Water Resour. Res., 12: 295-304.
HAZEN, A. (1914). Storage to be provided in impounding reservoirs for municipal water supply. Trans., ASCE, Pap. 1308, 77: 1547-1550.
HORTON, R.E. (1913). Frequency of recurrence of Hudson River floods. U.S. Weather Bureau Bull., 2, pp. 109-112.
HOSKING, J.R.M. (1990). L-moments: Analysis and estimation of distributions using linear combination of order statistics. J. Roy. Stat. Soc., Series B, 52(1), 105-124.
HOSKING, J.R.M. and WALLIS, J.R. (1987). Parameter and quantile estimation for the generalized Pareto distribution. Technometrics, 29(3), 339-349.
HOSKING, J.R.M., WALLIS, J.R. and WOOD, E.F. (1985a). An appraisal of the regional flood frequency procedure in the UK flood studies report. Hydrol. Sci. J., 30, 85-109.
HOSKING, J.R.M., WALLIS, J.R. and WOOD, E.F. (1985b). Estimation of the generalized extreme value distribution by the method of probability weighted moments. Technometrics, 27(3), 251-261.
HOUGHTON, J.C. (1978). Birth of a parent: the Wakeby distrbution for modeling flood flows. Water Resour. Res. 14: 1105-1110.
JAIN, D. and V.P. SINGH (1987). Estimating parameters of EV1 distribution for flood frequency analysis. Water Resour. Bulletin, 23(1): 59-145.
JENKINSON, A.F. (1955). The frequency distribution of the annual maximum (or minimum) of meteorological elements. Quarterly J. Royal Meteo. Soc., 81: 158-171.
JIN, M. and STEDINGER, J.R. (1989). Partial duration series analysis for a GEV annual flood

distribution with systematic and historical flood information. Dept. Civ. Eng., Penn. State Univ., PA. (unpublished paper).

KARR, A.F. (1976). Two extreme value processes arising in hydrology. J. Appl. Prob., 13: 190-194.

KAVVAS, M.L. (1982a). Stochastic trigger model for flood peaks. 1. Development of the model. Water Resour. Res., 18: 383-398.

KAVVAS, M.L. (1982b). Stochastic trigger model for flood peaks. 2. Application of the model to the flood peaks of Goksu-Karahacili. Water Resour. Res., 18: 399-411.

KIRBY, W. (1969). On the random occurrence of major floods. Water Resour. Res., 5: 778-784.

KIRBY, W. and MOSS, M. (1987). Summary of flood frequency analysis in the United States. J. Hydrol., 96: 5-14.

KLEMES, V. (1976). Comment on: Regional skew in search for a parent, by N.C. Matalas, J.R. Slack and J.R. Wallis. Water Resour. Res., 12(6), 1325-1326.

KONECNY, F. and NACHTNEBEL, H.P. (1985). Extreme value processes and the evaluation of risk in flood analysis. Appl. Math. Modelling, 9: 11-15.

KRITSKY, S.N. and M.F. MENKEL (1969). On principles of estimation methods of maximum discharge. Floods and their computation. I.A.H.S., No 84.

LANDWEHR, J.M., MATALAS, N.C. and WALLIS, J.R. (1979). Probability weighted moments compared with some traditional techniques in estimating Gumbel parameters and quantiles. Water Resour. Res., 15(5), 1055-1064.

LANGBEIN, W.B. (1949). Annual floods and the partial duration flood series. Transaction, Am. Geophys. Union, 30: 879-881.

LETTENMAIER, D.P. (1988). Evaluation and testing of flood frequency estimation methods. Proceedings, International Conference on Natural Hazards in European-Mediterranean Countries, Perugia, Italy, June.

MANN, H.B. and D.R. WHITNEY (1947). On the test of whether one of two random variables is stochastically larger than the other. Ann. Math. Statist., 18: 50-60.

MATALAS, N.C., SLACK, J.R. and WALLIS, J.R. (1975). Regional skew in search of a parent. Water Resour. Res., 11(6), 815-826.

MIQUEL, J. (1984). Guide Practique d'Estimation des Probabilités de Crues. Editions Eyrolles, Paris, 160 pp.

MOSONYI, E.F., EGGERS, H. & D. KOBERG (1975). Influence of Probability Analysis on Selection of Design Flood, paper presented at the Second World Congress of IWRA, New Delhi.

NACHTNEBEL, H.P. and KONECNY, F. (1987). Risk analysis and time-dependent flood models. J. Hydrol., 91: 295-318.

NERC (NATURAL ENVIRONMENT RESEARCH COUNCIL, LONDON) (1975). Flood Studies Report, Vols. 1-5, 1100 p.

NOETHER, G.E. (1976). Introduction to statistics: A nonparametric approach. 2nd edition. Houghton Mifflin, Boston.

NORTH, M. (1980). Time-dependent stochastic model of floods. J. Hyd. Div., 106: 649-665.

OUARDA, T. B.M.J., ASHKAR, F. and N. EL-JABI (1993). Peaks over threshold model for seasonal flood variations. Proceedings of the International Symposium on Engineering Hydrology, Hydraulic Division, ASCE, San Francisco, USA, July 25-30 (Chin Y. Kuo, Ed.): 341-346.

PHIEN, H.N. (1987). A review of methods of parameters estimation for the extreme value type 1 distribution. Journal of Hydrology, 90: 251-268.

PHIEN, H.N. and HIRA, M.A. (1983). Log-Pearson type 3 distribution: parameter estimation. J. Hydrol., 64, 25-37.

PICKANDS, J. (1975). Statistical inference using extreme order statistics. Ann. Stat., 3: 119-131.

POTTER, K.W. (1987). Research in flood frequency analysis: 1983-86. in J.R. Wallis (ed.), U.S. National Report to IUGG 1983-86, Reviews of Geophysics, 25(2), pp. 113-118 .

RAO, D.V. (1980). Log Pearson type 3 distribution: method of mixed moments. J. Hydraul. Div., ASCE, 106(6), 999-1019.

RASMUSSEN, P.F. and ROSBJERG, D. (1989). Risk estimation in partial duration series. Water Resour. Res., 25: 2319-2330.

RASMUSSEN, P.F. and ROSBJERG, D. (1991a). Evaluation of risk concepts in partial duration series. Stochastic Hydrol. Hydraul., 5: 1-16.

RASMUSSEN, P.F. and ROSBJERG, D. (1991b). Prediction uncertainty in seasonal partial duration series. Water Resour. Res., 27: 2875-2883.

RASMUSSEN, P.F. and ROSBJERG, D. (1991c). Application of Bayesian principles in regional flood frequency analysis. in G. Tsakiris (ed.), Advances in Water Resources Technology, Balkema, pp. 65-75.

RASMUSSEN, P., ASHKAR, F., ROSBJERG, D. and B. BOBÉE (1993). The POT method for flood estimation: A review. International Conference on Stochastic and Statistical Methods in Hydrology and Environmental Engineering, Waterloo, Ont., Canada, June 21-23.

RAYNAL, J.A. and J.D. SALAS (1986). Estimation procedures for the type 1 extreme value distrubution. Journal of Hydrology, 87: 315-336.

ROSBJERG, D. (1985). Estimation in partial duration series with independent and dependent peak values. J. Hydrol., 76: 183-195.

ROSBJERG, D., RASMUSSEN, P.F., and MADSEN, H. (1991). Modelling of exceedances in partial duration series. Proc. of the International Hydrology and Water Resources Symp., Perth, pp. 755-760.

ROSBJERG, D. and MADSEN, H. (1992). On the choice of threshold level in partial duration series. Proc. of the Nordic Hydrological Conference 1992, NHP-Report No 30, pp. 604-615.

ROSBJERG, D., MADSEN, H., and RASMUSSEN, P.F. (1992). Prediction in partial duration series with generalized Pareto-distributed exceedances. Water Resour. Res., 28: 3001-3010.

ROSSI, F., FIORENTINO, M. and VERSACE, P. (1984). Two-component extreme value distribution for flood frequency analysis. Water Resour. Res., 20(7), 847-856.

ROUSSELLE, J. and HINDIE, F. (1976). Incertitude dans les débits de crues: Approche bayesienne. J. Hydrol., 30: 341-349.

SHANE, R.M. and LYNN, W.R. (1964). Mathematical model for flood risk evaluation. J. Hyd. Div., 90: 1-20.

SMITH, J.A. (1987). Estimating the upper tail of flood distribution. Water Resour. Res., 23: 1657-1666.

SMITH, R.L. (1984). Threshold methods for sample extremes. In J. Tiago de Oliveira (ed.), Statistical Extremes and Applications, Reidel, pp. 621-638.

TAESOMBUT, V. and YEVJEVICH, Y. (1978). Use of partial flood series for estimating distribution of maximum annual flood peak. Hydrol. Paper 97, Colorado State University, Fort Collins, CO.

TAVARES, L.V. and DA SILVA, J.E. (1983). Partial duration series method revisited. J. Hydrol., 64: 1-14.

TODOROVIC, P. (1978). Stochastic models of floods. Water Resour. Res., 14: 345-356.

TODOROVIC, P. and ROUSSELLE, J. (1971). Some problems of flood analysis. Water Resour. Res., 7: 1144-1150.

TODOROVIC, P. and WOOLHISER, D.A. (1972). On the time when the extreme flood occurs. Water Resour. Res., 8: 1433-1438.

TODOROVIC, P. and ZELENHASIC, E. (1970). A stochastic model for flood analysis. Water Resour. Res., 6: 1641-1648.

U.S. WATER RESOURCES COUNCIL (USWRC) (1967). *A Uniform Technique for Determining Flood Flow Frequencies*, Bulletin No. 15 (Washington, DC).

VAN MONFORT, M.A.J. and WITTER, J.V. (1985). Testing exponentiality against generalized Pareto distribution. J. Hydrol., 75: 305-315.

VAN MONFORT, M.A.J. and WITTER, J.V. (1986). The generalized Pareto distribution applied to rainfall depths. Hydrol. Sci. J., 31:151-162.

W.M.O. (WORLD METEOROLOGICAL ORGANIZATION) (1989). Statistical distributions for flood frequency analysis. Operational Hydrology, Report N° 33. WMO Publ. N° 718, WMO, Geneva, Switzerland.

WALD, A. and WOLFOWITZ, J. (1943). An exact test for randomness in the non-parametric case based on serial correlation. Ann. Math. Stat. , 14, 378-388.

WALLIS, J.R. and WOOD, E.F. (1985). Relative accuracy of Log Pearson 3 procedures. J. Hydr. Eng., ASCE, 111(7), 1043-1057.

WANG, Q.J. (1991). The POT model described by the generalized Pareto distribution with Poisson arrival rate. J. Hydrol., 129: 263-280.

WAYLEN, P. and WOO, M.K. (1982). Prediction of annual floods generated by mixed processes. Water Resour. Res., 18: 1283-1286.

YEVJEVICH, V. (1972). Probability and statistics in hydrology. Water Resources Publications. Fort Collins, Colorado, 302 p.

YEVJEVICH, V. and TAESOMBUT, V. (1979). Information on flood peaks in daily flow series. in E.A. McBean, K.W. Hipel, and T.E. Unny (eds.), Inputs for Risk Analysis in Water Systems, Water Resources Publications, Fort Collins, Colo, pp. 171-192.

ZELENHASIC, E. (1970). Theoretical probability distributions for flood peaks. Hydrol. Paper 42, Colorado State Univ., Fort Collins, CO.

ESTIMATION OF EXTREME FLOODS WITH PARTICULAR REFERENCE TO SOUTHERN AFRICA

Professor D. Stephenson
Water Systems Research Group
Department of Civil Engineering
University of the Witwatersrand, JHB
Private Bag 3, Wits 2050
Republic of South Africa

ABSTRACT. The prediction of extreme floods associated with different recurrence intervals has always been a difficult task. Even the alternative of estimating maximum probable floods is fraught with uncertainty. The methodology is not yet standardized and the datums and factors to consider are not all incorporated in any single procedure for estimating these floods. The estimates are even more difficult in areas with limited data. Developing areas in particular have sparse hydrological networks and the spectrums of floods gauged in these areas provide limited, if not at times misleading, information. Rain data are often more reliable and more extensive and therefore many methods for estimating floods use rainfall data. The paper compares alternative methods such as the rational method, unit hydrograph method, soil conservation method, and empirical methods. Owing to wide variations in topography and climate over large catchments it is demonstrated that many of these methods cannot be scaled up or extrapolated. It appears catchment modelling using hydraulic techniques is the most reliable where extrapolation is required. Various empirical methods of estimating extreme floods are compared and it is shown that for large catchments available data illustrates the floods converge to common figures whichever data are used.

Background

Southern Africa is a region where rapid development is taxing the resourcefulness of hydrologists. Limited hydrological data have to be extrapolated to estimate river flows and floods. Owing to the vast variations in climate and landform, cross catchment correlations are unreliable. Regional flood formulae have been proposed for temperate zones but the extreme effects of cyclones on the one hand, and flood absorbing swamps or arid plains on the other, make hydraulic modelling a necessity for any flood estimation.

The area covers the Kalahari desert, the tropical rain forests of Zambia and Northern Mocambique, and the cyclonic east coast of Natal. In the later case despite reasonable gauging records, mathematical extrapolations have proved unreliable. In some other areas streamflow records are poor or non-existent. This sometimes applies to rainfall records which are often more readily available but not in ideal form. Methods of extrapolating limited data are presented. The method selected in each case should optimize the use of data available.

Methods used for estimating floods varies widely from country to country and the estimates are often open to question. A number of methods are based on restrictive assumptions. Many of these techniques are based on rainfall estimates over critical storm durations and monographs or equations for derivation of the flood from the rainfall data. Many of these techniques are sufficiently sophisticated to satisfy engineers' or

V. P. Singh and B. Kumar (eds.), Surface-Water Hydrology, 505–520.

hydrologists' sense of responsibility with regard to efforts in estimating floods ie. the method is often selected to suit either the funding or manhours considered necessary for the job or the engineer's conscience rather than on whether it is able to produce an adequate answer.

Unfortunately not only is there lack of hydrological data for estimating floods in many areas but also the technology for determining such floods is not developed. The extreme range in hydrology over relatively short distances makes extrapolation from one region to another dangerous and estimates are therefore unsubstantiated. It may take years to build up adequate data. Projections should take cognizance of the wide band of uncertainty in estimates and provision should be made in the design of bridges and waterways for revising designs if more accurate data comes to light requiring such revision. The significance of the selected design recurrence interval is also different in the environment considered. In rural areas flooding every few years is of lesser consequence than high bridge costs. (Structural designs on the other hand need to be based on higher recurrence intervals.)

The significance of an extreme event such a 1 000 year flood is often hidden in numbers. Although there is a chance a 1 000 year flood will occur in any year most mathematical methods of estimating its recurrence interval will give it a much lower recurrence interval. This is because historical records exist only for a much shorter period. For example if 100 years of flood records exist, the worst flood occurring in that period may be given a 100 year recurrence interval (using RI = rank/no. years record = 1/100). In fact there is a 10% chance that the 1 000 year flood occurred in that period, so many such assessments must be incorrect).

The recurrence interval calculated is supposed to represent the inverse of the probability of an event being exceeded under prevailing conditions. However the real probability of the event occurring in any extended period must account for climatic changes and change in catchment characteristics etc. Such changes can and do occur and should be considered in any hydrological projections if future flood risks are to be assessed. Catchment modelling would appear to be one way of allowing for such factors.

The PMF (probable maximum flood) provides a guide to the upper limit of floods. The PMF can be obtained empirically or deterministically. In the former case a plot of world flood peaks versus catchment area is given an envelope. This envelope may creep upwards as more data are obtained. For the deterministic method a catchment model can be used in conjunction with an estimate of PMP (probable maximum precipitation). PMF generated by rainfall modelling could account for changes in climate so it may be unrelated to projections of existing data. The WMO (1983) suggests two methods for simulating PMP - storm models, and storm maximization and transposition. The latter method is again empirical, so even if used with a deterministic flood model it results in a semi-empirical flood estimate.

The Unit hydrograph procedure is used extensively in South Africa for flood determination (Hydrological Research Unit, 1972; Pitman and Midgley, 1967). Regionally generalized 1-hour unitgraphs are multiplied by excess rainfall and summated over the assumed storm duration to obtain an S-curve. Hydrographs for any finite storm duration are then obtained by subtraction. The method assumes a linear excess rainfall runoff

TABLE 1 SIGNIFICANT FLOODS IN SOUTHERN AFRICA

River	Country	Catchment	Descript	MAP mm/a	MAR Mm3/a	Max Flood m3/s	Year	No of yrs
Limpopo	S Africa	201 000	Arid	600	2 600	21 000	1972	79
Sand	S Africa	7 703	Dry	500	20	2 550	1956	30
Klein Letaba	S Africa	3 250	Grass	670	205	3 510	1981	50
Komati	S Africa	21 600	Grass	950	640	7 800	1937	51
Usutu	S Africa	15 350	Grass	850	1 980	13 500	1984	200
Riet	S Africa	31 675	Grass	400	160	5 350	1988	76
Vaal	S Africa	193 000	Grass	800	4 400	7 800	1988	94
Orange	S Africa	342 000	Grass	500	11 000	16 200	1925	63
Senqu	Lesotho	7 950	Mountains	800	5 000	6 800	1976	200
Gouritz	S Africa	43 400	Cape,hilly	400	670	11 400	1981	40
Buffles	S Africa	4 001	Cape,arid	230	42	5 740	1981	200
Kei	S Africa	19 690	Cape,hilly	600	1 000	15 000	1874	114
Mbashe	S Africa	4 800	Cape,hilly	700	600	3 400	1959	33
Mzim-kulu	S Africa	6 630	Cape,hilly	800	1 000	6 520	1959	57
Mkuze	S Africa	2 647	Grass,hills	700	2 700	5 500	1984	60
Black Mfolosi	S Africa	3 396	Coast bush	840	370	10 000	1984	200
Mfolozi	S Africa	9 200	Coast bush	850	880	16 000	1984	200
Moko-manzi	S Africa	4 375	Coast bush	950	560	7 000	1856	132
Fish	Namibia	46 400	Desert	200	1 000	8 300	1972	27
Zambezi	Zimbabwe	525 000	Grass, bush	900	40 000	16 140	1958	80
Lundi	Zimbabwe	23 000	Grass	700	1 300	11 700	1973	50
Umniati	Zimbabwe	5 880	Grass	770	300	5 660	1953	40
Tati	Botswana	570	Arid	400	40	1 010	1972	17
Ntshe	Botswana	800	Arid	400	50	1 610	1972	17
Shashe	Botswana	7 810	Arid	500	500	1 750	1972	17
Save	Mocambique	102 000	Sub-trop	900	5 000	13 840	1973	30
Maputo	Mocambique	31 600	Sub-trop	900	2 800	16 000	1984	200
Pungwe	Mocambique	29 000	Bush,swamps	1 200	3 100	2 960	1977	18
Zambezi	Mocambique	1 200 000	Grass, bush	1 100	90 000	30 000	1959	50
Kafue	Zambia	153 000	Grass,swamp	1 100	7 400	1 982	1959	50
Luangwa	Zambia	141 000	Grass,swamp	1 200	8 000	4 500	1959	50

References - Kovacs (1988),Rodier and Roche (1988),Van der Leeden (1975) relationship and excess rain is taken as a proportion of rainfall. Further the unit hydrographs have to be transposed geographically within regions of hydrographic similarity. The basic time-area concept behind the rational method is therefore extended in the unit hydrograph method. Both rational and unit hydrograph methods do not account for hydraulic or physiographic characteristics and therefore may be unsafe to extrapolate for extreme events. (Stephenson 1983; Stephenson and Collins, 1988). This is particularly so in Southern Africa (Fig 1) with its wide variations in flood characteristics.

The assumption that the critical storm duration is equal to the concentration time of a catchment is used with the rational method, and a similar result appears from unit hydrograph application, on account of the assumption of linearity of the system. It can be shown by hydrodynamic methods (Stephenson and Meadows, 1985) that this can underestimate extreme floods.

Methods of estimating flood magnitudes

The various approaches used to estimate flood magnitudes or water levels are :

- I Deterministic, based on catchment models and known rain patterns.
- II Extrapolation from neighbouring catchments with spot gauging correlation.
- III Empirical methods based on extrapolation of limited time series data.
- IV Probabilistic, based on statistical fits to available data.

Where floods have been observed it is possible to correlate the flood peak with factors such as catchment area, rainfall rate or return period. The most common correlation is flood magnitude against catchment area. Here envelopes of extreme flood ie. the maximum ever recorded at any site are plotted and over the years as more data is gathered the line will invariably move upwards. Such plots generally have to be done on log paper owing to the wide range in catchment areas and the measurements may include floods from vastly differing catchment areas ie. world envelopes will cover all possible catchments, national envelopes may cover a limited range of catchments and more confined data may cover catchments with different topographies and configurations. The length of record or historical period of observation often varies from site to site as well as the method of estimating the flood.

Francou and Rodier (1967) proposed coefficients for applying to such curves with each curve associated with a certain probability that it is the maximum curve. The basis of such curves is not strictly according to physical principles and generally the curves only fit for larger catchments, whereas observations deviate from the curves for small catchments. For this reason the areas of inapplicability were referred to as the transition zone or non-flood zone. Equations have been fitted to the envelopes relating maximum probable flood to catchment area. These equations appear to be replacing the Creager rating curves (Creager et al, 1964) in practice.

Plots of flood magnitude versus recurrence interval for gauged sites can also be made apparently simply with special graph paper. Gumble's theory of extreme values was used to prepare graph paper which tends to straighten out most hydrological analyses. Log-log paper often tends to result in plots of flow versus probability or recurrence interval appearing a straight line for a large proportion of its length. Studies by Pegram and

Fig 1 Map of Southern Africa with rivers studied

Adamson (1988) indicate there may be two extreme value curves for data ie. the majority of the floods, or the more frequent floods, plot on one straight line and so-called outliers plot on another straight line. However whether they are only two lines or whether there is really not a straight line on any mathematically prepared paper cannot be proved. The complexity of the hydrological system cannot be accounted for on a two-dimensional curve except as an approximation ie. there are many different combinations of storms, coupled with physical limitations on the system such as wind directions and speeds, the amount of moisture possible in the atmosphere, topographical barriers and channel capacities. The latter factors affect even more significantly the probabilistic methods whereby extreme values are projected using some statistical equation.

The so-called deterministic methods of estimating floods also require projections ie. even a simple equation such as the rational equation whereby catchment area, design rainfall intensity and a multiplication factor are preassessed, requires extensive assumptions. Whether the method of transposing rainfall rate into a flood flow is based on such a simple equation or a more complicated numerical method such as unit hydrograph method or even a highly sophisticated and complex computer model, depends on the ability of the user as well as his urge to refine the method, but unfortunately less so on the soundness of the data.

Factors to consider in flood estimates

A model for estimating floods off any catchment must account for all the factors which affect the flood. It is not sufficient to transpose floods from one catchment to another in the same geographic region if hydrological factors differ. The following factors are known to affect runoff :

i) Rainfall rate
ii) Storm duration and temporal variation in the storm
iii) Spatial distribution of the storm
iv) Wind direction and therefore direction of movement of storm
v) Antecedent moisture conditions of the ground
vi) Permeability and other soil infiltration factors
vii) Topography
viii) Catchment cover
ix) Channel characteristics including flood plane characteristics for large flows
x) Channel bed and bank cover or alluvial characteristics

Whereas a long duration widespread storm may be the most critical in catchments such as the Kafue where there is large-scale storage and therefore a slow response time, in case of more arid regions such as Botswana, the long duration storms may result in a very small runoff. This is because infiltration rates are relatively high and can approach the precipitation rate. The short, sharp storm is more likely to result in a high runoff and for this reason the storm may not necessarily spread over the full catchment for the worst runoff. The aerial reduction factor is therefore likely to be less important because it is a smaller storm cell which is critical but storm movement is likely to influence considerably the runoff rate. On the other hand in mountainous regions such as Transkei the configuration of the catchment particularly near the mouth or the point where the flood hydrograph is to be estimated is important. Rapid concentration times from steep land may also result in short duration storms being critical. Steeper grounds are also usually rocky and infiltration is low, whereas antecedent moisture content may also be low. For this reason short low intensity storms may have a much lower runoff coefficient than for high intensity or longer duration storms or those with preceding small storms.

Methods of flood estimation

Direct methods of flood determination, whether they be empirical or probabilistic, rely on some observations of floods. Even indirect methods such as rainfall-runoff models rely on some flow measurements to calibrate the model or as a basis for obtaining runoff coefficients. In order of preference the following should be available :

Long-term continuous records at the site for direct methods
Records transposed from an adjacent site
Spot gaugings of floods for model calibration
Indirect flood measurements such as from depth, slope or width observations

Owing to the lack of streamflow data, and the difficulties of transposing flood data from whatever gauged sites there are, floods are more frequently synthesized from rain data. The rainfall-runoff model used may be any of the following :

Rational equation
Time area method (graphical or numerical)
Unit hydrograph with manual or computer calculations
Kinematic nomographs, equations or models
Hydrodynamic models

In an attempt to generalize results, simplistic correlations have been proposed by various authorities. eg. Institute of Hydrology for Eastern Botswana (1986) :

$$Q_p = 7.2\ A^{0.45}T^{0.36} \tag{1}$$

where Q_p is the flood peak in m^3/s with a return period T years off a catchment of area A km^2.

Op Ten Noort and Stephenson (1982) allowed for rainfall rate and produced equations of the form

$$Q_p = (0.0012M-0.21)\ A^{0.69}\ T^{0.49} \tag{2}$$

(for Eastern Botswana)
where M is the mean annual precipitation in mm. Their results were published in nomograph form (1982). Similar relationships were proposed for Zimbabwe (Mitchell, 1974) and Zambia (Balek, 1977).

RAINFALL DEPTH-DURATION-FREQUENCY RELATIONSHIP

Storm precipitation depth and average intensity are related to storm duration. Intense storms are generally of shorter duration than light rain and this is even more marked in tropical regions. There is a wide spectrum of storm intensities, and duration, in any record. In attempting to categorize storms, the duration must be defined and even this can vary between analyses because storms merge and generally have a non-uniform rainfall rate. Once storms of selected durations are identified, they can be ranked and assigned return periods such as has been done in Fig.2.

Unfortunately very few countries in the region keep continuous rainfall records on an extensive basis and one is lucky to obtain reliable daily, monthly or even annual totals. Whatever limited data are available could be plotted on log paper (eg. Fig. 3, for Eastern Botswana) and extrapolated. Equations of the following form could be fitted for example, by regression.

$$P = 1.5\ M^{0.5}\ T^{0.15}/t^{0.7} \quad (3)$$

where P is average storm intensity in mm/h, M is the mean annual precipitation in mm, T is the return period in years and t is the storm duration in hours. The equation is derived using point rainfall data from Bhalotra (1987) for Eastern Botswana.

Such data gives depths of point rain at a single raingauge and correlation with adjacent raingauges will show the peak does not occur simultaneously throughout a catchment, and in fact often does not occur during the same storm. Generalized areal reduction factors (eg. Pegram and Adamson, 1988) are therefore applied. However these may be inapplicable for restricted storms causing the worst runoff from quick response and flat areas, or for large-scale rain over flat slow-reacting basins.

COMMENTS ON CORRELATION OF FLOOD MAGNITUDE WITH CATCHMENT AREA

The major factor relating to flood magnitude is reputed to be catchment area (Kovacs, 1971). However unless basic characteristics are accounted for there is a wide ban in the plot of peak observed flood and catchment area (Fig. 4). For example the maximum observed flood of 1,929m^3/s on the Kafue (catchment 153,000 km^2) is only 12% of that on the Save of 13,800m^3/s (catchment area 102,000km^2), despite the fact that the mean annual precipitation is 1500 mm in the Kafue and 1100 mm on the Save catchment. The difference cannot be explained by rainfall extremes such as brought about by cyclones, since both catchments are bigger than catchments which are monopolized by such relatively short-term weather anomalies.

The floods in the Kafue are strongly affected by the polders or vast swamps along its reach, as revealed by catchment modelling. The attenuation of the flood plains spreading over the reed covered swamps is much greater than the routing effect of reservoirs on the river. The Kafue basin has both flatness and vegetation to alleviate floods. Extreme

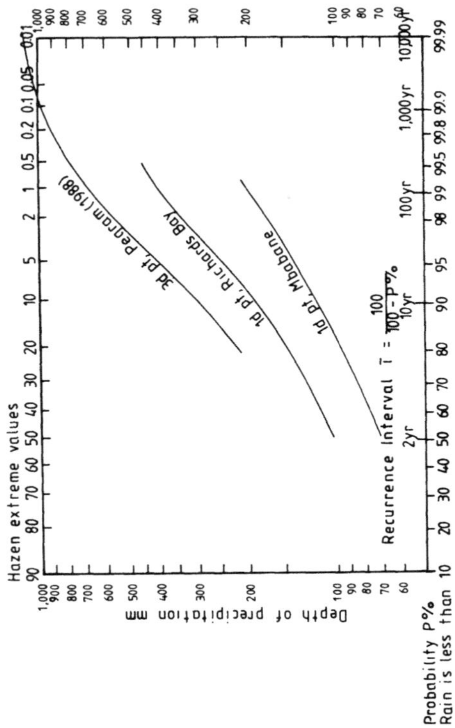

Fig. 2 Rainfall depth versus recurrence interval

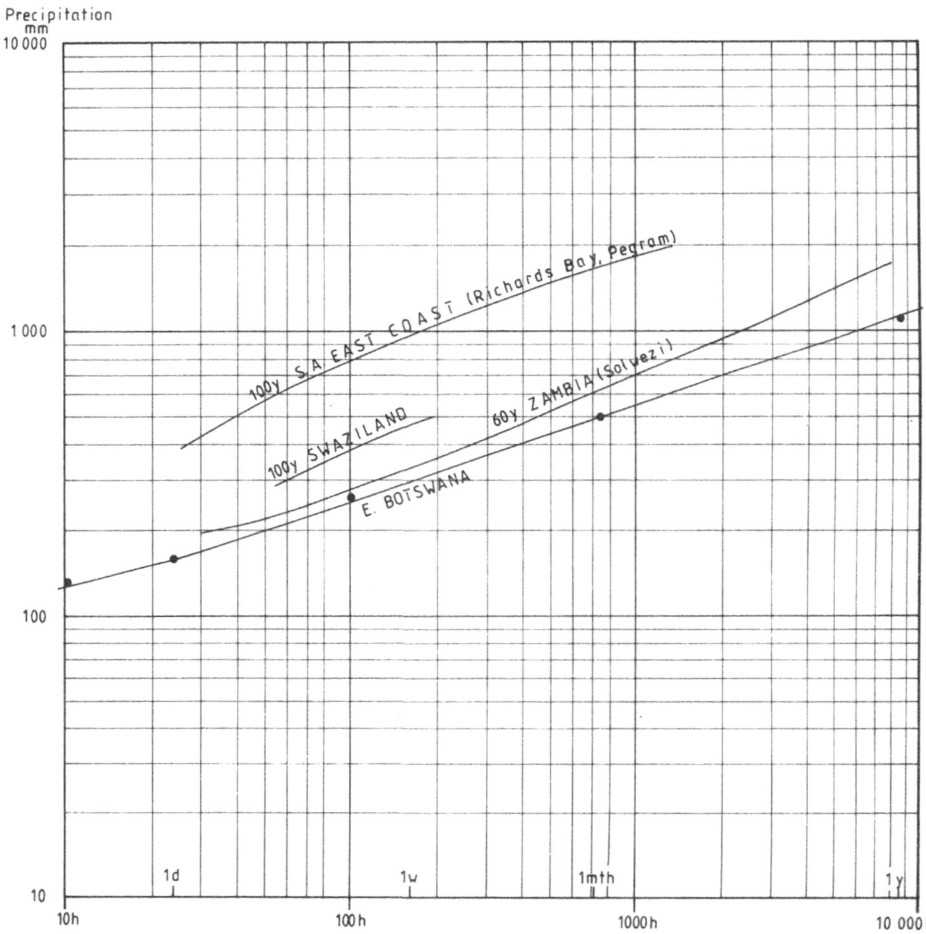

Fig. 3 Extreme Rainfall Depth - Duration Estimates

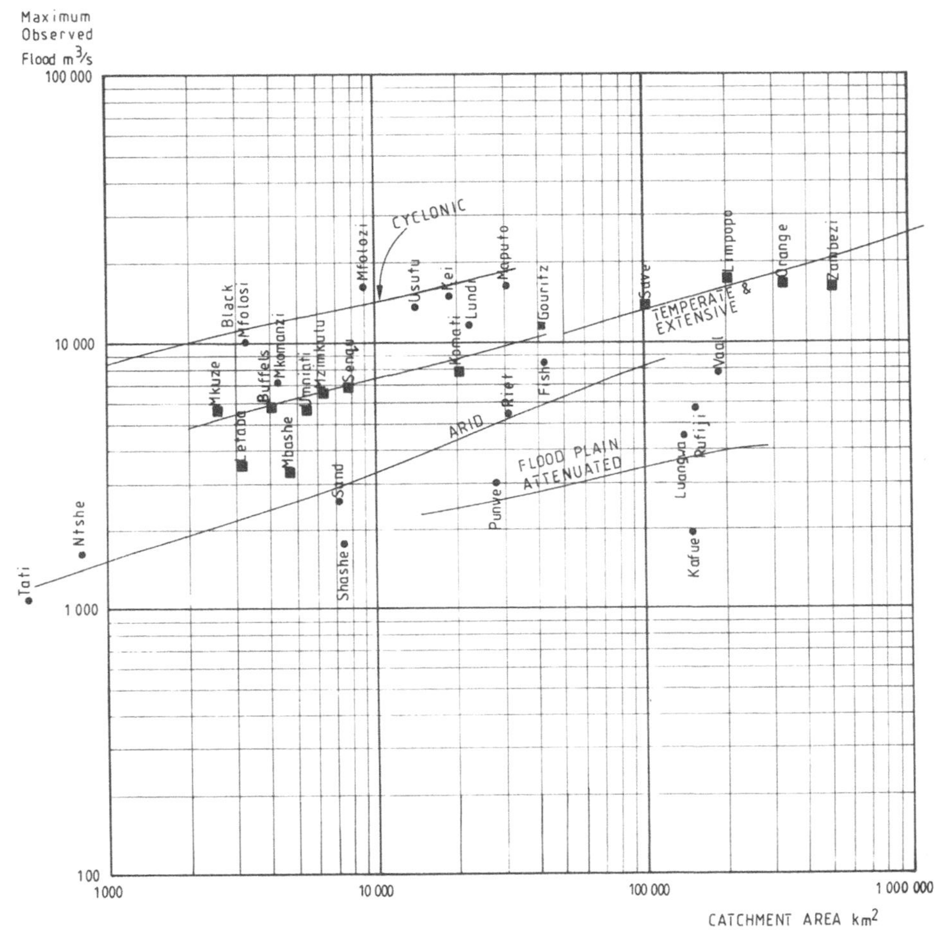

Fig. 4 Some maximum observed flood peaks in Southern Africa over the period 1900 - 1990 and fitted lines for various conditions.

projection however, should consider whether the vegetation could be affected by climatic changes.

The attenuating effect of swamps is also experienced by rivers through Mocambique. The Pungwe flows through swampy areas which could once have drained into the Zambezi river. Despite the attenuation in floods on the Pungwe flats by the wide flood plain water levels near the mouth of the river at the port of Beira can be strongly affected by tidal levels in the Indian Ocean, as the tidal fluctuation can be 7m.

TABLE 3

Table of selected catchments and the effect on flood peaks

River	Effect	CatchmentArea Km^2	Normalised max flood m^3/s	Max observed	Modelled 100 yr flood m^3/s
Orange	Long channel	342 000	18 000	16 000	15 000
Kafue	Wide marshes	153 000	14 000	2 000	3 000
Pungwe	Wide flood plane	2 900	9 000	3 000	3 000
Molhoutse	Flat arid	18 000	4 000	-	1 200
Shashe	Flat arid	7 800	3 000	1 800	
Mbashe	Steep lower reaches	5 000	6 000	3 000	5 000

Table 3 presents data showing the effect of catchment characteristics on flood peaks. Normalized flood refers to the value read of the respective line in Fig.4, i.e. expected maximum observed flood, for average catchments.

It will be observed from Table 3 that the effects of wide marshes or swamps is the most pronounced in attenuating floods (by a factor of 7 for Kafue). The factor reduces to 2 or 3 for flat arid planes (Motloutse, Shashe) and is less for long channels (Orange river).

It appears the effects of tropical cyclones experienced on the east coast of Southern Africa do not extend far inland. The upper envelope of extreme floods tapers off for catchments above 30 000 km^2. Floods off large catchments (greater than 100 000 km^2) conform to the temperate zone line. This is because no arid zones or cyclonic zones are large enough to influence the floods significantly. Even the arid flood zone appears to converge on the temperate line for extensive areas (Fig. 4). This accords with Francou and Rodier's (1967) theory that all lines converge for large catchments. This characteristic is absent on a plot of flood magnitude against recurrence interval (Fig. 5).

The fact that mean annual rainfall does not have as marked an effect on flood peaks as catchment characteristics is demonstrated by comparing for example floods in the Fish river (MAP 200 mm, area 464 000 km^2, max flood 8 300) with the Vaal river (MAP 800, area 19 300 km^2, max flood 7 800 m^3/s).

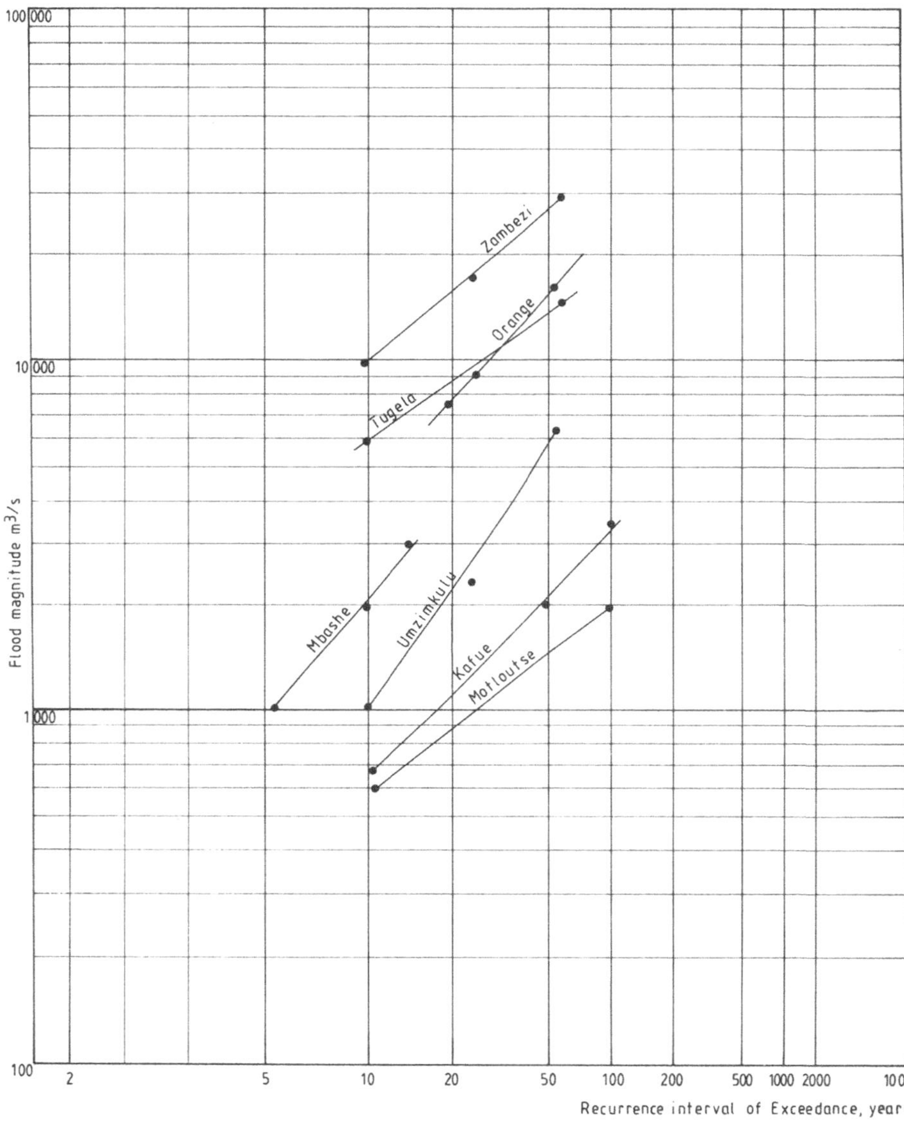

Fig.5 Flood Frequencies (F = years of record ÷ rank)

It is possible that extreme flood magnitudes off large catchments also converge on some asymptote which is a function of catchment area only. It is more likely that large catchments suffer limited volumes of precipitation owing to the restricted carrying capacity of the atmosphere. The catchment attenuation and spatial averaging effect become more likely, the larger the catchment, and topographic discontinuities and distance form the oceans restrict precipitation.

A knowledge of palaeofloods provides a further guide to probable maximum floods. Smith (1991) documented palaeofloods from geological evidence in South Africa. A comparison of these and historical recorded floods is made in table 2.

TABLE 2

Comparison of Palaeofloods and recorded floods in S.A.

River	Palaeofld. m^3/s	Recorded max	Date
Orange	40 000	10 000	1933
Umgeni	45 000	6 500	1856
Crocodile	9 000	3 300	1917
Sand	4 000	2 500	1956

A plot of flood magnitude versus frequency (eg. Fig.5) would indicate the Palaeofloods have recurrence intervals of the order of 500 years. Even so, Smith talks of climatic change affecting the palaeofloods. It is therefore disappointing that even higher floods have not been unearthed, but the method may hold potential when better methods are developed.

Conclusions

Conventional methods of storm analysis and flood determination often fall short in the case of extreme floods. It is not sufficient to extrapolate short duration records unless they include samples of extreme events.

Rainfall intensities can also not be extrapolated from short term records unless extremes such as cyclones are accounted for. In addition in projecting ahead for extreme events, climatic change and catchment changes must be considered.

Standard methods of estimating floods such as the rational method and unit hydrographs can often not be used to generate hydrographs for extreme storms because of the pre-conceived hydrographs' shapes, and the rainfall-runoff relationship is not linear. In addition hydrographs off one catchment do not necessarily conform to hydrographs off other catchments even in the same hydrographic region.

Channel characteristics can also influence the hydrograph and flood peak considerably. Marshy land can have a large routing effect such as in the Kafue and along the Pungwe flats. Alluvial channels such as those in Botswana experience a suspension of their beds to a considerable depth during extreme storms and therefore the flood water levels may not be as high as for stationary beds. On the other hand friction drag and therefore flow velocities are reduced by bed suspension but very little research has been conducted into these effects.

The critical storm duration is not necessarily related to the basin lag time as the runoff from short sharp storms is often more severe than for long duration storms in the case of flat arid land. The worst floods from swampy areas may be for storms with a duration considerably longer than the so-called concentration time or lag time of the basin.

The routing effect of reservoirs is small in the case of arid areas unless the reservoir is right at the point at which the hydrograph is required. This is because peak runoffs can occur for storms downstream of reservoirs and the spillway characteristics of reservoirs in these regions is usually such that there is not a large routing effect. On the other hand for the larger rivers in Southern Africa, reservoirs are frequently small in relation to the runoff volumes and the routing effect may not be large in comparison with the routing effect of flood plains and swamps along the channels.

Observation of flood hydrographs and autographic rainfall is required for improving flood estimates throughout the region. The data required specifically for calibrating models as unit hydrographs cannot be extrapolated. Owing to the limited resources of some countries in the region it may be necessary for richer countries to embark on this type of research. Most research at present is project orientated, and the costs of projects can be influenced considerably by the lack of regional flood research.

References

Balek, J. 1977. Hydrology and Water Resources in Tropical Africa. Elsevier.

Bhalotra, Y.P.R. 1987. Climate of Botswana, Part II Elements of Climate; 1; Rainfall. Dept. Met. Services, Botswana.

Creager, W.P., Justin, J.D. and Hinds, J., 1964. Engineering for dams, Wiley.

Francou, J. and Rodier, J.A. 1967. 'Essai de classification des crues maximales'. Proc. Leningrad Symp. on floods and their computation. UNESCO.

Hydrological Research Unit, 1972. Design Flood Determinations in South Africa, University of the Witwatersrand.

Institute of Hydrology, 1986. A review of flood estimates for Kolobeng and Metsemotlhaba dams. For Alexander Gibb and Partners, Botswana. Wallingford.

Kovacs, G. 1971. Relationship between characteristic flood discharges and catchment area. Symp. Role of hydrology in developing Africa II, WMO No. 301, pp 18-23.

Kovacs, Z.P. 1988. Regional maximum Flood Peaks in Southern Africa. TR137, Dept. Water Affairs, Pretoria.

Mitchell, T.B. Nov 1974. 'A study of Rhodesian floods and proposed flood formulae'. The Rhodesian Engineer.

Op Ten Noort, T. and Stephenson, D. 1982. Flood peak calculation in South Africa, Water Systems Research Group, University of the Witwatersrand.

Pegram, G. and Adamson, P. 1988. 'Revised risk analysis for extreme storms and floods in Natal'. Civil Engineer in S.A., Jan. pp 15-42. Discussion July, pp 331-6.
Pitman, W.V. and Midgley D.C. Aug. 1967. "Flood studies in South Africa : Frequency analysis of peak discharges". The Civil Engineer in S.A.
Rodier, J.A. and Roche, M. 1988. World catalogue of maximum observed floods. IAHS public **143**.
Smith, A.M. 1991. 'Extreme Palaeofloods, their climatic significance and the chances of similar floods re-occurring'. SA. Journal of Science, **87 (6)** 219-220.
Stephenson, D. July 1983. 'Hydrological Myths'. The Civil Engineer in SA. 337-383.
Stephenson, D. and Meadows, M.E. 1986. Kinematic Hydrology and Modelling. Elsevier.
Stephenson, D. and Collins, S.T. 1988. 'Problems due to inaccurate flood estimates at Collywobbles'. Proc. Intl. Congress on Large dams. San Francisco.
Van der Leeden, F. 1975. Water Resources of the World. Water Information Cr. NY.
World Meterological Organization (WMO) 1983. Guide to Hydrological Practices. Vol. II. Analysis. Forecasting and other Applications. 4th Edn. Geneva, WM0168.

FUNCTIONAL RELATIONSHIPS AND ASYMPTOTIC PROPERTIES OF DISTRIBUTIONS OF INTEREST IN HYDROLOGIC FREQUENCY ANALYSIS

F. Ashkar[1], B. Bobée[2] and T. B.M.J. Ouarda[3]
[1]Professor, [3]Research Associate, Faculty of Sciences, University of Moncton, Moncton, N.B. CANADA E1A 3E9.
[2]Professor, INRS-Eau, University of Québec, 2800 Einstein, Suite 105, Québec (Québec) CANADA G1X 4N8.

ABSTRACT. The procedure of hydrologic frequency analysis involves fitting a theoretical probability distribution to a series of flows, water levels or rainfall. The data series must meet the criteria of being independent and identically distributed (iid). The theoretical distribution must be adequately chosen to reflect the nature of the phenomenon and the characteristics of the data being modeled (positive or negative skewness, range, etc.). A number of statistical distributions, with various numbers of parameters, have been proposed and used in a number of countries for the fitting of samples of hydrologic flood data (maximum annual discharge, for example). These distributions are reviewed in this paper, and the main characteristics of each distribution are briefly discussed. The functional relationships between these most commonly used distributions are highlighted. Proofs are derived for all the relationships that are established, and all necessary transformations of variables are identified. Special cases for each distribution are also discussed. This paper includes also a study and classification of the distributions according to their asymptotic properties (characteristics of the right tail of the distribution). One-, two-, three-, and four-parameter distributions are considered in this study. The final results are summarized in a diagram outlining the functional relationship between the variates of the different distributions, and a table detailing the probability density function (pdf) and the different forms and characteristics of each distribution. Another table classifying the asymptotic properties of these distributions is also presented.

Key words: Asymptotic properties, hydrologic frequency analysis, flood data, statistical distribution, skewness, probability density function.

1. Introduction

Statistical hydrologic frequency analysis is the most common of the two main approaches to the estimation of design flood events (extreme high flood quantiles), the other approach being streamflow simulation from consideration of precipitation input (rainfall-runoff analysis). Statistical estimation of extreme high quantiles of a hydrologic variable such as river flow, water

V. P. Singh and B. Kumar (eds.), Surface-Water Hydrology, 521–535.

level, rainfall depth, has been a topic of extensive research during the last decades. In the present study we shall consider flood flows as the variable of interest (for the purpose of discussion) but the discussion bears interest to any other type of extreme hydrologic phenomenon.

Two general methodologies have been extensively used for probability estimation of extreme floods: the annual maximum flood (AMF) approach, and the peaks over threshold (POT) approach. A sample of size n of flood values constitutes the basis for a general flood frequency analysis method. In the AMF approach, these n values correspond to the maximum annual floods during an n-year period of record, whereas in the POT approach they signify all flood peaks exceeding a threshold level ("exceedances"). The focus in this work is on the AMF approach. After checking the flood sample for randomness, independence, homogeneity, stationarity (iid), and the absence of "outliers", the sample is fitted by an "appropriate" distribution (D), using a method of parameter estimation (E). This D/E-procedure is then used to make inferences about flood quantiles, and their associated confidence intervals.

A large number of D/E procedures for analyzing AMF series have been recommended. Although some of them have claimed a certain degree of physical justification, in reality none can be completely defended on physical grounds. According to Bobée et al. (1993), this large number of proposed D/E procedures has led to a state of confusion among practitioners as to what are the best methods to use.

Dealing with the flood frequency analysis problem is very complex: it requires a very good understanding of both statistics, and the physical phenomena involved. Considering the flood frequency analysis problem as one in statistical inference, two types of uncertainty are involved and need to be well investigated and understood: (1) sampling uncertainty and (2) model uncertainty. The practitioner is interested in the global uncertainty in the estimated flood quantiles, which is a combination of these two. It is important to note that global uncertainty is especially pronounced when flood quantiles beyond the observed sample need to be estimated ("extrapolation"; Ashkar and Bobée, 1987). This case is especially important in problems of hydraulic design. In this case, model error (error due to the choice of distribution) becomes relatively much more important than the sampling error. This illustrates the importance of looking for the best possible choice of distribution in a flood frequency analysis.

2. Choice Of Distributions For Fitting Flood Series

2.1. GENERAL CONSIDERATIONS FOR DISTRIBUTION CHOICE

Selecting a distribution (D) for fitting flood series involves decisions such as: choice of the number of distribution parameters, specification of the general "desirable" properties that a flood distribution should possess, and identification of the factors that determine the "descriptive" and "predictive" abilities of the distribution (Cunnane, 1987). In the case of AMF series, it is generally acknowledged (Morlat, 1956) that a distribution for fitting these series should have three parameters, in order to provide sufficient flexibility to the large variety of AMF data found in practice. It has been suggested by some investigators that the chosen distribution should have a distribution function and an inverse distribution function that can be expressed in closed form, and that the distribution must not be computationally complex. However, these distribution characteristics, although desirable, should not form a strict basis for choosing one type of flood distribution over another.

2.2. EFFECTS OF WRONG CHOICE OF DISTRIBUTION

The following comments by Lawless (1980) have special interest to the problem of flood frequency analysis (which has much in common with problems related to the use of "lifetime distributions"): "In any statistical work it is important to know the effects of departures from assumptions on the inferences one is making. This is especially important in estimation with lifetime distributions where one is often interested in the tails of the distribution. In such situations even modest departures from the assumed model can have sizeable effects on inferences. ... very often data are analyzed under the assumption of a particular model (such as the Weibull or lognormal) simply because (1) the model has been used before in similar situations, and (2) it fits the data on hand. This does not of course imply any absolute validity of the model, and one should ask the question whether estimates or other inferences will change much if some other similarly 'plausible' model is used instead. ... Unless there are very strong a priori reasons for believing in a particular model, a careful statistical analysis should include an examination of inferences under competing, plausible models."

Some countries have produced guidelines specifying particular distributions to be used by their government agencies and private enterprises for flood-flow frequency analysis. The log-Pearson Type 3 distribution (LP) has been proposed in the USA (Benson, 1968) and Australia (I.E.A., 1977), and the generalized gamma distribution (GG) has been recommended in the republics of the ex-USSR (Kritsky and Menkel, 1969), while the UK opted for the generalized extreme value distribution (GEV) (NERC, 1975). These guidelines can assist in seeking a certain uniformity in the management of water resources, but should be used with special care.

2.3. EFFECT OF TAIL SHAPE ON ESTIMATED QUANTILES

Early analyses of extreme hydrologic phenomena had assumed that the data being analyzed may be reasonably modeled by distributions which are somewhat "light-tailed". Loosely speaking, this assumption means that the tail of the pdf approaches zero in some sort of exponential manner. As noted by Ochoa et al. (1980), the well-known central limit theorems, which assert the asymptotic normality of sample means, assume that the distribution is at least sufficiently light-tailed to possess a finite variance. These authors also note that if the tail of the pdf is of such well-known forms as $x^a\exp(-bx)$ (gamma family) or $\exp(-ax^b)$ (Weibull family, normal), then all moments exist. In evaluating the distribution of extreme events, such as the maximum annual flood, Gumbel (1958) showed that for the most commonly used distributions, the distribution of the maximum of an independent and identically distributed (iid) set of random variables of large size, may be approximated by the type-I extreme-value distribution (EV1 or Gumbel). In practice, the hypothesis of this statement (iid, large size) are often violated. The derivation of the EV1 model also assumes that the distribution of individual observations has a tail behaviour roughly equivalent to the existence of all moments (Ochoa et al., 1980).

On the other hand, researchers in hydrology have also given much attention to distributions with tail behaviour of the form:

$$1\text{-}F(x) = ax^{-k} \qquad (1)$$

These "Paretian-tailed" distributions (whose name is deduced from the fact that their tail behaviour is similar to that of the Pareto family of distributions) possess no moments of order k or above (Ochoa et al., 1980). Among a set of independent random variables having a distribution with this tail behaviour, the maximum has a distribution approximated by the type-II extreme-value distribution (EV2). Referring to Jenkinson (1977), Ochoa et al. (1980) note that by using suitable linear transformations of variables the EV1 distribution may be regarded as a limiting case of the EV2. Alternatively, the EV2 may be regarded as a log-EV1 in the sense that, if $X \sim EV2$, then $Y = \log X \sim EV1$, as is explained in Bobée and Ashkar (1991). The type-II, or Paretian -tailed family of distributions has played a very important role in hydrology, especially that so much work has been made on the log-Pearson III distribution, which has a Paretian tail (see Bobée, 1975).

3. Purpose Of The Study

From the above discussion, it is seen that the hydrologist who needs to explore in detail what distributions are best fit for representing extreme empirical hydrologic data needs to have a very good knowledge and understanding of two things:

(1) what mathematical relations do exist between distributions (for example how a logarithmic transformation produces a "light-tailed" distribution from one that is "thick-tailed"?) Understanding these relations will allow a better assessment, for example, of the importance and impact of a type-II family vs. type-I (as alternatives) in estimating the frequency of extreme events.

(2) how can the tail behaviour of distributions be classified so that we can better assess the adequacy of the various tail shapes for representing extreme empirical data.

Consequently, the purpose of the present study is two-fold:

(1) to investigate the mathematical relationships between some of the most commonly used distributions in hydrology; and

(2) to classify the tail behaviour of these distributions in a systematic manner.

4. Relationships Between Distributions Commonly Used In Hydrology

The main characteristics of the distributions of interest in hydrologic frequency analysis are summarised in Table I. One-, two-, three- and four-parameter distributions are considered in this

TABLE I. The p.d.f. and some characteristics of distributions of interest in flood analysis (Continued in the next three pages)

Distribution	**Probability density function f(x), or cumulative distributin function F(x)**	**Range of variate (x) and value of parameters**
GG4(s,α,λ,m) Generalized Gamma 4-parameter	$f(x) = \frac{\lvert \alpha s \rvert e^{-[\alpha(x-m)]^s}[\alpha(x-m)]^{s\lambda-1}}{\Gamma(\lambda)}$	if $\alpha>0$, $x \geq m$ if $\alpha<0$, $x \leq m$ $\lambda>0$, $s \neq 0$, $-\infty < m < +\infty$
PI(a_1,a_2,m_1,m_2) Pearson Type I	$f(x)=K(x-a_1)^{m_1}(a_2-x)^{m_2}$	$a_1 < x < a_2$ $m_1 > -1$, $m_2 > -1$
GG3(s,α,λ) Generalized Gamma 3-parameter	$f(x) = \frac{\lvert \alpha s \rvert e^{-(\alpha x)^s}(\alpha x)^{s\lambda-1}}{\Gamma(\lambda)}$	if $\alpha>0$, $x \geq 0$ if $\alpha<0$, $x \leq 0$ $\lambda>0$, $s \neq 0$,
P (α,λ,m) Pearson Type 3	$f(x) = \frac{\lvert \alpha \rvert e^{-[\alpha(x-m)]}[\alpha(x-m)]^{\lambda-1}}{\Gamma(\lambda)}$	if $\alpha>0$, $x \geq m$ if $\alpha<0$, $x \leq m$ $\lambda>0$, $-\infty < m < +\infty$
PV(α,λ,m) Pearson Type 5	$f(x) = \frac{\lvert \alpha \rvert e^{\alpha(x-m)}[\alpha(x-m)]^{-(\lambda+1)}}{\Gamma(\lambda)}$	if $\alpha>0$, $x \geq m$ if $\alpha<0$, $x \leq m$ $\lambda>0$ $-\infty < m < +\infty$

WE3(s,α,m) Weibull 3-parameter	$f(x) = \lvert \alpha s \rvert e^{-[\alpha(x-m)]^s}[\alpha(x-m)]^{s-1}$
	if ($\alpha>0$ et $s>0$) ou ($\alpha<0$ et $s<0$):
	$F(x)=1-\exp\{-[\alpha(x-m)]^s\}$
	if ($\alpha>0$ et $s<0$) ou ($\alpha<0$ et $s>0$):
	$F(x)=\exp\{-[\alpha(x-m)]^s\}$
		if $\alpha>0$, $x \geq m$
		if $\alpha<0$, $x \leq m$
		$s \neq 0$,
		$-\infty < m < +\infty$

LP(α,λ,m) Log-Pearson Type 3	$f(x) = \frac{\alpha^{\lambda} e^{\alpha m} x^{-(\alpha+1)}(\ln x-m)^{\lambda-1}}{\Gamma(\lambda)}$	if $\alpha>0$, $x \geq e^m$ if $\alpha<0$, $0 \leq x \leq e^m$, $\lambda>0$
EV3(s,α,m) Extreme value Type 3	$f(x) = \lvert \frac{s}{m+x} \rvert \exp\{-e^{-s(\xi+x)}-s(x+\xi)\}$	$\alpha>0$, $s \neq 0$ $-\infty < m < +\infty$ $-\infty < x < +\infty$

EV2(s,α,m)
Extreme value
Type 2
	$f(x)=\|s\|\alpha^s(x-m)^{-(s+1)}\exp\{-\alpha^s(x-m)^{-s}\}$
	if $\alpha>0$ et $s>0$:
	$F(u)=\exp\{-[\alpha^s(x-m)^{-s}]\}$
	if $\alpha>0$ et $s<0$:
	$F(x)=1-\exp\{-[\alpha^s(x-m)^{-s}]\}$
		$\alpha>0$,
		$-\infty<m<+\infty$
		$-\infty<x<+\infty$

HA(m,α,ν) Halphen Type A (m,α,ν)	$f(x)=\frac{1}{2m^\nu K_\nu(2\alpha)}\exp\{-\alpha(\frac{x}{m}+\frac{m}{x})\}x^{\nu-1}$	$x>0$, $\alpha>0$, $m>0$, $-\infty<\nu<+\infty$
HB(m,α,ν) Halphen Type B (m,α,ν)	$f(x)=\frac{2}{m^{2\nu}ef_\nu(\alpha)}\exp\{-(\frac{x}{m})^2+\alpha\frac{x}{m}\}x^{2\nu-1}$	$x>0$, $m>0$, $\nu>0$, $-\infty<\alpha<+\infty$
HB^{-1}(m,α,ν) Halphen Type B^{-1} (m,α,ν)	$f(x)=\frac{2}{m^{-2\nu}ef_\nu(\alpha)}\exp\{-(\frac{m}{x})^2+\alpha\frac{m}{x}\}x^{-2\nu-1}$	$x>0$, $m>0$, $\nu>0$, $-\infty<\alpha<+\infty$
LN3(m,μ,σ) Log-normal 3-parameter	$f(x)=\frac{1}{\sigma(x-m)\sqrt{2\pi}}\exp\{-\frac{(\ln(x-m)-\mu)^2}{2\sigma^2}\}$	$x>m$, $-\infty<m<+\infty$ $-\infty<\mu<+\infty$ $\sigma>0$
GEV(α,λ,m)	$f(x)=\frac{1}{\alpha}[1-\lambda(x-\frac{m}{\alpha})^{\frac{1-\lambda}{\lambda}}]\exp\{-[1-\lambda(\frac{x-m}{\alpha})]^{\frac{1}{\lambda}-1}\}$ $F(x)=\exp\{-[1-\lambda(\frac{x-m}{\alpha})]^{\frac{1}{\lambda}}\}$	$\alpha>0$ if $\lambda<0$, $x\geq m+\alpha/\lambda$ if $\lambda>0$, $x\leq m+\alpha/\lambda$
PA(α,c) Pareto Type 1	Form A: $f(x)=\|\alpha\|e^{\alpha m}x^{-(\alpha+1)}$; $c=e^m$ $F(x)=\frac{1}{\alpha}(e^{-\alpha m}-x^{-\alpha})$ Form B: $f_x(x)=\beta^{-1}(1-\frac{kx}{\beta})^{\frac{1}{k}-1}$ $F_x(x)=1-(1-\frac{kx}{\beta})^{\frac{1}{k}}$ $k\neq0$	if $\alpha>0$, $x\geq c$ if $\alpha<0$, $0<x<c$ $-\infty<x<+\infty$ $\beta\neq0$
EX2(α,m) Exponential 2-parameter	$f(x)=\alpha e^{-\alpha(x-m)}$ $F(x)=e^{\alpha m}(1-e^{-\alpha x})$	if $\alpha>0$, $x\geq m$ if $\alpha<0$, $x\leq m$ $-\infty<m<+\infty$

G2(α,λ) Gamma 2-parameter	$f(x) = \frac{\|\alpha\| e^{-\alpha x}(\alpha x)^{\lambda-1}}{\Gamma(\lambda)}$	if $\alpha>0$, $x\geq 0$ if $\alpha<0$, $x\leq 0$ $\lambda>0$
ER(α,λ) Erlang	$f(x) = \frac{\alpha^{\lambda}}{\Gamma(\lambda)} x^{\lambda-1} e^{-\alpha x}$	if $\alpha>0$, $x\geq 0$ if $\alpha<0$, $x\leq 0$ λ positive integer
WE(s,α) Weibull 2-param	$F(x)=1-e^{-(\alpha x)^{s}}$	$x\geq 0$, $s\neq 0$, $\alpha>0$
EV1⁻(ξ,s) Extreme value Type 1, negative Skewness	$f(x)=\|s\|\exp\{-e^{-s(\xi-x)}+s(x-\xi)\}$	$-\infty<\xi<+\infty$ $s\neq 0$ $-\infty<x<+\infty$
EV1⁺(ξ,s) Extreme value Type 1, positive skewness	$f(x)=\|s\|\exp\{-e^{-s(\xi+x)}-s(x+\xi)\}$	$-\infty<\xi<+\infty$ $s\neq 0$ $-\infty<x<+\infty$
LGU(ξ,s) Log-Gumbel	$f(x)=\|s\| e^{-s\xi} x^{-(s+1)} \exp\{-x^{-s} e^{-s\xi}\}$	$-\infty<\xi<+\infty$ $x\geq 0$, $s\neq 0$
LG(α,λ) Log-gamma	$f(w)=\frac{\alpha^{\lambda} w^{-(\alpha+1)}(\ln w)^{\lambda-1}}{\Gamma(\lambda)}$	$-\infty<\alpha<+\infty$ $w>0$, $\lambda>0$
GT(α,λ) Standard Grassia-Tadikammalla	$f(x)=\frac{\alpha^{\lambda} x^{-(\alpha-1)}(\ln x)^{\lambda-1}}{\Gamma(\lambda)}$	$-\infty<\alpha<+\infty$ $-\infty<x<+\infty$ $\lambda>0$
LN2(μ,σ) Log-normal 2-parameter	$f(x)=\frac{1}{\sigma x\sqrt{2\pi}}\exp\{-\frac{(\ln x-\mu)^2}{2\sigma^2}\}$	$x>0$ $-\infty<\mu<+\infty$ $\sigma>0$
PV(α,λ) Pearson Type 5 (Gamma inverse)	$f(x) = \frac{\|\alpha\| e^{\alpha x}(\alpha x)^{-(\lambda+1)}}{\Gamma(\lambda)}$	if $\alpha>0$, $x\geq 0$ if $\alpha<0$, $x\leq 0$ $\lambda>0$
Beta(α,β)	$f(x)=\frac{1}{B(\alpha,\beta)} x^{\alpha-1}(1-x)^{\beta-1}$	$0<x<1$ $\alpha, \beta>0$
U(a,b) Uniform	$f(x)=\frac{1}{b-a}$ $F(x)=\frac{1}{b-a}x-\frac{a}{b-a}$	$a\leq x\leq b$, $a\neq b$ $-\infty<a<+\infty$ $-\infty<b<+\infty$

HN(α) Half-normal	$f(x) = \frac{2\|\alpha\|}{\sqrt{\pi}} e^{-(\alpha x)^2}$	if $\alpha > 0$, $x \geq 0$ if $\alpha < 0$, $x \leq 0$
SHN Standrad half-normal	$f(x) = \sqrt{\frac{2}{\pi}} e^{-\frac{x^2}{2}}$	$-\infty < x < +\infty$
$\chi^2(\nu)$	$f(x) = \frac{1}{\Gamma(\frac{\nu}{2}) 2^{\frac{\nu}{2}}} x^{(\frac{\nu}{2}-1)} e^{-\frac{x}{2}}$	$x \geq 0$, ν positive integer
EX(α) Exponential 1-parameter	$f(x) = \alpha e^{-\alpha x}$ $F(x) = 1 - e^{-\alpha x}$	if $\alpha > 0$, $x \geq 0$ if $\alpha < 0$, $x \leq 0$
G1(λ) Gamma 1-parameter	$f(x) = \frac{x^{\lambda-1} e^{-x}}{\Gamma(\lambda)}$	$x \geq 0$ $\lambda > 0$
RA(c) Rayleigh	$f(x) = 4c^2 x e^{-(c\sqrt{2}x)^2}$ $F(x) = 1 - e^{-(c\sqrt{2}x)^2}$	$x \geq 0$ $c > 0$
GL(λ) Logarithmic gamma	$f(x) = \frac{e^{\lambda x - e^x}}{\Gamma(\lambda)}$	$\lambda > 0$ $-\infty < x < +\infty$
Student(ν)	$f(x) = \frac{1}{\sqrt{\nu} B(\frac{1}{2}, \frac{1}{2}\nu)} (1 + \frac{x^2}{\nu})^{-\frac{1}{2}(\nu+1)}$	$-\infty < x < +\infty$ ν positive integer
Cauchy(a,b)	$f(x) = \frac{1}{\pi b [1 + (\frac{x-a}{b})^2]}$ $F(x) = \frac{1}{\pi} [\frac{\pi}{2} + Arctan(\frac{x-a}{b})]$	$-\infty < x < +\infty$ $-\infty < a < +\infty$ $b \neq 0$
Laplace	$f(x) = \frac{b}{2} e^{-b\|x-a\|}$	$-\infty < x < +\infty$ $b > 0$ $-\infty < a < +\infty$

work. Four-parameter distributions are recognized to have limited practical use in flood frequency analysis, nevertheless, they are included in this study to allow for a complete discussion of available distributions. Other characteristics, along with discussions of the advantages and disadvantages of each distribution are presented in Ashkar et al. (1993).

Figure 1. represents a single diagram illustrating systematically the relationships existing between the distributions of Table I. Proofs are derived for all the relationships that are established (Ashkar et al., 1993). All necessary transformations of variables, and special cases for all distributions are also identified. We present in the following sub-section a portion of these results for illustration purposes (for the two-parameter Gamma distribution).

4.1. THE TWO-PARAMETER GAMMA DISTRIBUTION G2(α,λ)

The probability density function of the distribution is:

$$f(x) = \frac{|\alpha| e^{-\alpha x} (\alpha x)^{\lambda-1}}{\Gamma(\lambda)} \tag{2}$$

This distribution is a special case of the Pearson Type 3 P(α,λ,m) when m=0.

4.1.1. Special Cases: The following are special cases of the G2(α,λ)

* $\alpha=1$, we obtain the one-parameter Gamma distribution G1(λ).
* $\alpha=1/2$, $\lambda=\nu/2$ and ν integer, we obtain the $\chi^2(\nu)$ distribution.
* λ is a positive integer, we obtain the Erlang distribution ER(α,λ).

4.1.2. Transformations:

* If Y follows a G2(α,λ) distribution, then $W=e^Y$ follows the Log-Gamma distribution LG(α,λ).

Proof:

$$W=e^{Y} \rightarrow Y=\ln w \rightarrow \left|\frac{dy}{dw}\right| = \frac{1}{w}$$

$$f(w) = f_{G2}(y) \left|\frac{dy}{dw}\right| \tag{3}$$

$$f(w) = \frac{|\alpha| e^{-\alpha y} (\alpha y)^{\lambda-1}}{\Gamma(\lambda)} \frac{1}{w} \tag{4}$$

$$f(w) = \frac{|\alpha| e^{-\alpha \ln w} (\alpha \ln w)^{\lambda-1}}{\Gamma(\lambda)} \frac{1}{w} \tag{5}$$

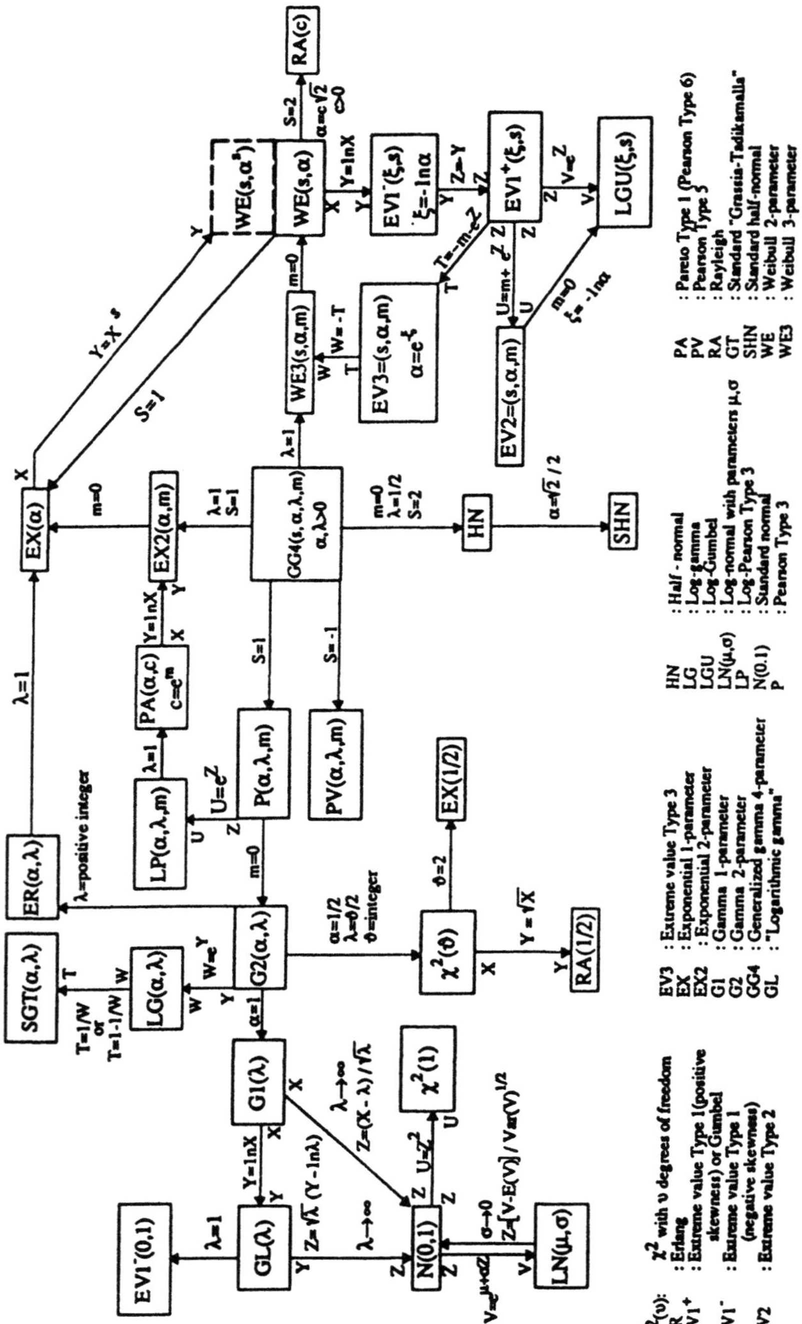

Figure 1. Relationships between distributions commonly used in flood analysis.

$$f(w) = \frac{\alpha^{\lambda}}{\Gamma(\lambda)} w^{-(\alpha+1)} (lnw)^{\lambda-1} \tag{6}$$

* If Y follows a G2(α,λ) distribution, then Z=Y+m follows the Pearson Type 3 distribution P(α,λ,m).

Proof:

$$Z=Y+m \Rightarrow \left|\frac{dy}{dz}\right| = 1$$

$$f(z) = f_{G2}(y) \left|\frac{dy}{dz}\right| \tag{7}$$

$$f(y) = \frac{\alpha^{\lambda}}{\Gamma(\lambda)} y^{\lambda-1} e^{-\alpha y} \tag{8}$$

$$f(y) = \frac{\alpha^{\lambda}}{\Gamma(\lambda)} (z-m)^{\lambda-1} e^{-\alpha(z-m)} \tag{9}$$

$$f(z) = \frac{|\alpha| e^{-[\alpha(z-m)]} [\alpha(z-m)]^{\lambda-1}}{\Gamma(\lambda)} \tag{10}$$

5. Asymptotic Properties Of Distributions Commonly Used In Hydrology

Table II represents a classification of the distributions reviewed in Section 4 according to their asymptotic properties. The importance of the right tail characteristics of the distribution is discussed in Section 2.3. Following is an example of the determination of the asymptotic behaviour for the three-parameter generalized gamma distribution.

5.1. THE THREE-PARAMETER GENERALIZED GAMMA DISTRIBUTION GG3(s,α,λ)

$$f(x) = \frac{|\alpha s| e^{-[\alpha x)]^{s}} [\alpha x]^{s\lambda-1}}{\Gamma(\lambda)} \tag{11}$$

TABLE II. Classification of distributions of interest in flood analysis according to their asymptotic properties (notations of distributions are the same as in TABLE I).

Class		Distribution	
I:	$x \sim T^K$	GG4(s,α,λ,m), s<0	K=-1/sλ
		GG3(s,α,λ), s<0	K=-1/sλ
		PV(α,λ,m)	K=1/λ
		WE3(s,α,m) s<0	K=-1/s
		PVII (c_0,c_2)	K=-c_2/(c_2+1)
		Student (ν)	K=1/ν
I.1:	$x \sim T^K$, K=1 ($x \sim T$)	LP (α,λ,m)	
		HB^{-1}(m,α,ν)	
		Cauchy (a,b)	
		PA(α,c) (de type II)	
		Beta (α,β)	
II:	$x \sim (\ln T)^K$	GG4(s,α,λ,m), s>0	K=1/s
		EV2(s,α,m)	K=1/s
		GG3(s,α,λ), s>0	K=1/s
		WE3(s,α,m), s>0	K=1/s
		LGU(ξ,s)	K=1/s
II.1:	$x \sim (\ln T)^K$, K=1 ($x \sim \ln T$)	HA(m,α,ν)	
		EV3(s,α,m)	
		P (α,λ,m)	
		Laplace (a,b)	
		PA(α,c) (de Type I)	
		EX2(α,m)	
		G2(α,λ)	
		$EV1^+$(ξ,s), s>0	
		$EV1^-$(ξ,s), s<0	
		G1(λ)	
		EX(α)	
II.2:	$x \sim (\ln T)^K$, K=1/2 ($x \sim \sqrt{\ln T}$)	HB(m,α,ν)	
		N(μ,σ)	
		RA(c)	
		HN(α)	
		SHN	
III:	$x \sim \exp\sqrt{(\ln T)}$	LN3 (m,μ,σ)	
		LN2 (μ,σ)	
IV:	$x \sim \ln(\ln T)$	$EV1^+$(ξ,s), s<0	
		$EV1^-$(ξ,s), s>0	
		GL(λ)	
V	$x \sim e^T$	GT(α,λ)	

$$\frac{df(x)}{dx} = K\exp\{-(\alpha x)^s\}\{(s\lambda-1)x^{s\lambda-2} - \alpha^s s x^{s(\lambda+1)-2}\} \tag{12}$$

From Gumbel (1958) we have:

$$T = -\frac{1}{f^2(x)}\frac{df(x)}{dx} = K\exp\{(\alpha x)^s\}\{-(s\lambda-1)x^{-s\lambda} + \alpha^s s x^{s(1-\lambda)}\} \tag{13}$$

$$\ln T = C + (\alpha x)^s + \ln\{-(s\lambda-1)x^{-s\lambda} + \alpha^s s x^{s(1-\lambda)}\} \tag{14}$$

$$* \;\; \text{If } s > 0 \Rightarrow \ln T \sim x^s \qquad x \sim (\ln T)^{\frac{1}{s}}$$

$$* \;\; \text{If } s < 0 \Rightarrow T \sim x^{-s\lambda} \qquad x \sim T^{-\frac{1}{s\lambda}}$$

For a complete development of the asymptotic properties of the distributions discussed in Table II the reader is referred to Ashkar et al. (1993).

6. Conclusions

The results of this study can be of special interest in flood frequency analysis, by providing the answers to some of the questions that arise when selecting the "best" statistical distribution to fit a data sample. The first part of the study deals with the relationship that exists between the distributions to choose from, or between one particular distribution and other distributions from the same family. The second part of this work provides a classification of the distributions according to their "tail shapes". This is obviously of special interest for the estimation of the probability of occurrence or return period of extreme events (flood analysis). The existence of five general types of asymptotic behaviour is established for the distributions of Figure 1, and several subclasses can be identified.

It must be reminded that hydrologic processes, such as river flows, are constrained by physical limitations (range, shape of distribution, skewness,...). Hence, the statistical distribution to be chosen for the representation of the process must conform to these constraints. In this context, Table I can be very useful in identifying distributions with characteristics that satisfy the natural conditions and requirements of the data. This information could be included in software designed to perform statistical frequency analysis of hydrological, climatological and meteorological phenomena, such as AJUSTE II (Perron et al.). In the same way, information concerning the tail behaviour of the distributions (Table II) can be incorporated in software performing the frequency analysis of extreme phenomena. It is obvious that, if a distribution was found to provide an inadequate estimation of extreme events related to a data set, than all other distributions from the same class (Table II) should also be discarded. Ultimately, all the information presented in the results of the present work, could be included in some type of hydrologic analysis expert system designed to aid in rational decision making.

7. References

ASHKAR, F., OUARDA T. B.M.J. AND B. BOBÉE (1993). Distributions statistiques utilisées en hydrologie: Transformations et propriétés asymptotiques. Scientific Report, Department of Mathematics, University of Moncton, NB.

ASHKAR, F. and B. BOBÉE (1987). The generalized method of moments as applied to problems of flood frequency analysis: some practical results for the log-Pearson Type 3 distribution. Journal of Hydrology, 90(3-4): 199-217.

BENSON, M. A. (1968). Uniform flood-frequency estimating methods for federal agencies. Wat. Res. Res. 4(5): 891-908.

BOBÉE, B. (1975). The log-Pearson type 3 distribution and its application in hydrology. Wat. Res. Res. 11(5): 681-689.

BOBÉE, B. and F. ASHKAR (1991). The gamma family and derived distributions applied in hydrology. Water Resources Publications, Littleton, Colo., 217 p.

BOBÉE, B., CAVADIAS, G., ASHKAR, F., J. BERNIER and P. RASMUSSEN (1993a). Towards a systematic approach to the comparison of distributions used in flood frequency analysis. Journal of Hydrology, 142: 121-136.

CUNNANE, C. (1987). Review of statistical models for flood frequency estimation. In V.P. Singh (ed.): Hydrologic Frequency Modeling, 49-95. D. Reidel Publishing Company.

GUMBEL, E.J. (1958). Statistics of extremes. Columbia University Press, New York, NY, 375 p.

I.E.A. (1977). Australian rainfall and runoff flood analysis and design. Institute of Engineers, Australia Camberra, A.C.T., 149p.

JENKINSON, A.F., (1977). The analysis of meteorological and other geophysical extremes. U.K. Meteorol. Off., Bracknell, Tech. Note 58.

KRITSKY, S.N. and M.F. MENKEL (1969). On principles of estimationmethods of maximum discharge. Floods and their computation. I.A.H.S., no 84.

LAWLESS, J.F. (1980). Inference in the generalized Gamma and Log Gamma distributions. Technometrics, 22(3): 409-419.

MORLAT, G. (1956). Les lois de probabilités de Halphen. Revue de statistique appliquée, 4(3): 21-46.

NERC (1975). Flood Studies Report, Vol. 1, Hydrological Studies, Natural Environment Research Council, 27 Charing Cross Road, London.

OCHOA, I.D., M.C. BRYSON AND H.-W. SHEN (1980). On the occurence and importance of paretian-tailed distributions on hydrology. Journal of Hydrology, 48: 53-62.

PERRON, H., BRUNEAU P., BOBÉE B. AND L. PERREAULT (1993). AJUSTE II, A software for hydrological data analysis. International conference on stochastic and statistical methods in hydrology and environmental engineering. Waterloo, Ontario, Canada, June 21-23.

FREQUENCY ANALYSIS OF UPPER CAUVERY FLOOD DATA BY L-MOMENTS

A RAMACHANDRA RAO
KHALED HAMED
School of Civil Engineering
Purdue University
West Lafayette, IN 47907 USA

ABSTRACT The objectives of the present study are to investigate the hydrological homogeneity of Upper Cauvery annual maximum flow data and to select a suitable distribution for the frequency analysis The L-moments method is used in this analysis The Upper Cauvery river basin is shown to be hydrologically heterogeneous The 3 parameter log normal and the generalized extreme value distributions are recommended for the frequency analysis of data in this region

1. Introduction

The Concept of Probability Weighted Moments (PWM) was introduced by Greenwood et al (1979) Since then, it has received considerable attention by Landwehr et al (1979 a,b), Hosking et al (1985), Hosking (1986), Hosking and Wallis (1987) and others PWM estimates are robust in the presence of outliers Parameter estimates from small samples computed by using the PWM method are sometimes more accurate than even the Maximum Likelihood (ML) estimates The PWM method is less complicated than the ML method With some distributions, such as the symmetrical Lambda and Weibull distributions, explicit expressions for the parameters are obtained by the PWM method, which cannot be done with either the ML or the method of moments (MOM)

Hosking (1986 and 1990) has defined the L-moments The L-moments are analogous to the conventional moments and are estimated by linear combinations of order statistics L-moments can also be expressed by linear combinations of PWM Thus the procedures based on PWM and L-moments are equivalent However, L-moments are more convenient because they are directly interpretable as measures of the scale and of the shape of probability distributions Hosking (1990) has used L-moment ratio diagrams to identify underlying parent distributions and L-moment ratios for testing hypotheses about forms of probability distributions Hosking and Wallis (1991) extended the use of L-moments and developed statistics that can be used in regional frequency analysis to measure discordancy, regional homogeneity and goodness-of-fit

The objective of the present work is to analyze the annual maximum flow data from the Cauvery River basin in south India by using the L-moment method Both regional and at-site parameter and quantile estimates are used and the differences between them are studied The L-moment ratios are used to identify candidate distributions and to evaluate the effectiveness of regional analysis

V P Singh and B Kumar (eds), Surface-Water Hydrology, 537–555

2. Theoretical Background

Probability weighted moments $M_{1,r,s}$ are defined by Greenwood et al. (1979) as in Eqs. 1 and 2 where F is the cumulative probability.

$$M_{1,r,0} = \beta_r = \int_0^1 x(F)\ F^r\ dF \tag{1}$$

$$M_{1,o,s} = \alpha_s = \int_0^1 x(F)\ (1-F)^s\ dF \tag{2}$$

Both α_s and β_r are linear in x and are of sufficient generality for parameter estimation. Also, α_s and β_r are related as in eqs. 3 and 4.

$$\alpha_s = \sum_{k=0}^{s} \binom{s}{k} (-1)^k\ \beta_k \tag{3}$$

$$\beta_r = \sum_{k=0}^{r} \binom{r}{k} (-1)^k\ \alpha_k \tag{4}$$

L-moments λ_{r+1} are defined by Hosking (1986 and 1990) in terms of PWMs α_s and β_r as in eq. 5,

$$\lambda_{r+1} = (-1)^r \sum_{k=0}^{r} p^*_{r,k}\ \alpha_k = \sum_{k=0}^{r} p^*_{r,k}\ \beta_k \tag{5}$$

where

$$p^*_{r,k} = (-1)^{r-k} \binom{r}{k} \binom{r+k}{k} \tag{6}$$

In particular,

$\lambda_1 = \alpha_0 \qquad = \beta_0 \qquad (7)$

$\lambda_2 = \alpha_0 - 2\alpha_1 \qquad = 2\beta_1 - \beta_0 \qquad (8)$

$\lambda_3 = \alpha_0 - 6\alpha_1 + 6\alpha_2 \qquad = 6\beta_2 - 6\beta_1 + \beta_0 \qquad (9)$

$\lambda_4 = \alpha_0 - 12\alpha_1 + 30\alpha_2 - 20\alpha_3 \qquad = 20\beta_3 - 30\beta_2 + 12\beta_1 - \beta_0 \qquad (10)$

For a given ordered sample $x_1 \leq .... \leq x_n$, $n > r$, the sample PWMs are calculated (Hosking, 1986) by eqs. 11 and 12,

$$a_s = \hat{\alpha}_s = \frac{1}{n} \sum_{i=1}^{n} (1-P_{i:n})^s\ x_i \tag{11}$$

$$b_r = \hat{\beta}_r = \frac{1}{n} \sum_{i=1}^{n} P_{i:n}^r\ x_i \tag{12}$$

where $P_{i:n}$ is a plotting position. The use of $P_{i:n} = (i-0.35)/n$ usually gives good results for the generalized extreme value distribution (GEV) (Hosking et al., 1985) and is recommended in general for analysis of hydrologic data (Cunnane, 1989). Sample L-moments (l_r) can be calculated by using eqs. 7-10 by replacing α_s or β_r by their sample estimates a_r and b_r. L-moment ratios, which are analogous to the conventional moment ratios are defined by Hosking (1986 and 1990) in eqs. 13 and 14,

$$\tau = \lambda_2 / \lambda_1 \tag{13}$$

$$\tau_r = \lambda_r / \lambda_2 \quad , \quad r \geq 3 \tag{14}$$

where λ_1 is a measure of location, τ is a measure of scale and dispersion (LC_v), τ_3 is a measure of skewness (LC_s), and τ_4 is a measure of kurtosis (LC_K) Sample L-moment ratios (t and t_r) are calculated by using eqs 13 and 14 and substituting l_r for their population values λ_r. The L-moment ratios (Hosking, 1990) offer an easy way to identify underlying distributions, particularly the skewed distributions The sample L-moment ratios plot as well separated groups for different distributions Therefore different distributions are easily discriminated by using them A distribution is considered to be suitable if the data spread consistently around it Hosking (1990) also suggests a test for normality against the skew alternative based on the statistic

$$N_n = v_n^{-1/2}\, t_3$$

where v_n is the variance of t_3 and is given by eq 15

$$v_n = 0\,1866 n^{-1} + 0\,8\, n^{-2} \tag{15}$$

The critical limits of N_n are obtained from the standard normal tables at the required significance level Hosking and Wallis (1991) have derived statistics to measure discordancy (D), regional heterogeneity (H) and goodness of fit (Z) A full description of these statistics is found in Hosking and Wallis (1991) A site i is considered to be unusual if D_i is large A suitable criterion for defining largeness is that D_i should be greater than 3 A region is declared heterogeneous if H is sufficiently large Hosking and Wallis suggest the region be regarded as *acceptably homogeneous* if H is less than 1, *possibly heterogeneous* if H is between 1 and 2, and *definitely heterogeneous* if H is greater than 2 Also, a given distribution is declared adequate if Z^{DIST} is sufficiently close to zero An acceptable criterion being that $|Z^{DIST}|$ is less than or equal to 1 64

As in the method of moments (MOM), parameter estimates are obtained by equating sample PWM or L-moments to the corresponding population values Depending on data availability, parameter and quantile estimates are obtained by using either data at a site or regional data or both (Cunnane, 1989) Flood estimates may be based on data at a site if the record is exceptionally long, or when regional data are not available, or when a region is very heterogeneous The advantage of joint use of at-site and regional data is that, in general, there is sufficient information in the combined set of data so that a multi-parameter distribution can be reliably used A method of combining regional data, which is used here, is the index flood method (Cunnane, 1989) The variate Q normalized by its mean is assumed to have the same distribution at each site The quantile Q_T at a site is estimated by eq 16,

$$\hat{Q}_T = \mu_i\, q_T \tag{16}$$

where q_T is the quantile estimate from the regional distribution and μ_i is the mean at the site The regional distribution parameters are obtained by using regional weighted averages of dimensionless L moments computed by dividing the moments by the mean μ_i of each station

3. Analysis of the Cauvery River Data

The annual maximum flow data from nine gauging stations on the tributaries to the Cauvery River in the state of Karnataka, India are used in this study. The locations of the gauging stations are shown in Figure 1. Details of the rivers and stations included in this study are given in Table 1.

The record length of each station is 63 years except for station 5 which has a record length of 57 years. Probability weighted moments as well as L-moments are calculated for data from each of the nine stations by using equations given in Section 2. Table 2 gives the standardized moments, which are the original moments computed by using the data divided by the mean at each station, as well as their weighted regional averages for these stations. The values of L-moment ratios LC_v, LC_s and LC_k are given in Table 3, as well as their weighted regional averages. The discordancy measures D_i for each of the nine sites are given in Table 3. The largest value of D_i is 1.93 for site 1, which is less than 3 recommended by Hosking and Wallis (1991). Consequently none of the nine sites may be considered to be unusual. The LC_v - LC_s moment ratio diagram for different rivers is shown in Figure 2.

The LC_s versus LC_k diagram for the data used in this study as well as for some of the common three parameter distributions is shown in Figure 3. The heterogeneity measures are given in Table 4. The standardized test value H_1 is 10.21 which is much higher than 2, the lower bound given by Hosking for a *definitely heterogeneous region.* Consequently the upper Cauvery river basin is heterogeneous, a conclusion supported by the information in Figures 2 and 3. The data in Figure 3 do not cluster around any distribution, but are scattered around all of them. The average value of the statistic H_2 (4.37) is again larger than 2. H_2 is shown later to represent the relationship between the at-site and regional estimates. Consequently the relationship between the regional and at-site estimates at different sites is quite diverse. The statistic H_3(0.38), which is shown later to represent the relationship between observed and fitted data indicates that, in general, there would be good agreement between at-site estimates and observed data.

The test of normality outlined in section 2 yields ν_n = 0.00316. The critical values of N_n at 5% significance level are ± 1.96. The normality assumption can be accepted if $|t_3| < 0.11$. Consequently, the normality hypothesis is rejected for all nine stations. Only skew distributions can be considered for these stations. In Table 5 the values of the goodness of fit measure Z^{DIST} for different distributions (Hosking, 1991) are given. Candidate distributions for which the $|Z^{DIST}|$ is less than 1.96 are the Generalized Logistic distribution (GLOG), the Generalized Extreme Value distribution (GEV) and the Log-normal (LN) distribution. Consequently they are used in further analysis. The Pearson-III (P-III) distribution is included for comparison purposes although it does not provide a good fit.

3.1 GEV Distribution

Parameter estimates of the GEV distribution are obtained from Hosking (1986 and 1990) and given in eqs. 17-20.

$$\hat{K} = 7.8590C + 2.9554C^2 \quad (17)$$

$$\hat{\alpha} = \frac{l_2 \hat{K}}{(1-2^{-\hat{K}})\,\Gamma(1+\hat{K})} \quad (18)$$

$$\hat{u} = l_1 + \frac{\hat{\alpha}}{\hat{K}}\left[\Gamma(1+\hat{K}) - 1\right] \quad (19)$$

$$C = \frac{2}{3+t_3} - \frac{\log 2}{\log 3} \quad (20)$$

Table 1 Details of Cauvery river and tributaries

Station	Name of the Station	Stream	Drainage area (km^2)	Annual Rainfall (mm)	Data Available	Q_{max} (m^3/s)
1	Chunchanakatte	Cauvery	2968	680	1918-1980	2879
2	Akkihebbal	Hemavathy	5198	742	1918-1980	2979
3	Unduwadi	Lakshmanathirtha	1502	763	1918-1980	1825
4	Nugu Dam	Nugu	984	920	1918-1980	877
5	Hullahalli	Kabini	4850	920	1918-1974	4129
6	Markonahalli Dam	Shimsha	4131	764	1918-1980	1952
7	Mangala Dam	Nagini	748	764	1918-1980	707
8	Kanva Dam	Kanva	344	839	1918-1980	998
9	Suvarnavathi Dam	Suvarnavathi	1437	676	1918-1980	2084

(source Ramesh et al., 1987)

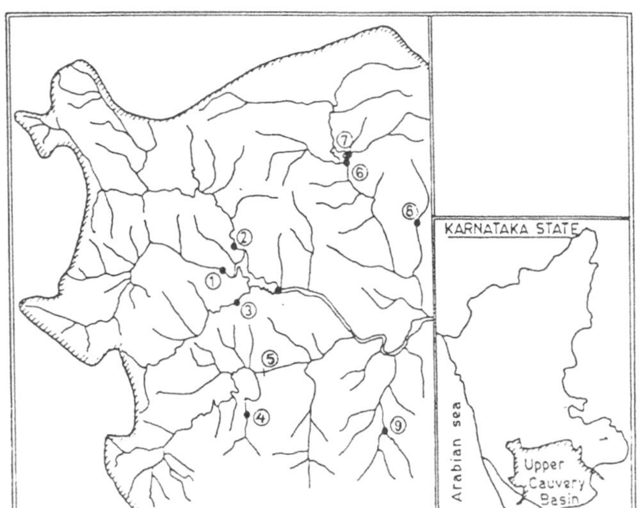

Figure 1. The Upper Cauvery Basin (Source: Ramesh et al. 1987)

Table 2. Standardized Probability Weighted Moments (PWMs) and L-moments

Station No.	b_0	b_1	b_2	b_3	b_4	a_0	a_1	a_2	a_3	a_4	l_1	l_2	l_3	l_4
1	1.0000	0.5952	0.4356	0.3477	0.2913	1.0000	0.4048	0.2452	0.1734	0.1330	1.0000	0.1904	0.0422	0.0296
2	1.0000	0.6144	0.4576	0.3705	0.3141	1.0000	0.3856	0.2288	0.1592	0.1204	1.0000	0.2288	0.0595	0.0534
3	1.0000	0.6926	0.5520	0.4681	0.4112	1.0000	0.3074	0.1668	0.1102	0.0806	1.0000	0.3852	0.1566	0.1122
4	1.0000	0.7343	0.6044	0.5217	0.4630	1.0000	0.2657	0.1358	0.0886	0.0654	1.0000	0.4686	0.2206	0.1133
5	1.0000	0.6227	0.4711	0.3869	0.3322	1.0000	0.3773	0.2258	0.1585	0.1208	1.0000	0.2454	0.0907	0.0757
6	1.0000	0.8042	0.6793	0.5911	0.5249	1.0000	0.1958	0.0709	0.0342	0.0195	1.0000	0.6084	0.2507	0.0935
7	1.0000	0.7397	0.6078	0.5238	0.4645	1.0000	0.2603	0.1283	0.0803	0.0569	1.0000	0.4794	0.2083	0.1205
8	1.0000	0.7414	0.5978	0.5055	0.4408	1.0000	0.2586	0.1150	0.0636	0.0398	1.000	0.4828	0.1382	0.0736
9	1.0000	0.7065	0.5628	0.4757	0.4165	1.0000	0.2935	0.1498	0.0932	0.0643	1.0000	0.4129	0.1378	0.1093
Average	1.0000	0.6953	0.5529	0.4665	0.4073	1.0000	0.3047	0.1623	0.1062	0.0774	1.0000	0.3906	0.1455	0.0869

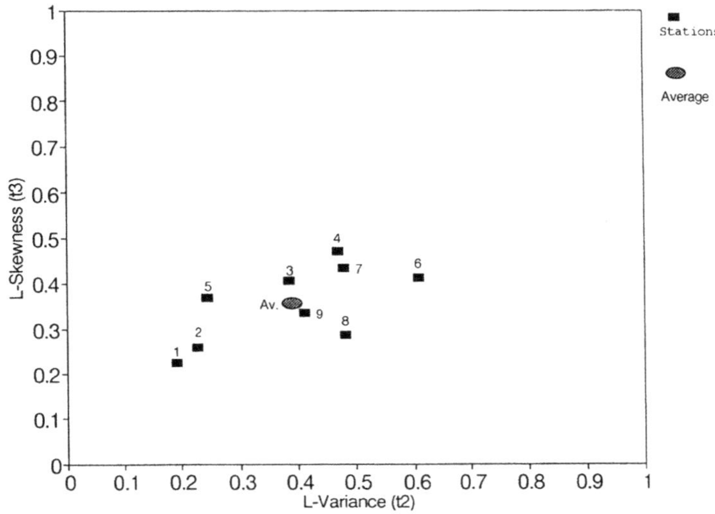

Figure 2. LC_v vs LC_s Moment Ratio Diagram

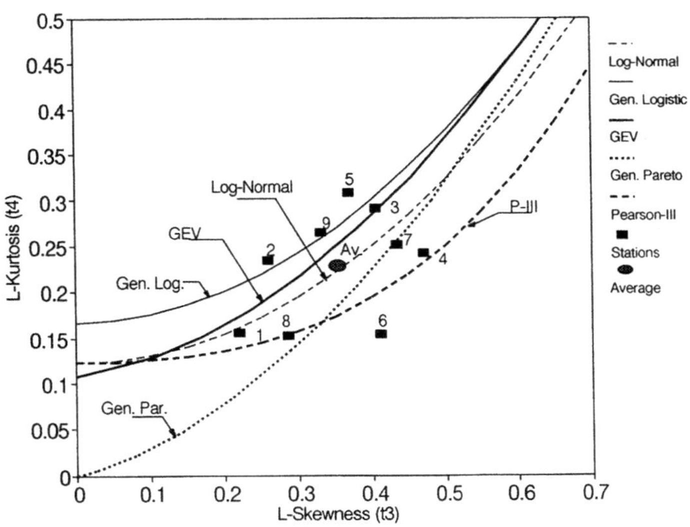

Figure 3. The LC_s vs LC_k Moment Ratio Diagram

Table 3. L-Moment Ratios and Discordancy Measures D_i

SITE	N	NAME	LC_v	LC_s	LC_k	D_i
1	63	SITE 1	0.1904	0.2217	0.1556	1.93
2	63	SITE 2	0.2288	0.2599	0.2332	0.62
3	63	SITE 3	0.3852	0.4065	0.2913	0.44
4	63	SITE 4	0.4686	0.4707	0.2417	1.18
5	57	SITE 5	0.2454	0.3696	0.3085	0.89
6	63	SITE 6	0.6084	0.4121	0.1537	1.25
7	63	SITE 7	0.4794	0.4345	0.2513	0.33
8	63	SITE 8	0.4828	0.2862	0.1524	1.17
9	63	SITE 9	0.4129	0.3338	0.2646	1.19
WEIGHTED MEANS			0.3906	0.3548	0.2272	

Table 4. Heterogeneity Measure H

Measure	Item	Value
H_1	OBSERVED S.D. OF GROUP LC_v	0.1323
	SIM. MEAN OF S.D. OF GROUP LC_v	0.0357
	SIM. S.D. OF S.D. OF GROUP LC_v	0.0095
	STANDARDIZED TEST VALUE	10.21
H_2	OBSERVED AVE. OF LC_v / LC_s	0.1394
	SIM. MEAN OF AVE. LC_v / LC_s	0.0689
	SIM. S.D. OF AVE. LC_v /LC_s	0.0162
	STANDARDIZED TEST VALUE	4.37
H_3	OBSERVED AVE. OF LC_s /LC_s	0.0944
	SIM. MEAN OF AVE. LC_s / LC_k	0.0866
	SIM. S.D. OF AVE. LC_s /LC_k	0.0206
	STANDARDIZED TEST VALUE	0.38

Number of Simulations = 500

Table 5. Goodness of Fit Measures Z^{DIST}

Distribution	LC_k	Z
GEN. LOGISTIC	0.272	1.34
GEN. EXTREME VALUE	0.250	0.53
LOG-NORMAL	0.222	-0.53
PEARSON TYPE III	0.174	-2.34
GEN. PARETO	0.184	-1.97

Parameter estımates at each statıon as well as theır regıonal average are gıven ın Table 6 Quantıle estımates are obtaıned usıng Equatıon 21

$$\hat{q}_T = \hat{u} + (\hat{\alpha}/\hat{K}) \left[1 - (-\log F)^{K}\right] \qquad (21)$$

Fıgure 4 (a d) shows at-sıte as well as regıonal quantıle estımates along wıth the correspondıng observed values for rıvers Cauverı, Lakshmanathırtha, Kabını and Shımsha

Table 6 Parameter Estimates for the GEV Distribution

Site	1	2	3	4	5
u	0 8319	0 7906	0 6117	0 5154	0 7573
α	0 2539	0 2863	0 3615	0 3783	0 2494
K	-0 0794	-0 1359	-0 3386	-0 4209	-0 2897

Site	6	7	8	9	Regional
u	0 3852	0 5108	0 5483	0 6003	0 6129
α	0 5639	0 4222	0 5768	0 4511	0 3945
K	-0 3460	-0 3750	-0 1740	-0 2408	-0 2937

3 2 Generalızed Logıstıc (GLOG) Dıstrıbutıon

Parameter estımates of the GLOG dıstrıbutıon are obtaıned from Eqs 22-25 (Hoskıng, 1986 and 1990)

$$\hat{K} = -t_3 \quad (22) \quad , \qquad \hat{\alpha} = \frac{l_2}{\Gamma(1+\hat{K})\ \Gamma(1-\hat{K})} \qquad (23)$$

$$\hat{u} = l_1 + (l_2 - \hat{\alpha}) / \hat{K} \qquad (24)$$

Parameter estımates for each of these statıons as well as theır regıonal average are gıven ın Table 7 Quantıle estımates are computed by usıng Equatıon 25

$$q_T = \hat{u} + \frac{\hat{\alpha}}{\hat{K}} \left[1 - \{ (1-F)/F \}^{K} \right] \qquad (25)$$

The regıonal quantıle estımates along wıth the at sıte observed data for Cauverı, Lakshmanathırtha, Kabını and Shımsha rıvers are shown ın Fıg 5

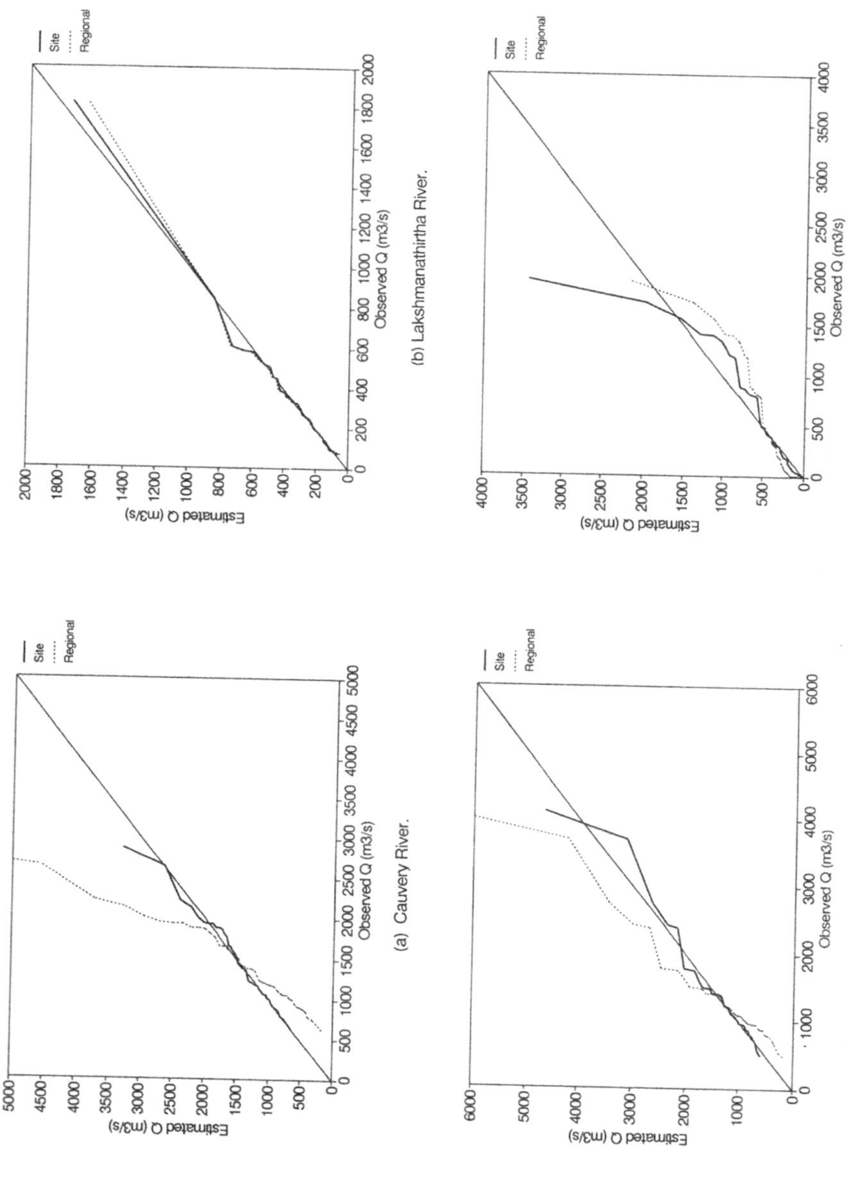

Figure 4. The Estimates and Observed Data for the GEV Distribution

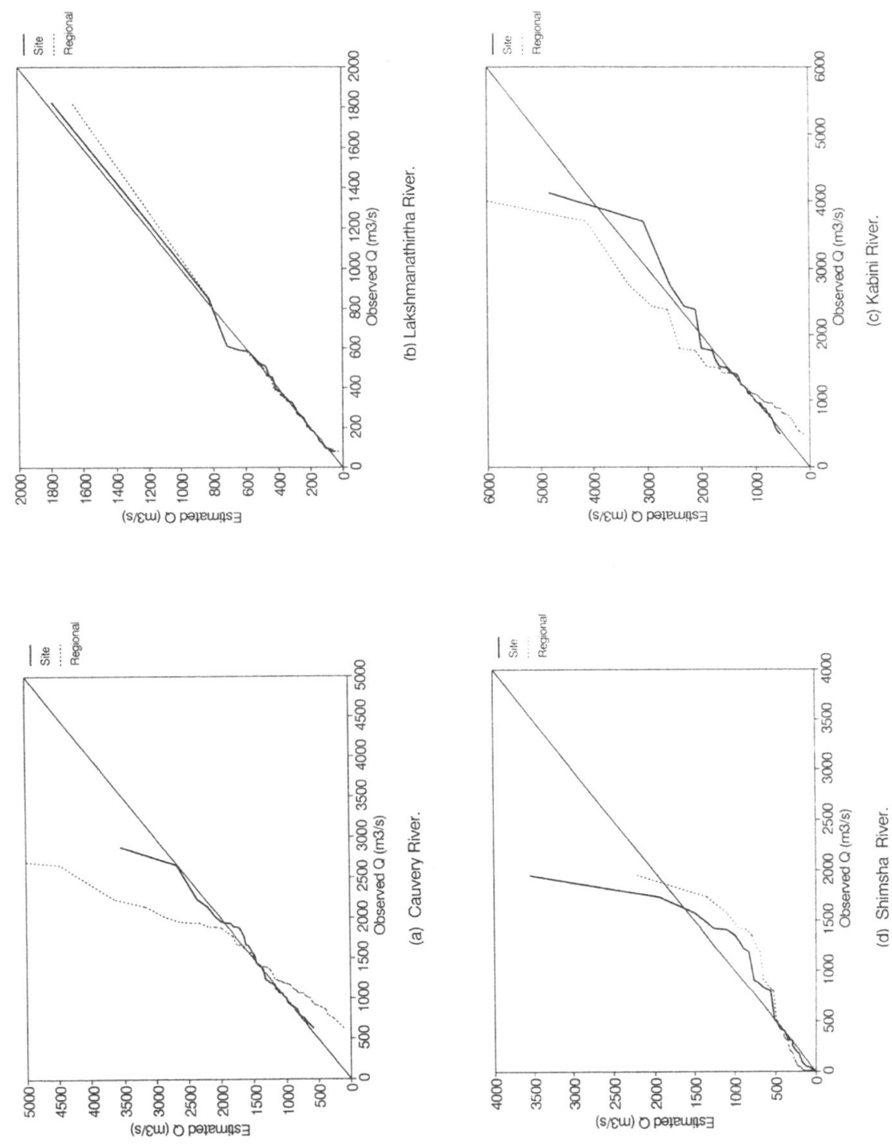

Figure 5. The Estimates and Observed Data for the GLOG Distribution

Table 7. Parameter Estimates for the GLOG Distribution

Site	1	2	3	4	5
u	0.9322	0.9504	0.7627	0.6748	0.8606
α	0.1754	0.2042	0.2887	0.3156	0.1939
K	-0.2217	-0.2599	-0.4065	-0.4707	-0.3696

Site	6	7	8	9	Regional
u	0.6208	0.6879	0.7817	0.7854	0.7857
α	0.4521	0.3438	0.4203	0.3413	0.3146
K	-0.4121	-0.4345	-0.2862	-0.3338	-0.3548

3.3. Log-Normal Distribution

Parameter estimates of the LN distribution are obtained from Eqs. 26-29 (Hosking, 1990),

$$\hat{\sigma} = 0.999281Z - 0.006118Z^3 + 0.000127Z^5 \qquad (26)$$

$$\hat{\mu} = \log\left[l_2 / erf(\hat{\sigma} / 2)\right] - \frac{\hat{\sigma}^2}{2} \qquad (27)$$

$$\hat{a} = l_1 - \exp\left(\hat{\mu} + \frac{\hat{\sigma}^2}{2}\right) \quad (28) \qquad Z = \sqrt{(8/3)}\ \Phi^{-1}\left(\frac{1+t_3}{2}\right), \qquad (29)$$

where Φ^{-1} is the inverse standard normal distribution function, and erf (.) is the error function, erf $(x) = 2\ \Phi\ (x\sqrt{2}) - 1$. Parameter estimates for data from each of the stations as well as their regional averages are given in Table 8. Quantile estimates are obtained using Equation 30.

$$q_T = \exp\left[\ \hat{\sigma}\ \Phi^{-1}(F) + \hat{\mu}\right] + \hat{a} \qquad (30)$$

Table 8. Parameter Estimates for the LN Distribution

Site	1	2	3	4	5
a	0.2504	0.2302	0.1624	0.1142	0.4150
μ	-0.3932	-0.4072	-0.5524	-0.6410	-0.8416
σ	0.4582	0.5396	0.8663	1.0195	0.7815

Site	6	7	8	9	Regional
a	-0.3056	0.0221	-0.4774	-0.0870	0.0312
μ	-0.1201	-0.4570	0.2124	-0.1623	-0.3116
σ	0.8794	0.9323	0.5965	0.7010	0.7482

The at-site as well as regional quantile estimates along with the observed data for Cauveri, Lakshmanathirtha, Kabini and Shimsha rivers are shown in Figure 6.

3.4 Pearson-III Distribution

Parameter estimates for the P-III distribution are obtained from Eqs. 31-34 (Hosking, 1991).

For $t_3 \geq 1/3$ then $t = 1 - t_3$ and

$$\beta = \frac{(0.36067t - 0.59567t^2 + 0.25361t^3)}{(1 - 2.78861t + 2.56096t^2 - 0.77045t^3)} \tag{31}$$

and for $t_3 < 1/3$ then $t = 3\pi\, t_3^2$

$$\beta = \frac{(1 + 0.2906t)}{(t + 0.1882t^2 + 0.0442\, t^3)} \tag{32}$$

$$\hat{\alpha} = \sqrt{\pi}\; l_2 \frac{\Gamma(\beta)}{\Gamma(\beta + 1/2)} \quad (33); \qquad \hat{\gamma} = l_1 - \hat{\alpha}\,\beta. \quad (34)$$

Parameter estimates as well as their regional average are given in Table 9. Quantile estimates can be obtained using the frequency factor formula in eq. 35, where K_T is computed using any of the formulae given by Bobée and Ashkar (1991) corresponding to a given probability F. The at-site as well as regional quantile estimates along with the observed data for Cauveri, Lakshmanathirtha, Kabini and Shimsha rivers are shown in Figure 7.

$$q_T = \hat{\alpha}\,\beta + \hat{\gamma} + K_T \sqrt{\hat{\alpha}^2\,\beta} \tag{35}$$

Table 9. Parameter Estimates for the P-III Distribution

Site	1	2	3	4	5
α	0.2388	0.3418	0.9970	1.5050	0.5592
β	2.2327	1.6363	0.6661	0.4844	0.8114
γ	0.4668	0.4407	0.3359	0.2709	0.5463

Site	6	7	8	9	Regional
α	1.6052	1.3644	0.8053	0.8272	0.8448
β	0.6470	0.5777	1.3535	0.9974	0.8815
γ	-0.0386	0.2117	-0.0899	0.1750	0.4668

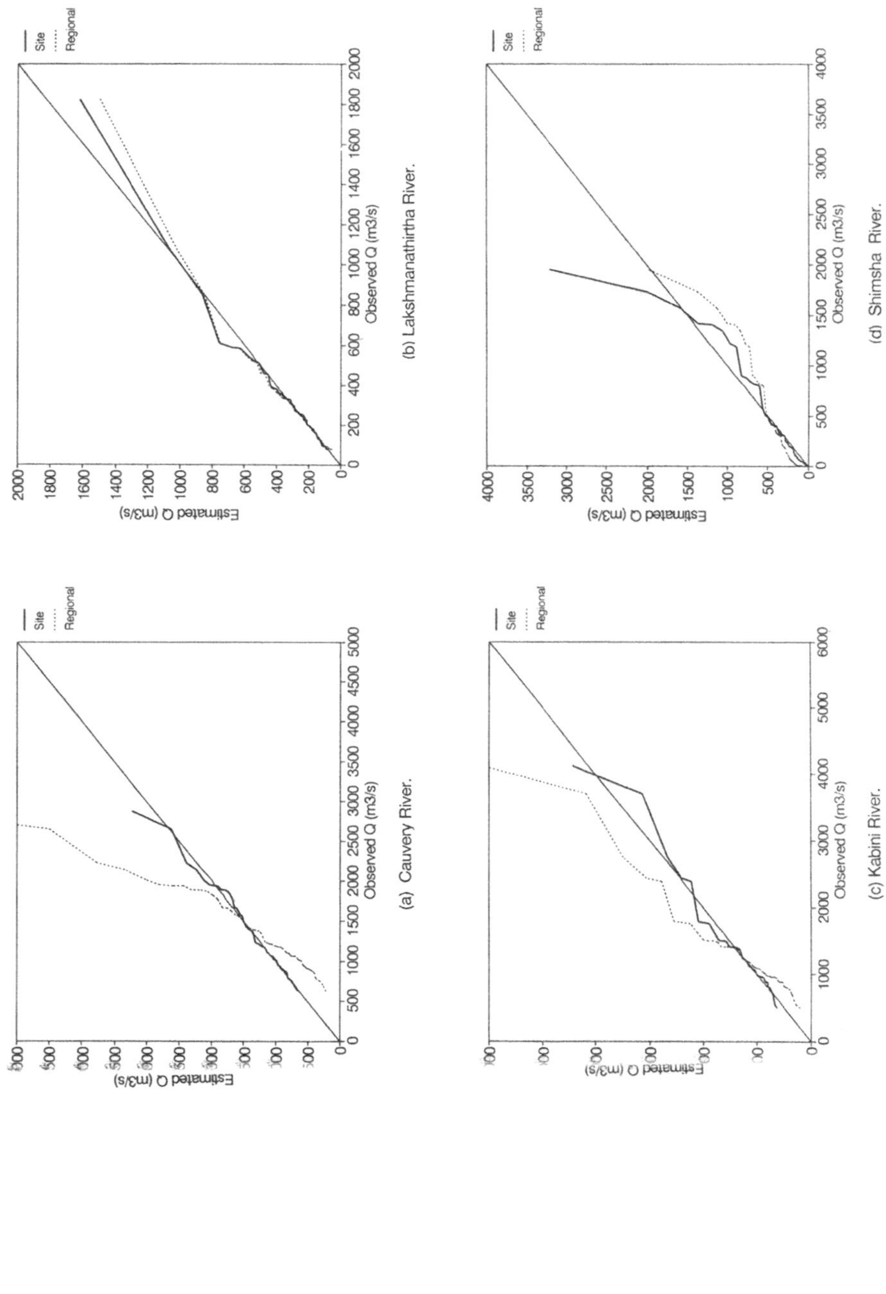

(a) Cauvery River. (b) Lakshmanathirtha River. (c) Kabini River. (d) Shimsha River.

Figure 6. The Estimates and Observed Data for the LN Distribution

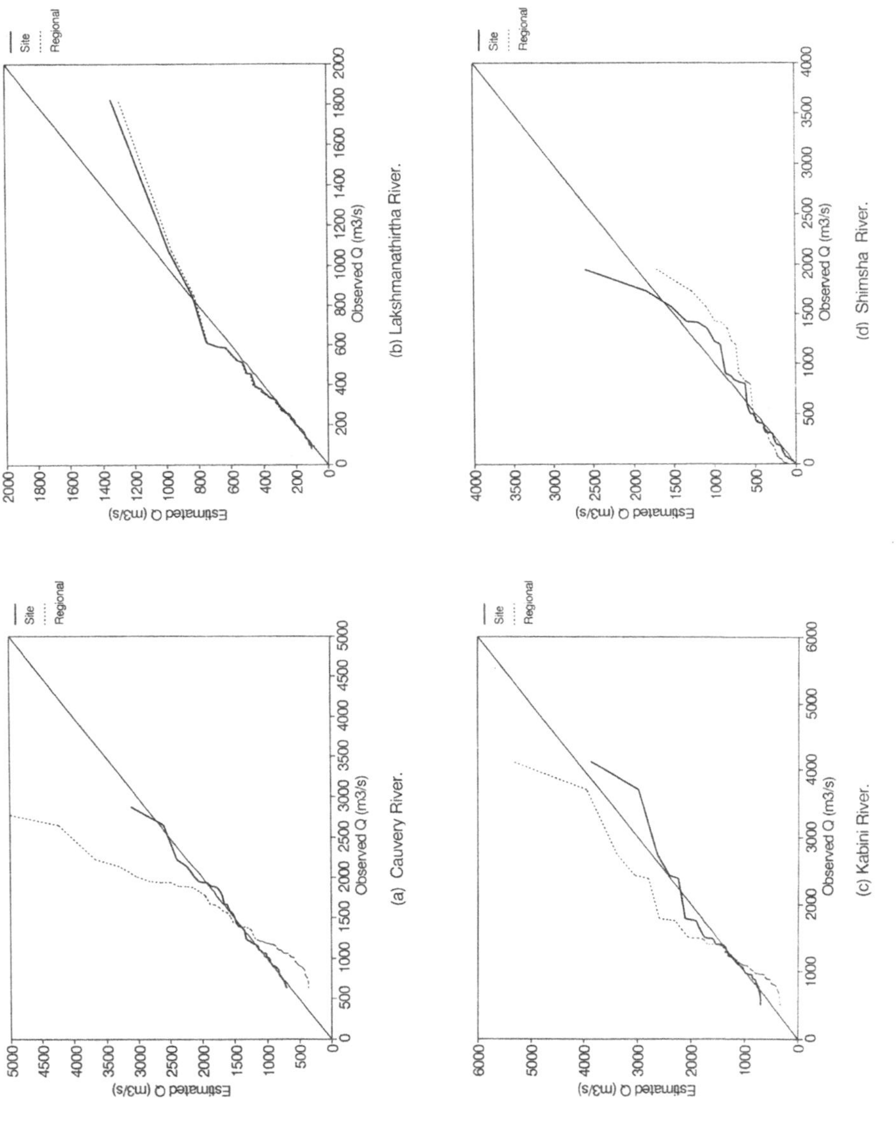

Figure 7. The Estimates and Observed Data for the P-III Distribution

4. Discussion of Results

From the results given in Figures 4-7, in general, three parameter distributions are acceptable for fitting the observed data. According to the goodness-of-fit measure (Z^{DIST}) the regional estimates obtained by the P-III distribution do not agree with the observed data as well as the other three distributions. However, it is hard to reach any conclusions by using only the graphical results. This is due to the fact that the difference in t_4 between the regional value and the fitted distribution in Figure 3, which is a measure for goodness of fit (Hosking and Wallis, 1991), is not that much larger for the P-III distribution than it is, for example, the GLOG distribution.

Examination of the results in Figure 4 shows that the difference between regional and at-site estimates, for the same distribution, depends on the site location, being maximum for station 1 and minimum for station 3. Results in Figure 3 offer no explanation for this observation. Figure 2 explains these results better because, for three parameter distributions, the parameter estimates depend on λ_1, λ_2 and t_3 which are included in Figure 2. Figure 3 contains t_3 and t_4, but only t_3 is used in parameter estimation. Therefore, the closer the statistics of a site is to the average in Figure 2, the closer is the agreement between regional and at-site estimates. In this sense, the regional heterogeneity measure H_2 based on the LC_v / LC_s ratio quantifies the average difference between regional and at-site estimates. For stations with t less than the average value in Figure 2 the regional estimates of high quantiles are greater than the at-site estimates, and the opposite is true for stations with t greater than the average value of t. This conclusion is supported, in general, by the results in Table 10. When at-site estimates are considered, each site is treated individually. The goodness of fit measure (Z) depends only on the difference between the regional average and fitted distribution value of t_4. The difference between at-site estimates and the observed data depends on the difference between the at-site and fitted distribution value of t_4, i.e. the location of the site statistic with respect to the distribution curves on the LC_s vs LC_k diagram given in Figure 3. The data from Station 3, according to Figure 3, must fit the GEV distribution much better than the data from Station 6. The Regional heterogeneity measure H_3, based on the LC_s/LC_k distance, gives an estimate of the average deviation between at-site estimates and the observed data.

Both the heterogeneity measures H_2, and H_3 based on LC_v/LC_s and LC_s/LC_k distances respectively, are important to select a good distribution for regional analyses. The measure H_2 indicates whether at-site and regional estimates are close to each other. The measure H_3 indicates whether the at-site estimates and the observed data are in agreement. A large value of H_2 usually indicates a large deviation between regional and at-site estimates and in turn between regional estimates and the observed data. Similarly, a large H_3 value indicates, in general, a large deviation between at-site estimates and the observed data, and in turn between regional estimates and the observed data, as is the case with large H_2. In the present study, the value of H_2 in Table 4 is large (4.37) whereas H_3 is small (0.38). As a result, at-site estimates are much closer to the observed data than the regional estimates as shown in Figures 4-7. Consequently, the region may have to be further subdivided to obtain better regional estimates.

Table 10. Higher Quantile Estimates for the LN Distribution

	F = 0.98		F = 0.99		F = 0.995		F = 0.998	
Station	R	S	R	S	R	S	R	S
1	4601	2651	5633	2960	6780	3277	8490	3714
2	3782	2472	4630	2824	5573	3194	6978	3715
3	1040	1082	1274	1357	1533	1673	1920	2159
5	4395	3276	5380	3928	6476	4659	8109	5759
4	494	632	605	829	728	1064	912	1442
6	1383	2051	1693	2640	2083	3318	2553	4366
7	381	479	466	616	561	777	703	1029
8	736	800	901	959	1085	1130	1358	1373
9	1233	1257	1510	1528	1818	1826	2276	2264

R: Regional Estimate S: At-Site Estimate

Finally, quantile estimates for a probability F = 0.998 are given in Table 11. For both regional and at-site estimates, the estimates for a certain station in descending order are given by GLOG, GEV, LN and P-III. These results conform to the relative position of these distributions in Figure 3. For a given station, the appropriate distribution is the one which is closest to that station in Figure 3. If the statistic for a station plots higher than all distributions such as station 5, then higher quantiles will tend to be underestimated. If a station plots lower than other distributions such as station 6, then the flood estimate tends to be overestimated. If the station plots within the distributions (station 1), the distributions below will tend to underestimate and the distributions above will tend to overestimate flood magnitudes.

Table 11. Q_s Estimates (F = 0.998) at Different Sites

Station	Regional				At-Site			
	GLOG	GEV	LN	P-III	GLOG	GEV	LN	P-III
1	10628	10176	8490	6833	4389	3845	3714	3483
2	8735	8364	6978	5617	4487	3946	3715	3393
3	2403	2301	1920	1545	2703	2513	2159	1628
4	1142	1093	912	734	1797	1712	1442	987
5	10150	9719	8109	6525	7094	6526	5759	4650
6	3195	3059	2553	2054	5523	5131	4366	3209
7	879	842	703	565	1292	1214	1029	740
8	1700	1628	1358	1093	1715	1501	1373	1201
9	2848	2727	2276	1832	2834	2545	2264	1866

5. Conclusions

On the basis of results presented herein, the following conclusions are presented.

1. The Upper Cauvery basin is hydrologically heterogeneous. It will have to be subdivided into smaller regions to get hydrologically homogeneous regions.

2. As a result of conclusion 1 above, the at-site estimates are better than the regional estimates for the Upper Cauvery river data.

3. The lognormal or the generalized extreme value distributions is recommended for use in the Upper Cauvery river basin.

6. Acknowledgement

Dr. J.R.M. Hosking most graciously sent us his computer program and reports for our use. We are grateful to him. We would like to thank Dr. R. Prasad for sending us the data used in this study.

7. References

Bobée, B. and Ashkar, F. (1991) "The Gamma Family and Derived Distributions Applied in Hydrology". *Water Resources Publications*, Colorado.

Cunnane, C.(1989) "Statistical Distributions for Flood Frequency Analysis." WMO Operational Hydrology Report No. 33, WMO-No.718.

Greenwood, J.A., Landwehr, J.M. and Matalas, N.C. (1979) "Probability Weighted Moments: Definition and Relation to Parameters of Several Distributions Expressible in the Inverse Form." *Water Res. Research*, 15(5), 1049-1054.

Hosking, J.E.M. (1991) "Fortran Routines for Use with the Method of L-Moments, Version 2". IBM Research Report RC 17097, IBM Research, Yorktown Heights, NY.

Hosking, J.R.M. and Wallis, J.R. (1991) "Some Statistics Useful in Regional Frequency Analysis". *Research Report* RC 17096 (#75863), IBM Research, Yorktown Heights, NY.

Hosking, J.R.M. (1986) "The Theory of Probability Weighted Moments." Research Report RC12210 (#54860), *IBM Research*, Yorktown Heights, NY.

Hosking, J.R.M. (1990) "L-Moments: Analysis and Estimation of Distributions Using Linear Combination of order Statistics." *J. royal Stat. Soc. B,* 52(1), 105-124.

Hosking, J.R.M. and Wallis, J.R. (1987) "Parameter and Quantile Estimation for the Generalized Pareto Distribution." *Technometrics,* 29, 339-349.

Hosking, J.R.M., Wallis, J.R. and Wood, E.F. (1985) "Estimation of the Generalized Extreme Value Distribution by the Method of Probability Weighted Moments." *Technometrics,* 27, 251-261.

Landwehr, J.M., Matalas, N.C. and Wallis, J.R. (1979a) "Probability Weighted Moments Compared With Some Traditional Techniques in Estimating Gumble Parameters and Quantiles." *Water Res. Research,* 15(5), 1055-1064.

Landwehr, J.M., Matalas, N.C. and Wallis, J.R. (1979b) "Estimation of Parameters and Quantiles of Wakeby Distribution, Parts 1 and 2." *Water Res. Research,* 15(6), 1361-1379.

Ramesh, M., Murthy, M.C.S. and R. Prasad (1987) "Analysis of Flood Frequencies in the Cauvery Valley", in "Hydrologic Frequency Modelling", Proc. Int. Symp. on Flood Frequency and Risk Analysis, D. Reidel Pub. Co., Dordrecht, Holland, pp. 505-513.

FLOOD PREDICTION BY USE OF PARTIAL DURATION SERIES

ARIE BEN-ZVI
Israel Hydrological Service
P.O.Box 6381
Jerusalem 91063
ISRAEL

ABSTRACT. Partial duration series are series composed of all and only the high flow events which exceed a pre-determined threshold value. As such they provide the exact information for computation of recurrence intervals of large magnitudes. Recurrence intervals of rare magnitudes are predicted by the exceedance probability of these magnitudes in the series, divided by the mean frequency of occurrence of events in that series. The frequency of occurrence is usually described by the Poisson distribution, although in some cases the negative binomial distribution fits the data better. The distribution of magnitudes is described, in most cases, by the negative exponential distribution, but a number of other distributions are also applied. The criteria for selection of a distribution are based upon theoretical reasoning, goodness-of-fit, regional uniformity, and also on the trend observed for the higher values. Threshold values are selected through empirical, physical or statistical reasoning. Most predictions are made for peak discharges. Some others are made for volumes of events, and a recent introduction is of volumes beyond design discharges. The combination of selected distributions, their parameters, and the threshold values for individual stations in a region provides a basis for development of a regional model.

DEFINITIONS

Flood is commonly defined as an unusually wide overbank extension of streamflow. Official declarations of flood areas are related to danger of lives and to damage to property. In science, flood is defined as a streamflow event which attains extremely large magnitudes. As such, floods are understood as occurring infrequently. Yet, there is no clear-cut distinction between floods and other events of large magnitudes.

Flood prediction is an assignment of occurrence probabilities to magnitudes of floods which might occur. In delivering the results to their cliens, hydrologists transform these probabilities into recurrence intervals. The recurrence interval, of a magnitude x of a variable X, is the mean time elapses between successive occurrences of events for which $X \geq x$. The transformation is done through the mean occurrence frequency

V. P. Singh and B. Kumar (eds.), Surface-Water Hydrology, 557–565.

of the considered events.

A reasonable assignment of probabilities requires a fit of a probability distribution to observed magnitudes. Owing to the limited duration of quantitative hydrological observations, and to the infrequent occurrences of flood events, there are available, in almost all cases, too few floods on record to which a good fit can be made. Hydrologists overcome this difficulty by fitting the distribution to a series composed of a sufficient number of recorded large events. Two methods are practiced for the construction of this series, one is of the maximal value recorded during every hydrological year, and the other is of all the values which exceeded a pre-selected threshold value. A series obtained by the former method is called annual maxima series, while that obtained by the latter method is called partial duration series. The present work deals only with partial duration series.

MAGNITUDES

An objective prediction of floods is made through a probability distribution which is assumed to describe the distribution of possible magnitudes of floods. In most of the works, the negative exponential distribution is selected for this purpose (e.g. Todorovic and Zelenhasic, 1970; Cunnane, 1973; North, 1980; Ashkar and Rousselle, 1983; Nachnebel and Konecny, 1987; Choulakian et al., 1990; Rasmussen and Rosbjerg, 1991; Rasmussen et al., 1994). Yet, the lognormal distribution is considered in some other works (e.g. Hiemstra et al., 1976; Hiemstra and Francis, 1981; Ben-Zvi, 1982; Rosbjerg, 1987). The criteria for selection of a distribution are goodness-of-fit to recorded data, regional consistency and mathematical convenience. Recently, an additional criterion, about the thickness of upper tail of data, is also being considered (Rosbjerg, 1992; Ben-Zvi, 1993, 1994). Rosbjerg et al. (1992) recommend a use of the negative exponential distribution for long (thick) tailed samles, and of the correct distribution for short (thin) tailed samples.

Noting that the magnitudes, which are of interest in many hydrologic predictions, are larger than those on records, a number of researchers prefer distributions which adhere to trends in the upper tails of recorded data (e.g. Bryson, 1974; Picands, 1975; Smith, 1987). Bryson (1974) defines the trend as being heavy when the observed density distribution approaches zero more slowly than the negative exponential distribution does, and as being light when the distribution approaches zero more rapidly. He tests the rapidity of approach through a variable called conditional mean exceedance, CME, which is defined by:

$$CME_x = E(X - x \mid X \geq x) \tag{1}$$

where X is a considered variable and x is a pre-selected value of X.

Smith (1987) proposes a graphical display for the determination of tail trend, where CME_x is plotted against x. An example for this display is shown in Fig. 1. An unbounded thick (i.e. heavy) tail would render, on the display, an upwards trend, whereas a bounded thin (i.e. light)

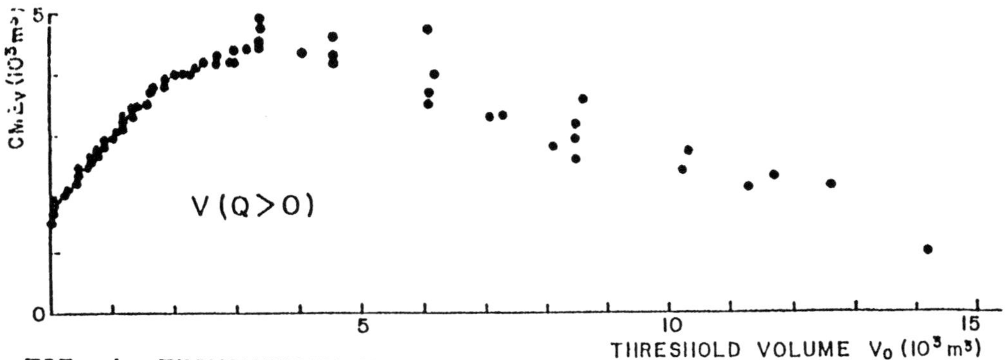

FIG. 1: EXAMINATION OF TAIL TREND

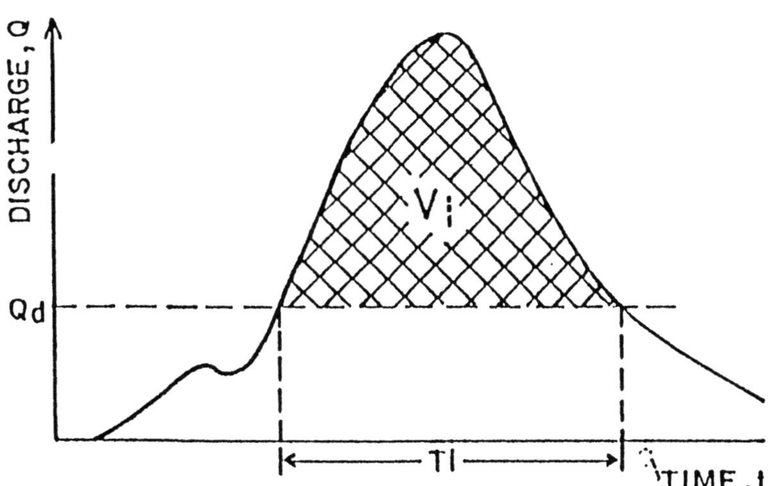

FIG. 2: ILLUSTRATIVE SKETCH

tail would render a downwards trend.

Trends of upper tails have been examined, so far, for a few cases of annual maxima series of peak discharges, for two case studies of peak discharges in partial duration series (Rosbjerg et al., 1992; Ben-Zvi, 1994) and for a case of volumes beyond design discharges (Ben-Zvi, 1993). Yet, it should be noted here that an extrapolation of a bounded light tail might lead to a statistical, rather than a physical, estimation of maximal probable values. This situation is not desired, at least in cases where the maximal probable magnitude is not considerably higher than the maximal observed one.

FREQUENCY

Consideration of the frequency of occurrence of flow events is required for two purposes. One is the transformation of occurrence probabilities in the series of magnitudes into recurrence intervals, and the other is for generation of synthetic series of events where occurrence times are required. The former purpose requires the mean frequency of occurrence of the events which compose the partial duration series while the latter one requires an application of a probability distribution.

The following equation is applied for the transformation:

$$T(x) = P_{e}(X \geq x)/n \qquad (2)$$

in which $T(x)$ is the recurrence interval of x (in years), $P_{e}(X \geq x)$ is the exceedance probability of x in the series, and n is the mean number of events per year ($year^{-1}$).

The occurrence distribution is generally considered as being Poissonian (e.g. Shane and Lynn, 1964; Todorovic and Zelenhasic, 1970; Ashkar and Rousselle, 1987; Cruise and Arora, 1990; Rasmussen and Rosbjerg, 1991; Rosbjerg et al., 1992; Rasmussen et al., 1994). Some refined versions of this distribution are also considered (e.g. Todorovic and Rousselle, 1971; Taesombut and Yevjevich, 1978; Kavvas, 1982; Nachtnebel and Konecny, 1987). The refined versions can improve the predictions in cases of seasonal variations or differences in types of floods, and in case of presistance in the occurrence of floods. Ben-Zvi (1991) has shown an advantage for the negative binomial over the Poisson distribution for partial duration series of runoff in Israel.

THRESHOLD VALUE

Partial duration series are being practiced in hydrology for more than sixty years (Jarvis et al., 1936). These series contain all and only evets of which the examined magnitude exceeds a pre-determined threshold value (called also base level). The determination of threshold values is usually done through subjective considerations. Obviously, the higher the threshold the fewer are the events in the selected series. This results in a lower data handling and computational load, a higher effect of sampling variations and, in many cases, a lower dependence between the selected events (Bhuiya and Yevjevich, 1968).

The original USGS guidelines for selection of the base level is that it should be equal to the lowest annual flood on record, but in a long record the level is raised to include, on the average, three or four events per year (Langbein, 1949). Chow (1953) proposes that the sample size be equal to the number of years on record. Cunnane (1973) finds an advantage when the sample size is at least 1.65 times the number of years on record. Taesombut and Yevjevich (1978) find the advantage when the sample size is at least 1.95 times the number of years on record. Ashkar and Rousselle (1987) prefer samples for which the variance of the occurrence frequency is close enough to the mean, which, in their case study, results in sizes two to three times larger than the number of years on record. Kavvas (1982) elaborates on the physical and engineering considerations for the selection of thresholds, whereas Ashkar and Rousselle (1983) prefer statistical considerations for the selection. We prefer statistical considerations for regional works (e.g. Ben-Zvi, 1991) and physical considerations for particular predictions (e.g. Ben-Zvi and Cohen, 1983).

Ben-Zvi (1991) has examined the effect of threshold value on the fit of the Poisson and the negative binomial distributions to recorded occurrences of events. He concluded that the fit of either distribution to samples with frequent events does not guarantee the fit to sub-samples composed of events of lower frequency only. Owing to shortage in data for events exceeding high thresholds, he could not test the fit of those distributions to purely flood samples (i.e. exceedance frequency ≤ 0.1 year^{-1}). Yet, indirect considerations indicate a good fit of either distribution, and particularly of the negative binomial, to the occurrence of such flood samples.

The cited statistical selections of threshold values consider the occurrence frequency of events in the samples. Another criterion, of goodness-of-fit, has recently been proposed (Ben-Zvi, 1993, 1994). This criterion is applied when the selected distribution is fitted to nested sub-samples of the upper tail. The threshold associated with the best fit is selected. It is believed, though not shown, that this selection would contribute to the credibility of the predictions.

VARIABLES

Having selected a threshold value and fitted probability distributions to occurrences and magnitudes in the partial duration series (i.e. obtained the most suitable values for the parameters), one can estimate the time intervals between given magnitudes, and magnitudes of events to which given recurrence intervals are assigned. Most of the flood predictions are made for the peak discharge, which is the maximal momentary discharge during the flood. The other ones are made for the flood volume or the flood duration. A recent introduction is of flood volume beyond the design discharge (Ben-Zvi, 1993). This variable describes the volume of innundation upstream from river constrictions, such as bridges and culverts, and is closely related to the damage inflicted there by flood waters.

VOLUME BEYOND DESIGN DISCHARGE

The volume beyond design discharge is a new variable in hydrology, and therefore we shall elaborate here on its determination. Let us consider a given cross section on a river, which is constricted to a design discharge Q_d, with an associated recurrence interval T_d. This cross section allows for a free passage of all flood events of which the discharges are lower than Q_d. On the average, once in T_d, there would occur an event i, in which, for a certain duration T_i, the approaching discharge, $Q(t)$, $t \in T_i$, is higher than Q_d. For every elementary time interval, $dt \in T_i$, the extra volume $dV(t) = (Q(t) - Q_d)dt$ is accumulated upstream from the constricted cross section. The maximal volume of extra water for the i'th event, V_i, is equal to the integral of $dV(t)$ throughout T_i. The determination of V_i from a recorded hydrograph is illustrated in Fig. 2. Upon substitution of V_i into the local geometry, and with respect to the physical properties of the site, one can compute the depth and areal extent of the resulted innundation and estimate the damage inflicted by the i'th event.

The series of volumes beyond design discharge is a peculiar case in hydrology. Its occurrences are those of a partial duration series of peak discharges associated with a threshold value of Q_d, and its magnitudes are of volumes determined with respect to that discharge. The recurrence interval of a magnitude, v, in this series, T(v), is the ratio of the recurrence interval of the design discharge, T_d, to the exceedance frequency of v in the series of extra volumes, $P_e(V \geq v)$:

$$T(v) = T_d/P_e(V \geq v) \tag{3}$$

REGIONALIZATION

Predictions for ungauged sites are based on regionalization of results obtained for gauged sites. For this purpose, records are collected for stations located within the considered region and its vicinity, and distribution functions are fitted to partial duration series of each one of these stations. Threshold and parameter values, as well as mean occurrence frequency are determined for each one of the stations. These are regressed against measurable, or clearly definable, properties of the watersheds of the gauging sites. An altertnative regression against the watershed properties is of magnitudes to which given recurrence intervals are assigned. Next, a decision is taken about the geographic, or otherwisely defined, region through which the regression equations are applicable. These equations, together with the distribution functions, form a model for prediction of magnitudes and frequency of floods at any site within the defined region.

PREPARATION

Preparation of predictions is involved with objective computations and subjective decisions. The results serve, in most practical cases, as quantitative inputs for engineering and other water resources management decisions. Many of the users are not sufficiently aware of the role of subjective decisions in the preparation process. In order to provide them with a notion of these aspects, and to raise the credibility of the prediction, a clear report on the preparation process should be an integral portion of the supplied prediction. Another contributer to the value of prediction is a logical and consistent methodology of preparation.

Preparation of a prediction begins with a definition of the purpose for which the prediction is made. This may be for a particular engineering project at a given site, for a multisite project, for a number of alternatives of a given project, for flood zone mapping, etc. In most cases, the required recurrence intervals are specified by the clients. Other cases are of specified design discharges or of deferred specification untill the economic consequences of the prediction are evaluated. An important decision, about the prediction variable, is taken by the hydrologist at this stage.

The second stage is a survey of available data on which the prediction would be based. The survey should cover the entire geographic region where the prediction is required, as well as its vicinity. Attention is paid to duration, quality, completeness, range, and geographical distribution of the records. This stage is concluded by a decision on the basis for the regional model (i.e. which stations and durations provide the most suitable data, is a regional model required or a direct application of station results would suffice, can a regional model be formulated such as to cover the entire project area, etc.).

Following these decisions, proper data are collected. These are occurrence times of flow events, magnitudes of selected flow variable(s) for these events, watershed properties for the gauging sites and for the application sites, and the hydrologic regime of the study area. The last item would support delineation of the region's borders and the other subjective decisions.

The principal subjective decisions are taken next. These are selection of candidate probability distributions, methods for fitting distributions to data, criteria for selection of the most approprite distributions, threshold values (or criteria for their selection), candidate variables for the regressions, etc. Having made these decisions, the hydrologist would run the principal computations which are described in the preceding chapters, and take additional decisions in accordance with the chosen criteria.

For each one of the stations, the computations result in a vector of magnitudes and their assigned recurrence intervals, and a list of parameter and threshold values. These data are regressed against watershed properties. Formulation of the equations and selection of variables are mostly done through a trial and error process which is involved with successive subjective decisions. This stage results in a model through which the required predictions is made. A number of hydrologists add to the model an estimation of accuracy of the predicted values. This task is not discussed here.

The last stage includes an application of the model and reporting of the results. As it is mentioned above, a good and clear report, in which all achievements and difficulties are described, raises the prediction credibility. Furthermore, the need for such a report elevates the quality of work throughout all of its stages and provides a good basis for improvements in future works.

ADVANTAGES

Use of partial duration series is more economical and simpler than that of complete series, but it is more complicated than that of annual maxima series. Partial duration series exactly suits to the definition of recurrence interval and to the assessment of economic and other consequences of floods, whereas annual maxima series provides only an approximate information on these subjects. Partial duration series suits better than annual maxima series to investigations on flood causes because it is composed of all and only the large events. In applications for arid areas, annual maxima series loses its advantage of indepedence of events from one another because the events occur there rarely and sporadically.

REFERENCES

Ashkar, F. and Rousselle, J. (1983) 'Some remarks on the truncation used in partial duration series models', Water Resour. Res. **19**, 477-480.

Ashkar, F. and Rousselle, J. (1987) 'Partial duration series modeling under the assumption of Poissonian flood count', J. Hydrol. **90**, 135-144.

Ben-Zvi, A. (1982) 'Flow events in the Negev - a regional quantitative model', Water International **7**, 127-133.

Ben-Zvi, A. (1991) 'Observed advantage for negative binomial over Poisson distribution in partial duration series', Stoch. Hydrol. Hydraul. **5**, 135-146.

Ben-Zvi, A. (1993) ' Distribution of flood volumes beyond design discharges', Proc. Symp. Engrg. Hydrol., ASCE, San Francisco, CA, USA, 97-102.

Ben-Zvi, A. (1994) 'Fit of probability distributions to upper sub-samples of partial duration series', Proc. Conf. Stoch. and Statist. Methods in Hydrol. and Environ. Engrg., Waterloo Univ. Ontario, Canada.

Ben-Zvi, A. and Cohen, O. (1983) 'Maximal discharges in the lower Yarkon River', Rep. 4/83, Israel Hydrol. Service, Jerusalem, Israel (in Hebrew).

Bhuiya, R.K. and Yevjevich, V. (1968) 'Effects of truncation on dependence in hydrological time series', Hydrol. Paper 31, Colorado State Univ., Fort Collins CO, USA.

Bryson, M.C. (1974) 'Heavy-tailed distributions: properties and tests', Technometrics **16**, 61-68.

Choulakian, V., El-Jabi, N. and Moussi, J. (1990) 'On the distribution of flood volume in partial duration series analysis of flood

phenomena', Stoch. Hydrol. Hydraul. **4**, 217-226.
Chow, V.T. (1953) 'Frequency analysis of hydrologic data with special application to rainfall intensities', Bull. **414**, Engrg. Exp. Station, Univ. of Illinois, Urbana IL, USA.
Cruise, J.F. and Arora, K. (1990) 'A hydroloclimatic application strategy for the Poisson partial duration model', Water Resour. Bull. **26**, 431-442.
Cunnane, C. (1973) 'A particular comparison of annual maxima and partial duration series methods of flood frequency prediction', J. Hydrol. **18**, 257-271.
Hiemstra, L.A.V. and Francis, D.M. (1981) 'Run hydrographs for prediction of flood hydrographs', J. Hydr. Div. ASCE **107**, 759-775.
Hiemstra, L.A.V., Zucchini, W.S. and Pegram, G.G.S. (1976) 'A method of findind the family of runhydrographs for given return periods', J. Hydrol. **30**, 95-103.
Jarvis, C.S. and "others" (1936) 'Floods in the United States magnitude and frequency', Water Supply Paper **771**, USGS, Washington DC, USA.
Kavvas, M.L. (1982) 'Stochastic trigger model for flood peaks, I. Development of the model', Water Resour. Res. **18**, 383-398.
Langbein, W.B. (1949) 'Annual floods and the partial duration flood series', Trans. AGU **30**, 879-881.
Nachtnebel, H.P. and Konecny, F. (1987) 'Risk analysis and time-dependent flood models', J. Hydrol. **91**, 295-318.
North, M. (1980) 'Time dependent stochastic model of floods', J. Hydr. Div. ASCE **106**, 649-665.
Pickand, J. (1975) 'Statistical inferences using extreme order statistics', Anals Statis. **3**, 119-131.
Rasmussen, P.F., Ashkar, F., Rosbjerg, D. and Bobee, B. (1994) 'The POT method for flood estimation: a review', Proc. Conf. Stoch. and Statis. Methods in Hydrol. and Environ. Engrg., Waterloo Univ., Ontario, Canada.
Rasmussen, P.F. and Rosbjerg, D. (1991) 'Evaluation of risk concept in partial duration series', Stoch. Hydrol. Hydraul. **5**, 1-16.
Rosbjerg, D. (1987) 'Partial duration series with log-normal distributed peak values', in V.P. Singh (ed.), Hydrologic Frequency Modeling, Riedel, 117-129.
Rosbjerg, D., Madsen, H. and Rasmussen, P.F. (1992) 'Prediction in partial duration series with generalized Pareto-distributed exceedances', Water Resour. Res. **28**, 3001-3010.
Shane, R.M. and Lynn, W.R. (1964) 'Mathematical model for flood risk evaluation', Proc. Hydr. Div. ASCE **90**, 1-20.
Smith, J.A. (1987) 'Estimating of the upper tail of flood frequency distributions', Water Resour. Res. **23**, 1657-1666.
Taesombut, V. and Yevjevich, V. (1978) 'Use of partial flood series for estimating distribution of annual flood peaks', Hydrol. Paper **97**, Colorado State Univ., Fort Collins CO, USA.
Todorovic, P. and Rousselle, J. (1971) 'Some problems of flood analysis', Water Resour. Res. **7**, 1144-1150.
Todorovic, P. and Zelenhasic, E. (1970) 'A stochastic model for flood analysis', Water Resour. Res. **6**, 1641-1648.

COMPARISON OF FIVE DISTRIBUTIONS FOR SUITABILITY IN FLOOD FREQUENCY ANALYSIS IN BANGLADESH

Md. Abdul Karim
Department of Civil Engineering
Bangladesh University of Engineering and Technology (BUET)
Dhaka-1000, Bangladesh.

and

Jahir Uddin Chowdhury
Institute of Flood Control and Drainage Research
BUET, Dhaka-1000, Bangladesh.

ABSTRACT. A study has been conducted to compare five distributions used in flood frequency analysis in Bangladesh. The distributions are Log-normal, Gumbel, Generalized Extreme Value, Pearson Type 3 and Log-pearson Type 3. Comparisons have been based upon at-site root mean square error in fit, at-site probability plot correlation coefficient, regional L-moment ratio diagrammme, regional distribution of L-coefficient of skewness and regional distribution of largest normalized discharges. The study indicates that the Generalized Extreme Value distribution represents the statistical behaviour of observed annual maximum discharges best. Pearson Type 3 and Gumbel distributions are not suitable for frequency analysis of discharge data in Bangladesh.

1. Introduction

Different organizations employ different probability distribution functions in the frequency analysis of flood flows in Bangladesh. A comparative study has been made to determine which distribution represents the statistical characteristics of observed flood data best. This paper discusses the methods of comparison and the results.

2. Discharge Data

Annual maximum (AM) discharge data from 31 gage stations have been used. The length of data record varies from 16 to 24 years. The stations have been selected by performing statistical tests for stationarity of mean and standard deviation (SD) of flood samples.

3. Candidate Distributions

Five frequency distribution functions have been included in the comparison. These distributions are used in Bangladesh. The distributions are: Log-normal (LN) with parameters estimated by maximum likelihood method, Gumbel with probability weighted moment (PWM) method, Generalized Extreme Value (GEV) with PWM method, Pearson Type 3 (P3) with method of moment (MM) and Log-pearson Type 3 (LP3) with MM in log-space. The PWM method employed is after Greenwood et al.(1979) and Hosking et al.(1985).

V. P. Singh and B. Kumar (eds.), Surface-Water Hydrology, 567–575.

4. Methods of Comparison

4.1. APPROACH OF THE METHODOLOGY

Five methods of comparison have been employed. Two of them are goodness-of-fit study and they are based upon root mean square error (RMSE) in fit and probability plot correlation (PPC) coefficient. Remaining three are statistical behaviour studies and they are based upon regional L-moment ratio diagrammme, regional distribution of L-moment ratio and regional distribution of largest normalized discharges. A report prepared by Cunnane (1989) for the World Meteorological Organization emphasizes the value of statistical behaviour studies in choosing between distributions. These methods examine whether candidate distributions are capable of producing random samples having the same statistical characteristics as observed samples. These methods require generation of random data from candidate distributions such that sizes of generated samples are similar to observed samples. In this study, data have been generated by setting the parameters of a parent distribution equal to the regional average of estimated at-site parameters. The methods of comparison are briefly discussed below.

4.2. ROOT MEAN SQUARE ERROR IN FIT

The RMSE of a fitted distribution to the observed discharge data at a station has been computed as

$$\mathrm{RMSE} = \left[\frac{1}{n}\sum_{i=1}^{n}\left(\frac{y_i - w_i}{y_i}\right)^2\right]^{1/2} \qquad (1)$$

where, n is the sample size at the station, y_i is the observed discharge of a given probability and w_i is estimated discharge for the same probability. Mean and standard deviation (SD) of RMSEs at 31 gage stations for every distribution are given in Table 1.

4.3. PROBABILITY PLOT CORRELATION COEFFICIENT

The adequacy of a fitted distribution can be evaluated by the PPC coefficient which is essentially a measure of the linearity of probability plot. It gives the correlation ρ between the ordered observations y_i and the corresponding fitted quantiles w_i determined by plotting position for each y_i. If $\bar{y}$ denotes the average value of the observations at a station and $\bar{w}$ denotes the average value of the fitted quantiles, then the PPC coefficient of the fitted distributuion at that station is given by

$$\rho = \frac{\sum_{i=1}^{n}(y_i - \bar{y})(w_i - \bar{w})}{\left[\sum_{i=1}^{n}(y_i - \bar{y})^2 \sum_{i=1}^{n}(w_i - \bar{w})^2\right]^{1/2}} \qquad (2)$$

Mean and SD of PPC coefficients at 31 gage stations for every distribution are also given in Table 1.

It may be mentioned that a simple goodness-of-fit test is the PPC coefficient test developed by Filliben (1975). Application of this test to frequency analysis of hydrologic data has been investigated by Vogel (1986, 1987), Vogel and Kroll (1989), Vogel and McMartin (1991) and Chowdhury et al.(1991).

TABLE 1. Statistics of RMSE and PPC coefficient of distributions fitted to AM discharge data from 31 gage stations

Goodness-of-fit Criteria		LN	Gumbel	GEV	P3	LP3
RMSE	Mean	0.092	0.173	0.075	0.018	0.078
	SD	0.065	0.140	0.058	0.018	0.060
PPC cioefficient	Mean	0.970	0.873	0.981	0.989	0.979
	SD	0.020	0.060	0.011	0.012	0.013

4.4. MOMENT RATIO DIAGRAMME

Moment ratio diagramme is useful as a guide in selecting a distribution for describing a set of random variables. In the recent Handbook of Hydrology, Stedinger et al.(1993) discuss importance of L-moment ratio diagrammes in the selection of appropriate frequency distribution function. A moment ratio diagramme is a plot of one moment ratio of a given order as a function of a lower order one, such as coefficient of skewness (CS) versus coefficient of variation (CV) and Kurtosis versus CS. Moment ratio diagramme can be constructed using either the conventional product moment ratios or the L-moment ratios.

4.4.1. L-moments. The L-moments, introduced by Hosking (1990), are another way of summarizing the statistical properties of hydrologic data. L-moments can be expressed as linear combinations of PWMs. Details of L-moments are also given in Stedinger et al.(1993).

The PWM of order r can be defined as (Greenwood et al., 1979)

$$\beta_r = E\{X[F(X)]^r\} \tag{3}$$

where, $E\{\bullet\}$ denotes expectation, $F(X)$ is the cumulative distribution function (cdf) of X.

An unbiased estimator of β_r is (Landwehr et al.,1979)

$$b_r = \frac{1}{n}\sum_{i=1}^{n}\frac{(i-1)(i-2)\ldots.(i-r)}{(n-1)(n-2)\ldots.(n-r)}y_i \tag{4}$$

where, y_i are the ordered values so that $y_1 \le \ldots.. \le y_2 \le \ldots.. \le y_n$. Then the estimates of L-CV($\hat{t}_2$), L-CS($\hat{t}_3$) and L-kurtosis($\hat{t}_4$) are (Hosking, et al.,1990)

$$\hat{\tau}_2 = \frac{2b_1 - b_0}{b_0} \tag{5a}$$

$$\hat{\tau}_3 = \frac{6b_2 - 6b_1 + b_0}{2b_1 - b_0} \tag{5b}$$

and

$$\hat{\tau}_4 = \frac{20b_3 - 30b_2 + 12b_1 - b_0}{2b_1 - b_0} \tag{5c}$$

4.4.2. Regional L-moment Ratio Diagramme. Wallis (1988), Cunnane (1989) and Hosking (1990) illustrated that compared to the product moment ratio diagramme, the L-moment ratio diagrammes possessed better ability to discriminate between distributions. In this study, $\hat{\tau}_4$ versus $\hat{\tau}_3$ diagrammes have been examined. Figure 1(a) shows plot of $\hat{\tau}_4$ versus $\hat{\tau}_3$ constructed from observed discharge samples while Figs.1(b) and 1(c) are from random regional samples generated respectively from GEV and P3 parent distributions with constant regional parameters. In fact five plots of $\hat{\tau}_4$ versus $\hat{\tau}_3$ were obtained from generated data from five distributions. Among them the diagramme (Fig.1b) from the GEV distribution gives best similarity with that from observed data (Fig.1a) while the diagramme (Fig.1c) from P3 distribution has no similarity.

4.5. REGIONAL DISTRIBUTION OF L-CS

Examination of the sampling distribution of moment ratios in a region is useful in the inference of distribution for AM series for the region (Cunnane, 1989). Values of moment ratio calculated from flood samples from several gage stations in a region define an observed sampling distribution of the moment ratio. This sampling distribution of a moment ratio can be used as a criterion to be reproduced by Monte Carlo derived samples from candidate distributions. For example Rossi et al.(1984) used the observed sampling distribution of CS obtained from 39 Italian AM series for comparing candidate distributions.

In this study, the regional distribution of $\hat{\tau}_3$ has been examined. Observed and generated values of $\hat{\tau}_3$ are displayed in Fig.2(a) by using a Gumbel probability axis and Gringorten plotting positions.

It may be mentioned that Chowdhury et al.(1991) derive small sample sampling variance of $\hat{\tau}_2$, $\hat{\tau}_3$ and $\hat{\tau}_4$ as a function of shape parameter for the GEV distribution to provide a powerful test whether a particular data set is consistent with a GEV distribution with a regionally estimated value of shape parameter, or a regional shape parameter and CV.

4.6. REGIONAL DISTRIBUTION OF LARGEST NORMALIZED DISCHARGES

A sample of largest discharges in a region can be formed by taking the largest value from every normalized AM flood sample of equal size for a concurrent period from several gage stations in that region. Cunnane (1989) suggests that comparison of observed sampling distribution of largest discharges with the Monte Carlo derived distribution of largest discharges can be helpful in identifying a distribution whose sample extremes behave similarly to observed AM extremes.

Among 31 gage stations, there are 13 stations at which AM discharge sample of equal size (n = 16) covering the period of 1972 to 1987 are available. A new sample of size n = 13, is formed by taking the largest discharges after normalizing the flood samples. Figure 2(b) displays regional distributions of largest normalized discharges from observed data and generated data from five distributions.

5. Discussion

Performance of five candidate distributions are summarized in Table 2 based upon goodness-of-fit analysis and statistical behaviour studies. The table indicates that the GEV distribution is the best among five candidate distributions in representing the statistical characteristics of AM discharge data in Bangladesh.

Although the $\hat{t}_3$ - $\hat{t}_4$ plot [Fig.1(b)] obtained from the GEV distribution is not exactly equal to the plot in Fig.1(a) obtained from observed data, the L-CS of generated samples are quite close to those of observed samples. Of course the L-kurtosis of generated samples are slightly smaller.

TABLE 2. Performance of candidate distributions

Method of Comparison	LN	Gumbel	GEV	P3	LP3
Goodness-of-fit analysis	S	N	G	G	G
Regional L-kurtosis versus L-CS diagramme	S	G	G	N	S
Regional distribution of L-CSs	S	S	G	N	S
Regional distribution of largest normalized discharges	S	N	G	G	G

[G: Good; S: Satisfactory; N: Not satisfactory]

Figures 1(c) and 2(a) indicate that the P3 distribution is unable to represent the statistical behaviour of AM discharge data in Bangladesh. Figure 2(b) indicates that the Gumbel distribution is also unable to reproduce the distribution of largest discharges. Besides, the RMSE is the largest and the PPC coefficient is the smallest when Gumbel distribution is compared with other four distributions as can be seen from Table 1. When

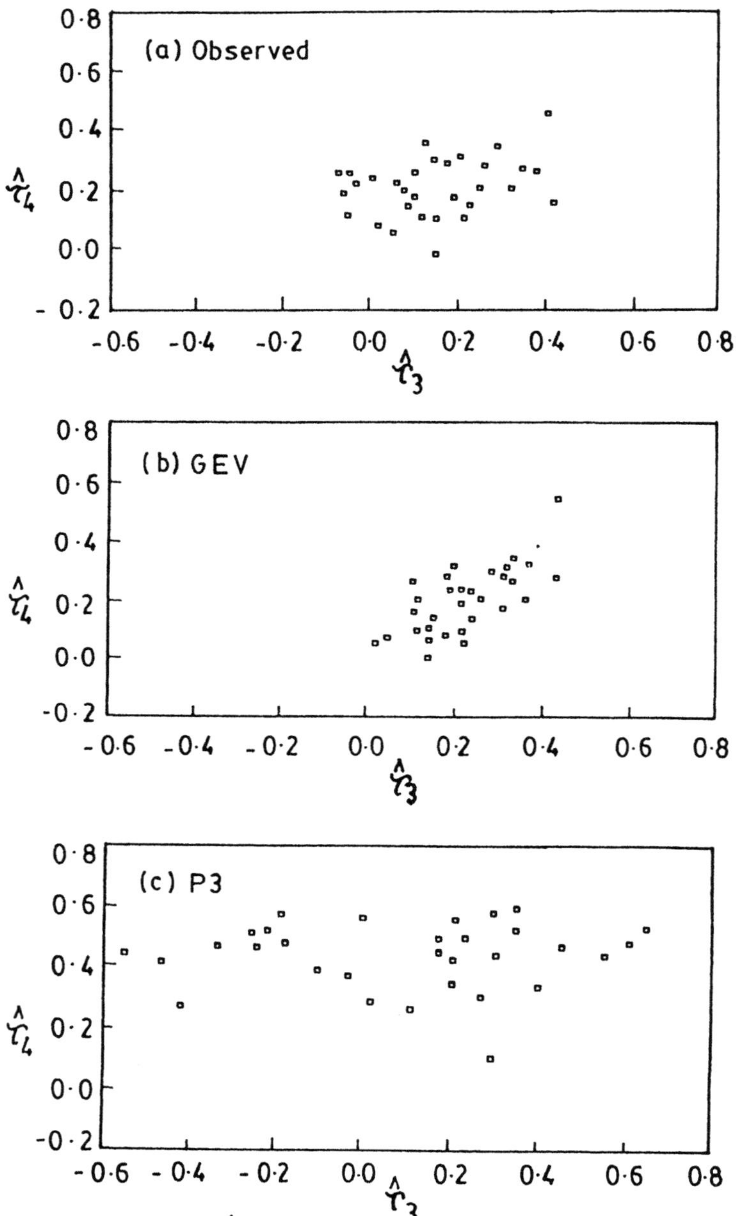

Figure 1. Plot of $\hat{\tau}_4$ versus $\hat{\tau}_3$ generated from (a) observed data (b) GEV distribution and (c) P3 distribution.

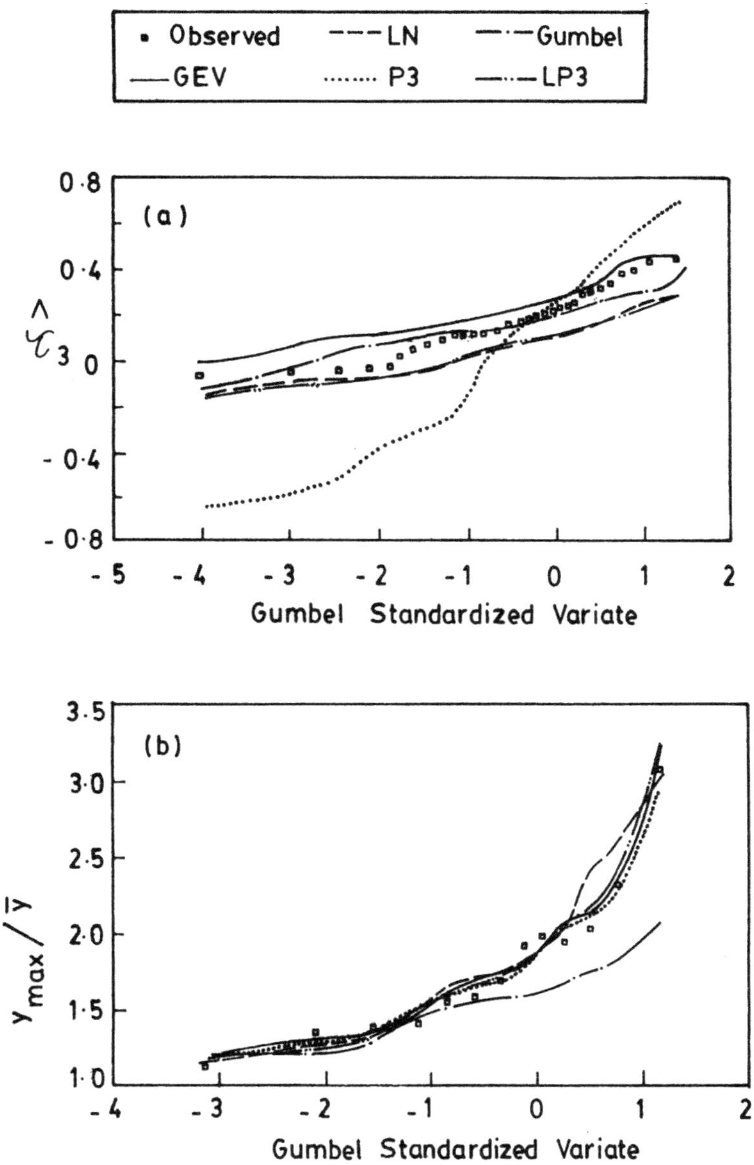

Figure 2. Comparison of regional distributions of (a) L-CS and (b) normalized largest discharges using Gringorten plotting position on a Gumble scale.

the PPC coefficient test is applied after Vogel (1986, 1987), fitted Gumbel distribution at 19 out of 31 gage stations is rejected at 5% significance level. It may be mentioned that this is the most widely used distribution in Bangladesh. It appears from Table 2 that performance of the LP3 distribution is better than the LN distribution.

It is seen from mean and SD values in Table 1 that the RMSEs and PPC coefficients of three parameter distributions (GEV, P3 and LP3) are very close and appear to be within sampling variability. Although the P3 distribution is unable to model statistical behaviour of AM series, its RMSE is the lowest and PPC coefficient is the highest as can be seen from Table 1. This is because three parameter distributions have greater flexibility which allows close fit to the data.

6. Conclusions

The GEV distribution is the best among five candidate distributions in representing the statistical behaviour of AM discharge series in Bangladesh. Gumbel and P3 distributions are not suitable for AM discharge data. The LP3 distribution is better compared to the LN distribution.

RMSE and PPC coefficient are not good guide to choosing between three parameter distributions. The statistical behaviour studies are of help in this respect. The value of behaviour studies in the selection of a distribution for a country or region has been emphasized by Cunnane (1989).

7. Acknowledgement

This study was done at the Institute of Flood Control and Drainage Research of BUET, Dhaka, Bangladesh. We appreciate the useful comments given by J.R. Stedinger.

8. References

Chowdhury, J.U., Stedinger, J.R. and Lu, L.H.(1991), Goodness-of-Fit Tests for Regional Generalized Extreme Value Flood Distribution, Water Resour. Res., 27(7), 1765-1777.

Cunnane, C. (1989), Statistical Distributions for Flood Frequency Analysis, Operational Hydrology Report No. 33, WMO - No. 718, World Meteorological Organization , Geneva, Switzerland.

Filliben, J.J. (1975), The probability plot correlation test for normality, Technometrics, 17(1), 111-117.

Greenwood, J.A., Landwehr, J.M., Matalas, N.C. and Wallis, J.R.(1979), Probability weighted moments: Definition and relation to parameters of several distributions expressible in inverse form, Water Resour. Res., 15(5), 1049-1054.

Hosking, J.R.M.(1990), **L-moments: analysis and estimation of distribution using linear combinations of order statistics**, J. of Royal Stat. Soc., B, 52(1), 105-124.

Hosking, J.R.M., Wallis, J.R. and Wood, E.F.(1985), **Estimation of generalized extreme-value distribution by method of probability weighted moments**, Technometrics, 251-261.

Landwehr,J.M., Matalas, N.C and Wallis, J.R.(1979), **Probability weighted moments compared with some traditional techniques in estimating Gumbel parameters and quantiles**, Water Resour. Res., 15(5), 1055-1064.

Rossi, F., Fiorentino, M. and Versaece, P.(1984), **Two component extreme value distribution for flood frequency analysis**, Water Resour. Res., 20(7), 847-856.

Stedinger, J.R., Vogel, R.M. and Georgiou, E.F.(1993), **Frequency Analysis of Extreme Events**, Chap. 18, Handbook of Hydrology, ed. D.R. Maidment, McGraw-Hill Inc., New York.

Vogel, R.M.(1986), **The probability plot correlation coefficient test for the normal, log-normal and Gumbel distributional hypothesis**, Water Resour. Res., 22(4), 587-590.

Vogel, R.M.(1987), **Correction of the probability plot correlation coefficient test for the normal, log-normal and Gumbel distributional hypothesis**, Water Resour. Res., 232(10), 2013.

Vogel, R.M. and Kroll, C.N.(1989), **Low flow frequency analysis using probability plot correlation coefficients**, J. of Water Resources Planning and Management, ASCE, 115(3), 338-357.

Vogel, R.M. and McMartin, D.E.(1991), **Probability plot goodness-of-fit and skewnwss estimation procedures for pearson type 3 distribution**, Water Resour. Res., 17(12), 3148-3158.

Wallis, J.R.(1988), **Catastrophes, computing and containment: living in our restless habitat**, Speculation in Science and Technology, 11(4), 295-315.

INVESTIGATION OF POTENTIAL FLASH FLOODS FOR UNGAGED REGION USING THE KRIGING METHOD

Tiao J. Chang
Civil Engineering Dept.
Ohio University
Athens, Ohio 45701
USA

Hong Y. Sun
R.D. Zande Associates
1237 Dublin Road
Columbus, Ohio 43215
USA

ABSTRACT. For a small rural community situated on a flood plain, it is not economically justified for any flood protective structures. A community warning system by the use of existing precipitation gage stations could be an alternative. This paper investigated the regional distribution of potential flash floods based on several existing precipitation gages in the studied area. Daily truncation levels of 99.9 and 99.99% were used to derive potential flash floods, where a 99.9% level means that 0.1% of historically recorded daily precipitation are larger than or equal to the truncated value. The higher the truncation level, the more severe the potential flash flood. Daily precipitation records from twenty-one gages in the Scioto River Basin were selected for the study. The kriging method based on the minimum unbiased estimation was used to estimate spatial interpolation for ungaged areas. These estimations will form the regional distribution of potential flash floods at various truncation levels. Contour lines of these potential flash floods at each truncation level were constructed and associated errors were estimated. The result shows that estimation errors associated with constructed contour lines were reasonably small.

INTRODUCTION

Fatalities and damage in the 1990 Shadyside flash flood in Ohio raised a serious question of flood protection for an ungaged small watershed. The Shadyside area, lying on two tributaries of the Ohio River, Wegee and Pipe Creeks, has no rain gages. The only rainfall estimate in the watershed for this particular event came from a resident of upper Wegee Creek where daily accumulation in a child's wading pool was about 100 to 130 mm. The county sheriff had an unofficial report of 70 mm rainfall in a hour in the town 30 kilometers northwest of the Shadyside (NOAA, 1991). While neither of these estimates were reliable, 26 known fatalities and enormous property damage in the region are certain.

Small rural communities that are not economically justified for any flood protective structures are extremely vulnerable to floods. An early warning system for such communities may reduce possible flood damage. Based on the method of truncation level, this study used daily records of precipitation to derive potential flash

V. P. Singh and B. Kumar (eds.), Surface-Water Hydrology, 577–589.

floods at a gaged location. Then, the kriging method was applied to obtain regional distributions of flash floods for a possible early warning in ungaged areas.

The method of truncation level for deriving potential flash floods has been used for the construction of hydrologic droughts (Yevjevich, 1967, 1972; Chang and Stenson, 1990; Chang and Kleopa, 1991; Chang, 1987, 1991; Chang and Teoh, 1992). The kriging method for obtaining a regional distribution of potential flash floods was successfully performed by Delhomme (1978) and Karlinger and Skrivan (1980) in the study of spatial precipitation.

POTENTIAL FLASH FLOODS

Based on partial duration series, potential flash flood events are derived from daily records of precipitation. By applying a specified truncation level to a time series of daily precipitation, the sequence of consecutive days whose precipitations exceed the specified truncation level is defined as a potential flood event.

To obtain truncation levels, historical data of daily precipitation are sorted in a descending order. Then, a Y% truncation level is the value that corresponds to the ith rank of the sorted data with a total record length N, where the relationship between i^{th} rank and Y% truncation level can be expressed as:

$$i=\frac{(100-Y)}{100}N \tag{1}$$

Consequently, at the Y% truncation level, Y% of the historic data are less than this truncation level. A potential flood event occurs whenever historical records of precipitation are greater than a specified truncation level.

Furthermore, potential flash floods were also derived based on annual duration series. Extreme storm events of 50-year and 100-year return periods were estimated by the Gumbel Method. Values of these storms served as truncation levels for deriving potential flash floods. The following equation was used to estimate such truncation levels:

$$X=X_m+(0.7797Y-0.45)\sigma_x, \tag{2}$$

where X_m is the mean of annual potential flood duration series and σ_x is its standard deviation; Y is the reduced variate, which is a function of probability by the following relation:

$$Y=-\ln[-\ln(1-P)], \tag{3}$$

where P is the probability of a given value being equaled or exceeded, P = 1/T, T is the return period; e is the base of napierian logarithms.

REGIONAL DISTRIBUTION OF FLASH FLOODS

To obtain regional distribution of potential floods, the kriging method is used which is based on the linear minimum variance unbiased estimation (Kitanidis, 1983). Given n potential flood measurements of Z at spatial locations, x_1, x_2, ... x_n, an estimator using a linear combination of n variables is considered as:

$$Z_o^* = \sum_{i=1}^{n} \alpha_i Z(x_i), \tag{4}$$

where Z^*_o is an estimation of a true value Z_o; α_i is the weight of the observed value at location x_i. The goal is to find a set of weights which give the best estimation so that the estimation is unbiased with a minimum variance. The unbiased means that

$$E[Z_o^* - Z_o] = 0, \tag{5}$$

where E is the expectation operator. The minimum variance means that

$$E[(\sum_{i=1}^{n} \alpha_i Z(x_i) - Z_o)^2] = minimum \tag{6}$$

In order to satisfy the unbiased condition, coefficients $\alpha_1, \alpha_2, ..., \alpha_n$, must be selected so that

$$\sum_{i=1}^{n} \alpha_i - 1 = 0 \tag{7}$$

Based on Equation 7, the estimation error can be written as follows:

$$Z_o^* - Z_o = \sum_{i=1}^{n} \alpha_i Z(x_i) - Z_o \tag{8}$$

Therefore, the problem of kriging technique is reduced to minimize Equation 6 subject to the constraint of Equation 4. Using the Lagrange multiplier, the necessary conditions for the problem of minimization yield n+1 equations with as many unknowns in the following:

$$\sum_{j=1}^{n} \alpha_j \tau(x_i - x_j) = \tau(x_i - x_o) + \mu, \; i = 1, 2, \ldots, n, \tag{9}$$

$$\sum_{j=1}^{n} \alpha_j = 1, \tag{10}$$

where τ is the semi-variogram, which is the expectation of $(Z(x_i)-Z(x_j))^2$; μ is a Lagrange multiplier. By simultaneously solving these equations, the optimum weights α_j can be obtained. The variance of this estimation is given by

$$\sigma^2 = -\sum_{i=1}^{n}\sum_{j=1}^{n} \alpha_i \alpha_j \tau(x_i - x_j) + 2\sum_{i=1}^{n} \alpha_i \tau(x_i - x_o), \tag{11}$$

where the variance can be interpreted as the estimation error associated with each kriged estimate. When x_o coincides with a data location at x_i, the system solution yields $\alpha_i = 1$. Therefore, this results in $Z_o^* = Z(x_i)$ at location x_i and $\sigma^2 = 0$, that is, the kriging estimator is an exact interpolator at the observed point.

If the semi-variogram can not be estimated without bias due to nonstationality, Delhomme (1978) provided the extended solution by using the higher order differences to filter out the nonstationality. Assuming that G(z) is a linear operator, then G(z) is called the generalized increment of order k if it can filter out polynomials of order k, that is, for every polynomial $P_k(x)$ of order k,

$$G(z + P_k) = G(z) \tag{12}$$

Furthermore, the estimator Z(x) satisfies the following two conditions:

$$E[G(z)] = 0, \tag{13}$$

and

$$E[G(z)^2] = \iint K(x_1-x_2)\, G(dx_1)\, G(dx_2)\,, \tag{14}$$

where $K(x_1\text{-}x_2)$ is the generalized covariance function; x_1 and x_2 are two spatial locations; the estimator Z(x) is also called the intrinsic function of order k. Therefore, the theory of kriging can be extended by the use of linear estimator of the form

$$L^*(z) = \sum_{i\in I} \alpha_i Z(x_i)\,, \tag{15}$$

where I is a subset of the data points. The requirement for the location-invariance is

$$L^*(Z(x+h)) = \sum_{i\in I} \alpha_i Z(x_i+h)\,, \tag{16}$$

for every distance h. Thus, the kriging problem is to minimize $E[(L^*(z)\text{-}L(z))^2]$ subject to $E[L^*(z)\text{-}L(z)] = 0$. The solution to the problem can be obtained following the procedure described in Equations 9 and 10.

APPLICATIONS

Twenty-one precipitation gage stations were selected in the Scioto River Basin as shown in Figure 1 for this study. Truncation levels for deriving potential flash floods, i.e., 99.9%, 99.99%, and values of 50-year and 100-year returned periods, are estimated and listed in Table 1 along with their monthly means and record lengths. It is shown that the truncated values increase as their corresponding levels increase, which implies an increase of potential flood severity. Based on the 21 truncated values at each severity level of potential flash floods, the kriging method was applied to obtain a regional distribution to determine the potential threat of flash floods in the ungaged region.

A computer program developed by Kafritsas and Bras (1981) was used to perform the regionalization analysis at each severity level of potential flash floods. First, identify the order of intrinsic function, where models are ranked according to their errors and the model with the lowest average rank is the best. Secondly, determine coefficients of the generalized covariance function for the intrinsic function. This is to obtain all qualified generalized covariance models with the same order of intrinsic function at the first step. Next, select the best generalized covariance model by the jacknife estimator. This is done by dividing the group of data

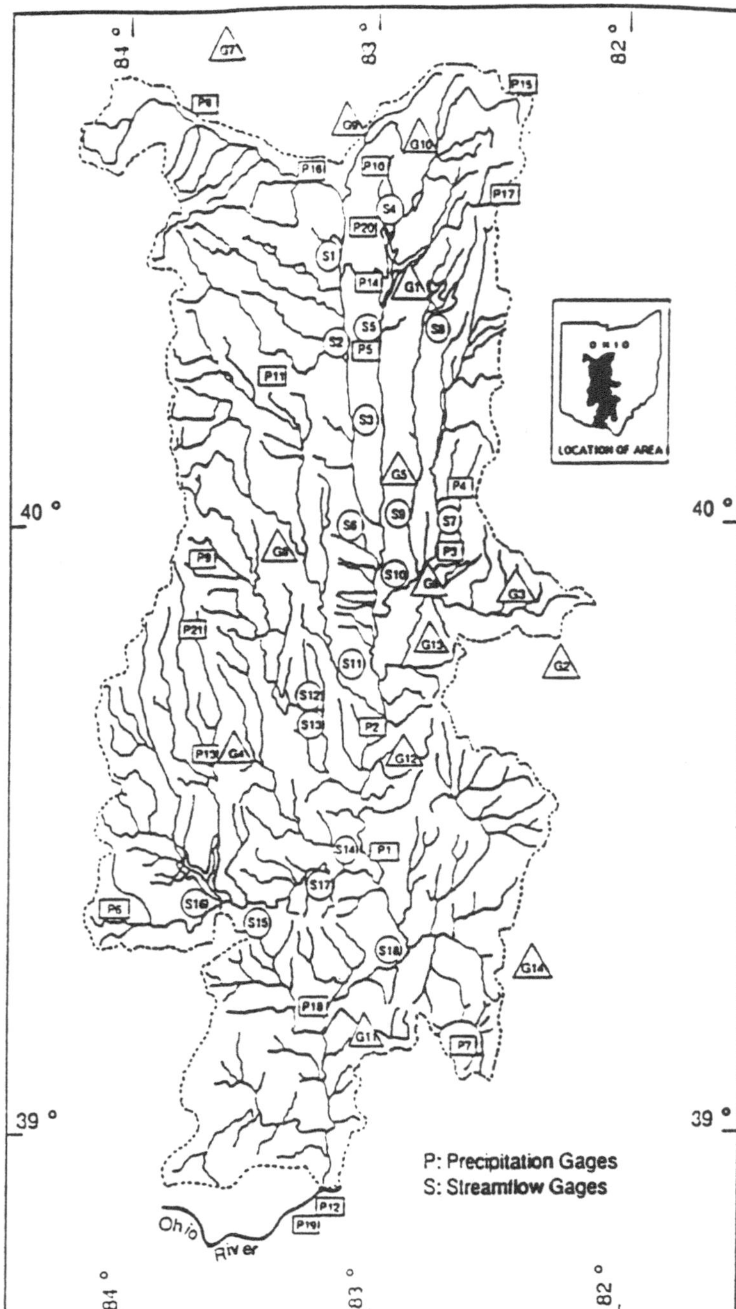

Figure 1. Gage stations in the Scioto River Basin

points, J, into two subgroups J_1 and J_2 and calculating

$$r_1 = \frac{\sum_{j \in J_1} K_j^2(z)}{\sum_{j \in J_1} \sigma_j^2} \tag{17}$$

and

$$r_2 = \frac{\sum_{j \in J_2} K_j^2(z)}{\sum_{j \in J_2} \sigma_j^2} \tag{18}$$

where the numerators are the sums of square of kriging errors and the denominators are the theoretical variances of errors for each corresponding subgroup. Then a jacknife value J_e is computed as follows:

$$J_e = 2r - \frac{n_1 r_1 + n_2 r_2}{n_1 + n_2} \tag{19}$$

where

$$r = \frac{\sum_{j \in J} K_j^2(z)}{\sum_{j \in J} \sigma_j^2} \tag{20}$$

and n_1 and n_2 are numbers of data points in subgroup J_1 and J_2, respectively. The best model is selected by the criterion that J_e is approximately equal to one.

Finally, kriging estimates and their associated errors are estimated using the best generalized covariance model. Based on these kriging values, contour lines are constructed for each severity level of potential flash floods. Figures 2 and 3 show contour lines for the examples of 99.9% and 99.99% potential flash floods, while Figures 4 and 5 give regional distributions for potential floods of 50-year and 100-year return periods. It is noted that associated errors for each kriging estimation are less than 10%.

CONCLUSION

Based on both partial duration series and annual duration series, truncation levels for deriving potential flash floods are defined and estimated from daily precipitation time series. This is to reflect the severity of potential flash floods since

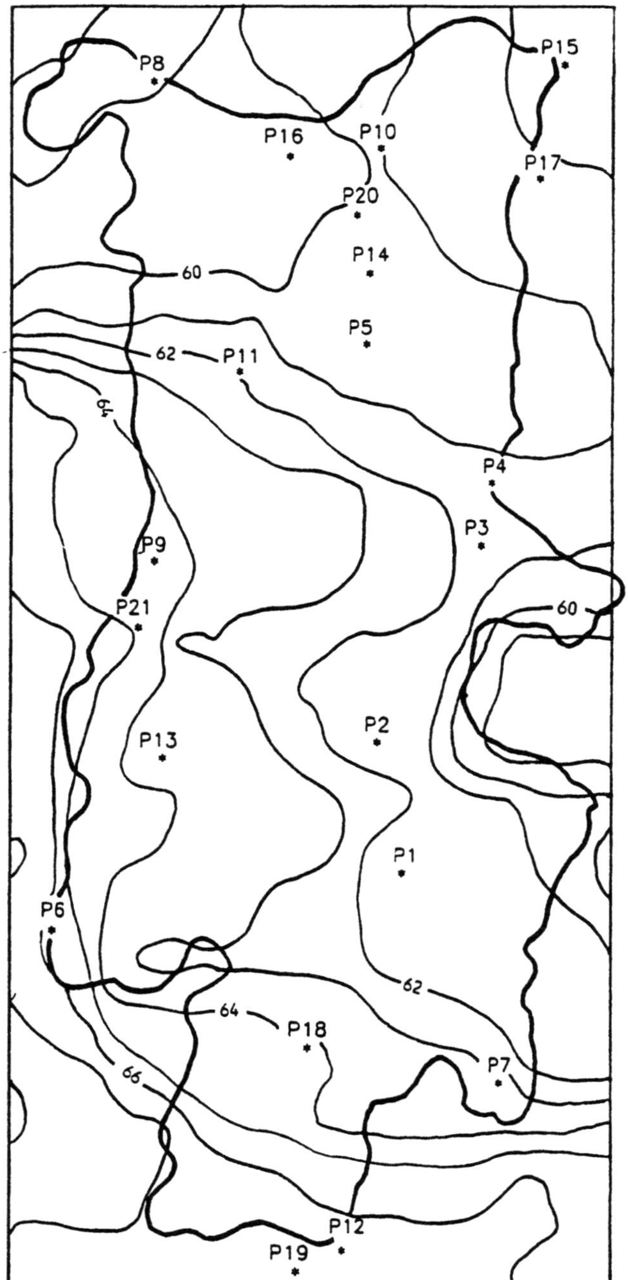

Figure 2. Regional distribution of potential flash floods derived by 99.9% truncation level

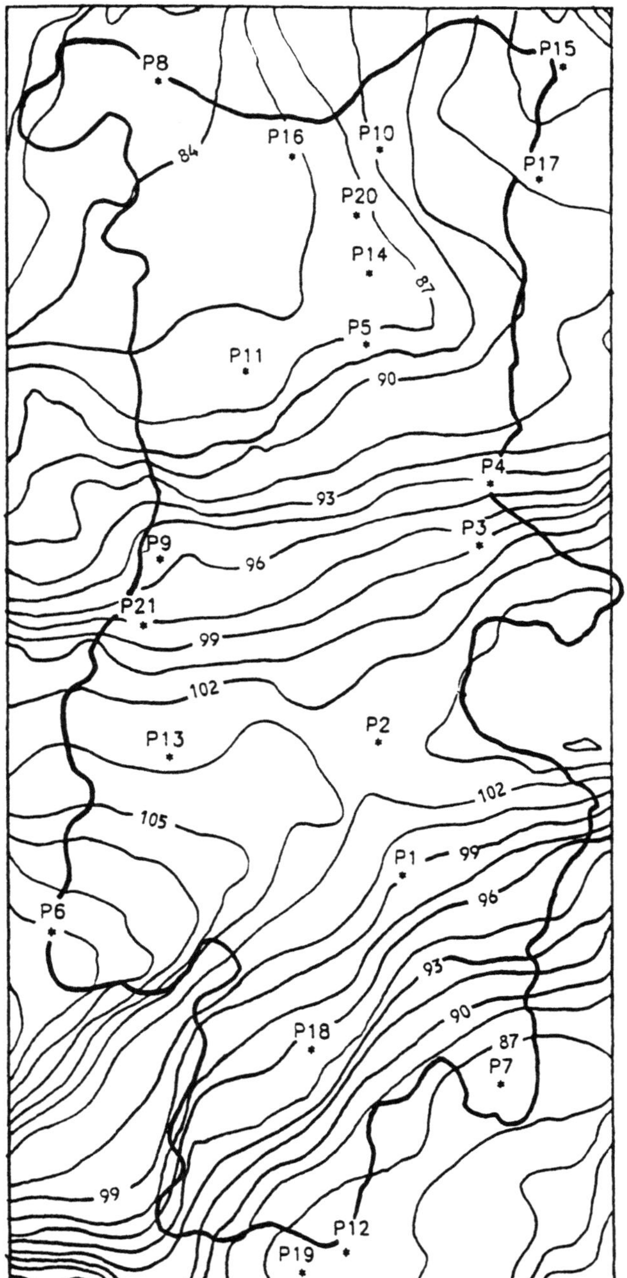

Figure 3. Regional distribution of potential flash floods derived by 99.99% truncation level

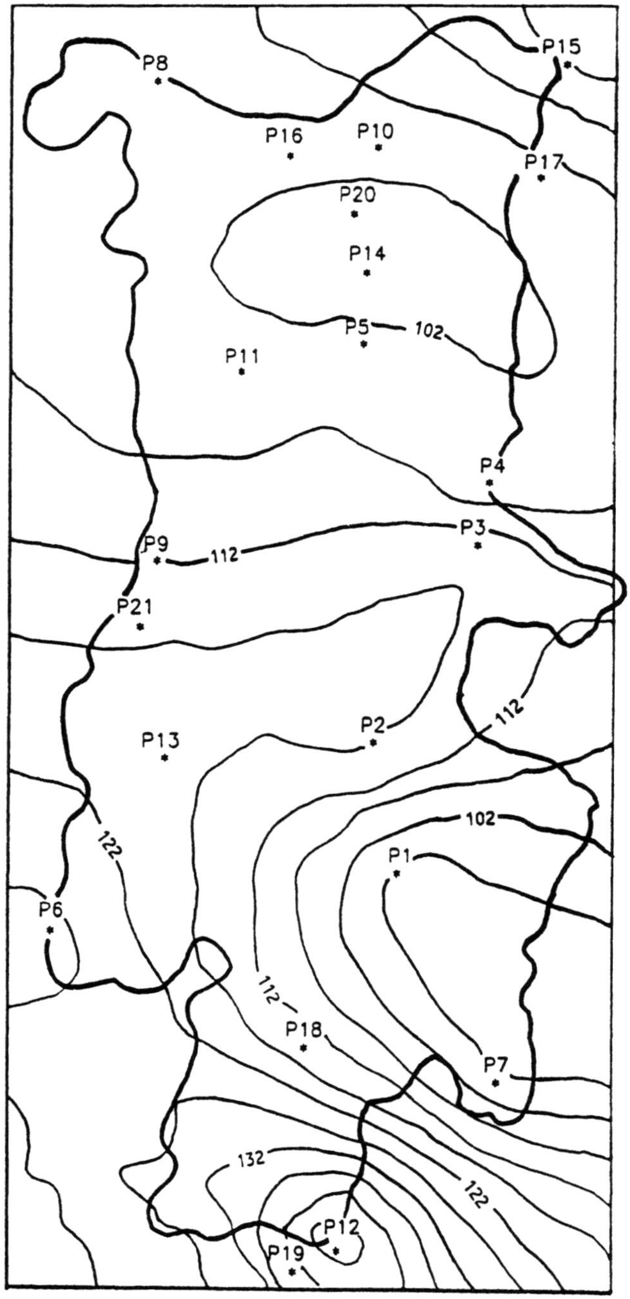

Figure 4. Regional distribution of potential flash floods derived by 50-year daily storm

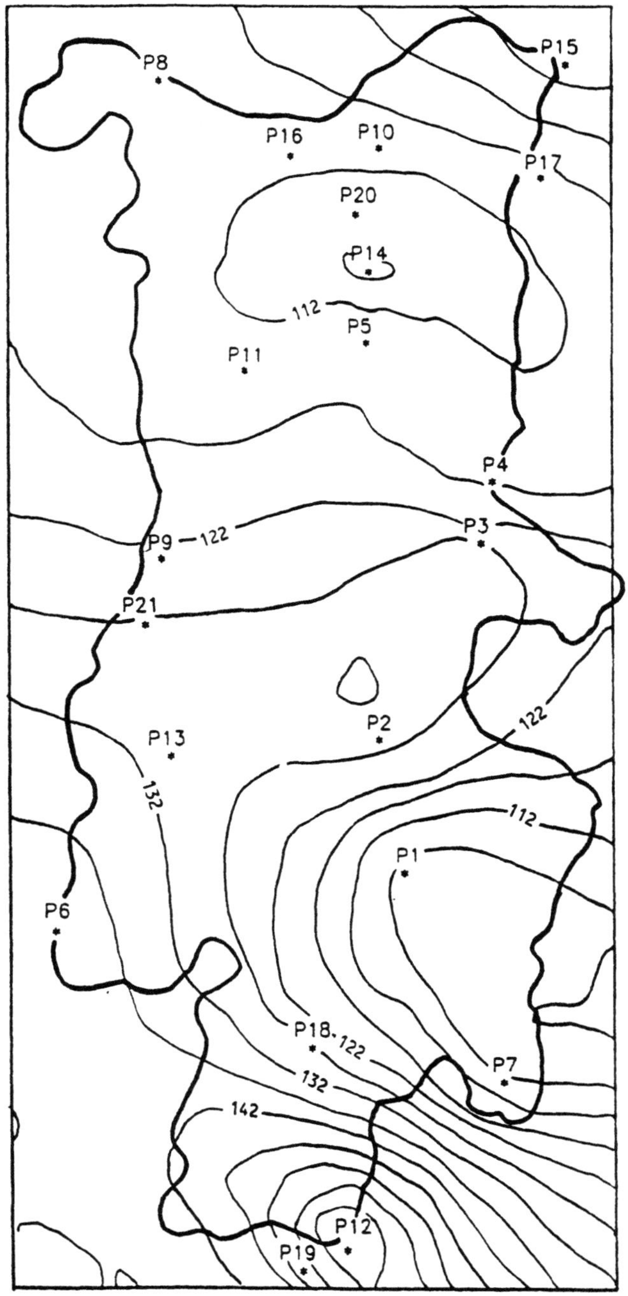

Figure 5. Regional distribution of potential flash floods derived by 100-year daily storm

severe rainfall storms result in flash floods in the studied region. Table 1 shows that daily truncated values for deriving potential flash floods are mostly greater than their corresponding monthly mean values. The truncated values increase as their associated levels, i.e. an increase of potential threat of flash flood.

The severity level of potential flash flood obtained from historic records at a gage station is strictly a localized measure since the daily precipitation is observed only at a certain location. Based on the truncation levels, used to derive potential flash floods, from the selected 21 gage stations, a regionalization can be made through the use of kriging method. This results in a regional distribution of potential threat of flash floods for ungaged areas. The estimate errors for all the kriging estimations are relatively small. This increases the confidence for using such a regional distribution to determine the potential threat of flash floods for a possible early warning in ungaged areas.

Table 1. Precipitation gage stations in the Scioto River Basin and their Statistics

No.	Station	Index No.	Record Length	Monthly Mean (mm)	Potential Flash Floods (daily values in mm)			
					99.9%	99.99%	50-yr	100-yr
1	Chilicothe	152809	1973-86	81	52	103	93	102
2	Circleville	159205	1951-86	80	61	111	120	133
3	Columbus Valley	178305	1949-86	80	64	116	117	130
4	Columbus WS	178605	1949-86	79	58	86	101	111
5	Delaware	211901	1936-86	78	62	99	107	117
6	Hillsboro	375808	1940-86	91	69	119	129	143
7	Jackson 2 NW	400409	1956-86	87	65	75	95	103
8	Kenton	418904	1973-86	76	56	94	104	115
9	London Water Works	468105	1936-86	82	66	90	111	121
10	Marion	494205	1951-86	74	58	97	103	114
11	Marysville	497905	1936-86	77	61	87	102	112
12	Portsmouth	678109	1936-86	87	73	106	162	181
13	Washington Court H.	879405	1936-86	82	62	104	118	131
14	Delaware Lake	212405	1950-86	76	62	79	93	102
15	Galion	302102	1953-86	81	71	107	119	131
16	La Rue	440905	1949-86	97	60	85	103	113
17	Mt. Gilead Lakes P.	553505	1953-86	80	60	93	104	114
18	Piketon Aec P. sta.	663009	1949-86	81	67	102	113	125
19	Portsmouth US Gt.	678609	1957-86	88	71	92	130	144
20	Prospect	686105	1957-86	79	58	90	100	111
21	Sedalia	753805	1949-86	83	71	96	117	129

REFERENCES

Chang, Tiao J.,1987. "Analysis and Simulation of Three-Component Floods in the Ohio River Basin," Paper in the Hydrologic Frequency Modeling, edited by V.P. Singh, D. Reidel Publishing Co., pp. 583-594.

Chang, T.J. and Jennifer R. Stenson. 1990. "Is It Realistic to Define a 100-Year Drought for Water Management?" Water Resources Bulletin, 26(5):823-829.

Chang, T.J., 1991. "Investigation of Precipitation Droughts by the Use of Kriging Method, "Journal of Irrigation and Drainage Engineering, ASCE, Vol. 117, No. 6, pp. 935-943.

Chang, T.J. and X. Kleopa, 1991. "A Proposed Method for Drought Monitoring,

"Water Resources Bulletin, AWRA, Vol. 27, No. 2, pp. 275-281.

Chang, T.J. and C.B. Teoh, 1992. "Study of Groundwater Availability in case of Drought," Water Forum'92, edited by M. Karamouz, ASCE, pp. 130-137.

Delhomme, J.P., 1978. "Kriging in Hydrosciences", Advances in Water Resources, Vol.1, No.5, pp.251-266.

Kafritsas, J., & R.L. Bras, 1981. The Practice of Kriging, Report 263, Ralph M. Parson Lab., Massachusetts Institutes of Technology, Cambridge, Massachusetts.

Karlinger, M.R., and Skrivan J.A. 1980. "Kriging Analysis of Mean Annual Precipitation, Power River Basin, Montana and Wyoming," U.S. Geological Survey Report 80-50, Reston, 25p.

Kitanidis, P.K., 1983. "Statistical Estimation of Polynomial Generalized Covariance Functions and Hydrological Applications," Water Resources Research, Vol.19, No.4, pp.909-921.

NOAA, 1991. Natural Disaster Survey Report: Shadyside, Ohio, Flash Flood June 14, 1990, National Weather Service, National Oceanic and Atmospheric Administration, Department of Commerce, Silver Spring, Maryland, 125p.

Yevjevich, V.M., 1967. "An Objective Approach to Definitions an Investigations of Continental Hydrologic Droughts," Hydrologic Paper 23, Colorado State University, Fort Collins, Colorado.

Yevjevich, V.M., Probability and Statistics in Hydrology, Water Resources Publications, Fort Collins, Colorado, 1972.

SECTION 8:

DROUGHT ANALYSIS

A STOCHASTIC CHARACTERIZATION OF PALMER DROUGHT SEVERITY INDEX

G.V. Loganathan[1], S. Mostaghimi[2], M.K. Tchaou[2], V.K. Lohani[1]

Departments of Civil[1] and Agricultural[2] Engineering respectively
Virginia Tech
Blacksburg, VA 24061, USA

ABSTRACT. The Palmer Drought Severity Index (PDSI) combines a set of key meteorologic and hydrologic variables to assign a numerical value for drought severity which can be used as a unified drought scale for comparison among different geographical regions. The index values are further put in several drought severity classes with highest class being the severest drought. This class assignment is taken advantage of, to formulate a time varying (non-homogeneous) Markov chain approach for characterizing the underlying drought stochastic process. The approach identifies drought prone geographical regions, persisting drought classes, periods of return to a particular drought class, and yields short term predictions for future droughts. A time homogenous Markov chain is also formulated. The average of the statistics taken over all months from the non-homogeneous Markov chain agrees quite well with that of the time homogeneous chain. Monthly statistics from the non-homogeneous chain also compare very well with the empirical results obtained with the aid of 1152 months of data for the years 1895-1990 belonging to two climatic divisions in Virginia, USA, namely the Tidewater Region the Southwest Mountains Regions. The Tidewater Region is short of water as opposed to the Southwest Mountains Regions which is water rich.

1. Introduction

The analysis of droughts including the identification of drought prone regions is necessary for deciding long term policies at planning stage. In an operational mode, short term prediction, combined with appropriate regulations for meaningful allocation of water among uses is needed. In this study short term prediction is taken to mean a prediction over a crop growing season. The farmer should be advised of the most likely state of water surplus or shortage over the crop growing period. In this paper the Palmer Drought Severity Index (PDSI) is used to classify various wet and dry states or classes. The PDSI takes into consideration meteorologic and hydrologic variables in a comprehensive manner to assign a quantitative measure of dryness/wetness (Palmer, 1965). Well detailed analysis of the formulation of the PDSI is given in Alley (1984). Karl (1986) provides another comprehensive analysis and recommends the use of Palmer Hydrologic Drought

V. P. Singh and B. Kumar (eds.), Surface-Water Hydrology, 593–608.

Index (PHDI) which skips the backtracking step involved in the PDSI computation so that it will be applicable in real time. Detailed reviews are also given in Guttman (1991) and Johnson and Kohne (1993). Based upon the frequency distribution of PDSI across all months and climatic divisions Karl (1986) has grouped the PDSI values into seven classes (states) with class 1 (PDSI greater than or equal to 4.00) being the wettest and class 7 (PDSI less than or equal to -4.00) being the driest state. Table I provides Karl's delineation of wet/dry spell states.

Large scale development of the irrigation sector in Virginia, has been reported by various researchers. Cox et al. (1981) project that by the year 2000, the state will have an irrigated area of the order of 51,591 acres as against the actual irrigated area of 28,275 acres in 1974. Ross et al. (1982) report that the amount of irrigation equipment available, and used on farms, in dry years have risen steadily from the 1930's to 1982. The feasibility of irrigation schemes, however, depends on availability of water from primary sources like wells, impoundments and streams. It is important to develop appropriate methodologies on the nature of droughts in the State. Van Bavel and Lillard (1957) analyzed the historical pattern of drought in the State and observed that drought was not serious until June and from June through September (part of the growing season), the probability of frequent periods of moisture deficiency occurred at least 3 years in 10 and was a very real problem throughout Virginia. Van Bavel and Lillard reported the results only for the months April through September as the growing season. Later Vellidis et al. (1985) have concurred with the observations of Van Bavel and Lillard (1957). In a recent study conducted by the State Water Control Board (SWCB, 1990) nine drought years have been identified in the State during the study period of 1957-87 which is again about 30% probability.

In this paper two climatic divisions CD1 - the Tidewater Region and CD6 - the Southwest Mountains Regions of the state of Virginia are considered. A methodology which assesses (i) the probabilities of occurrence of the various drought classes for an area, (ii) the duration of a particular drought class, (iii) the expected passage time to go from one drought state to another, (iv) the expected recurrence time of a particular drought state, and (v) the expected drought states for future periods, is presented. The class delineation of PDSI is utilized in a Markov Chain formulation. The monthly variations in weather behavior is well represented by formulating a time varying monthly Markov chain. Based upon the non-homogeneous Markov Chain approach a complete characterization of the underlying wet/dry state stochastic process is obtained. The theoretical results are supported by the empirical analyses of ninety six years of monthly data for the period 1895-1990 for the two climatic divisions.

2. Non-homogeneous Monthly Markov Chain Analysis

The class delineation of the PDSI provided by Karl (1986) is utilized to formulate a Markov Chain analysis of droughts. Let X_n be the random variable representing the drought (wet) state for month n. For example, $X_1 = X_{Jan} = 1$ represents the occurrence of state 1 in January. The underlying stochastic process is completely described by a

Markov Chain if the transition probabilities denoted by $p_{i,j}^{n,n+1}$ for moving from state i in month n to state j in month n+1 and the initial state vector, $f^{(0)}$, describing the probabilities of the seven states for the beginning month, are prescribed. In the present study twelve monthly (non-homogeneous) transition probability matrices describing the state transfers from months January to February; February to March; ..., and December to January are formulated. It is hypothesized that these matrices are cyclic in the sense that there is no yearly variation; that is, the transition probabilities depend only on the month and not on the year. The transition probability

$$p_{i,j}^{(n,n+1)} = P[X_{n+1} = j \mid X_n = i], \text{ for } i, j = 1, 2, ..., 7 \text{ and } n = 1, 2, ..., 12 \tag{1}$$

is computed by

$$p_{i,j}^{(n,n+1)} = \frac{N_{i,j}^{(n,n+1)}}{N_i^{(n)}} \tag{2}$$

in which: $N_{i,j}^{(n,n+1)}$ = number of transfers from state i in month n to state j in month n+1; $N_i^{(n)}$ = number of times state i is realized in month n. If $\mathbf{N_i^{(n)}}$ is zero for some, i, we define $p_{i,j}^{(n,n+1)} = \frac{1}{7}$ for all $j = 1, 2, \ldots, 7$.

2.1 MONTHLY STEADY STATE PROBABILITIES

It is of interest to know what the probability is that for any given month a particular drought (wet) state will be realized. The drought proneness of a region is reflected in a large probability for the drought states. Let $f^{(k)}$ be the state probability vector after k transitions. Using the initial state probability row vector $f^{(0)}$ and the monthly transition matrices we can write

$$f^{(k)} = [f^{(0)}]\,[P_1]\,[P_2] \cdots [P_k] \tag{3}$$

in which: P_1 = (7 x 7) monthly transition matrix associated with the starting month, say January to February, i.e., $P_1 = P^{(1,2)} = P^{(\text{Jan., Feb.})}$. Of course, the transitions do not have to use January as the starting month. Also, note that because of the cyclic nature of these matrices, the transition matrix for months 14 to 15 denoted by $P^{(14,15)}$ is the same as $P^{(2,3)} = P^{(\text{Feb., Mar.})}$, the February-March transition matrix.

For long term, that is as $k \to \infty$ we would like to know whether $f^{(\infty)}$ has steady state probabilities independent of $f^{(0)}$. This will be true if the product of the transition matrices $[P_m]$ through $[P_k]$ denoted by $\phi^{(m,k)}$ called the composite matrix

$$\phi^{(m,k)} = [P_m]\,[P_{m+1}] \cdots [P_k] \tag{4}$$

is a constant stochastic matrix with identical rows (Isaacson and Madsen, 1976) for large k. For such a constant stochastic matrix it follows from Eq. (3) that $f_m^{(k)}$ will be independent of $f^{(0)}$; furthermore, each state has a steady state probability value corresponding to that state (column) constant probability of $\phi^{(m,k)}$. However, because the beginning month, m, influences the value of $\phi^{(m,k)}$ the steady state probabilities of $f_m^{(k)}$ will depend on m. To interpret $f_m^{(k)}$ as $k \rightarrow \infty$, consider Eq. (4) as follows. Let the constant (identical rows) stochastic matrix

$$\begin{aligned}\phi^{(1,\infty)} = [\mathrm{Jan}] = \{&[P_1]\,[P_2] \cdots [P_{11}]\,[P_{12}]\} \\ \{&[P_1]\,[P_2] \cdots [P_{11}]\,[P_{12}]\} \cdots\end{aligned} \tag{5}$$

Because [Jan] is a constant stochastic matrix it follows

$$\text{row } [\mathrm{Jan}] = f_1^{(\infty)} \tag{6}$$

Now consider

$$\begin{aligned}\phi^{(2,\infty)} = [\mathrm{Feb}] = \{&[P_2]\,[P_3] \cdots [P_{11}]\,[P_{12}]\,[P_1]\} \\ \{&[P_2]\,[P_3] \cdots [P_{11}]\,[P_{12}]\,[P_1]\} \cdots\end{aligned} \tag{7}$$

and we obtain

$$[\mathrm{Feb}] = [P_2]\,[P_3] \cdots [P_{11}]\,[P_{12}]\,[\mathrm{Jan}]\,[P_1] \tag{8}$$

from which upon manipulation it follows

$$[\mathrm{Feb}] = [\mathrm{Jan}]\,[P_1] \tag{9}$$

and therefore

$$\text{row } [\mathrm{Feb}] = f_2^{(\infty)} = \text{row } [\mathrm{Jan}]\,[P_1]$$

similarly we can show

$$\begin{aligned}\text{row } [\mathrm{Mar}] &= f_3^{(\infty)} = \text{row } [\mathrm{Feb}]\,[P_2] \\ \text{row } [\mathrm{Apr}] &= f_4^{(\infty)} = \text{row } [\mathrm{Mar}]\,[P_3] \\ &\cdots\cdot \\ \text{row } [\mathrm{Dec}] &= f_{12}^{(\infty)} = \text{row } [\mathrm{Nov}]\,[P_{11}]\end{aligned} \tag{10}$$

$$\text{row [Jan]} = f_1^{(\infty)} = \text{row [Dec]} \, [P_{12}]$$

In general we obtain, *row* [Month n+1] = *row* [Month n] [P_n]. Eq. (10) provides a means to evaluate the monthly drought/wet steady state probabilities. It is a system of linear equations in terms of the monthly steady state probabilities. These steady state probabilities for the Tidewater region are listed in Table II. Also, shown in Table II are the steady state probabilities obtained from a mean monthly transition matrix which does not consider in which months the transitions occur. This analysis is elaborated in the section on mean monthly homogeneous Markov chain analysis. Table II also contains Karl's (1986) empirical estimates of the steady state probabilities computed for the entire USA across all the months. It is of interest to note that for the Tidewater Region (CD1) for the months of August through October the probability of occurrence of a drought state (classes 5, 6, or 7) is 0.3 which agrees with the results of Van Bavel and Lillard (1957). Wallis (1993) reports that the probability of PDSI being in state 7 for the month of July for CD1 is between 0.00 and 0.05 and the present analysis yields 0.0208.

2.2 EXPECTED UNINTERRUPTED RESIDENCE TIME

The expected uninterrupted residence time of the process in a state indicates the actual duration of that state for that region. The process stays in state i without migrating to another state for 'm' time periods when the following event occurs

$$\{X_1 = i = X_2 = \ldots = X_{m-1} \mid X_0 = i\} \tag{11}$$

The probabilities of events specifying uninterrupted stay for different time periods in a particular state, i, can be computed as follows. For example, one month duration of stay denoted by m=1 starting with the month of January is given by

$$P[X_{Feb} \neq i \mid X_{Jan} = i] = P[m = 1 \mid X_{Jan} = i] = 1 - p_{i,i}^{1,2} \tag{12}$$

where: $p_{i,i}^{1,2}$ = probability of moving from state i in January to the same state i, in February. For m=2 starting from the month of January the probability can be computed by

$$P[m = 2 \mid X_{Jan} = i] = P[X_{Mar} \neq i, X_{Feb} = i \mid X_{Jan} = i] \tag{13}$$

$$= P[X_{Feb} = i \mid X_{Jan} = i] \, P[X_{Mar} \neq i \mid X_{Feb} = i] \tag{14}$$

$$= p_{i,i}^{1,2} \left(1 - p_{i,i}^{2,3}\right) \tag{15}$$

Likewise, the probabilities of events defining consecutive stay for higher number of time steps can be computed. For example, for staying 12 time steps consecutively in state i, starting in January, the probability will be

$$P\left[m = 12 \mid X_{Jan} = i\right] = p_{i,i}^{1,2}\, p_{i,i}^{2,3} \cdots p_{i,i}^{10,11}\, p_{i,i}^{11,12}\left(1 - p_{i,i}^{12,1}\right) \tag{16}$$

It is readily observed that the computation of probabilities for various events defining uninterrupted stay in state i involves the multiplication of ith row and i^{th} column entries (diagonal elements) of the consecutive transition matrices. If any one of these entries is zero, the computation stops at that point because all the remaining probabilities for higher durations of stay go to zero. Once the probabilities for uninterrupted stay for various time periods are computed, the expected uninterrupted residence time for state j, $E[R_{uj}]$, is given by

$$E[R_{uj} \mid \text{starting month}] = \sum_k k\, P(m = k \mid \text{starting month}) \tag{17}$$

It is seen that the uninterrupted residence time is sensitive to the starting weather state. The expected uninterrupted residence times for various starting months and states are given in Table III. The longest residence time is for the state 4 which implies that the weather tends to stay normal for relatively longer period for the Tidewater Region, CD1. The average continuous residence times for the drier states (5, 6, and 7) are higher than the wet states (1, 2, and 3) which indicates that a drought once occurs would stay for a relatively longer period than a wet spell. From Table III it is realized that if state 7 occurs in the month of October it may last for 5 months. Also, if the extreme dry state 7 occurs in September, it is likely to persist in the dry state.

2.3 EXPECTED FIRST PASSAGE TIMES

The expected first passage time is defined as the expected time period taken for the process to go to a state, j, for the first time starting from some state, i, and is denoted by $m_{i,j}$. For a non-homogenous chain the starting month, n, is crucial in deciding the expected first passage time and therefore we let $m_{i,j}^{n}$ as the first passage time for a process to reach state, j, starting from state, i, in month, n. Mathematically it attains the form

$$m_{i,j}^{(n)} = (1)\, p_{i,j}^{(n,\,n+1)} + \sum_{k \neq j} p_{i,k}^{(n,\,n+1)} \left(m_{k,j}^{(n+1)} + 1\right) \tag{18}$$

which states that the state, j, in month (n+1) can be reached in one step or in one step some state $k \neq j$ is reached in month (n+1) and it takes a further $m_{k,j}^{(n+1)}$ steps to reach j. This equation is simplified to

$$m_{i,j}^{n} = 1 + \sum_{k \neq j} p_{i,k}^{(n,\,n+1)}\, m_{k,j}^{(n+1)} \tag{19}$$

For example, for n=1 we obtain

$$m^{1}_{i,j} = m^{Jan}_{i,j} = 1 + \sum_{k \neq j} p^{(1,2)}_{i,k}\, m^{Feb}_{k,j} \tag{20}$$

in which: $p^{(1,2)}_{i,k}$ = January-February transition probability for reaching state, k, from state, i. The solution of system of linear equations in Eq. (19) yields the expected first passage times. The *mean recurrence time* $m^{(n)}_{i,i}$ can be computed from the results of Eq. (19) by

$$m^{(n)}_{i,i} = (1)\, p^{(n,n+1)}_{i,i} + \sum_{k \neq i} p^{(n,n+1)}_{i,k} \left(m^{(n+1)}_{k,i} + 1\right) \tag{21}$$

which simplifies to

$$m^{(n)}_{i,i} = 1 + \sum_{k \neq i} p^{(n,n+1)}_{i,k}\, m^{(n+1)}_{k,i} \tag{22}$$

The expected first passage times for the state 4 for the Tidewater Region (CD1) are given in Table IV. The averages of the expected first passage times taken over the twelve months agree with the mean passage times computed from the mean monthly matrix.

2.4 MONTHLY DROUGHT STATE PREDICTION

It has been shown in the above sections that Markov Chain with non-homogenous transition matrices can describe monthly weather transition satisfactorily. Therefore, these matrices can also be used to make predictions of weather states in the future months. At the end of a month, the weather state in that month is known. Let $E[X_{Feb} \mid X_{Jan}]$ denote the expected weather state for February, given that during the month January the weather state observed is, i. Then the conditional expected state for the month of February is given as

$$E\left[X_{Feb} \mid X_{Jan} = i\right] = \sum_{j=1}^{7} j P(X_{Feb} = j \mid X_{Jan} = i) = \sum_{j=1}^{7} j\, p^{(1,2)}_{i,j} \tag{23}$$

where: $p_{i,j}^{(1,2)}$ is the conditional probability of being in state j in February given that in January the state is, i, which is the jth column element of the ith row of the transition matrix $P^{(1,2)}$. For the transition from February to March, X_{Feb} = NINT $(E[X_{Feb} \mid X_{Jan}])$, where: NINT = the nearest integer. Therefore,

$$E[X_{Mar} \mid X_{Feb}] = \sum_{j=1}^{7} j\, p^{(2,3)}\,(\text{NINT}\,(E[X_{Feb} \mid X_{Jan}]), j) \tag{24}$$

In a similar manner the prediction for subsequent months can be made. This approach is used to forecast the weather states during years 1930, 1931, and 1978 in the Tidewater climatic region. The forecasts are based on known March weather state. The years 1930 and 1931 correspond to the extreme drought situation in the region whereas the year 1978 corresponds to the wet weather conditions. The forecasts and comparison with observed values are given in Table V. It is seen that the predicted values compare quite well with the observed values for the first 3-4 months. However, for longer lead months the prediction is in general not accurate. This is attributed to the lack of updating of information. However, keeping in mind that many irrigation applications may not need a forecast of more than 4 months of lead time, the predictions can be considered satisfactory.

3. Mean Monthly Homogeneous Markov Chain Analysis

In order to characterize the mean monthly transition behavior of weather, transition probabilities giving emphasis on transitions among states, irrespective of the months in which these take place during the year are computed. If N_T is the total number of data points used and N_{ij} is the number of times the process transits from state i to state j, regardless of the month, and N_i the total number of times the process is in state i, then

$$p_{ij} = \frac{N_{ij}/N_T}{N_i/N_T} = \frac{N_{ij}}{N_i} \tag{25}$$

denotes the probability of transition from state i to state j. The mean monthly transition matrix is then defined as $P = [p_{ij}]$ for $i, j = 1, 2, ..., 7$. It is observed that the mean monthly matrix represents a closed communicating class of all weather states (1, 2, ..., 7). This is logical since in reality any state of weather is possible in a mean monthly transition. The absence of any transient or absorbing states indicates that neither there exists a weather state from which the system disappears forever nor there exists a permanent weather state in which the system is trapped.

3.1 STEADY STATE PROBABILITIES

The mean monthly matrix represents an irreducible aperiodic Markov chain which has limiting probabilities which are independent of starting state. These steady state probabilities denoted by the vector λ are computed using the equation

$$\lambda = \lambda\,[P] \tag{26}$$

These steady state probabilities are given in Table II.

3.2 MEAN RECURRENCE TIME

The mean recurrence time, m_{jj}, for a state j is defined as the expected number of transitions until a Markov chain, starting in state j, returns to that state. Since, on the average, the chain will spend 1 unit of time for every m_{jj} units of time in state j, it follows that

$$m_{jj} = \frac{1}{\lambda_j} \tag{27}$$

The mean recurrence times for various states in both climatic regions are given in Table VI. From Table VI it is observed that on an average the driest state reoccurs at shorter time periods as compared with the wettest state for the Tidewater Region, CD1. However, for the Southwestern Mountains region, CD6 the reverse is true. This indicates that the Tidewater Region is expected to experience severer drought events than the South Mountain region during the same period. This conclusion is in agreement with the results reported by the State Water Control Board (1990) which has found that during 1957-1987 the CD1 region experienced 10 drought years while CD6 region had only 5 drought years.

3.3 MEAN PASSAGE TIME

The mean monthly transition matrix can be analyzed to the find the mean passage time which gives an estimate of the number of transitions the process takes, on the average, to go from one particular state to another for the first time. The mean passage time is computed as

$$m_{ij} = 1 + \sum_{k \neq j} p_{ik}\, m_{kj} \tag{28}$$

where: m_{ij} = mean passage time to go from state i to j; p_{ik} = one step transition probability to go from state i to k; m_{kj} = mean passage time to go from state k to j. Table VII gives the mean passage times in months. It is seen that the process takes relatively higher number of transitions to go away from the normal weather state to either dry or wet states for the first time than to return to the normal state. It is also noticed that the mean number of transitions to go from the normal state (4) to the extreme weather states (1 or 7) are about 13-19 times higher than to come back to the normal state.

4. Summary

The non-homogeneous Markov chain analysis provides useful drought characteristics such as most likely state of drought severity at a location, drought durations, and periods of return to various drought classes. Such a comprehensive characterization of droughts is extremely useful for planning purposes. The method is also extended to predict future drought states contingent on the present month's observed state. It should be useful in situations where short term predictions are needed such as in irrigation applications.

5. References

Alley, W.A. (1984). "The Palmer Drought Severity Index: Limitations and Assumptions", *Journal of Climate and Applied Meteorology*, vol. **23**, pp. 1100-1109.

Cox, W.E., L.A. Shabman and J.W. Looney (1981). "Virginia's Water Resources: Policy and Management Issues", Dept. of Agric. Econ. & Civil Engrg., Virginia Tech., Blacksburg.

Guttman, N.B. (1991). "A Sensitivity Analysis of the Palmer Hydrologic Drought Index", *Water Resources Bulletin*, vol. **27**, no. 5, pp. 797-807.

Isaacson, D.L., and R. Madsen (1976). *Markov Chains*: *Theory and Applications*, John Wiley, New York.

Johnson, W.K., and R.W. Kohne (1993). "Susceptibility of Reservoirs to Drought Using Palmer Index", *Journal of Water Resources Planning and Mgmt.*, ASCE, vol. **119**, no. 3 pp. 367-387.

Karl, T.R. (1986). "The sensitivity of the Palmer Drought Severity Index and Palmer's Z Index to Their Calibration Coefficients Including Potential E.T.", *Journal of Climate and Applied Meteorology*, vol **25**, pp. 77-86.

Ross, B.B., F.S. Wright, N.L. Powell, and V.O. Shanholtz, (1982). "Status of Irrigation Development in Virginia", *File Report*, Dept. of Agric. Engrg., Virginia Tech, Blacksburg, VA.

Palmer, W.C. (1965). "Meteorological Drought", *Research Paper No. #5*, U.S. Weather Bureau, Washington, D.C.

State Water Control Board (1990). "Drought Indicator Parameters", *Information Bulletin No.* 582, Richmond, VA.

Tchaou, M.K. (1992). "A Markov chain approach for analyzing Palmer Drought Severity Index", *M.S. Thesis*,Virginia Polytechnic Institute and State University, Blacksburg.

Van Bavel, C.H.M. and J.H. Lillard (1957). "Agricultural Drought in Virginia", *Va. Agric. Exp. Station, Tech. Bulletin, No.* 128, Blacksburg, Va.

Vellidis, G., B.B. Ross and D.B. Taylor (1985). "Agricultural Drought in Virginia 1930-83", *Info. Series 85-3*, College of Agric. and Life Sc., Virginia Tech, Blacksburg, VA.

Wallis, J.R. (1993). "The National Drought Atlas", *Proc. 20th Anniversary Conference on Water Management in the '90s, ASCE*, Seattle, WA, pp. 455-462.

Table I. PDSI values and corresponding stochastic states.

PDSI value	Weather Spell	Stochastic State
4.00 or more	extremely wet	1
3.00 to 3.99	severely wet	2
1.50 to 2.99	mild to moderate wetness	3
-1.49 to 1.49	near normal	4
-1.50 to -2.99	mild to moderate drought	5
-3.00 to -3.99	severe drought	6
-4.00 or less	extreme drought	7

Table II. Monthly Steady State Probabilities (CD1 Region)

Wet/Dry States / Month	1	2	3	4	5	6	7
Jan.	.0104	.0520	.2198	.5191	.1249	.0312	.0426
Feb.	.0104	.0313	.2193	.5518	.1248	.0312	.0312
March	0	.0626	.1773	.5521	.1560	.0208	.0312
April	0	.0732	.1668	.5520	.1352	.0416	.0312
May	.0417	.0313	.1667	.5628	.1455	.0416	.0104
June	.0209	.0314	.1666	.5730	.1353	.0624	.0104
July	0	.0627	.2086	.4578	.2084	.0417	.0208
Aug.	.0104	.0730	.1982	.4162	.2397	.0625	0
Sept.	0.104	.0937	.1356	.4587	.2184	.0624	.0208
Oct.	.0209	.0626	.2082	.3958	.2604	.0313	.0208
Nov.	.0208	.0521	.2499	.4063	.1979	.0417	.0313
Dec.	0	.0729	.2082	.4688	.1771	.0417	.0313
Twelve Month Average	0.0121	0.0582	0.1937	0.4928	0.177	0.0425	0.0235
Mean Monthly Matrix	.0122	.0582	.1937	.4926	.1773	.0425	.0234
Karl (1986)	.05	.06	.17	.45	.17	.06	.04

Table III. Expected Uninterrupted Residence Times (months) (Tidewater Region, CD1)

State / Starting Month	1	2	3	4	5	6	7
Jan.	2	1.19	2.68	5.64	3.89	3.11	3.99
Feb.	1.0	1.0	2.52	5.39	3.85	3.17	3.99
March	1.17	2.05	2.91	5.17	3.43	3.26	3.00
Apr.	1.21	1.26	2.70	4.71	3.31	2.27	2.00
May	1.5	1.81	2.72	4.27	3.00	2.51	3.00
June	1.0	2.44	2.50	3.92	3.51	2.02	2.00
July	1.14	2.16	2.39	4.47	3.63	2.03	1.00
Aug.	1.0	1.74	2.14	4.92	3.50	2.07	1.5
Sept.	2.5	1.73	2.72	4.90	3.83	2.13	3.67
Oct.	1.5	2.17	3.19	5.36	2.97	2.25	5.33
Nov.	1.0	2.35	2.92	5.92	3.08	1.89	4.33
Dec.	1.28	1.69	2.88	5.99	3.29	1.78	5.00

Table IV. Mean Monthly First Passage Times To State 4

From State / Month	1	2	3	4	5	6	7
Jan.	8.20	4.80	4.69	1.53	5.64	7.06	11.31
Feb.	7.20	4.14	4.95	1.77	6.18	9.08	10.71
Mar.	6.57	6.20	4.72	1.83	6.22	9.02	9.71
April	6.19	6.24	4.23	1.53	5.66	8.02	8.71
May	7.17	3.52	4.46	1.65	6.06	7.97	7.19
June	8.56	7.57	4.34	1.86	6.13	7.26	6.19
July	6.66	7.56	4.59	2.99	6.71	6.31	5.19
Aug.	7.39	6.87	4.47	2.64	5.93	8.39	6.56
Sept.	6.80	6.39	5.03	2.14	5.81	8.77	6.10
Oct.	5.80	6.19	4.84	2.34	5.05	10.45	10.20
Nov.	3.54	6.05	4.71	2.27	5.14	9.57	9.20
Dec.	6.95	5.08	4.93	1.91	5.15	8.41	12.31
Average	6.75	5.88	4.66	2.04	5.81	8.36	8.62

Table V. Predicted and Observed Weather Spells for Virginia's Tidewater Region

Month	1930		1931		1978	
	Predicted	Observed	Predicted	Observed	Predicted	Observed
January	4	4	7	7	2	2
February	4	4	7	7	3	3
March	5*	5	7*	7	2*	2
April	5	5	7	7	2	2
May	5	5	6	6	2	1
June	5	5	6	5	3	2
July	5	6	6	5	3	2
August	5	6	5	5	3	2
September	5	7	5	5	3	3
October	5	7	5	5	3	4
November	5	7	5	5	3	4
December	5	7	5	6	3	4

*Observed state, not predicted

Table VI. Mean Recurrence Times (months) (CD1 Region)

State	Tidewater Region (CD1)	Southwestern Mountains (CD6)
1	81.96 (94.47)*	56.82
2	17.18 (18.13)*	28.49
3	5.16 (5.16)*	5.32
4	2.03 (2.04)*	1.95
5	5.64 (5.59)*	5.29
6	23.53 (21.84)*	22.22
7	42.74 (50.98)*	82.64

( )* indicates average over all months from the non-homogeneous chain

Table VII. Mean Passage Times (months) (CD1 Region)

State	1	2	3	4	5	6	7
1	81.3	11	9.9	6.2	24.1	76.4	207.0
2	117.3	17.5	6.8	5.9	23.8	76.1	206.8
3	145.3	31.9	5.2	4.7	23	74.9	205.5
4	157.3	43.3	13.6	2.03	18	70.2	200.8
5	163.9	48.2	19.1	6.1	5.6	52.3	182.8
6	164.3	48.2	20.5	8.2	11.4	23.2	130.3
7	165.1	50	21.4	8.5	18.1	33.9	41.7

GPSR Compliance

The European Union's (EU) General Product Safety Regulation (GPSR) is a set of rules that requires consumer products to be safe and our obligations to ensure this.

If you have any concerns about our products, you can contact us on ProductSafety@springernature.com

In case Publisher is established outside the EU, the EU authorized representative is:

Springer Nature Customer Service Center GmbH
Europaplatz 3
69115 Heidelberg, Germany

Batch number: 10607320

Printed by Printforce, the Netherlands